Biogeochemistry of Trace Metals

Edited by

Domy C. Adriano

CRC Press is an imprint of the
Taylor & Francis Group, an **informa** business

First published 1992 by CRC Press
Taylor & Francis Group
6000 Broken Sound Parkway NW, Suite
300 Boca Raton, FL 33487-2742

Reissued 2018 by CRC Press

A Library of Congress record exists under LC control number: 91042965

Publisher's Note
The publisher has gone to great lengths to ensure the quality of this reprint but points out that some imperfections in the original copies may be apparent.

ISBN 13: 978-1-138-50666-4 (hbk)
ISBN 13: 978-1-138-55772-7 (pbk)
ISBN 13: 978-1-315-15026-0 (ebk)

Visit the Taylor & Francis Web site at http://www.taylorandfrancis.com and the CRC Press Web site at http://www.crcpress.com

ADVANCES IN TRACE SUBSTANCES RESEARCH

Series Preface

The need to synthesize, critically analyze, and put into perspective the ever-mounting body of information on trace chemicals in the environment provided the impetus for the creation of this series. In addition to examining the fate, behavior and transport of these substances, the transfer into the food chain and risk assessment to the consumers, including humans, will also be taken into account. It is hoped then that this information will be user-friendly to students, researchers, regulators, and administrators.

The series will have "topical" volumes to address more specific issues as well as volumes with heterogeneous topics for a quicker dissemination. It will have international scope and will cover issues involving natural and anthropogenic sources in both the aquatic and terrestrial ecosystems. To ensure a high quality publication, volume editors and the editorial board will subject each article to peer review.

Thus, **Advances in Trace Substances Research** should provide a forum where experts can discuss contemporary environmental issues dealing with trace chemicals that hopefully can lead to solutions resulting in a cleaner and healthier environment.

Domy C. Adriano
Editor-in-Chief

ADVANCES IN TRACE SUBSTANCES RESEARCH

Preface

Trace metals are an important environmental contaminant that affect all the ecosystem components, both in the aquatic and terrestrial systems. There has been an ever-increasing body of information that indicates trace metals are continually accumulating in the food chains. Classical examples are toxic effects in the food chain from the itai-itai disease for cadmium and the Minamata disease for mercury, both in Japan. And now, lead poisoning in children is becoming a social and health problem not only in the United States but in most of the industrialized countries. The present book *Biogeochemistry of Trace Metals* is yet another compendium of the most recent information on the effects of trace metals in soil quality and its potential threat on the transfer of these contaminants to the consumers.

Most of the papers were presented during the First International Conference on the Biogeochemistry of Trace Elements (formerly Metals in Soils, Plants, Waters and Animals) held in Orlando, Florida in May of 1990. There are two chapters on background levels of metals in soils and/or plants — one for western Europe and one for the temperate, humid zone of Europe; six chapters are devoted to the metal cycling and transfer in the food chain in agroecosystems; two chapters are devoted to uptake and accumulation of metals by bacteria, fungi and invertebrate animals; two chapters are devoted to mechanistic aspects of metals — one on metal tolerance in plants, the other on metal-humic substance complexes; and two special papers are on the microbial aspects of soil selenium losses and manganese sorption on soil constituents.

The book should be especially interesting to researchers and practitioners in the field. One other beneficial aspect of the book is its international coverage and how various special cases from the application of municipal sewage sludge onto land with respect to cadmium transfer in the food chain are highlighted. Guidelines in soil levels and diets for cadmium and other metals are also discussed.

Domy C. Adriano

Domy C. Adriano received his Ph.D. in Soil Science at Kansas State University in 1970. He is currently Professor and Head of the Division of Biogeochemical Ecology at the University of Georgia's Savannah River Ecology Laboratory located near Aiken, South Carolina. He has research stints at the University of California, Riverside and at Michigan State University. He was an Associate Editor for the *Journal of Environmental Quality* in 1980–1985 and is Editor-in-Chief of Advances in Trace Substances Research and Editor of Advances in Environmental Science. He founded and organized the International Conference on the Biogeochemistry of Trace Elements (formerly Metals in Soils, Water, Plants and Animals), the first one held in Orlando, Florida, April 28–May 3, 1990. He is currently the Chairperson of its International Executive Committee. He has received several professional awards, including Fellow of the Soil Science Society of America and of the American Society of Agronomy.

Dr. Adriano's professional interests include the biogeochemistry of trace contaminants in the soil-plant system from the application of hazardous wastes, trace element biogeochemistry, waste management, environmental restoration, and soil remediation.

Acknowledgments

This volume contains information presented and discussed by the plenary speakers for the International Conference on Metals in Soils, Waters, Plants and Animals held in Orlando, Florida, April 30–May 3, 1990.

The conference would not have been successful without the assistance of certain key individuals: J. C. Corey, Westinghouse Savannah River Laboratory; I. P. Murarka, Electric Power Research Institute; M. Chino, University of Tokyo; B. Davies, University of Bradford; I. Iskandar, U.S. Army Corps of Engineers; J. Ryan, U.S. Environmental Protection Agency; R. Severson, U.S. Geological Survey; C. L. Strojan, M. C. Newman and P. M. Bertsch of the Savannah River Ecology Laboratory; W. J. Birge, University of Kentucky; J. Huckabee, Electric Power Research Institute; F. A. M. de Haan, Wageningen, The Netherlands; P. M. Giordano, Tennessee Valley Authority; T. C. Hutchinson, University of Toronto; D. Sauerbeck, Braunschweig, Germany and I. Thornton, Imperial College of Science and Technology. The following organizations and agencies also deserve special acknowledgment for their financial contributions: the Association of Metropolitan Sewerage Agencies; the Electric Power Research Institute; the Hitachi Kiden Kogyo Ltd. of Japan; Showa Engineering Co., Tokyo; the Tennessee Valley Authority; the University of Kentucky; the U.S. Army Corps of Engineers and Wageningen Agricultural University.

I am especially grateful for the excellent secretarial assistance of Ms. Edith Towns and Ms. Brenda Rosier, who flawlessly handled the reviews of the chapters in this volume. Undoubtedly the following reviewers enhanced the quality of the manuscripts with their rigorous criticisms. The reviewers included:

Alloway, B. J.	University of London
Angle, J. S.	University of Maryland
Baker, A. J. M.	University of Sheffield
Baligar, V. C.	USDA Appalachian Soil & Water Conservation Research Laboratory
Bloom, P. R.	University of Minnesota
Candelaria, L. M.	University of California-Riverside
Carlson, C. L.	Bowling Green State University
Chang, A. C.	University of California-Riverside
Chen, J.	Peking University
Chino, M.	University of Tokyo
Cronan, C. S.	University of Maine
Crossley, D. A.	University of Georgia
Dudka, S.	Institute of Soil Science & Plant Cultivation (Poland)
Evans, D. A.	Clemson University
Gambrell, R. P.	Louisiana State University
Gilmour, C. C.	The Academy of Natural Science-Benedict Estuarine Research Laboratory
Harsh, J.	Washington State University
Hunter, D.	University of Georgia-SREL

Johnston, C. T.	University of Florida
Juste, C.	I.N.R.A. Station d'Agronomie (Bordeaux)
Kirkham, M. B.	Kansas State University
Klubek, B. P.	Southern Illinois University-Carbondale
Korcak R. F.	USDA, ARS, Fruit Laboratory
Kuo, S.	Washington State University
Marschner, H.	University of Hohenheim (Germany)
McGrath, S. P.	Rothamsted Experimental Station
Mikkelsen, R. L.	North Carolina State University
Miller, W. P.	University of Georgia
Nutter, W. L.	University of Georgia
Pacyna, J. M.	Norwegian Institute for Air Research
Pierzinski, G.	Kansas State University
Porcella, D. B.	Electric Power Research Institute
Roth-Holzapfel, M.	University of Ulm (Germany)
Sauerbeck, D. R.	Institute of Plant Nutrition & Soil Science (Germany)
Severson, R. C.	USDOI Geological Survey (Denver)
Smith, J. A. C.	University of Oxford
Steinnes, E.	University of Trodheim (Norway)
Taylor, R. W.	Alabama A & M University
Watson, A. P.	Oak Ridge National Labortory

Contributors

Brian J. Alloway
Environmental Science Unit
Department of Geography
Queen Mary and Westfield College
University of London
Mile End Road
London E1 4NS, U.K.

Massimo Angelone
ENEA
C.R.E. Casaccia
Labortory Geochemistry
Via Anguillarese, 301
00060 Rome, Italy

Claudio Bini
Department of Crop Production
University of Udine
V. Fagagna, 208
33100 Udine, Italy

Zueng-Sang Chen
Department of Agricultural Chemistry
National Taiwan University
Taipei, Taiwan 10764, ROC

Anna Chlopecka
Institute of Soil Science & Plant Cultivation
Osada Palacowa
24-100 Pulawy, Poland

Jonathan R. Cumming
Department of Botany
Marsh Life Science Building
University of Vermont
Burlington, VT 05405

Brian E. Davies
Department of Environmental Science
University of Bradford
Bradford, West Yorkshire BD7 1DP, England

Stanislaw Dudka
Institute of Soil Science & Plant Cultivation
Osada Palacowa
24-100 Pulawy, Poland

William T. Frankenberger, Jr.
Department of Soil & Environmental Sciences
University of California
Riverside, CA 92521

Teresa Gawinowska
Institute of Soil Science & Plant Cultivation
Osada Palacowa
24-100 Pulawy, Poland

Karl Harmsen
Institute for Soil Fertility Research
P.O. Box 30003
9750 RA Haren, The Netherlands

R. B. Harrison
University of Washington
College of Forest Resources
AR-10
Seattle, WA 98195

Philip M. Haygarth
Institute of Environmental & Biological Sciences
Lancaster University
Lancaster LA1 4YQ, U.K.

Charles L. Henry
University of Washington
College of Forest Resources
AR-10
Seattle, WA 98195

U. Karlson
Department of Soil & Environmental Sciences
University of California
Riverside, CA 92521

Riaz A. Khattak
Department of Soil Science
NWFP Agricultural University
Peshawar, Pakistan

Andrew P. Jackson
Environmental Resources Limited
Eaton House
Wallbrook Court
North Hinksey Lane
Oxford OX2 0QS, U.K.

Kevin C. Jones
Institute of Environmental & Biological Sciences
Lancaster University
Lancaster LA1 4YQ, U.K.

Christian Juste
I.N.R.A. — Station d'Agronomie
Centre de Recherches de Bordeaux, BP 81
F-33883 Villenave d'Ornon cedex, France

Alina Kabata-Pendias
Institute of Soil Science & Plant Cultivation
Osada Palacowa
24-100 Pulawy, Poland

Nicholas W. Lepp
School of Natural Sciences
Liverpool Polytechnic
Byrom Street
Liverpool L3 3AF, U.K.

Bernd Markert
Systems Research Group
University of Osnabrück
POB 4469
4500 Osnabrück, Germany

Michel Mench
I.N.R.A. — Station d'Agronomie
Centre de Recherches de Bordeaux, BP81
F-33883 Villenave d'Ornon cedex, France

Albert L. Page
Department of Soil & Environmental Sciences
Geology 2247
University of California
Riverside, CA 92521

Mechthild Roth
University of Ulm
Ecology & Morphology of Animals (Biology III)
Albert-Einstein-Allee 11
D-7900 Ulm/Donau, Germany

Nicola Senesi
Istituto di Chimica Agraria
Universita di Bari
Via Amendola, 165/A
70126 Bari, Italy

A. Brian Tomsett
Department of Genetics and Microbiology
University of Liverpool
Liverpool L69 3BX, U.K.

Contents

1

Trace Metals in the Environment: Retrospect and Prospect*

Brian E. Davies

Department of Environmental Science, University of Bradford, Bradford, West Yorkshire, BD7 1DP, England

ABSTRACT

Progress made in understanding trace metals in the environment is reviewed, especially over the last 30 years. The main focus is research in soils, plant uptake, and implications for human health. Until the 1960s the major concern was micronutrient research. Since then, the subject has broadened to encompass excess concentrations of both micronutrients and toxic trace metals. Despite a great increase in our understanding of these metals, there are still important questions unanswered. These include establishing a sound basis for sampling soils, sample treatment, and data handling. There is a need to establish background concentrations of trace metals locally, regionally, and nationally. Our understanding of the composition and dynamics of the soil solution is far from satisfactory, especially in terms of its role as a supplier of nutrient ions. Present models for predicting plant uptake from soils are too simple, and more complex models are needed. More generally, trace metal research is still dominated by

---

* Keynote address at International Conference on Metals in Soils, Waters, Plants and Animals. Grenelefe Resort and Conference Center, Orlando, Florida.

fact-finding, and insufficient emphasis is given to hypothesis testing and to predictive modeling. There is a lack of agreement on a rigorous definition of technical terms in common use. Although much is now known of the health effects of environmental levels of cadmium and lead, the more general impact of environmental concentrations of trace metals on health has been neglected.

INTRODUCTION

Trace metals, or more generally, trace elements, are now extensively researched in the life, agricultural, and environmental sciences. But widespread interest in trace metals has emerged from academic obscurity only over the last 25 to 30 years; essentially it is a post-World War II (1939–1945) subject. This shift in emphasis can be exemplified by comparing the 7th edition of Russell's classic text, *Soil Conditions and Plant Growth* (1937)[1] with the current 11th edition.[2] The former described, in a few paragraphs, boron, manganese, iron, and other trace metals as only having "stimulative and prophylactic effects" on plant growth, whereas the latter devotes a whole chapter (34 pages) to "Micronutrients and Toxic Elements."

It is important to understand something of the historical development of any subject in order to identify its major research topics, to describe their development, and, hence, to suggest the priorities of the future. We are probably too close in time to many of the major developments in trace metal research for a balanced history to be written; nonetheless, by trying to understand where we have come from we can better plan where we might go. I am primarily concerned in this chapter with trace metals in soils and plants since this is my own subject. My knowledge of trace elements in aquatic ecosystems or in animal or human health is not sufficiently detailed to allow me to attempt a valid synopsis: for animals and people a useful review has been provided by Mertz.[3] Furthermore, the account which follows is not intended as an exhaustive review of the literature. It is the personal (perhaps idiosyncratic) view of someone who has been active in this field of research for some 25 years and has had the privilege of meeting many of the workers whose papers are cited. It is intended as a contribution to debate.

HISTORICAL RETROSPECT

The 1920s and 1930s were the years when the essentiality of trace metals for higher plants was of great interest for physiologists. Indeed, the rules for designating an element as essential were drawn up by Arnon and Stout[4] at the end of that period. In the same decades, agronomists were describing the plant diseases or growth defects which were seen when elements such as boron or zinc were in restricted supply in agricultural or horticultural soils. But it was probably the wartime demands for greatly increased crop production that led to more and more areas of soil being identified where low concentrations of trace elements were limiting yields. The first major text devoted to trace elements in plants and animals was that of Stiles.[5] Writing in the early 1940s, he observed that no copper deficiency diseases of plants were known to occur in Britain, yet within a few years that was no longer so. Caldwell[6] described how the first case of copper deficiency in English agricultural crops was encountered in cereals in 1946 and 1947 on a reclaimed deep peat. He wrote, "In subsequent years many areas of deficiency were found ... in all the fenland counties." Thus we see how the imperatives for much increased food production in the 1940s led to the reclamation of peat lands with inherently low copper contents, the more intensive use of other soils made heavy demands on the ability to provide nutrients, and the marginal supply situation for micronutrients became evident. Now, copper problems for both crops and animals are widely recognized throughout the British Isles, as are deficiencies of other micronutrients. Tinker[7] suggested that trace element deficiencies were a considerable but not excessive problem for British agriculture, and his rough estimate of the total annual cost of micronutrient fertilizers and sprays was between 2.5 and 3 million pounds sterling, "which is appreciable, even if far less than that of the major elements."

There are nine trace elements which are commonly accepted as micronutrients for higher plants, namely, boron, chlorine, copper, iron, manganese, molybdenum, nickel, sodium, and zinc. The situation for cobalt, iodine, silicon, and vanadium is equivocal. The essentiality for six was established before 1940: Mn by McHargue,[8] B by Warrington,[9] Zn by Sommer and Lipman,[10] Cu by Lipman and MacKinney[11] and independently by Sommer,[12] and Mo by Arnon and Stout.[4] In the postwar period, the essentiality for Cl was demonstrated by Broyer et al.,[13] for Na by Brownell and Wood,[14] and for Ni by Eskew et al.[15] It is curious that, despite the rapidly growing interest in trace metals in the environmental and life sciences, little effort is apparently being put into deciding the essentiality or otherwise of other elements. Moreover, although nearly 70 years have lapsed since boron was established as an essential micronutrient, we still have no clear understanding of its role in plant metabolism. For bacteria and fungi, micronutrient and trace metal research are still in their infancy. The situation for plants contrasts with that for animals, where 14 trace elements (Si, V, Cr, Mn, Fe, Co, Ni, Cu, Zn, As, Se, Mo, I, and F) are recognized as essential; 7 were identified in the period 1950 to 1980.[3,16]

The postwar period has seen a growing concern with excess concentrations of a number of elements in the environment as a consequence of anthropogenic emissions. Many of these elements are conveniently grouped together as *heavy metals*, although the term is not precisely defined (often authors adopt a lower limiting metallic density of 6 g cm^{-3}; see Davies and Jones in Wild[2]) and the origin of its usage is obscure. Figure 1 is based on a simple citation count for the trace metals Cd, Co, Cu, Ni, Pb, and Zn for soils and plants, using the abstracting journal *Soils and Fertilizers* as a source. The graph should not be interpreted too rigorously, since no attempt was made to guard against multiple citations. It does, however, demonstrate how publication rates have risen markedly since 1950.

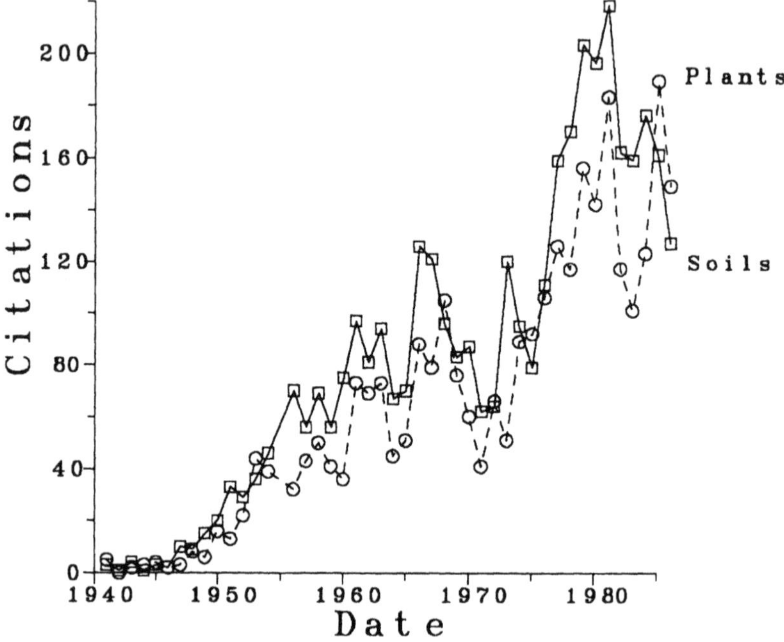

FIGURE 1. Trace metal publications.

A critical historical account has yet to be written for heavy metal research, but it is nonetheless possible to note some seminal papers of the 1960s which have stimulated major areas of work in many countries over the following 20 to 30 years. This is not to deny the importance of earlier papers. For example, Griffith[17] wrote a comprehensive account of the adverse consequences of fluvial dispersal of lead-rich mine wastes in west Wales immediately after World War I. A decade later Bertrand and Okada[18] described elevated concentrations of lead in French soils.

Several overlapping developments had to take place before trace element research in the agricultural and life sciences became more than a local or minor interest. One was possibly a change of attitude in health care. Prior to the discovery and wide provision of antibiotics in the 1940s to cure bacterial infections and the simultaneous introduction of the insecticide DDT, which controlled the insect vectors of diseases, medical attention was largely preoccupied with the infectious diseases; the importance of other influences on health was less clear and often discounted.

Rachel Carson's book *Silent Spring*, eutrophication of lakes following river pollution by phosphates and nitrates from detergents and fertilizers, or the problems of combating smog and photochemical smog all awakened the public conscience to environmental degradation. Minamata (methyl mercury poisoning) and Itai-itai (cadmium) diseases in Japan helped focus attention on pollution by toxic metals. These all created a climate of thought in which it was acceptable to consider a role for trace metals in human health. Now, an ill-defined environmental influence is recognized as important in the etiology of diseases such as cancer, and there is some evidence to support a role for environmental and dietary trace metals.

The influence of new analytical techniques should also be recognized since they have made relatively easy, the rapid, and reliable analysis of trace metals. For example, Bertrand and Okada[18] obtained their lead data from a hydrochloric acid extract of calcined soil which was saturated with hydrogen sulfide gas and allowed to stand for 24 hr in a sealed flask. This was followed by filtration and the precipitate was washed with more H_2S-saturated acid. The soil was re-extracted, retreated with H_2S and, ultimately, the precipitates were bulked. Purification steps followed to remove other metal sulphides and, eventually, the pure, dry lead sulfate yield was weighed. One can only admire the perseverance in obtaining data for an element in soil in which no one at that time was really interested.

In the postwar period, spectrophotometry and in particular the dithizone method, become widely available for lead and other trace metals. Warren and Delavault[19] were active in Canada studying trace elements in relation to biogeochemical prospecting. They wrote, "Surprisingly little is known about the distribution of lead in soils." They went on to comment that extraneous contamination should be taken into account in biogeochemical prospecting for new metal ores, and they named two previously unrecognized sources. One was contamination by petrol exhaust fumes (from tetraethyl lead as an anti-knock agent in the fuel) and the other was orchard insecticide spray (lead arsenate). They concluded that "industrial salting" provided a more widespread and serious problem than was anticipated. Subsequently, Cannon and Bowles[20] published their classic paper on lead in roadside soils and plants in Colorado and thereby launched a myriad of investigations by others over the next 28 years. The wider problem of urban contamination has also been a prolific research area, and this can be traced to a short report by Purves[21] concerning contamination of soils in Edinburgh and Dundee (Scotland) by copper and boron. In recent years the problem of sewage sludge has occupied many workers. This topic also originated in the 1960s when Le Riche[22] published a paper on metal contamination of soil and the appearance of zinc-induced chlorosis in crop plants in the Woburn (England) market-garden experiment where sewage sludge had been used as a manure substitute over many years.

It is less easy to track studies on metal uptake by plants back to one or a few sources since contaminating metals such as copper or zinc are also essential micronutrients and have therefore been the subject of continuing investigation since early in the century. A pioneer publication was a color atlas of deficiency symptoms published by Wallace.[23] One should also note the report by Mitchell and Reith[24] which first demonstrated a marked seasonality in the lead content of pasture grasses. Similarly, papers by Marten and Hammond[25] and MacLean et al.[26] were among the first to discuss the relatively low bioavailability of lead and the soil factors which control the plant uptake of the element. Motto et al.,[27] Page et al.,[28] and Lagerwerff et al.[29] in the U.S. opened up the ongoing debate concerning the relative contributions of leaf-deposited and root-translocated lead to the total plant content and the possibility that foodstuffs might contain excess lead. During the 1960s cadmium received barely a mention. Yet, in the 1970s, this element, together with lead, dominated the trace metal research world. Shacklette[30] provided one of the earliest reviews of cadmium in plants. Figure 1 illustrates the dramatic rise in trace metal publications in the 1970s, stimulated by the discoveries of the 1960s.

PROSPECT: SOME RESEARCH UNCERTAINTIES AND NEEDS

There are now many thousands of journal papers, dozens of books, and hundreds of official reports all concerned with aspects of trace metals in soils, waters, plants, and animals. It is a good time to take stock, to review what is known, and to suggest future destinations. The decline in publication rates in recent years, shown in Figure 1, is not because most of the answers are now known. Sadly, most of the questions have still to be asked. Officially perceived research priorities have shifted, e.g., to acid precipitation or the greenhouse effect (global warming), and funding for trace element research has become more difficult to obtain. What follows is this author's personal choice of topics.

Sampling and Analysis

It might be supposed that the basic problems of environmental sampling have been adequately studied and satisfactory techniques evolved after 30 or more years of work. Sadly, this is not necessarily so. Of course there has been progress. Water and blood sampling are arguably the most advanced in this respect, since sampling methods have been well researched and most practitioners now adopt broadly similar sampling protocols. In geochemical prospecting, stream sediment sampling is used worldwide, and Webb,[31] in the introduction to the Wolfson Geochemical Atlas, has described standardized procedures. Warren[32] has given an account of appropriate methodologies when sampling twigs from trees in biogeochemical prospecting. For soils, however, there is a lack of sampling guidance based on careful research.

Davies and Wixson,[33] working in the lead and copper mining county of Madison, Missouri, collected soils on a regular 1 km grid over an area 8 × 11 km. A special feature of their work was that they duplicated the field sampling. Samples were taken by bulking auger cores from around a circle of 2 m in diameter. A second sampling circle was established 10 m away from the center of the first and the sampling repeated. These cores were analyzed separately. Subsequently, the duplicated data were combined and the average values used to characterize the whole area. For lead, the relative standard deviation for the investigational area was 223%. But for the duplicated field samples the relative standard deviation was 73% and the correlation coefficient for the duplicates was $r = 0.85$ ($p < 0.001$) (Figure 2). It was concluded that the local variability was less than the area variability and duplicated field sampling was not necessary.

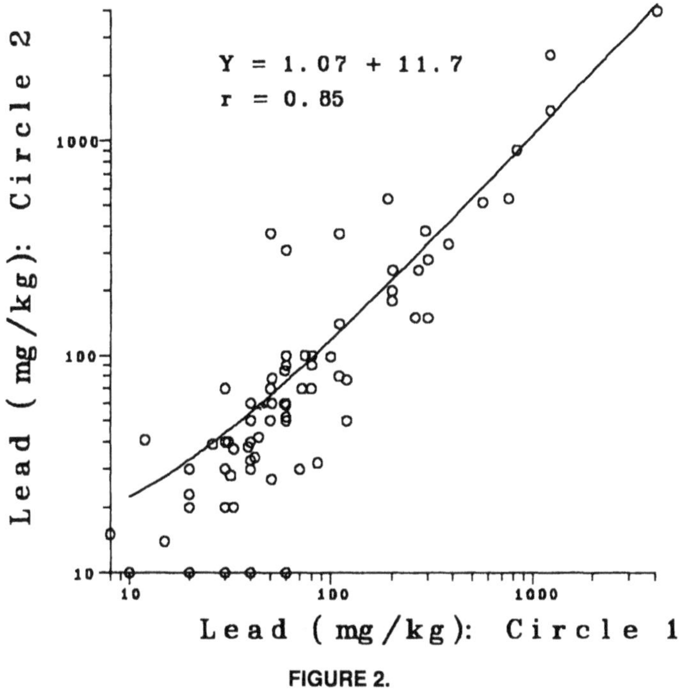

FIGURE 2.

Jackson et al.[34] have shown that simple random soil sampling over an area 1 km square gave relative standard deviations for lead and copper of 36 and 15%, respectively. When the area was stratified into woodland, agricultural land, and land within 50 m of roads, the relative standard deviation for lead improved to 21% and to 11% for copper. In general, however, far too little work has been done on the lateral variability of trace metals. Research is needed for soil heterogeneity at various scales such as 1 m, 10 m, 100 m, and 1000 m.

There are other uncertainties. There has been no comparative evaluation of different sampling tools such as augers. Sample preparation practices vary widely. Workers with a soil science background usually grind soil to pass a sieve of 2 mm aperture, but those coming from the earth sciences usually prepare much finer fractions (<100 μm or <64 μm). Since, in many soils, the finer fractions tend to have higher concentrations of some trace metals, comparing data obtained by different workers is difficult. Even the terminology of sample preparation may be confusing. Many authors quote their sieve aperture sizes in terms of mesh numbers and are apparently oblivious to the fact that these are differently defined in Great Britain and the U.S. There are also problems at the analysis and interpretation stages. Is it really necessary to grind further the <2 mm soil fine earth before digestion? How should plant material be washed so as at once to remove surface contaminants yet minimize loss of labile constituents? Or later, at the interpretation stage, what is achievable and acceptable analytical quality for soils plants in *routine* analysis? Trace element data sets are frequently bedevilled by some values being below the analytical detection limit. Most authors use the detection limit as the value in subsequent statistical interpretation, although this leads to bunching and truncation at lower values. Is there not a valid weighting procedure to overcome this problem?

These few comments outline the real difficulties in obtaining reliable soil, sediment, and, often, plant data. Yet the final interpretation can be no better than the original data. If these were not representative, the conclusions can hardly be valid.

Baselines, Backgrounds, Averages, and Contamination

In contrast with the 1960s, there is now a plethora of trace metal data for soils, sediments, and plants. The data are not very representative in global terms. Most case studies have taken place in the industrialized world, and the intertropical areas have been particularly neglected. Furthermore, attention has often focused on the "interesting" areas (geochemically anomalous or industrially polluted areas) and data tend to be scarce for the clean and normal.

Fundamental to the *pedological* approach to soil science is the concept that the soil landscape can be classified into soil *series*, each formed from a mosaic of *pedons*, the individual soil bodies. A soil series consists of soils of similar parentage and formed under similar environmental conditions as evidenced by their having similar assemblages of soil horizons. Such a classification is based upon comparisons of observable properties such as texture, color, or the nature of horizon boundaries. It is attractive to suppose that soils within a given series would also have a similar chemical composition. Indeed, this is inherent in the distinction between some *groups* of soil: rendzinas and rankers are both relatively undifferentiated shallow soils, but the former are calcareous whereas the latter are not. One can hypothesize that the trace metal composition of a given horizon in a given soil series could be described in terms of a mean and a range of values. Another horizon in that soil or the same horizon in a different soil could be distinguished by a comparison of means and ranges.

In geochemistry a long period of debate began when Ahrens[35] published a paper suggesting that, for trace elements in igneous rocks, their distribution was *log-normal* not *normal* and that this was a fundamental law of geochemistry. The paper was based on the premise that there was a need for an accurate determination of the abundances of elements in rocks, and disputed the view of some "that the distribution of an element is likely to be so complex and irregular in geological materials that the mathematical handling of the observations would be almost hopelessly difficult and of little or no practical significance." Essentially, this is where trace metal research in soils, plants, waters, and animals is today, and we too must ask whether our element distributions can be described in straightforward statistical terms.

These comments may seem somewhat abstruse, but there is a real practical dilemma. Unless trace metal concentrations in different areas can be estimated reliably and a valid estimation of background levels made, legislators may introduce mandatory or guideline levels which are unrealistically low. A problem with the published data is the generally unsatisfactory way in which they have been reported and interpreted. Too often arithmetic means are used when inappropriate. Too often comparisons are made with other data by uncritical selection and inappropriate techniques.

These problems of data interpretation are especially pertinent when *contamination* or *pollution* are discussed. The two terms are often used interchangeably, and equally often in a way that suggests the contaminant is harmful. This usage is not satisfactory, since a substance which is harmful to one organism may not be harmful to another at that concentration. Science depends for its progress on careful and reliable measurement, careful and reliable interpretation, and rigorous definition of terms.

It is much preferable to define a contaminant as a substance found in a given medium at a concentration higher than that which one would expect from other considerations and where the source of the additional concentration of the substance appears to be human activity. Essential to such a definition is the need for reliable values for background or baseline concentrations. Again, these are terms which are not satisfactorily defined. One would like to restrict baseline concentrations to descriptions of environments unaffected by human activity and background concentrations to evaluations of the general concentrations of trace metals found in a particular area.

It is a second and separate question to ask whether the observed concentration of the contaminant is enough to cause harm to some organism. If it is, then the contaminant is also a pollutant.

Of course, these definitions can lead to apparent absurdities: a farmer limes a naturally acid field, and the additional calcium is therefore a contaminant but is not a pollutant. Any attempt to define a contaminant as something capable of doing harm founders on the question, harmful to what?

It would be of great interest to establish the concentrations of many trace elements, especially pollutants such as Cd or Pb, in environmental materials in areas where human activity has been minimal. The suspicion is that human activity tends to generate widespread, low level contamination, and, therefore, even at locations far from known industrial sources the metal contents of soils, sediments, or plants are not necessarily "natural."

Davies[36] made a graphical and statistical analysis of soil lead data from four areas of England and Wales and concluded that the (geometric) mean content of apparently uncontaminated agricultural soils was 42 mg Pb/kg. In a later study for Wales where soils were sampled on a regular grid across the country the geometric mean lead content was also 42 mg/kg (B. E. Davies, unpublished results). In contrast, Ure and Berrow[37] reported the geometric mean for Scottish soils as 14 mg Pb/kg. Shacklette et al.[38] reported a geometric mean of 16 mg Pb/kg for 863 soil samples in the U.S. In a lead mining county (Madison) of southeast Missouri, the geometric mean was 60 mg/kg.[33] The mean content of 4970 reported soil values worldwide has been reported as 29.2 mg/kg.[37] There is general agreement that the lead content of crustal rocks is approximately 14 mg/kg and that the lead content of soils in remote areas is also approximately 15 to 25 mg/kg.

A number of reports have suggested that lead levels are higher than might be expected in rural areas of industrialized countries or states and that long distance transportation of lead and other pollutants is a contributor to local soil lead contents. Reiners et al.[39] reported that the lead contents of soils in remote parts of New England were higher than expected and still rising. Johnson et al.[40] analyzed buried organic horizons from forests in the northeastern U.S. and concluded that the lead content of the forest floor had risen five- to tenfold in the past century. Moyse and Fernandez[41] studied a montane ecosystem in western Maine and interpreted their soil lead data as being in agreement with "the notion of a gradient of decreasing metal deposition with distance from the southern New England urban corridor."

Elwood et al.[42] sampled garden soils at 56 cottages on three remote British islands where petrol-driven vehicles have never, or only very occasionally, been used and where industry in its accepted sense is entirely absent. The mean lead contents of the garden soils on the three islands were 62, 87, and 59 mg/kg. The conclusion drawn here is that human habitation in its own right tends to raise trace metal contents in soils around that habitation.

These local studies support the global assessment by Nriagu and Pacyna[43] that man-induced mobilization of trace metals into the biosphere is now very significant. They calculated that if total metal inputs were dispersed uniformly over the global cultivated land area of 16×10^{12} m^2 the annual rates of metal application would be 1.0 g ha^{-1} for Cd and Sb to about 50 g ha^{-1} for Pb, Cu, and Cr.

Chemical Forms, Speciation, and Bioavailability of Trace Metals

The last 15 years or so have seen an increasing interest in the chemical form of trace metals in environmental media. Determinations of the total metal content of soils or sediments have a valuable place in characterizing the properties of those materials in an absolute sense, but it has long been understood that only the more soluble metal fractions have any biological significance. Agricultural scientists of the 19th century attributed the different "feeding powers" of plants to differences in the total acidity of the root sap. This led J. B. Dyer, in the late 19th century, to propose the use of 1% citric acid as an extractant for soil phosphate; a useful account of this and other early work is to be found in Russell.[1]

From these origins sprang the soil scientist's approach to investigating the chemical form of a trace metal in soil. Inherent in this approach is to suppose that the plant root derives nutrients from the soil solution and that the composition of the soil solution is maintained through dissolution of other soil phases. This has led to the use of a variety of soil extractants, each of which selectively leaches the soil; the amount of a metal thereby extracted is a reflection of the labile soil pool which replenishes the soil solution. Two of the most popular reagents are EDTA and DTPA. EDTA (ethylene diamine tetraacetic acid) was proposed by Viro[44,45] for assessing the Ca, Cu, and Zn status of acid forest soils. DTPA (diethylenetriaminepentaacetic acid) was introduced by Lindsay and Norvell[46] as a soil test for Zn, Fe, Mn, and Cu. It must be remembered, however, that the objective of using these extractants is to identify soils where fertilizer supplementation is required: "A DTPA soil test was developed to identify near neutral and calcareous soils with insufficient available Zn, Fe, Mn, or Cu for maximum yields of crops.[46] Often the results from these soil tests are used to estimate the required amount of fertilizer applications.

Environmental and toxicological interest in trace metals is concerned with the concentration of the metal in the tissues of food or fodder crops. There is a need to develop extractants which will predict uptake. For the most part researchers have used the same extractants as those developed by agronomists in identifying responsive soils. There should be no *a priori* supposition that an extractant which successfully predicts fertilizer need will also predict metal uptake.

Early work has been based on the assumption that uptake effectively depends on supply. Data are then interpreted by linear regression analysis. It has long been understood that uptake is modified by many factors such as soil pH or competing ions. The last decade has seen a growing interest in more complex multivariate linear models where these other factors are taken into account in the soil content vs. uptake model.[47–50] This approach merits more attention.

Another complexity arises from the nature of plant uptake of trace metals. The models used so far, even those which allow for the influence of, e.g., soil pH or cation exchange capacity, assume a linearity of uptake. But this is not so. Baker[51] has proposed that plants generally fall into one of three groups: accumulators, where plants concentrate metals from high or low soil levels; indicators, where tissue concentrations reflect the soil concentrations and a linear relationship is found; excluders, where tissue concentrations are maintained at a low level until, at higher soil concentrations, the restrictive mechanism breaks down and unrestricted transport results.

It is to be hoped that in future years more complex, and hence more realistic, models will be suggested. It is also hoped that statisticians might help in suggesting techniques other than multiple linear regression which is not always applicable.[52]

When one considers the importance of the soil solution and the root/solution interface, the relative lack of research is striking. The reason probably lies in the difficulties experienced in obtaining the soil solution and satisfactorily analyzing, even for total metal content, the low concentrations of trace metals which are obtained. Generally, authors have chosen a method for obtaining soil solution on grounds of practicability. Campbell et al.[53] have reviewed the techniques available for obtaining the soil solution from field-moist soils and selected for their studies a centrifugation method and an immiscible liquid displacement method. They concluded that the natural variation in the chemistry of soil solutions over a few days or weeks was small in relation to other sources of variation, that many of the major solutes showed a cyclical seasonal variation, and there were usually significant differences between soil solutions from different soil series as well as between areas of different land use. These conclusions are supportive of the hypothesis advanced above, that the pedological classification of the soil landscape into series is perhaps a useful basis for characterizing the chemical composition of soils. But before these ideas can be explored further, it will be necessary to compare directly the various methods available for obtaining the soil solution with a view to standardization. Some progress has been made here by Van den Ende[54,55] and Sonneveld et al.[56] who have compared the usefulness of saturation extracts, pressure extracts, and fixed ratio water extracts for routine testing of glasshouse soils. It must be noted, though, that these several reports are concerned with major nutrient ions, and no similar work has been reported for trace elements.

The root itself is not a passive absorber of nutrients from the soil solution but actively alters the chemical conditions in the rhizosphere. Experimental plant physiology has made good progress in this topic. For example, Hauter and Mengel[57] have described the use of an antimony electrode of diameter 0.5 mm to determine pH at a root surface. In a noncalcareous, sandy soil the root surface pH value was 1 unit lower than the value found 0.5 to 1 cm away from the root. Gijsman[58] measured rhizosphere pH along the root axis of Douglas fir using glass microelectrodes. The experimental soil was strongly acid (pH 3.9). Supplying fertilizer nitrogen in the ammonium form resulted in acidification of the rhizosphere (–0.3 pH mean compared with bulk soil) along nearly the entire root, whereas with nitrate supply alkalization (+0.9 pH compared with bulk soil) extended over the entire growth zone of the root. It was concluded that nitrate nutrition enabled the plant locally to protect its most essential root zone from extreme acidity by raising rhizosphere pH. Nye[59] has published a detailed mathematical discussion of the estimation of nutrient uptake by roots releasing solubilizing agents. He has applied the theory to phosphorus uptake by rape which releases hydrogen ions from its roots and thereby mobilizes phosphate in otherwise deficient conditions.

A root does not dangle naked in the soil solution. Numerous plant species are embedded in a mucilaginous layer, a feature first reported by Jenny and Grossenbacher.[60] Morel et al.[61] have described how mucilage binds trace metals in the order Pb > Cu > Cd, and the resulting modification of the flux of metals to the root would decrease the transport of Pb and Cu whereas Cd would reach the root relatively unimpeded.

Just as a root does not dangle bare in the soil solution, an ion does not float naked. There is a wide range of physicochemical forms (species) and our knowledge of them comes more from water chemistry than soil chemistry. The toxicity of a metal depends very much on its chemical form. Minamata disease was a vivid example of this since methyl mercury, being lipid soluble, is rapidly absorbed but not excreted. An interesting development in recent years has been the writing of computer programs to predict the stable species in aqueous solutions given certain conditions. Sposito[62] has described the use of GEOCHEM in this respect. Detailed studies of the soil solution are now greatly needed.

Not only is it important to know more about metal species in solutions, it is also necessary to know more about the chemical form of these metals in soils and sediments. The general approach here is to extract the soil or sediment with several extractants, separately or sequentially, each of which is presumed to attack only a limited number of pools, or even one pool of the metal. An early soil fractionation scheme was that of McLaren and Crawford,[63] whose scheme distinguished five fractions for copper: soil solution and exchangeable copper, copper weakly bound to specific sites, organically bound copper, copper occluded by oxide material, and residual copper. There are many such schemes, but all have one defect in common. They are largely empirical, and the forms which they extract are not defined. In particular they lack a detailed investigation of their mode of operation. Here, too, is a major area which needs rescuing from its present empiricism.

Modeling

The foundation of good science is a systematic and careful acquisition of basic data followed by a rigorous and objective interpretation of those data. In the experimental sciences it is important that any experiment is described in sufficient detail that it can be repeated by another research group in order to test the published conclusions. In much environmental research this repetition of work is rarely possible because of the expense and difficulties inherent in repeating another group's survey. In reality, of course, most experimental work is never duplicated except as student class exercises. Published work, however, should lead to hypotheses which can be tested and, hence, predictive models derived. This is where much of trace metal research is weakest. Too many publications are still focused on data gathering, and testable models are still in their infancy.

Water engineers are developing models for sediment transport in rivers and estuaries, and these show promise for application to contaminated sediments. On a smaller scale, Marcus[64,65] has attempted to model the dispersion of copper and other trace metals in streams. The physics of atmospheric plume dispersal are also well understood, but our understanding of the dynamics of trace metal movement through the food chain is still very primitive. The challenge of the 1990s for environmental science is to move beyond simple data gathering and start to be predictive. Only then can we hope to manage and not mismanage our environment.

Trace Metals and Health

Finally, a move away from soils and plants to, perhaps, the ultimate question. How do trace metals affect human health? Again, one goes back to the late 1950s to see where the story started. In a stimulating book, Voisin[66] suggested that human cancer might be linked to the geochemical environment, especially through copper. Since then, numerous authors have compared disease maps with geochemical maps and have commented on resemblances (e.g., Piispanen[67]).

Research in environmental geochemistry and health over the last 25 years has been dominated by the toxicology problems of lead and cadmium, especially lead. This is not an appropriate occasion to attempt any sort of review of the past work on these elements, but the outline story is familiar to any worker in trace metals. It has been shown that lead in air derived from the combustion products of alkyl lead added to petrol and released in vehicle exhaust gases raises body burdens. Enzyme activity, especially for amino-levulinic acid dehydrase (ALAD), which catalyzes one stage in the formation of haeme, is increasingly depressed as blood lead increases. There is a body of, admittedly disputed, evidence that relatively low blood levels of lead can cause detectable mental impairment. The consequences of these reports have included a reduction in the regulatory or advisory levels in many countries for lead in blood, a phasing out or outright banning of lead in petrol and paints, and a general move to the use of food tins manufactured without lead solder. These have had the desired effect of reducing environmental and blood lead levels in countries which have enforced these controls. Cadmium is less of a success story. Despite the years of effort it is still possible to argue that it is an element looking for a disease to cause in terms of environmental exposure.

Perhaps now that the main peak of health-related studies on lead has passed, it is time to return to the original questions of the 1950s and 1960s, namely, do environmental trace metals affect human health? The debate and the research have not stopped over the years. Warren has consistently argued for a role for lead in the etiology of multiple sclerosis, and in a recent paper[68] has reviewed the possible health implications of Hg, I, Cd, Cu, Al, Pt, Se, Cu, As, Zn, and Au. Aluminium, in particular, is the subject of much concern since it may have a causative role in Alzheimer's disease. Irvine et al.[69] have studied multiple sclerosis clusters in Saskatchewan and have proposed a link with excess Pb, Ni, and Zn in soil. One should not claim that environmental trace elements necessarily cause these diseases. It is enough to show they may have an influence, since this could lead to public health measures which, at least, could reduce our health problems.

REFERENCES

1. Russell, E. J. *Soil Conditions and Plant Growth.* 7th Ed. (London: Longmans, Green and Company, 1937).
2. Wild, A. *Russell's Soil Conditions and Plant Growth.* 11th Ed. (Harlow, England: Longman Scientific and Technical, 1988).
3. Mertz, W. "The Essential Trace Elements," *Science* 213:1332–1338 (1981).
4. Arnon, D. I. and P. R. Stout "Molybdenum as an Essential Element for Higher Plants," *Plant Physiol.* 14:599–602 (1939).
5. Stiles, W. *Trace Elements in Plants and Animals* (Cambridge, England: Cambridge University Press, 1946).
6. Caldwell, T. H. "Copper Deficiency in Crops. I. Review of Past Work," in Trace Elements in Soils and Crops: Maff Tech. Bull. 21, London: HMSO, 1971.
7. Tinker, P. B. "Trace Elements in Arable Agriculture," *J. Soil Sci.* 37:587–601 (1986).

8. McHargue, J. S. "The Role of Manganese in Plants," *J. Am. Chem. Soc.* 44:1592–1598 (1922).
9. Warrington, K. "The Effect of Boric Acid and Borax on the Broad Bean and Certain Other Plants," *Ann. Bot.* 37:629–692 (1923).
10. Sommer, A. L. and C. B. Lipman. "Evidence on the Indispensable Nature of Zinc and Boron for Higher Green Plants," *Plant Physiol.* 1:231–249 (1926).
11. Lipman, C. B. and G. Mackinney. "Proof of the Essential Nature of Copper for Higher Green Plants," *J. Pomol.* 10:593–599 (1932).
12. Sommer, A. L. "Copper as an Essential for Plant Growth," *Plant Physiol.* 7:339–345 (1932).
13. Broyer, T. C., A. B. Carlton, C. M. Johnson, and P. R. Stout. "Chlorine — a Micronutrient Element for Higher Plants," *Plant Physiol.* 29:526–532 (1954).
14. Brownell, P. F. and J. G. Wood. "Sodium as an Essential Micronutrient Element for Atriplex Vesicaria, Heward," *Nature (London)* 179:635–636 (1957).
15. Eskew, D. L., R. M. Welch, and E. E. Cary. "Nickel: An Essential Micronutrient for Legumes and Possibly All Higher Plants," *Science* 222:621–623 (1983).
16. Schwarz, K. "Recent Dietary Trace Element Research, Exemplified by Tin, Fluorine, and Silicon," *Fed. Proc.* 33:1748–1757 (1974).
17. Griffith, J. J. "Influence of Mines Upon Land and Livestock in Cardiganshire," *J. Agric. Sci. (Cambridge)* 9:366–395 (1919).
18. Bertrand, G. and Y. Okada. "Sur l'Existence du Plomb dans la Terre Arable," *C. R. Acad. Sci. Paris* 196:826–828 (1933).
19. Warren, H. V. and R. E. Delavault. "Observations on the Biogeochemistry of Lead in Canada," *Trans. R. Soc. Can.* 54:11–20 (1960).
20. Cannon, H. L. and J. M. Bowles. "Contamination of Vegetation by Tetraethyl Lead" *Science* 137:765–766 (1962).
21. Purves, D. "Contamination of Urban Garden Soils with Copper and Boron," *Nature (London)* 210:1077–1078 (1966).
22. Le and H. H. Riche. "Metal Contamination of Soil in the Woburn Market Garden Experiment Resulting from the Application of Sewage Sludge," *J. Agric. Sci. (Cambridge)* 71:205–208 (1968).
23. Wallace, T. "The Diagnosis of Mineral Deficiencies in Plants by Visual Symptoms." London: HMSO, 1943.
24. Mitchell, R. L. and J. W. S. Reith. "The Lead Content of Pasture Herbage," *J. Sci. Food Agric.* 17:437–440 (1966).
25. Marten, G. C. and P. B. Hammond. "Lead Uptake by Bromegrass from Contaminated Soils," *Gronomy J.* 58:553 554 (1966).
26. MacLean, A. J., R. L. Halstead, and B. J. Finn. "Extractability of Added Lead in Soils and Its Concentration in Plants," *Can. J. Soil Sci.* 49:327–334 (1969).
27. Motto, H. L., R. H. Daines, D. M. Chilko, and C. K. Motto. "Lead in Soils and Plants: Its Relationship to Traffic Volume and Proximity to Highways," *Environ. Sci. Tech.* 4:231–238 (1970).
28. Page, A. L., T. J. Ganje, and M. S. Joshi. "Lead Quantities in Plants, Soil and Air Near Some Major Highways in Southern California," *Hilgardia* 41:1–31 (1971).
29. Lagerwerff, J. V., W. H. Armiger, and A. W. Specht. "Uptake of Lead by Alfalfa and Corn from Soil and Air," *Soil Sci.* 115:455–460 (1973).
30. Shacklette, H. T. "Cadmium in Plants." Geological Survey Bulletin 1314-G Washington, D.C.: U.S. Government Printing Office, 1972.

31. Webb, J. S. *The Wolfson Geochemical Atlas of England and Wales* (Oxford: The Clarendon Press, 1978).
32. Warren, H. V. "Biogeochemistry, Trace Elements, and Mineral Exploration," in *Applied Soil Trace Elements*, Davies, B. E., Ed. (Chichester: John Wiley and Sons, 1980).
33. Davies, B. E. and B. G. Wixson. "Trace Elements in Surface Soils from the Mineralised Area of Madison County, Missouri, U.S.A.," *J. Soil Sci.* 36:551–570 (1985).
34. Jackson, K. W., I. W. Eastwood, and M. S. Wild. "Stratified Sampling Protocol for Monitoring Trace Metal Concentrations in Soil," *Soil Sci.* 143:436–443 (1987).
35. Ahrens, L. H. "The Lognormal Distribution of the Elements," *Geochim. Cosmochim. Acta* 5:49–73 (1954).
36. Davies, B. E. "A Graphical Estimation of the Normal Lead Content of Some British Soils," *Geoderma* 29:67-75 (1983).
37. Ure, A. M. and M. L. Berrow. "The Elemental Constituents of Soil," in *Environmental Chemistry Vol. 2* (Bowen H. J. M. London: Royal Society of Chemistry, 1982).
38. Shacklette, H. T., J. C. Hamilton, J. G. Boerngen, and J. M. Bowles. "Elemental Composition of Surficial Materials in the Conterminous United States," Geological Survey Professional Paper 574-D. Washington, D.C.: U.S. Government Printing Office, 1971.
39. Reiners, W. A., R. H. Marks, and P. M. Vitousek. "Heavy Metals in Subalpine and Alpine Soils of New Hampshire," *Oikos* 26:264–275 (1975).
40. Johnson, A. H., T. G. Siccama, and A. J. Friedland. "Spatial and Temporal Patterns of Lead Accumulation in the Forest Floor in the Northeastern United States," *J. Environ. Qual.* 11:577–580 (1982).
41. Moyse, D. M. and I. J. Fernandez. "Trace Metals in the Forest Floors at Saddleback Mountain, Maine in Relation to Aspect, Elevation and Cover Type," *Water Air Soil Pollut.* 34:385–397 (1987).
42. Elwood, P. C., R. Blaney, R. C. Robb, A. J. Essex-Cater, B. E. Davies, and C. Toothill. "Lead Levels on Traffic-less Islands," *J. Epidemiol. Comm. Health* 139:256-258 (1985).
43. Nriagu, J. O. and J. M. Pacyna. "Quantitative Assessment of Worldwide Contamination of Air, Water and Soils by Trace Metals," *Nature (London)* 333:134–139 (1988).
44. Viro, P. J. "Use of Ethylene Diaminetetraacetic Acid in Soils Analysis. I. Experimental," *Soil Sci.* 79:459–465 (1955).
45. Viro, P. J. "Use of Ethylene Diaminetetraacetic Acid in Soil Analysis. II. Determination of Soil Fertility," *Soil Sci.* 80:69–74 (1955).
46. Lindsay, W. L. and W. A. Norvell. "Development of a DTPA Soil Test for Zinc, Iron, Manganese and Copper," *Soil Sci. Soc. Am. J.* 42:421–428 (1978).
47. Gough, L. P., J. M. McNeal, and R. C. Sverson. "Predicting Native Plant Copper, Iron, Manganese and Zinc Levels Using DTPA and EDTA Soil Extractants, Northern Great Plains," *Soil Sci. Soc. Am. J.* 44:1030–1036 (1980).
48. Iyengaar, S. S., D. C. Marten, and W. P. Miller. "Distribution and Plant Availability of Soil Zinc Fractions," *Soil Sci. Soc. Am. J.* 45:735–739 (1981).
49. Soon, Y. K. and T. E. Bates. "Chemical Pools of Cadmium, Nickel and Zinc in Polluted Soils and Some Preliminary Indications of Their Availability to Plants," *J. Soil Sci.* 33:477–488 (1982).

50. Tills, A. and B. J. Alloway. "An Appraisal of Currently Used Soil Tests for Available Copper with Reference to Deficiencies in English Soils," *J. Sci. Food Agric.* 34:1190–1196 (1983).
51. Baker, A. J. M. "Accumulations and Excluders — Strategies on the Response of Plants to Heavy Metals," *J. Plant Nutr.* 3:643–654 (1981).
52. Webster, R. "Is Regression What You Really Want?" *Soil Use Manage.* 5:47–53 (1989).
53. Campbell, D. J., D. G. Kinniburgh, and P. H. T. Beckett. "The Soil Solution Chemistry of Some Oxfordshire Soils: Temporal and Spatial Variability," *J. Soil Sci.* 40:321–339 (1989).
54. Van den Ende, J. "Estimating the Chemical Composition of the Soil Solution of Glasshouse Soil. 2. Relationships Between the Compositions of Soil Solution and Aqueous Extracts," *Neth. J. Agric. Sci.* 37:323–334 (1989).
55. Van den Ende, J. "Estimating the Chemical Composition of the Soil Solution of Glasshouse Soil. 1. Compositions of Soil Solution and Aqueous Extracts," *Neth. J. Agric. Sci.* 37:311–322 (1989).
56. Sonneveld, C., J. Van den Ende, and S. S. De, Bes. "Estimating the Chemical Composition of Soil Solutions by Obtaining Saturation Extracts or Specific 1:2 by Volume Extracts," *Plant Soil* 122:169–175 (1990).
57. Hauter, R. and K. Mengel. "Measurement of pH at the Root Surface of Red Clover (Trifolium Pratense) Grown in Soils Differing in Proton Buffering Capacity," *Biol. Fertil. Soils* 5:295–298 (1988).
58. Gijsman, A. J. "Rhizosphere pH Along Different Root Zones of Douglas-fir (*Pseudotsuga menziesii*), as Affected by Nitrogen Source," in *Plant Nutrition–Physiology and Application*, van Beusichem, M. L., Ed. (Amsterdam: Kluwer Academic Publishers, 1990).
59. Nye, P. H. "On Estimating the Uptake of Nutrients Solubilized Near Roots or Other Surfaces," *J. Soil Sci.* 35:439–446 (1984).
60. Jenny, H. and K. Grossenbacher. "Root Soil Boundary Zone as Seen by Electron Microscope," *Soil Sci. Soc. Am. Proc.* 27:273–277 (1963).
61. Morel, J., M. Mench, and A. Guckert. "Measurement of Pb, Cu, Cd Binding with Mucilage Exudates from Maize (*Zea mays* L.) Roots," *Biol. Fertil. Soils* 2:29–34 (1986).
62. Sposito, G. "The Chemical Form of Trace Metals in Soils," in *Applied Environmental Geochemistry*, Thornton, I., Ed. (London: Academic Press, 1983).
63. McLaren, R. G. and D. Crawford. "Studies in Soil Copper: I. The Fractionation of Copper in Soils," *J. Soil Sci.* 24:172–181 (1973).
64. Marcus, W. A. "Dilution Mixing Estimates of Trace Metal Concentrations in Suspended Sediments," *Environ. Geol. Water Sci.* 14:213–219 (1989).
65. Marcus, W. A. "Copper Dispersion in Ephemeral Stream Sediments," *Earth Surf. Proc. Landforms* 12:217–228 (1987).
66. Voisin, A. *Soil, Grass and Cancer* (London: Crosby Lockwood and Son, Ltd., 1959).
67. Piispanen, R. "Geochemical Interpretation of Cancer Maps of Finland," *Environ. Geochem. Health* 11:145–147 (1989).
68. Warren, H. V. "Geology, Trace Elements and Health," *Soc. Sci. Med.* 8:923–926 (1989).
69. Irvine, D. G., H. R. Schiefer, and W. J. Hader. "Geotoxicology of Mutiple Sclerosis: The Henribourg, Saskatchewan, Cluster Focus. II. The Soil," *Sci. Total Environ.* 77:175–188 (1988).

2

Trace Elements Concentrations in Soils and Plants of Western Europe

M. Angelone[1] and C. Bini[2]

[1] ENEA, C.R.E. Casaccia, Geochemistry Laboratory, Via Anguillarese, 301 — 00060 Rome, Italy
[2] Department of Crop Production, University of Udine, Via Fagagna, 208 — 33100 Udine, Italy

ABSTRACT

The literature on trace element distribution in soils of western Europe is reviewed.

Existing data on elemental concentrations in soils, micronutrient uptake by plants, and the relationships with human health are widely scattered in the literature, and usually refer to only a few elements. Lead, Cd, Cu, Cr, Ni, and Zn are the most studied, since most of them are toxic to humans, while others may be phytotoxic or deficient to plants.

The levels of most elements in soils of western Europe fall within the range considered "normal" for unpolluted soils. A slight contamination, however, occurs in many areas, especially in the neighborhood of industrial factories and along roads with intensive traffic.

Lead (Belgium, Netherlands, Scotland: 94, 270, 1000 mg/kg, respectively), Cd (France, Germany, Netherlands, Spain, Scotland: 2, 2.9, 5.3, 4.0, 2.4 mg/kg, respectively), Zn (Germany, Netherlands, Sweden, Scotland: 492, 1020, 318, 987 mg/kg, respectively), Cu (Netherlands, Spain, Scotland: 110, 400, 2500 kg, respectively), and Ni (Belgium, Spain, Scotland: 151, 300, 5000 mg/kg, respectively) are the most hazardous trace elements in the environment.

Trace elements contents in plants frequently indicate some enrichment with Pb (Austria, Italy: 665 and 4036 mg/kg, respectively); Zn (Austria, Belgium, Germany, Italy: 1537, 90, 94, 1606 mg/kg, respectively); Cu (Germany, Italy: 46, 360 mg/kg, respectively); Cd (Germany: 1.45 mg/kg); B (Finland 37.7 mg/kg); and sometimes with B, Cr, and Co. A good correlation for the mean of certain elements (Cu, Zn, Mo, Se) between soil content and vegetation is recorded too.

Health hazards caused by food contamination with heavy metals have also been reported (e.g., cadmiosis, Pb, and Hg intoxication), although the mechanism governing the interactions between trace elements and human health are not well known.

INTRODUCTION

Trace elements have received increasing attention in recent years in western Europe. Many questions arise concerning their presence in soils, plants, water, and the food chain. Information concerning trace element distribution in the environment, micronutrients uptake by plants, and the relationships with human health, however, is still lacking in several countries.

Existing data on elemental concentrations in soils and plants are widely scattered in the literature, and usually refer to only a few elements. Pb, Cd, Zn, Cu, Cr, and Ni are the most studied, since some of them are toxic to humans, while others may be phytotoxic or induce deficiency symptoms in cultivated plants.

Unfortunately, many data, especially on minor elements, are of doubtful value due to the lack of adequate analytical techniques, precision, and accuracy. For many trace elements literature data are missing, or they are hardly available because of linguistic problems and difficulties in obtaining reports and papers published in local journals with limited circulation within the scientific community.

In order to avoid such difficulties in the preparation of this review, only papers published in readily available international journals or presented at international symposia were considered. A circular letter was distributed among European colleagues in order to facilitate the availability and circulation of papers or to attain data from different countries. Fifteen Western countries were recorded, listed in alphabetical order.

The search for data included only western European studies published over the past 15 years. If a research group published many reports from the same sites, only the major reports were included. In many cases, reports or single papers related to restricted areas have been considered only to avoid a total lack of information on such areas.

Table 1. Trace Element Content in Soils and Plants of Austria (mg/kg)

	Cu	Zn	Ni	Cr	Pb	Cd	Fe	Mn
(a)	17	65	20	20	15	0.20	13,300	310
Range								
m	10	6	1	1	2	0.01	7600	230
M	300*	8900*	57	50	3190*	0.71	19,980*	1300
(b)	12	1710*	—	—	670*	—	103	81
(c)	—	1537*	—	—	665*	—	654	82

Note: (a) Total levels of trace elements in soils; (b) average levels of extractable trace elements in soils; (c) average levels of heavy metals in plants. Samples from mined areas, indicated by asterisks, were excluded from the means.

From Aichberger,[1] Kazda and Glatzel,[3] Glatzel and Kazda,[4] and Sieghart.[2,5]

The objectives of this literature review are to:

1. compile and evaluate literature data on trace elements in western Europe,
2. attain data on metal toxicity in soils and plants,
3. determine the present level of pollution where an effect can be detected, and
4. identify fields where more research is needed.

AUSTRIA

Trace elements in soils and plants of Austria were investigated during the 1980s, especially with regard to land and forest degradation. Austria is a country where major industrial installations are lacking. Potential local pollutants are thus limited to occasional mills, farms, households, and to heavy traffic. However, atmospheric pollution from remote industrial settlements of conterminous regions may be of significance by affecting soils and forests, major land resources for this small country.

On an early soil survey Aichberger[1] and Sieghardt[2] identified many soil profiles developed from different parent materials (braunerde, pseudogley, rendzinas, and podzols) inclusive of some metalliferous dumps. In the same soil samples, the content and distribution of heavy metals Cu, Zn, Ni, Cr, Pb, and Cd were determined. The average contents and the ranges for these elements are reported in Table 1. Lead and Cd concentrations were higher in the topsoil and showed a positive correlation with the organic matter content. Copper, Zn, Ni, and Cr were distributed uniformly in the soil horizons, so that a dependence of their contents upon the parent material could be assumed, irrespective of the soil development. The data recorded did not show any metal accumulation in the soils investigated. More recent surveys in forest soils at different beech stands pointed to pollution phenomena occurring in the topsoil, especially near the tree stems.[3,4] It was concluded that significant amounts of heavy metals are deposited with stemflow, and strong soil acidification enhances heavy metal solubility, promoting leaching into deeper soil horizons.

The distribution of heavy metals in different organs of plants (roots, leaves) showed that the absorption of Pb is mostly confined to primary roots, and Pb reaches the above-ground plant tissues in the form of lead-chelates.[2,5] For zinc, a preferential accumulation in leaves was observed. Absorption and transport of Cu and Cd in plants was limited, since these metals have low concentrations in the soil. A dramatic increase in foliar levels of Mn was recorded together with a decrease of the Ca/Al ratio. Moreover, plant growth was reduced on the acidic substrates, and root damage could be observed. The data presented suggest that heavy metal compartmentation may be part of a tolerance mechanism diminishing a surplus of potentially toxic metals to plants.

BELGIUM

Research on trace element distribution in soils and their availability and uptake by plants in Belgium was stimulated in the early 1980s, following the increasing use of sewage sludge in agriculture and the consequent soil contamination hazard.[6] Moreover, air and soil pollution resulting from both automobile exhaust fumes and emissions from industrial plants has been recognized as a major problem.

Belgium is nearly entirely covered by sedimentary rocks, and more than 65% of the land consists of agricultural soils.[7] The Belgian soils are relatively coarse textured and acidic, with problems of erosion and plant nutrition.[8] Heavy fertilization increased the levels of macronutrients in soils even more than five times the international mean contents.

Among the micronutrients, the B, Cu, Zn, Pb, Mn, and Mo contents of soils and plants (Table 2) correspond more or less to the normal international level,[189] while the concentrations of Cr (average 90 mg/kg) and V (average 148 mg/kg) are close/over the levels considered phytotoxically excessive in surface soils.[189]

The selenium content of selected Belgian soils and its uptake by plants was investigated by Robberecht et al.[9] Total Se ranged from 0.04 to 0.27 mg/kg (average 0.11 mg/kg), and the levels of Se in plants ranged from 0.05 to 0.11 mg/kg (average 0.07); both are low levels in comparison to data from other countries and regarding the classification of Hamdy and Gissel-Nielsen (1976) quoted in Robberecht.[9]

According to De Temmerman et al.,[10] the concentration of trace elements for Belgian soils increases with increasing clay content, especially in soils under forest. This is particularly true for Pb, Hg, and Sb in the top layers of forest soils, where trace metal accumulation results from a long range transport and/or from industrial processes. Similar results were obtained by Albasel and Cottenie[11] for soils and plants collected near major highways, industrial areas, and urban

Table 2. Trace Element Content in Soils and Plants of Belgium (mg/kg)

	Cu	Zn	Ni	Cr	Pb	Cd	Fe*	Mn	Co	Mo	V	Rb	Sr	Se	B	Ga	Sn	Hg	As
a)	17	57	33	90	38	0.33	15.9	335	14.0	2.6	148	75	830	0.11	32	14	5		
Range																			
m	8	14	1	10	15	0.12	3.1	12	0.3	0.2	25	24	10	0.04	4	1	1		
M	30	130	151	300	94	0.54	35.0	1100	40.0	5.0	300	120	300	0.27	100	35	15		
(b)	17	34	14	—	42	0.25	—	120	5.3	0.21	47	—	—	—	0.52	—	—	0.08	6
(c)	9	90	—	—	50	—	—	102	—	0.32	—	—	—	0.07	4.4	—	—		

Note: (a) Averages and ranges of total concentrations of trace elements in soils; (b) averages of extractable trace elements in soils ; (c) averages of trace elements in plants.

* 10^3

From De Temmerman et al.,[7] Sillanpaa,[8] Robberecht et al.,[9] Albasel and A. Cottenie,[11] and Willaert and M. Verloo.[13]

gardens of Belgium. A remarkable contamination of soils with Zn, Pb, and Mn was recorded, while the uptake of trace elements by plants varied considerably as a function of soil conditions and the distance from the contaminant source. For instance, the Pb content of the plants was 3 to 10 times higher than the content of the reference plants, sampled at least 100 m from the roadside.

A striking difference in transfer of heavy metals from soil to plant was observed by Kiekens et al.[12] between the sandy and the heavy clay soils amended with sewage sludge. Plant uptake and accumulation proved to be higher for Zn and Cd than for Ni and especially Cu, which are associated with soil organic matter.

Recent observations on the extractability and plant uptake of heavy metals in Belgian soils (Willaert and Verloo[13]) suggest that Zn, Cu, and Cd fractions linked to clay and sesquioxides are more plant available than the fractions bound to organic matter and free Mn-oxides.

The results given in Table 2 emphasize that no hazardous accumulation of some heavy metals (with exception of Cr and V) occur in soils of Belgium. The nonessential trace elements such as Pb and Cd may be taken up by plants where they are present as soil contaminants. Therefore, their entrance into the food chain may be a long-range ecological and health hazard.

DENMARK

The concentrations of trace elements in Danish soils are not well recognized. During the 1970s a survey of the content of trace elements in 53 representative arable soils was carried out with the aim to record the levels of trace elements, which are of primary importance in evaluating the effects of increasing pollution, and to describe concentrations of the metals in relation to the other soil parameters (e.g., clay content, CEC, and organic matter) and to plant uptake.[14-16] Based on background levels recorded previously, applied trace element research received attention during the 1980s, provided that soils continuously receive heavy metals through atmospheric deposition and application of fertilizer, manure, and urban sludges.[17] Indeed, most heavy metals are accumulating in the topsoil of Denmark as well as all over western Europe, thus causing increasing metal concentrations in the environment and the food chains.

Several authors[18–23] focused their attention on heavy metals in agricultural soils, their sorption, and uptake by plants. These soils are formed mainly on slightly weathered quaternary deposits, derived from sediments of different geological age.[14] Data reported in Table 3 indicated that in the soils examined, no significant problems of heavy metal contamination were observed. Lead showed increasing soil concentrations over time in approximately half of the soils, with an annual increase of the order of 0.3% with respect to the soil content. It is suggested that most of the Pb input could be atmospheric deposition from leaded gasoline.

Table 3. Mean Levels and Ranges of Total Trace Elements in Danish Soils (mg/kg)

	Cu	Zn	Ni	Cr	Pb	Cd	Fe*	Mn	Co	Li	Sr
Average	11.1	7	6.9	21	16	0.24	12.0	233	2.35	10.4	75
Range											
m	1.5	7	1.5	10	2	0.10	4.2	104	1.60	3.2	17
M	24.0	76	25.0	39	31	0.84	22.4	315	3.20	19.4	118

Note: * $\times 10^3$

From Tjell and M. F. Hovmand[14] and Christensen and Tjell.[2]

Only a quarter of the soils studied showed increasing Zn soil concentrations over time, with annual increases of the order of 1%. The input of zinc is believed to originate from atmospheric deposition and its use as a micronutrient added to commercial fertilizers. Copper showed increasing soil concentrations in more than half of the soil samples, with an order of 1% of the soil content, and the main input is supposed to be from commercial fertilizers. Cadmium concentrations increased in approximately half of the soils investigated at a rate of 1 µg/kg/year, as a result of higher inflows (mainly from the atmosphere and P-fertilization).

Among the elements investigated, only Ni showed slightly decreasing soil concentrations, of the order of 0.5% of the soil content, probably because of a greater leaching from the topsoil than the other heavy metals.

Ion balance studies were implemented by Rasmussen[24] under the first observations of a forest dieback more severe than usually seen in Denmark, in comparison with similar plant damage in Sweden and the Netherlands.

In Greenland, in the absence of polluting industries, Pb and Cd concentrations were shown to be at the same level as those found in western Europe, suggesting that the source of pollution could possibly be the influence from the great industrial centers of the northern hemisphere, as long-distance transport of aerosols. Therefore, the influence from the environment is regarded a significant factor in the specific pattern of disease, particularly as regards the risk from Pb and Cd exposure.

FINLAND

Since 1978 a large amount of geochemical data have been collected in Finland and in Scandinavia as well, upon the first modest beginning of geochemical prospecting. [25]

Within the subject of soil science there have also been interest in many of the elements included in the geochemical prospecting, especially after the discovery of natural heavy metal poisoning in soils and the food chain.[26,27] A comprehensive registration of the chemical composition of Finnish food has given valuable support for geomedical evaluation in this country.[28,29] The geochemical data

Table 4. Mean Levels and Ranges of Extractable Trace Elements in Soils (a) and Plants (b) of Finland (mg/kg)

	Cu	Zn	Fe	Mn	Mo	Se	B
Soils	4.3	2.64	563	21.4	0.54	0.08	0.56
Range							
m	0.9	0.34	130	1.0	0.08	0.07	0.14
M	16.5	13.13	1995	63.2	2.07	0.10	1.88
Plants	6.1	34.50	31	45.5	0.37	—	8.98
Range							
m	3.6	17.40	19	18.0	0.01	—	3.48
M	13.1	47.20	57	423.0	1.88	—	37.30

From Marjanen,[31] Sillampaa,[8] and Nayha.[30]

attained from Finland (Table 4) showed relatively little influence by industrial activity and pollution for most of this country. Some elements showed significant geographical distribution patterns, others may have hazardous effects upon flora and fauna. In comparison with other nordic countries, low concentrations of Se have been observed.[8,30,31]

The average B content of Finnish soils is slightly lower with respect to the content of surface soils of the world, since it has been improved during the 1970s by addition of some B to most of the fertilizers used in Finland. The same is true of Cu. Previous investigations showed that Cu deficiency occurred quite frequently in Finland, especially in the peat soils and in the coarse textured mineral soils, and indicated that the possibility of response to Cu fertilization was somewhat more likely than in other countries.

Mean levels of Mn in Finnish soils and plants are lower than the international levels, being directly proportional to soil acidity. However, cases of "primary" deficiency (due to low total Mn content) are very rare compared to "secondary" deficiency (low availability mainly due to high soil pH). The average Mo content of Finnish soils is somewhat lower than the international average, though the present status does not indicate any serious Mo problem in Finland.

Mean levels of total Zn in both soils and plants are higher than the international average contents, and this has been found to inhibit the uptake of Mn. Moreover, large amounts of phosphorus added to soils with fertilizers may cause higher Zn contents in plants and grains.[8]

Trace elements (Mn, Se, Co, Zn) have recently been an object of interest among cancer researchers in Finland. Their significance in human metabolism has been investigated[31,32] and a synergism has been found among them. A deficiency in either Se or Mn, as well as an excess of Zn, for instance, weakens the function of the respiratory chain. The abundance of Zn in Finnish foodstuffs may prevent the absorption of Se, and a shortage of Mn in many areas may reduce the utilization of Se in the cells.

FRANCE

Since the past decades, the INRA (Institute National de la Recherche Agronomique) has paid attention to the problem of trace elements in soils and the agriculture in France, with special regard to deficiency-toxicity symptoms in plants. Several problems were encountered in the evaluation of the micronutrient availability in soils. Plant deficiency and toxicity symptoms have been reported in France since 1970.[33–35] Iron chlorosis of vines and fruit trees, Mn and Cu deficiencies in cereals (mainly wheat), B on sunflowers, Zn on maize, and Mo in several crops were most of the problems that appeared during the last decade and received practical solutions.[35,36]

Attention was focused especially on:

- better defining the role and the fate of trace elements in the soil,[37,38]
- quantifying the possible flows of metals trough the soil, their retention and transfer,[39,40]
- defining the tolerance threshold of trace elements to soils, particularly for the elements Cd, Hg, Pb (toxic to humans and animals), Zn, Cu, Ni (toxic to plants),[41,42] and
- studying the sorption and uptake of trace elements by plants and animals in relation to fertilizer and sewage sludge application.[43–48]

A summary of data on trace elements in French soils is reported in Table 5.

Human activities have been recently regarded as responsible for metal content rise in soils and plants in France as well as in other countries. In specific cases, repeated applications of fertilizers and sewage sludges were demonstrated to increase the average content of Cd, Cu, Zn, and Hg up to toxic levels.[42–44,46,48] Soil pollution from vehicle fumes and industrial plants was shown as well.

The general conclusion is that the French soils affected by metal pollution cover around one million hectares. The levels of heavy metals (Zn, Cu, Cd, Pb) do increase mostly upon fertilization and sewage sludge applications on agricultural soils. Metal bioavailability also was increased as a result of soil acidification, with the exception of Cd. Total inflow of Cd would be mainly due to phosphate fertilizers and atmospheric deposition. Contributions from sludge application may be important to inflows of Cd as well as Mn, Zn, Ni, and Cr. Contribution from industrial plants and vehicles may account for Zn, Cd, Cu, Cr, and Pb inflows in the environment.

GERMANY

Investigations connected with trace and toxic metals in soils have been conducted in Germany since the mid 1970s, in order to both assess the micronu-

Table 5. Mean Levels of Total Trace Elements in French Soils (mg/kg)

	Cu	Zn	Ni	Cr	Pb	Cd	Mn	Co	Se	B	Hg
Average	13	16	35	29.0	30	0.74	538	8	0.03	21.0	0.04
Range											
m	3	5	24	2.5	22	0.05	38	1	0.02	8.5	0.03
M	20	38	56	56.0	40	2.00	1700	14	0.10	35.0	0.05

From Aubert and Pinta,[48a] Moré and Coppenet,[43] Lemaire et al.,[44] Dejou et al.,[36] and Juste and Solda.[48]

trient availability and point out the continuing pollution of soils by heavy metals. The recognition of the "new type" of forest decline observed all over central Europe[49] and the associated hazards connected with trace element deficiency or toxic metal enrichment in the food chain resulted in a variety of hypotheses and opinions for possible causes.[50] Particular attention was given to Al, as it has been proposed that soil acidification caused by acid deposition in Europe has led to the release of toxic levels of Al in soil solution, mortality of fine roots, and ultimately forest dieback. [51]

In the late 1970s and early 1980s, the attention of German soil scientists, was focused on the trace element distribution in soil profiles, aimed at understanding their background levels and mobility within the soil.

Schilichting and Elgala[52] recorded positive correlations of Mn, Zn, Cu, and Co with clay and CEC (respectively $r = 0.54$–0.60; 0.76–0.70; 0.68–0.78; 0.76–0.67 for clay and CEC) and also with iron in soil profiles differing considerably in horizon pattern and parent materials. Deviations for all elements are ascribed to lithogenic differences in soils. The mobility of the heavy metals (EDTA or NH_4 Ac-extractable) in topsoil could be related to the organic matter and the pH. Similar results were obtained by Fassbender and Seekamp,[53] who found significant correlations between ion activities and NH_4Cl-extractable fractions of Cd, Co, Ni, and Pb ($r = 0.88$, 0.89, 0.97, and 0.79, respectively).

Murad and Fischer[54] investigated the trace element distribution in soils of a rural region in Western Germany and subdivided the elements investigated into two groups: (i) those not affected by industrial pollution (Rb, Sr, Zr, Y) and (ii) those who are characteristic constituents of industrial effluents.

Of all the investigated elements, a linear correlation exists between Pb and organic matter content ($r = 0.75$), which confirms the affinity of this element for O.M. and, possibly, extraneous additions as a result of long-distance environmental pollution.

Schwertmann et al.[55,56] examined the trace element distribution (Fe, Cu, Zn, Pb, Cd, Co, and Ni) in pedosequences from different parent materials and found that the total trace element amounts, irrespective of the soil development, lay within the normal ranges of these elements, except a few samples with high concentrations of Zn and Cd. In the topsoil, the elements investigated were accumulated in the order Cu < Zn < Cd < Pb, partly due to pedogenetic factors and partly to atmospheric input. An increase in the mobile fractions in the solum was generally recorded. With regard to the affinity toward the organic matter, Cd and Pb appear to behave similarly. Moreover, the total element balance indicates a slight gain of Pb and loss of Zn, whereas the total amounts of the other elements were essentially unchanged.

The inputs and outputs of trace elements in different ecosystems (soil-water-plant) and which processes are primarily responsible for redistribution of these elements were the items discussed in the early and mid 1980s in many papers.[57–62] All the authors point to heavy annual inputs of elements such as Cu, Zn, Cd, and Pb from the atmosphere. Air pollution from industry, home firing, and motor vehicles is the probable source. Instead, other elements (Be, Cr, Fe, Co, and Ni) have small atmospheric inputs, suggesting that, besides input with precipitation and dry deposition, the requirements for biomass buildup must be supplied mainly by internal sources, probably weathering of minerals. A summary of the trace element concentrations in soils of Germany is reported in Table 6.

Other investigations[51,63–66] show that the mobility and availability of heavy metals is high at strongly acid soil reaction. Therefore, a high input of heavy metals may lead to toxic environmental effects of the accumulated elements. The relative mobility of heavy metals in acid soils is shown to decrease in the order: Cd > Ni > Zn > Mn > Cu > Pb > Hg. Since the organic matter possesses a high binding capacity for several heavy metals at strongly acid soil reaction, it can reduce the toxic effects of heavy metals in contaminated acid soils. The recorded accumulation of metals in the humic layer of acid forest soils is explained in part by this assumption. On the other hand, Zottl and Huettl,[49] and Hantschel et al.[67] found low contents of Mg, K, Zn, Mn, Pb, and Cd in declining spruce stands of southwest Germany. Significant correlations between the ionic composition of soil extracts and the nutrient contents of spruce needles were recorded. It is suggested that the observed decline symptoms may be due to nutrient deficiencies (Mg, K, and Zn) at different site conditions (e.g., elevation, rainfalls, and forest practices). Leaching of the more mobile elements from the topsoil by acid atmospheric deposition is considered a predisposing stress factor.

The levels of trace elements available to plants are shown in Table 6.

The general conclusions of the quoted German literature agree with the opinion that high winds and precipitation, and frequent cloudiness with horizontal interception of aerosols, are responsible for high deposition rates. The observations of the behavior of particular elements and comparisons with the results of the quoted studies support the statement made above.

GREECE

The recent trace element research in Greece was aimed especially at assessing the toxic effects of heavy metals to plants and the environment. Previous studies on micronutrients concerning the relationships of bioclimatic zones of Greece with various characteristics of forest soils concluded that the concentrations of most macro- and micronutrients in the topsoil (humic horizon) were several times higher than those in the mineral portion of the soil profile.[68,69] Moreover,

Table 6. Mean Levels of Total (a) and Extractable (b) Trace Elements in Soils and Plants (c) of Germany (mg/kg)

	Cu	Zn	Ni	Cr	Pb	Cd	Fe	Mn	Co	V°	Zr°	Y°	Be°
(a)													
Topsoil	22.0	83	15	55	56.0	0.52	11.14*	806	6.7				
										37	240	20	8
Subsoil	28.0	96	13	53	33.0	0.38	13.18*	788	10.6				
Range													
m	1.0	13	4	46	12.0	0.06	1.60*	25	0.1	11	160	10	3
M	84.0	492	70	59	174.0	2.90	66.00*	1800	23.0	63	350	30	14
(b)													
Topsoil	3.6	10.4	—	—	19.6	0.25	121.00	111	0.29				
Subsoil	1.8	4.5	—	—	6.2	0.15	47.00	61	0.32				
Range													
m	0.1	1.5	—	—	2.9	0.03	7.00	7	0.15				
M	8.0	33.0	—	—	85.0	1.50	700.00	270	0.65				
(c)													
Plants	24.5	41.5	12	11	19.0	0.38	236.00	818	0.66				
Range													
m	3.0	11.0	8	8	0.6	0.06	50.00	160	0.50				
M	46.0	94.0	16	14	73.0	1.45	560.00	1900	0.90				

Note: ° Indicates that no distinction between top and subsoil has been made.

* $\times 10^3$

From Zöettle and Huettl,[49] Zöettle et al.,[49a] Schilichting and Elgala,[52] Murad and Fischer,[54] Schwertmann et al.,[55] Schwertmannet al.,[63] König et al.,[65] and Neite.[66]

there is a concentration gradient from the Mediterranean to the mountainous zone for most elements (e.g., Fe, Mn, and Zn) irrespective of the soil type.

Karataglis and Babalonas[70] and Babalonas et al.[71] examined the concentrations of several trace elements in plants and soils of northern Greece and noted that both in soils and plants the levels of Co, Ni, Cr, and Cd were considerably low. Instead, the contents of Zn, Pb, Cu, and, in some cases, Mn were considerably high approaching toxic levels in plants as well as in soils. In most instances, the relative abundances of these elements in soils decrease (Zn > Pb > Cu) and such behavior was found in the related plants. Copper, which is more toxic than Zn, forms more stable compounds and is thus available to plants in small amounts, though in some cases it exceeds 500 mg/kg (e.g., in *Rumex acetosella* 817 mg/kg Cu), a value that is considered quite high for the normal development of the plant. Also the zinc concentrations in plants exceed 500 mg/kg at some sites, *Cistus incanus* being the one with the highest value (1079 mg/kg Zn).

Concerning Pb, a case study carried out in the Thessaloniki region in northern Greece (Karataglis et al.[71]) found the Pb content of the soil samples to range from 710 mg/kg to 1500 mg/kg in the immediate neighborhood of a factory, while in the fields at some distance from the factory (>1000 m) the lead levels ranged between 29 and 120 mg/kg. Moreover, the greatest Pb concentrations was found in plant roots (up to 459 mg/kg), with the shoots and leaves following (up to 66 mg/kg). A close relationship between lead levels in the soil and the plant was recorded.

A summary of the trace element distribution in soil at different sites of Greece, including soils of mining areas, is reported in Table 7.

Since soils and plants are affected by wet and dry depositions of anthropic origin, Pb being one of the most common components of these depositions, the following classification of lead-polluted areas is proposed:

- soils with background (physiological) levels: Pb < 100 mg/kg;
- soils slightly affected: Pb 100–200 mg/kg;
- soils affected: Pb 200–400 mg/kg;
- soils markedly affected: Pb 400–800 mg/kg;
- soils strongly affected: toxic concentrations, Pb > 800 mg/kg.

ITALY

In the last ten years a systematic survey on the distribution of geochemically and toxicologically relevant trace elements in the soils of Italy was carried out with the aim to collect basic information on the background levels of trace elements, their geochemical behavior, and the relationship between weathering and pedogenetic processes.[73–85] Soil samples were collected mostly from areas least exposed to pollution, and only very few samples were collected in arable lands or near industrial areas.[74,90]

Table 7. Mean Levels of Total Trace Elements in Soils of Greece, Including Mining Area (mg/kg)

	Cu	Zn	Ni	Cr	Pb	Cd	Mn	Co
Average	1588	1038	101	94	398	7.4	1815	21
Range								
m	117	80	33	17	23	5.0	240	12
M	4509	10,547	212	228	7349	45.0	6485	45

From Nakos,[69] Karataglis and Babalonas,[70] Babalonas et al. 1987,[71] and Karataglis et al.[72]

The resulting trace element concentrations are generally similar to the world soil averages (Table 8). Because of local phenomena of contamination, however, the values of Pb, Zn, and Cd are somewhat higher in certain areas, as well as those of Ni, Cr, and Co which are released by mafic minerals during weathering.[84–86]

The data relative to the trace elements distribuited in the individual soil horizons provide the following indications:

1. variation in the trace element content of the different horizons of the same profile were limited, with the exception of Pb, As, and Mn which are usually higher at surface as a consequence of biogenetic processes;
2. conversely, leaching processes are responsible for the decrease of most elements in eluviated horizons of leached soils;[83,84]
3. in agricultural soils,[74,90,91] Cu and Zn levels are high in surface horizons as a consequence of fertilizer application. Cadmium level may also increase following P fertilization;
4. in mining areas very high contents of Pb, Zn, and Cd were recorded,[88,89] as well as for Ni, Co, Cu, and Cr in ophiolitic areas;[80,84–86]
5. sensible variations in minor element contents were found among the different soil profiles, which could be correlated to both the pedogenetic processes and the dominant clay mineral association.

Data on extractable trace elements in Italian soils[84–86] provide useful information on their bioavailability:[84–86]

1. the total metal content in soils correlates well with the EDTA-complexed forms;
2. manganese is the most readily extractable element, followed by Cu, Co, Ni, and Zn which are accumulated at surface, in connection with the organic matter content;
3. a general decrease with depth was recorded for most extractable elements, which is consistent with the weathering pattern of rocks;
4. the mean levels of extractable elements, inclusive of heavy metals, fall in the ranges which are considered "not toxic" to plants. Therofore, little contamination, if any, occurs in the soil investigated.

Table 8. Mean Levels of Total (a) and Ranges (b) of Extractable Trace Elements in Soils and Total Levels in Plants (c) of Italy (mg/kg)

	Cu	Zn	Ni	Cr	Pb	Cd	Fe*	Mn	Co	Mo	V	As	Sn	Hg
(a)	51	89	46	100	21	0.53	37	900	16	0.9	87	16	5.5	0.4
(b) Range														
m	1	<0.2	<0.1	0.1	<0.2	—	4	1	0.1	<0.1	0.2	—	—	—
M	96	48	52	1	21	—	690	819	13	1.3	15	—	—	—
(c) Range														
m	2	10	—	—	8	1.00	—	13	—	0.04	—	—	—	—
M	360	1600	—	—	4036	26.00	—	268	—	5.00	—	—	—	—

* $\times 10^3$

From Bini et al.,[74–77,79] Angelone et al.,[86] Leita et al.,[88,89] and Polemio et al.[90,91]

A few papers are available on trace elements in plants of Italy since the 1980s.[88,89,91a] From the data recorded, no symptoms of toxicity or deficiency appear to affect plants; however, Mo deficiency was observed in exceptionally acid soils.

The trace element distribution in Italian soils is largely determined by the mineralogical and chemical composition of the rocks they are derived from.

Slight contamination from atmospheric deposition (Pb, Cd, Zn, and Hg) and/or agricultural practices (Cu and Zn) occurs locally.

Potential hazard to plants may arise in some ophiolitic or mined areas, whose infertility could be a consequence of a synergism due to combined effect of major and trace elements, especially heavy metals.

NETHERLANDS

In the Netherlands, the governmental policy for soil protection takes the concept of so-called multifunctionality as the principal point of departure in land use.[92] Specific soil functions are:

- bearing function,
- food production function,
- filter function for surface water and ground water, and
- ecological function.

With respect to the last function, the contribution of soil to global element cycling may be mentioned. Especially Cd is taken as a guide element, since it is one of the most hazardous environmental pollutants faced at present. Indeed, it received extensive attention with respect to its behavior in soil systems (uptake, sorption, and transfer from crops to animals and humans).

The heavy metal contents of Dutch topsoils originating from natural reserves and from agricultural lands were assessed by Van Driel and Smilde[93] and Lexmond and Edelman.[94] In their study, the clay content was considered to dominate the original heavy metal content in areas that may be considered as "not heavily polluted". Clay as abiotic factor was found to be the dominant factor, followed by the organic matter content,[95] in decreasing the toxicity of heavy metals (especially Cd and to a lesser extent Zn, Cr, Cu, Ni, and Pb) in soil microbial respiration.

A summary of trace element levels in Dutch soils is reported in Table 9.

The accumulation of toxic trace metals in crops is receiving attention in the Netherlands, as well as throughout western Europe, in relation to environmental pollution and human health hazard. The mobility of heavy metals and uptake by crops was assessed by Gerritse et al.,[96,97] Wiersma et al.,[98] and Van Lune[99] in different Dutch soils. According to these authors, the uptake of a trace element by a plant is coupled to a "chemiosmotic" process across the membranes of active root cells. Accumulation will continue until transport is balanced by increased

Table 9. Mean Levels of Total Trace Elements in Soils, Plants, and Livestock of The Netherlands (mg/kg)

	Cu	Zn	Ni	Cr	Pb	Cd	Mn	Hg	As
Soils	18.6	72.5	15.6	25.4	60.2	1.76	435		
Range									
m	4.0	9.0	2.0	2.0	6.0	0.10	57		
M	110.0*	1020.0*	39.0	76.0	270.0	5.30	963		
Plants					0.33	0.06			
Livestock									
Milk									
m					2.0	0.15		< 0.5	< 1
M					7.0	0.26		2.3	
Kidney									
m					420.0	270.00		5.0	17
M					510.0	300.00		9.0	53
Liver									
m					110.0	60.00		3.0	8
M					210.0	65.00		7.0	12

* Indicates data not included in the mean.

From Doelman and Haanstra,[85] Gerritse et al.,[96] Van Lune,[99] and Vreman et al.[100,101]

uptake by the plant at a higher concentration level. Ionic strength, composition and pH of the soil solution have a significant effect on the solubility of heavy metals. Although it is impossible to generalize the results, it can be said that for all crops similar trends for Zn, Cd, and Cu were found. Zinc is most likely to compete with Cd for uptake, while no significant correlation with plant uptake could be established for Pb. Particularly as regards concentrations in soils of possible contaminated areas,[99] Cd and Pb were 1.5 and 2.5 times as high as those found by Wiersma et al.,[98] while median Cd and Pb concentrations in the crops from these soils were rather normal.

Recent trends of research on trace elements in the Netherlands point to the intake of heavy metals by domestic animals. Vreman et al.[100,101] paid attention to the transfer of Cd, Pb, Hg, and As from feed into milk and various organs of cows and bulls. Liver and kidney were the primary sites of element accumulation, but only for Cd, the (proposed) levels of tolerance were exceeded, suggesting that the pathological lesions observed were not related to the sorption of heavy metals.

NORWAY

Studies on the distribution of trace elements in soils, plants, and animals (including humans) were carried out in Norway during the last decade, particularly with regard to bioavailability and toxicity for man.[102] An initiative to organize geochemical investigations in Norway was intended to start in early 1970s within a program for Man and the Biosphere (MAB), and symposia on

Table 10. Mean Levels of Total Trace Elements in Norwegian Soils (mg/kg)

	Cu	Zn	Ni	Cr	Pb	Cd	Co	V	As	Se
	19	60	61	110	61	0.95	13	37.9	4.4	0.63
Range										
m	10	40			10	0.4				
M	40	100			140	2.0				

From Steinnes,[103] Låg and Bøvilken,[105] Låg,[107] Låg and Steinnes,[108,109] and Schalin.[111]

geomedical aspects in present and future research were organized by the Norwegian Academy of Science and Letters in recent years.[103,104] Indeed, geomedical investigations showed that environmental factors influenced several "endemic" diseases (e.g., goiter and lack of iodine; dwarfism and lack of zinc; caries and lack of fluorine).

In Norway, poisoning from Pb was found in several areas and naturally Cu-poisoning patches have also been detected.[105] Industrial pollution with Zn, Pb, Cd, Hg, Cu, As, Se, and F was found in food plants in the immediate vicinity of industrial areas,[106] and high contents of Ni, Cu, Hg, and As were detected in soils from areas in the neighborhood of abandoned mining areas.[107] In recent years, attention was given to the distribution of halogens, Se and As in natural soils of Norway,[108,109] since Se toxicity for livestock is well known, as well as because of its reputation as a carcinogen agent and multiple-schlerosis inhibitor.[110,111]

The average concentrations of some trace elements in natural surface soils from Norway are reported in Table 10. The average composition of the samples is similar to the crustal average. Most elements (e.g., Se, Pb, and Cd) have a geographic distribution indicating the importance of an airborne supply of these elements to the soil, and a high correlation with annual precipitation ($r = 0.83$ for Se). Moreover, high correlation coefficients for some element pairs (Se–Pb; Se–Cl; As–Pb) were found.

Since the investigated areas are mainly rural, with relatively little influence by industrial activity and pollution, the geochemical data recorded are assumed to reflect the natural geochemical background. Some of the elements showing significant geographical distribution patterns are essential to plants or animals. Others may have hazardous effects upon flora or fauna or even humans if present in sufficiently high concentrations.

PORTUGAL

A considerable effort for the elaboration of a Soil Trace Micronutrients Map of Portugal has been done recently by the Pedology Department of Portuguese Agricultural Reasearch Station.[112–117] Special attention was given to the trace elements (B, Co, Cu, Fe, Mn, Mo, Zn, Cr, and Ni) in soils developed on many different parent materials (mafic rock,[113] limestones,[114] sandstones,[118–122] and granitic rocks[123,124]).

Soils developed from serpentine rocks, outcropping in northeastern Portugal, were studied by Sequiera et al.[113] (Table 11a). These soils are rocky and shallow and have a scanty vegetation cover. They are characterized by high levels of total and extractable Ni and high total Cr content (average 2400 mg/kg), although the extractable fraction of this element is not much higher than 1 mg/kg. The available Cu level is below the toxicity level (30 mg/kg), indicated by Berrow and Reaves,[183] and the available Mn level falls within the normal range for this element.[189]

Soil samples from granitic and schist rocks were collected in the region of Nord East,[123,124] which represents about 1/10 of the country. Sampling sites included both cultivated and uncultivated areas at different elevation and morphology. The total trace element contents are strictly related to underlying rocks. The large variability of some elements is thus related to the percentage of mineral species in which they are present. Concerning granitic rocks, available B (Table 11b) results are low. The concentrations of Co and Mo are very low; for these elements deficiency phenomena cannot be excluded.[118] The concentrations of Cu and Zn seem to be related to the soil management, with phytotoxicity hazard in cultivated soils as a consequence of fertilizer applications. The available Mn levels are generally low, as well as those of total Mn.

The distribution of total and extractable Fe, Mn, and Zn in old alluvial calcareous soils in southern Portugal[114] showed some micronutrient deficiency (Zn, Fe, and Mn), apparently as a result of a nutritional disequilibrium characteristic of large areas of Portugal in which calcareous soils predominate. Data referred to total and extractable concentration in these soils are reported in Table 11c.

Studies on sandy soils in the neogene region, south of the Tagus River, were reported in many papers.[118–122,125] The distribution of major and trace elements in these soils was generally related to the soil mineralogy, particularly for Mo, Co, and B. Total Zn and Cu were well correlated with the fine fraction of soil. Some biological accumulation of Cu occurs at the surface. Total levels of Mn (mean 50 mg/kg) are low (Table 11d); leaching, which is enhanced by organic acids and the high permeability of soils, seems to have contributed to Mn decrease. Data on extractable Zn, Cu, Mo, Co, Mn, and B (Table 11d) show a generalized deficiency for these elements. The highest levels of extractable Zn, Cu, and Mn in surface horizons indicate that these elements are bioaccumulated in the form of chelates.

With the aim to evaluate the levels and distribution of Cu and Zn in two vineyard soils after many years of fungicides application, Jaghalaes et al.[126] studied soil profiles developed on sedimentary rocks within an area modified by anthropic activity in central Portugal. The levels of total and extractable Cu were higher at surface (118 and 72 mg/kg, respectively) and decreased with depth. The accumulation of Cu in soils, was in good agreement with the estimated amount of Cu in fungicides. No significant accumulation of Zn occurred (ranges: total Zn 67 to 49 mg/kg; extractable Zn 2.8 to 0.2 mg/kg), probably due to the more recent incorporation of this metal in fungicides.

Table 11. Mean Levels of Total and Extractable Trace Elements in Soils of Portugal (mg/kg)

	Cu	Zn	Ni	Cr	Fe	Mn	Co	Mo	B
Soils on ultra basic rocks									
Total			2353	2400		1552	112.6		
Range									
m			1021	772		624	19.0		
M			5893	4017		2886	199.0		
Extractable	9.7		62.4	1.3		19.7			
Range									
m	2.2		11.0	1.2		2.0			
M	30.4		245.0	1.6		78.0			
Soils on Granite									
Total	13.2	85.0				337.0	3.7	0.63	34.0
Range									
m	4.6	57.0				93.0	1.0	0.13	1.0
M	21.8	110.0				876.0	11.0	1.09	158.0
Extractable	2.4	0.9				6.5	0.2	0.09	0.17
Soils on Schist									
Total	56.7	98.4				585	9.5	0.43	67.20
Extractable	4.6	1.2				11.0	0.7	0.13	0.17
Alluvial soils									
Total		41.7			29,951	340			
Range									
m		5.0			3180	50			
M		95.0			102,500	726			
Extractable		1.1			14	19.5			
Range									
m		0.05			2	<1			
M		7.5			57	92			

Sandy soils										
Total	3.5	8.6					50	7.9	0.14	50.8
Extractable	0.16	0.23					6	0.05	0.016	0.3
General average										
Total	24.5	58.4					328[a]	33.3	0.30	59.0
Extractable	4.1	0.8				12.5	0.3	0.08	0.21	

[a] Serpentinitic soils are excluded from the mean.

From Santhos Couthinho et al.,[114] Castelo Branco et al.,[118] Sequieira,[119–122] Vieira E. Silva and Domingues,[123] Da Silva Teixeira et al.,[124,125] and Megalhaes et al.[126]

The general conclusion of the above-mentioned authors is the following: the total trace element distribution in soils of Portugal is generally related to the characteristics of parent materials they were derived from; data on extractable elements show that some deficiency may occur. In the vineyards, the long-term application of Cu-rich fungicides is an important source of soil contamination.

SPAIN

Moderate attention was paid to trace elements in Spanish soils during the past few decades. Increasing interest was given first to trace elements related to soil genesis and distribution,[127–129] since the total content of trace elements in soils is of great importance in geochemical studies.[130–132] Knowledge of soluble forms of trace elements received subsequent consideration, as it can provide basic information for agronomic practices.[133–136] Indeed, the available level of a trace element is most certainly of greater environmental significance than the total level.[137] Moreover, the increasing application of sewage sludges and urban refuse compost as fertilizers has brought about many studies aimed at evaluating the fertilizing capacity of these products.[138–141] Indeed, richness in essential plant micronutrients is one of the main characteristics of compost, but it may produce undesirable effects due to an increase in the amount of certain essential micronutrients to toxic levels (e.g., boron) or to contaminating processes (e.g., entering the food chains and affecting humans and animals; contaminating by migration to groundwaters).

Particularly as regards agriculture and the environment, possible heavy metal (Pb, Cd, Cu, Zn, Ni, and Cr) contamination in different plain soils of Spain was investigated by Cala Rivero et al.[130] According to these authors, some heavy metal levels (Pb, Cd, and Zn) in the soils investigated were higher than the recommended levels, though some differences among the various pedological zones were recorded. Moreover, based on the international toxicity index referring to the zinc level ("Zn equivalent" = mg/kg Zn + 2mg/kg Cu + 8 mg/kg Ni > 250 mg/kg), a potential toxicity was recorded for more than 35% of the soils investigated. Furthermore, a very good correlation between ionic radii and element solubility was found, with the following order: Pb > Cd > Cu > Zn > Ni > Cr. On the other hand, a soil survey in the province of Valencia[131] proved levels of heavy metals (Cd, Co, Cr, Cu, Ni, Pb, and Zn) were within the ranges quoted in the literature,[189,190] with the exception of Pb and Cd in fluvisols. Hence, no evident signs of contamination were recorded in the region investigated, regardless of the soil type.

The elemental content of agricultural soils in the Valderas area were investigated recently by Aller and Deban.[133] The levels found in this study were lower than those reported in the literature, and the ranges were those expected in an uncontaminated zone, though some precautions are necessary for those elements that are potential contaminants, such as Cd, Cu, and Hg, or have potential deficiency problems, such as Mo and Zn. Hence, in the area studied there were no toxicity problems for plants, but a few deficiency problems can be expected with Cu, Mo, and Zn.

Although some discrepancies occur in the Spanish literature concerning the criteria of soil sampling (e.g., topsoil or full profiles) and the related trace element distribution, the data reported in Table 12 can be considered representative for Spanish soils. Levels of most elements fall within the ranges considered "normal" for soils, with exceptions for mined areas[128] which were excluded from the means (Ni, Cr, Co, and Cu); slight contamination with Pb and Cd from atmospheric inputs or industry was observed in soils of restricted areas; potential toxicity to plants was recorded for about 30% of the soils studied; potential deficiency may be expected for some elements such as Cu, Mo, and Zn in agricultural soils.

SWEDEN

A few data on trace elements are available from Sweden in current literature. During the past decade the stress upon the environment has raised demands for better knowledge about the anthropogenic impact on the geochemical and hydrochemical cycles,[143] especially in relation to forest soil acidification from air pollution.[144–147] Indeed, the prevailing soil acidification will increase the release and leachability of many elements in soil, especially the topsoil, where the biological activity proved to be highly sensitive to heavy metal pollution.[148–150] The average total heavy metal concentrations in soils, reported in Table 13, is from a review of Swedish literature about this subject. The high levels reported for Zn, Fe, and Ti are mostly dependent on the parent material composition.

Sweden appears to still be one of the less polluted regions in Europe, as compared, for example, with Germany, even if some airborne (wet and dry) deposition may contribute to increasing pollution and acidification. Most metals are to some extent accumulated in the A horizon. Toxic cations Cd and Ni have negative budgets in the soil, but the problem with high Cd concentrations in acidic soil solutions will become more serious with increasing soil acidification and may cause plant root damages.

Attention has recently been paid to trace element uptake and toxicity to plants. Several studies have been carried out in Sweden on this subject. Baath[151] reviewed the effects of Cd, Cu, Zn, and Pb on soil microorganisms and concluded the following degree of toxicity to be the most commonly found, irrespective of the organic matter content of soils: Cd > Cu > Zn > Pb. Balberg-Pahlsson[148] noted that, among the heavy metals, Zn is the least toxic to vascular plants, and that the degree of toxicity is influenced by biological availability of the metals in the soil and the nutritional status and age of the plant.

Tyler[145] studied the uptake, retention, and toxicity of heavy metals in lichens and noted interspecies differences in sensitivity and the development of extreme tolerance in certain species.

Table 12. Mean Levels of Total (a) and Extractable (b) Trace Elements Content in Soils and in Plants (c) of Spain (mg/kg)

	Cu	Zn	Ni	Cr	Pb	Cd	Co	Mo	Sr	Se	Hg	Li
(a)	14	59	28	38	35	1.70	0.6	1.4	155.0	0.30	0.2	25.5
Range												
m	4	10		12	6	8	0.04	5.0	< 1.0			
M	400	109	300	3640	58	4.00	24.0	2.0				
(b)	3.8	0.7	2.5	4	0.4	0.47		0.13	7.7	0.1	0.18	1.6
Range												
m	0.4	0.3	1.0	0.4	0.3	0.03						
M	41.0	35.0	375.0	41.0	30.0	0.76						
(c)	17.4	—	22.7	21.6								

From Lopez and Guitian,[128] Cala Rivero et al.,[130] Boluda et al.,[131] Aller and Deban,[132] and Nogales.[139]

Table 13. Mean Levels of Total Heavy Metals in the Topsoil of Podzols from Sweden (mg/kg)

	Cu	Zn	Ni	Cr	Pb	Cd	Mn	Fe	Ti
	8.5	182	4.4	2.3	69	1.2	770	6300	3600
Range									
m	7.4	100	1.7	1.3	51	0.9			
M	39.8	318	10.0	8.0	150	2.0			

From Melkerud [143] and Bergkvist et al.[150]

The quoted authors concluded that at the present status of knowledge it is difficult to propose a limit for toxic concentrations of heavy metals, since there are ambiguous methods of determining the biologically available fraction of the metals in soils. Indeed, some studies give the metal concentrations in the leaf tissues as toxic limit values; others point to metal accumulation in the roots; moreover, the synergic effects of heavy metals and other pollutants such as SO_2 have to be stressed.[152]

SWITZERLAND

Switzerland is a very small country largely covered by forests. Recently attention has been paid to trace metals in relation to environmental pollution problems, since forestry and the environment are considered important land resources. In 1981 the federal government established regulations that set limits for the content of heavy metals in sewage sludge and for the amount that may be applied per year.[153] In the meantime, the rapid decline in the health of many forests in central Europe[154] has been suggested as a possible effect of acidic precipitations. However, existing data on elemental concentrations in soils and plants of Switzerland are very scarce.

Muller[154a] examined 80 soil profiles from alpine grass habitats in the upper Engadine and concluded that weathering, soil genesis, and element release are strongly influenced by the site characteristics, especially the podzolizing potential of the plant litter. Indeed, soil acidification enhances the element release, availability, and uptake by plants.

Keller[155, 156] (quoted in Bergkvist[150]) reported average levels for lead (39 mg/kg) and cadmium (0.8 mg/kg) in soils of the Zurich region, with ranges between 18 and 60 mg/kg for Pb and 0.2 to 1.4 mg/kg for Cd, indicating that the area investigated is still under the threshold of pollution according to Kabata-Pendias.[189]

Wyttenbach et al.[154] reported data from 31 forestry sites of an area centered around the city of Winterthur and noted that the material trapped by the spruce needles has roughly the same composition as soil, with very small enrichment coefficients. Only Na, Cl, and Br at sites in the vicinity of motor highways show increased concentrations, possibly originating from burning of gasoline and use of NaCl as deicing salt.

Schmitt and Sticher[153] studied the availability of Cd, Pb, and Cu in soils and concluded that the presence of competing metals increased the general availability especially for cadmium.

The levels of trace elements in Switzerland are low enough to exclude heavy pollution phenomena. However, as suggested by Schmitt and Sticher,[153] the threshold values for heavy metals in the sewage sludge should be decreased, since the retention of these metals by the soil cannot be supposed for a long time.

UNITED KINGDOM

Many reports and papers on trace element distribution in soils of Great Britain are available in the recent literature. Many of these are on soils of England and Wales, while others deal with Scottish soils. In this review the distinction has been maintained.

A broad distinction may be made among different studies carried out in England and Wales. Background levels of trace elements in soils and historical mining areas were investigated by Bradley et al.,[157] Davies,[158–160] Jones,[161,162] McGrath,[163,164] and Archer and Hodgson.[166] Several studies were focused on pedological evolution,[159–165] statistical elaboration,[158,165] heavy metal distribution,[157,161,164,168,169] mineralized areas,[161,168,171] and rural and urban areas.[170,172–174,176] Many of these data were produced by the Soil Survey of England and Wales, and mainly refer to topsoil and subsoil, with samples collected at 5 km intervals on a national square grid.

Data are usually presented as means, median or geometric mean; consequently no further statistical elaboration of these informations has been done here. Lead, Cu, Zn, Cd, and Ni were the most studied elements, in consideration of their potential toxicity to plants and man.

A collection of these data is reported in Table 14. According to the references quoted in this table, a correlation with parent rock is evident for most of the elements investigated (e.g., Be, Rb, Sr, Zr, Ni, Cr, and Co). Moreover, a strong correlation between Hg and Pb, and to a lesser extent Cu and Zn, was found, with Hg levels up to 1.78 mg/kg in contamined soils.

Silver in uncontamined soils ranged between 0.01 and 0.17 mg/kg, a very low level, but in mine spoils areas its amount increased up to 7.0 mg/kg. The levels of most elements, however, fall in the range expected for unpolluted or slightly contaminated soils. Occasionally high levels of trace elements (Cu, Pb, and Zn) were recorded, suggesting that some sites might be polluted as a consequence of ore processing, smelting, and industrial pollution.

Levels of extractable elements in soils were examined by Musa and Bridges[169] and Archer and Hodgson,[166] who concluded that in England and Wales problems of micronutrient deficency (e.g., for B, Cu, and Zn) do arise in less than 5% of the soils investigated; on the contrary, some mining areas may have phytotoxic concentrations of Cu, Zn and Pb. Moreover, abnormal levels of total Mo (range 2.5–13.2 mg/kg) in Welsh soils were recorded; the available Mo, however, proved less critical to plants and animals, ranging from 0.10 to 0.16 mg/kg.

Comprehensive information on background levels of trace elements in Scottish soils is given in many papers.[177–179,181–186] A summary of the substantial data is reported in Table 15. From the above quoted literature, a wide range of variability is observable for the elements Cr, Co, Cu, and Pb, mostly depending on the different parent materials.

Particular attention was paid to the distribution of total and extractable Pb and Cu in the soil profiles.[183–186] Both these elements are accumulated at surface, forming metal chelates with organic matter. The derived mean ($x^* = \text{antilog } x'$;

Table 14. Mean Levels of Total (a) and Extractable (b) Elements in Soils of England and Wales (mg/kg)

	Cu	Zn	Ni	Cr	Pb	Cd	Fe	Mn	Co	Mo	Rb	Hg	As	Sr	Br	Zr	Y	Ag
(a)	15.6	78.2	22.1	44	48.7	0.7	30.5*	1405	11.5	—	94	0.25	17.5	85	50	275	25	0.18
(b)																		
EDTA	4.8	5.4	1.6	—	11.2	0.2	—	—	0.34	—								
HOAc	1.0	5.8	1.0	—	1.6	0.2	—	—	0.62	0.24								

Note: Data from mined or polluted areas are excluded from the mean.

* $\times 10^3$

From Bradley,[157] Davies,[159,160,168] Jones et al.,[161,162] McGrath et al.,[163] McGrath,[164,165] Archer and Hodgson,[166] and Bradley et al.[167]

Table 15. Mean Levels of Total Trace Elements in Scottish Soils (mg/kg)

		Cu	Zn	Ni	Cr	Pb	Cd	Mn	Co	Mo	Hg	V	As	Ga	Se
		23	58.0	37.7	150.0	19	0.47	830	13	1.50	0.12	76.5	8.4	17	0.17
Range	m	<1	0.7	0.5	0.5	1	0.01	2	0.4	0.05	0.01	0.5			
	M	2.5*	987.0	5.0*	10.0*	1*	2.40	10*	400	40.00*	1.71	500.0			

* $\times 10^3$

From Ure et al.,[178] Berrow et al.,[179] and Berrow and Reaves.[181–183] where x′ = arithmetic mean)[186] for total Pb is 15 mg/kg (range 2.5 to 85 mg/kg) and that for extractable Pb 0.24 mg/kg (range 0.016 to 3.4 mg/kg). The derived means for total and extractable Cu are 10 mg/kg and 0.85 mg/kg, respectively (ranges 0.93 to 110 and 0.08 to 9.8 mg/kg, respectively).

The recorded data suggest that concentration of extractable Cu is inadequate to supply the nutritional requirement of cereal crops, and approximately 70% of soils contain metal levels inadequate to support herbage. There is evidence, however, that a slight contamination and possible phytotoxicity may occur at some mined sites.

Referring to the authors who studied trace element distribution in the United Kingdom,[159,160,163,164,166,167,178,181,182] we may conclude that the variability of trace elements in soils of the U.K. is principally caused by the variety of underlying rocks of different ages and lithology. Besides, the consequence of past mining activities and heavy and light industry have led to a widespread level of contamination (sometimes very high) of Cd, Pb, Zn, Cu, and Ni. Many data from the national grid[171] also contain some very high metal concentrations as a consequence of environmental pollution. Data on the extractable elements are not so numerous as those on total element concentration. However, if some mineralized areas are excluded, the threshold toxicity for some heavy metals (Cu, Pb, Ag, and Mo) is rarely reached; also, deficiency phenomena are extremely rare.

CONCLUDING REMARKS

During the past few decades a considerable amount of data on trace elements in the environment (especially heavy metals) have accumulated in western Europe. A thorough knowledge on trace element concentration in certain countries is still lacking, however.

The long-term influence of acid deposition, sewage sludges, and heavy fertilization on soils is an important item in present and future research. Knowledge of these long-term effects is urgently needed, not only from a scientific point of view but from a practical standpoint as well, since it has considerable practical implications in, for example, agriculture, the food chain, and health hazards. Moreover, there is a lack of information about metal speciation and washing out in the system soil-plant with regard to micronutrients and phytotoxic metal uptake.

With our present knowledge it is difficult to propose a limit for a toxic concentration of trace elements to plants and animals. There is no completely unambiguous method of determining the biologically available fraction of the metals in soils. Neither is it possible to use the trace element contents in different parts of the plants (e.g., roots, leaves), since response to metals depends on the plant itself, and most of metals are not biologically active or can be transported up to the above-ground parts. Moreover, the toxic effect may be synergistic, and the critical limit reduced, where external conditions (e.g., soil pH, Eh, nutritional status, drought) may play an important role when assessing the plant tolerance. Indeed, this depends upon the soil metal content and pH and is very specific for each element.

Bearing in mind that every report reviewed here has its own assumptions and aims, that methods vary greatly, and that results have been acquired in different ways, it may not be correct to draw any general conclusion from the studies reviewed. There is, however, a considerable amount of information in support of the implication pertaining to the metal content and availability in soils and plants.

The levels of most elements in soils of western Europe are consistent with the values considered "normal" for unpolluted soils, with exception for mined areas. In Table 16, we have reported the values relative to western European countries and, for comparison, the means of U.S., Canadian, and world soils.

Although the mean levels do not show any particular enrichment of trace elements, the higher values of the reported ranges (see previous tables) show that a slight soil contamination occurs in many areas, especially in the neighborhood of industrial factories and along roads with intensive traffic. High winds, intense cloudiness, and precipitation are considered responsible for high deposition rates of trace metals. Lead (Belgium, Netherlands: 94 and 270 mg/kg, respectively), Cd (France, Germany, Netherlands, Spain: 2, 2.9, 5.3, 4 mg/kg, respectively), Zn (Germany, Netherlands, Sweden: 492, 1020, 318 mg/kg, respectively), Cu (Netherlands, Spain: 110 and 400 mg/kg, respectively), and Ni (Belgium, Spain: 151 and 300 mg/kg, respectively) are the most hazardous trace elements in the environment, and potential toxicity to plants was recorded in some areas investigated. Therefore, there is no land in western Europe where the local balance for most heavy metals in the ecosystem is not strongly influenced by atmospheric pollution. We would expect, on a long term, the build-up of toxic concentration levels and adverse effect on plant growth.

Trace elements contents in plants of some countries of western Europe are reported in Table 17. The data are consistent with those of soils, since any particular enrichment is observed. However, the ranges recorded for some elements (see also Tables 4, 6, 8, and 12) point to some enrichment of Pb (Austria, Italy), Zn (Austria, Belgium, Germany, Italy), Cu (Germany, Italy), Cd (Germany), Mn (Germany), and sometimes with B, Cr, and Co.

The effects of trace elements excess may appear as phytotoxicity and inhibition of plant growth. Due to the low phytoxicity of some elements, however (most plants are quite tolerant to, e.g., Pb, Cu, Zn), the consumption of contamined plants by humans may be only apparently safe. Health hazards caused by food pollution with mineral elements have already been reported (e.g., Hg and Pb intoxication and cadmiosis). More research is therefore needed in this important field of environmental medicine (or geomedicine) in order to attain better knowledge of the mechanisms governing the interactions between trace elements and human health.

Table 16. Mean of Some Total Trace Elements in Soils of Western Europe, in Comparison with the Contents of U.S., Canadian, and World Soils (mg/kg)

Country	Cu	Zn	Ni	Cr	Pb	Cd	Fe	Mn	B
Austria	17	65	20	20	150	0.20	13,300	310	—
Belgium	17	57	33	90	38	0.33	1,638	335	32
Denmark	11	7	7	21	16	0.24	1,236	315	—
France	13	16	35	29	30	0.74	—	538	21
Germany	22	83	15	55	56	0.52	1,147	806	—
Greece	1,588	1,038	101	94	398	7.4	—	1,815	—
Italy	51	89	46	100	21	0.53	37,000	900	—
Netherlands	18.6	72.5	15.6	25.4	60.2	1.76	—	—	
Norway	19	60	61	110	61	0.95	—	—	—
Portugal	24.5	58.4	—	—	—	—	—	328	59
Spain	14	59	28	38	35	1.70	—	—	—
Sweden	8.5	182	4.4	2.3	69	1.20	6,300	770	—
England & Wales	15.6	78.2	22.1	44	48.7	0.70	3,141	1,405	—
Scotland	23	58	37.7	150	19	0.47	—	830	—
Calculated Average									
+ Greece	131.6	137	32.7	55.9	66.7	1.30	9,108	732	37
- Greece	19.5	68	27	52.7	39	0.79		633	
World Soils	20	50	40	200	10	0.30	—	850	
U.S. Soils	25	54	20	53	20	0.50	—	560	
Canadian Soils	22	74	20	43	20	0.30	—	520	
Excessive levels in Soils	100	250	100	100	200	5	—	1500	30

From Aubert and Pinta,[48a] Vinogradov,[187] McKeague and Wolynetz,[188] and Kabata-Pendias.[189] For western Europe references see previous tables.

Table 17. Mean Levels of Trace Elements in Plants of Some Western European Countries (mg/kg)

	Cu	Zn	Ni	Cr	Pb	Cd	Mn	Fe	Mo	Se	Co
Austria		1537			665		82	654			
Belgium	9	90			50		102		0.32	0.07	
Finland	6.1	34.5					45	31		8.98	0.66
Germany	24.5	41.5	12	11	19	0.38	818	236			
Netherlands					0.33	0.06					
Spain	17.4		22.7	21.6							
General average											
+ Austria	14.2	425.7	23.3	16.3	183.6	0.22	262	307	NA	4.56	NA
- Austria		55.3			23.1						

Note: Mined areas are excluded from the mean.

REFERENCES

1. Aichberger, K. "Schewermetallgehalte einiger bodenprofile oberösterreichs," *Bodenkultur* 31(3):215–228 (1980).
2. Sieghardt, H. "Schwermetall und Nahrelementgehalte von Pflanzen und Bodenproben Schwermetalllaltiger Halden im Raum Bleiberg in Karnte (Osterreich)," *Z. Pflanzenernaehr. Bodenkd.* 150:129–134 (1987).
3. Kazda, M. and G. Glatzel. "Schewermetallaureicherung und Sckwermetallvergugbarkeit im einsiekerungsbereich von Stammablaunfawsser in Buchenwaldern (*Fagus sylvatica*) des wienerwaldes," *Z. Pflanzenernaehr. Bodenkd.* 147:743–752 (1984).
4. Glatzel, G. and M. Kadza. "Wachstum und mineralstoffernahrung von Buche (*Fagus sylvatica*) und spitzahorn (*Acer glatanoides*) auf versanertem und schwermetallelastetem bodenmaterial aus dem Einsickermgsbereich von Stammabflußwasser in Buchenwaldern," *Z. Pflanzenernaehr. Bodenkd.* 148:429–438 (1985).
5. Sieghardt, H. "Heavy Metal Uptake and Distribution in *Silene vulgaris* and *Minuartia verna* Growing on Mining-Dump Material Containing Lead and Zinc," *Plant Soil* 123:107–111 (1990).
6. Guns, M. S. "Reserche sur les Teneurs on Metaux Lourds dans les Composts Urbains Belges," *Rev. Agric.* 35(4):2837–2848 (1982).
7. De Temmerman, L. O., J. R. Istas, M. Hoenig, S. Dupire, G. Ledent, Y. Van Elsen, H. Baeten, and A. De Majer. "Définition des Teneurs "normales" des Éléments en Trace de Certains Sols Belges en Tant Que Critère de Base Pour de la Pollution des Soles en Genéral," *Rev. Agric.* 35(2):1915–1944 (1982).
8. Sillanpaa, M. "Micronutrients and the Nutrient Status of Soils: A Global Study," F. A. O. Soils Bull. 48, Rome (1982), p. 444.
9. Robberecht, H., D. Vandenberghe, and R. Van Grieken. "Selenium in the Belgian Soils and its Uptake by Ryegrass," *Sci. Total Environ.* 25:61–69 (1982).
10. De Temmerman, L. O., M. Hoenig, and P. O. Scokart. "Determination of 'Normal' Levels of Trace Elements in Soils," *Z. Pflanzenernaehr. Bodenkd.* 147:687–694 (1984).
11. Albasel, N. and A. Cottenie. "Heavy Metal Contamination Near Major Higways, Industrial and Urban Areas in Belgian Grassland," *Water Air Soil Pollut.* 24:103–109 (1985).
12. Kiekens L., A. Cottenie, and G. Van Landschoot. "Chemical Activity and Biological Effect of Sludges-Borne Heavy Metals and Inorganic Metals Added to Soils," *Plant Soil* 79:89–99 (1984).
13. Willaert, G. and M. Verloo. "Extractability and Plant Uptake of Zn, Cu and Cd in Belgian Soils," in *Heavy Metals in the Hydrological Cycle* (London: Selper Ltd., 1988) pp. 319–326.
14. Tjell, J. C. and M. F. Hovmand. "Metal Concentrations in Danish Arable Soils," *Acta Agric. Scand.* 28:81–89 (1978).
15. Møberg, J. P., L. Petersen, and K. Rasmussen. "Constituents of Some Widely Distributed Soils in Denmark," *Geoderma* 42:295–316 (1988).
16. Gissel-Nielsen G. and A. A. Hamdy. "Plant Uptake of Se and Se Values in Different Soils," *Z. Pflanzenernaehr. Bodenkd.* 141:67–75 (1978).

17. Jørgensen, S. E. "Do Heavy Metals Prevent the Agricultural Use of Municipal Sludge?" *Water Res.* 9:163–170 (1975).
18. Assad, F. and J. D. Nielsen. "A Thermodynamic Approach for Copper Adsorption on Some Danish Arable Soils," *Acta Agric. Scand.* 34:377–385 (1984).
19. Tjell, J. C. and T. H. Christensen. "The Retrospective Development in Metal Concentrations in Arable Soils (in Danish)," Dept. of Environmental Engineering. Technical University of Denmark, Lyngby, February 1985.
20. Tjell, J. C. and T. H. Christensen. "Evidence of Increasing Cd Contents of Agricultural Soils," in *Proc. Int. Conf.: Heavy Metals in the Environment* (Athens, September 1985), pp. 391–393.
21. Christensen, T. H. "Cadmium Sorption Onto Two Mineral Soils," Report 80–1, Dept. of Sanitary Engineering, Technical University of Denmark, Lyngby. 1980.
22. Anderson, P. R. and T. H. Christensen. "Distribution Coefficients of Cd, Co, Ni and Zn in Soils," *J. Soil Sci.* 39:15–22 (1988).
23. Christensen, T. H. and J. C. Tjell. "Long-Term Trends in Heavy Metals Contents of Agricultural Soils," in *Proc. Int. Conf.: Heavy Metals in the Environment* (Geneva, September 1989).
24. Rasmussen, L. "Potential Leaching of Elements in the Danish Spruce Forest Soils," *Water Air Soil Pollut.* 31:377–383 (1986).
25. Kvalheim, A., Ed. *Geochemical Prospecting in Fennoscandia*, (New York: John Wiley & Sons, 1967) p. 350.
26. Oksanen, H. E. and M. Sandholm. "The Selenium Content of Finnish Forage Crops," *Maat. Tiet Aikak.* 41(4):250–253 (1970).
27. Sippola, J. "Selenium Content of Soils and Timothy in Finland," *Ann. Agric. Fenn.* 18:182–187 (1979).
28. Koivistonen, P., Ed. "*Mineral Element Composition of Finnish Food, Acta Agric. Scand.*, Suppl. 22, p. 171 (1980).
29. Saari, E. "Finnish Research Project on Mineral Elements in Soils, Plants and Foodstuffs," in *Geochemical Aspects in Present and Future Research*, J. Låg., Ed. (Oslo: Universitatesforlaget, 1980) pp. 167–171.
30. Nayha, S. "Mortality in Northern Finland: Temporal Trends and Regional Variations," in *Geochemical Aspects in Present and Future Research*, J. Låg, Ed. (Oslo: Universitatesforlaget, 1984) pp. 63–69.
31. Marjanen, H. "On the Relationship Between the Contents of Trace Elements in Soils and Plants and the Cancer Incidence in Finland," in *Geochemical Aspects in Present and Future Research*, J. Låg, Ed. (Oslo: Universitatesforlaget., 1980) pp. 149–166.
32. Marjanen, H. "Possible Causal Relationship Between the Easily Soluble Amount of Manganese on Arable Mineral Soil and Susceptibility to Cancer in Finland," *Ann. Agric. Fenn.* 8:326–334 (1969).
33. Coppenet, M. "Les Oligoéléments en France. Rappel Chronologique des Principaux Travaux Effectués en France," *Ann. Agron.* 21(5):469–482 (1970).
34. Coppenet, M. "Les Oligoéléments en France. Examples de Problèmes Régionaux. 1. Le Massif Armoricain," *Ann. Agron.* 21(5):587–601 (1970).
35. Loue, A. "Déficiences en Oligoélémentes Actuellement Reconnues sur les Plantes Cultivées en France (Fe, Mn, Zn, Cu, B, Mo) au Cours de la Dernière Décennie," *Sci. Sol* 2:89–107 (1983).

36. Dejou, J., F. X. Montard, M. Lamand, and J. Bellanger. "Effects de l'Application de Cu et Zn en Sol Volcanique sur les Teneurs du Sol et D'un Ray-Grass Anglais en ces Deux Oligo-Éléments," *Agronomie* 5(19):841–850 (1985).
37. Pedro, G. and A. B. Delmas. "Les Principes Géochimiques de la Distribution des Éléments-Traces dans les Sols," *Ann. Agron.* 21(5):483–518 (1970).
38. Henin, S. "Les Éléments-Traces dans le Sol," *Sci. Sol* 2:67–71 (1983).
39. Guillet, B., E. Jeanroy, C. Rougier, and B. Souchier. "Le Cycle Biogeochimique et la Dynamique du Comportement des Éléments-Traces (Cu, Pb, Zn, Ni, Co, Cr) dans les Pédogenèses Organiques Acides," Note Techn. et Sci. Centre de Pedologie, 27 (Nancy, 1980) p. 49.
40. Godin, P. "Le Sources de Pollution des Sols: Essai de Quantification des Risques dus aux Éléments-Traces," *Sci. Sol* 2:73–87 (1983).
41. Ministere de L'Environment. "Séminaire Élément-Traces et Pollution des Sols" (Paris, 1982) 1 vol.
42. Juste, C. "Problèmes Posés por L'èvalutation de la Disponibilité pour la Plante des Éléments-Traces du Sol et de Certains Amendements Organiques," *Sci. Sol* 2:109–122 (1983).
43. Morè, E. and M. Coppenet. "Teneurs au Se des Plants Fourrageres. Influence de la Fertilization et des Apport de Sélénite," *Ann. Agron.* 31:297–317 (1980).
44. Lemaire, F., R. Morichon, and J. Meignan. "Egrondage de Bones de Station d'Épuration sur un Sol Sableux Supportant une Rotation Fourragère on une Rotation Légumière en Région Sautmuroise,"*Sci. Sol* 2:123–134 (1983).
45. Trichet, J. "Rétention des Métaux Lourds (Cu, Mn, Zn, Pb, Cd, Ni, Co, as) par les Sols Volcaniques. Incidences sur le Transfer de ces Métaux par les Plantes, le Bétail et les Eaux," Ministere de L'Enviroment, c.R. SDS/81–219 (Paris 1983), p. 29.
46. Morel, J. L. and A. Guckert. "Evolution en Plein Champ de la Solubilité dans DTPA des Metaux Lourd du Sol Introduits par des Égoudages des Bones Urbaines Chaulées," *Agronomie* 4 (4):377–386 (1984).
47. Juste, C. and J. Tauzin. "Evolution du Contenu en Métaux Lourds d'un Sol de Limon Maintenue en Jachère Nue après 56 Annèes d'Application Continue de Divers Engrais et Amendements," c.R. Acad. Agric. Fr, 72/9–739–746 (1986).
48. Juste, C. and P. Solda, "Influence de l'Addition de Differentes Matières Fertilusantes sur la Biodisponibilité du Cd, Mn, Ni et Zn Contenus dans un Sol Sabbleux Amendé par des Nones de Station d'Épuration,"*Agronomie* 8(10)897–904 (1988).
48a. Aubert, M. and M. Pinta. *Trace Elements in Soils* (Amsterdam: Elsevier, 1977) p. 395.
49. Zöettl, H. W. and R. F. Huettl. "Nutrient Supply and Forest Decline in Southwest Germany," *Water Air Soil Pollut.* 31:449–462 (1986).
49a. Zöettl, H. W., K. Stahr, and K. Keilen. "Spurenelementverteilung in einer bodengesellstkoft in Barkaldegranitgebriet (Sudsckawarzwald),"*Mitt. Dtsch. Bodenk. Geellsch.* 25:143–148 (1977).
50. Ulrich, B. "Natural and Anthropogenic Components of Soil Acidification," *Z. Pflanzenernaehr. Bodenkd.* 149:702–717 (1986).
51. Davies, M. R. "Chemical Composition of Soil Solutions Extracted from New Zeland Beech Forest and West Germany Beech and Spruce Forest," *Plant Soil* 126:237–246 (1990).

52. Schilichting, E. and A. M. Elgala. "Schwermetallverteilung und Tongehalte in Böden," *Z. Pflanzenernaehr. Bodenkd.* 6:563–571 (1975).
53. Fassbender, H. W. and G. Seekamp. "Fraktionen und löslichkeit der schwermetalle Cd, Co, Cr, Cu, Ni und Pb in Böden," *Geoderma* 16:55–69 (1976).
54. Murad, E. and W. R. Fischer. "Mineralogy and Heavy Metal Contents of Soils and Stream Sediments in a Rural Region of Western Germany," *Geoderma* 21:133–145 (1978).
55. Schwertmann, U., W. R. Fischer, and H. Fechter. "Spurenelemente in Bödensequenzen. I. Zwei Braunerde-Podsol sequence aus Tonschieferschutt," *Z. Pflanzenernaehr. Bodenkd.* 145:161–180 (1982).
56. Schwertmann, U., W. R. Fischer, and H. Fechter. "Spurenelemente in Bödensequenzen. II. Zwei Pararendzina-Pseudogley-sequenzen aus Löß," *Z. Pflanzenernaehr. Bodenkd.* 145:181–196 (1982).
57. Stahr, K., H. W. Zöttl, and F. R. Hadrich. "Transport of Trace Elements in Ecosystems of the Bärhalde Watershed in the Southern Black Forest," *Soil Sci.* 130(4):217–224 (1980).
58. Heinrichs, H. and R. Mayer. "The Role of Forest Vegetation in the Biogeochemical Cycle of Heavy Metals," *J. Environ. Qual.* 9:111–118 (1980).
59. Ulrich, B. "Ökologische gruppierung von böden mach ikrem chemiscken Bödensustand," *Z. Pflanzenernaehr. Bodenkd.* 144:289–305 (1981).
60. Matzner, E. and B. Ulrich. "Raten der deposition der internen Produktion und des umsatzes von protonen in zwei waldökosystemen," *Z. Pflanzenernaehr. Bodenkd.* 147:290–308 (1984).
61. Zech, W., T. Suttner, and E. Popp. "Elemental Analyses and Physiological Response of Forest Trees in SO-Polluted Areas of NE Bavaria," *Water Air Soil Pollut.* 25:175–183 (1985).
62. Bartels, U. and J. Block. "Ermittlung der gesamtsäuredeposition in nordrhein Westfälischen fichten-und Buchenbeständen," *Z. Pflanzenernaehr. Bodenkd.* 148:689–698 (1985).
63. Bruemmer, G. W., J. Gerth, and U. Herms. "Heavy Metal Species, Mobility and Availability in Soil," *Z. Pflanzenernaehr. Bodenkd.* 149:382–398 (1986).
64. Fischer, W. R. "Property and Heavy Metal Complexation by Aqueous Humus Extracts," *Z. Pflanzenernaehr. Bodenkd.* 149:399–410 (1986).
65. König, N., P. Baccini, and B. Ulrich. "Der einfluß der natürlichen organischen substausen auf die Metalverteilung Zwischen Boden und Bodenlösung in einen souren Waldboden," *Z. Pflanzenernaehr. Bodenkd.* 149:68–82 (1986).
66. Neite, H. "Zum einfluß von pH un organischen kohlenstoffgehalt auf die löslicheit von eisen, blei, mangan und zink in waldböden," *Z. Pflanzenernaehr. Bodenkd.* 152:441–445 (1989).
67. Hantschel, R., M. Kaupenjohann, J. Gradl, and W. Zech. "Ecologically Important Differences Between Equilibrium and Percolation Soil Extracts, Bavaria," *Geoderma* 43:213–227 (1988).
68. Apostolakis, C. G. and C. E. Douka. "Distribution of Macro and Micronutrients in Soil Profiles Developed on Lithosequences and Biosequences in Northern Greece," *Soil Sci. Soc. Am. Proc.* 34:290–296 (1970).
69. Nakos, G. "Relationships of Bioclimatic Zones and Lithology with Various Characteristics of Forest Soils in Greece," *Plant Soil* 79:101–121 (1984).
70. Karataglis, S. and D. Babalonas. "The Toxic Effects of Copper on the Growth of *'Solanum lycopersicum'* Collected from Zn and Pb-soils," *Angew. Bot.* 59:45–52 (1985).

71. Babalonas, D., S. Karataglis, and V. Kabassakalis. "Zinc, Lead and Copper Concentration in Plants and Soils from Two Mines in Chalkidiki, North Greece," *J. Agron. Crop Sci.* 158:87–95 (1987).
72. Karataglis, S. S., B. Kabasakalis, and C. Alexiadis. "Environmental Lead Pollution," in *Proc. Int. Conf. on Environmental Pollution* (Thessaloniki, Greece, 1981) pp. 203–211.
73. Bellanca, A., A. Di Caccamo, and R. Neri. "Mineralogia e geochimica di alcuni suoli della Sicilia Centro-Occidentale: studio delle variazioni composizionali lungo i profili pedologici in relazione ai litotipi d'origine," *Mineral Petrogr. Acta* 24:1–15 (1980).
74. Bini, C., M. Dall'Aglio, R. Gragnani, and V. Papagni. "Distribuzione e Circolazione degli Elementi in Traccia nei Suoli. Studio di una Zona Agricola del Chianti," *Rend. Soc. Ital. Mineral. Petrol.* 38(2):803–816 (1982).
75. Bini, C., O. Ferretti, E. Ghiara, and R. Gragnani. "Distribuzione e Circolazione Degli Elementi in Traccia nei Suoli. Suoli della Regione Puglia," *Rend. Soc. Ital. Mineral. Petrol.* 39(1):281–296 (1984).
76. Bini, C., E. Ghiara, and R. Gragnani. "Distribuzione Degli Elementi in Traccia nei Suoli. Pedologia e Geochimica di una Toposequenza sul Versante Occidentale della Cima Vertana (Alto Adige)," *Rend. Soc. Ital. Mineral. Petrol.* 39(1):281–296 (1984).
77. Bini, C., O. Ferretti, E. Ghiara, and R. Gragnani. "Distribuzione e Circolazione degli Elementi in Traccia nei Suoli. Pedogenesi, Mineralogia e Geochimica dei Suoli dell'Emilia Occidentale," *Rend. Soc. Ital. Mineral. Petrol.* 41(1):95–112 (1986).
78. Bini, C., O. Ferretti, C. Orlandi, and S. Torcini. "Distribuzione e Circolazione degli Elementi in Traccia nei Suoli. Una Sequenza Altimetrica di Suoli su Rocce Carbonatiche del M.te Terminillo (Rieti)," *Rend. Soc. Ital. Mineral. Petrol.* 41(2):297–309 (1986).
79. Bini, C., M. Dall'Aglio, O. Ferretti, and R. Gragnani. "Background Levels of Trace Elements in Soils of Italy," *Environ. Geochem. Health* 10(2):63–69 (1988).
80. Bini, C., R. Gragnani, and G. Ristori. "Soil Genesis and Evolution from Mafic and Ultramafic Rocks in the Northern Apennines," *Ofioliti* 9(3):337–352 (1984).
81. Bini, C., N. Coradossi, A. M. Froio, and R. Gragnani. "Mineralogia e Geochimica dei Suoli Sviluppati su Formazioni Mio-plioceniche dell'Appennino Tosco-Romagnolo," *Mineral. Petrogr. Acta* 30:181–202 (1986–1987).
82. Angelone, M., C. Bini, L. Leoni, C. Orlandi, and F. Sartori. "I Suoli del bacino del Brasimone (Bo): Pedogenesi, Mineralogia, Geochimica," *Mineral Petrogr. Acta* 31:217–241 (1988).
83. Angelone, M., C. Bini, R. Gragnani, and G. G. Ristori. "Mineralogical and Geochemical Evolution of Two Podzolic Soils on Granitic Rock (Eastern Alps, Italy)," *Chem. Erde* 50:279–295. (1990).
84. Bini, C., O. Vaselli, N. Coradossi, M. G. Pancani, and M. Angelone. "Clay Mineral Formation from Mafic Rocks in Temperate Climate, Central Italy," in *Proc. 9th Int. Clay Conference* (Strasbourg, 1989).
85. Bini, C., N. Coradossi., O. Vaselli., M. G. Pancani, and M. Angelone. "Weathering and Soil Mineral Evolution from Mafic Rocks in Temperate Climate," *Proc. 14th Int. Congress of Soil Science* (Kyoto, August 1990) (7): p. 54.

86. Angelone, M., O. Vaselli, C. Bini, N. Caradossi, and M. G. Pancani. "Total and EDTA-Extractable Element Contents in Ophiolitic Soils from Tuscany (Italy)," *Z. Pflanzenernaehr. Bodenkd.* 154:(1990).
87. Bini, C. "Contributo per la Mineralogia e la Geochimica dei Suoli dell'Appennino," *Atti Accad. Naz. Lincei* (Roma, 1990).
88. Leita, L., M. De Nobili, and P. Sequi. "Content of Heavy Metals in Soils and Plants Near Cave del Predil, Udine, Italy," *Agrochimica* 32(1):94–97 (1988).
89. Leita, L., M. De Nobili, G. Pardini, F. Ferrari, and P. Sequi. "Anomalous Contents of Heavy Metals in Soils and Vegetation of a Mine Area in S.W. Sardinia, Italy," *Water Air Soil Pollut.* 48:423–433 (1989).
90. Polemio, M., N. Senesi, and S. A. Bufo. "Soil Contamination by Metals. A Survey in Industrial and Rural Areas of Southern Italy," *Sci. Total Environ.* 25:71–79 (1982).
91. Polemio, M., S. A. Bufo, and N. Senesi. "Minor Elements in South-east Italy Soils," *Plant Soil* 69:57–66 (1982).
91a. Bargagli, R., G. Borchigiani, B. F. Sieghel, and S. M. Sieghel. "Accumulation of Mercury and Others Metals by Lychen *Parmelia Sulcata*, of Italy Minesite and a Volcanic Area," *Water Air Soil Pollut.* 45(3–4):315–327 (1989).
92. De Haan, F. A. M., S. Van der Zee, and W. H. Van Riemsdijk. "The Role of Soil Chemistry and Soil Physics in Protecting Soil Quality: Variability of Sorption and Transport of Cd as an Example," *Neth. J. Agric. Sci.* 35:347–359 (1987).
93. Van Driel, W. and K. W. Smilde. "Heavy Metal Contents of Dutch Arable Soils," *Landwirtsch. Forsch. Sonderh.* 38:305–313 (1982).
94. Lexmond Th. M. and Th. Edelman. "Current Background Values of a Number of Heavy Metals and Arsenic in Soils (in Dutch)," in *Handboek voor Miliebeheer*, Vol. 4. Bodembefcherming, chap. D4100, Samsen, Alphen AAN DEN RIJN (1987) p. 32.
95. Doelman, P. and L. Haanstra. "Short-Term and Long-Term Effects of Cd, Cr, Cu, Ni, Pb and Zn on Soil Microbial Respiration in Relation to Abiotic Soil Factors," *Plant Soil* 79:317–327 (1984).
96. Gerritse, R. G., R. Vriesma, J. W. Dalenberg, and H. P. de Roos. "Effect of Sewage Sludge on Trace Element Mobility in Soils," *J. Environ. Qual.* 11:359–364 (1982).
97. Gerritse, R. G., W. Van Driel, K. W. Smilde, and B. Van Luit. "Uptake of Heavy Metals by Crops in Relation to Their Concentration in the Soil Solution," *Plant Soil* 75:393–404 (1983).
98. Wriesma, D., B. J. Van Goor, and N. G. Van der Veen. "Cd, Pb, Hg, As Concentration in Crops and Corresponding Soils in The Netherlands," *J. Agric. Food Chem.* 34:1067–1074 (1986).
99. Van Lune, P. "Cd and Pb in Soils and Crops from Allotment Gardens in The Netherlands," *Neth. J. Agric. Sci.* 35:207–210 (1987).
100. Vreman, K., N. G. Van der Veen, E. J. Van der Molen, and W. G. de Ruig. "Transfer of Cd, Pb, Hg, and As from Feed into Milk and Various Tissues of Dairy Cows: Chemical and Pathological Data," *Neth. J. Agric. Sci.* 34:129–144 (1986).
101. Vreman, K., N. G. Van der Veen., E. J. Van der Molen, and W. G. de Ruig. "Transfer of Cd, Pb, Hg, As, from Fuel into Tissues of Felterning Bulls: Chemical and Pathological Data," *Neth. J. Agric. Sci.* 36:327–338 (1988).

102. Låg, J. *Geomedical Aspects in Present and Future Research* (Oslo: Universitatsforlagets, 1980) p. 226.
103. Steinnes, E. "Some Geographical Trace Element Distribution of Potential Geomedical Relevance," in *Geomedical Research in Relation to Geochemical Registrations*, J. Låg, Ed. (Oslo: Univeritatsforlagets, 1984) pp. 175–186.
104. Bølviken, B. and R. T. Ottesen. "Geochemistry in the Nordkalott Project, Northern Finland, Norway and Sweden," in *Geomedical Research in Relation to Geochemical Registrations*, J. Låg, Ed. (Oslo: Univeritatsforlagets, 1984) pp. 17–25.
105. Låg, J. and B. Bøvilken. "Some Naturally Heavy-Metal Poisoned Areas of Interest in Prospecting Soil Chemistry and Geomedicine," *Nor. Geol. Unders.* 304:73–96 (1974).
106. Låg, J. "Innhol av tungmetaller og enkelte andre stoffer i noen prøver av kulturjord og matvekster fra odd-omradet (English Summary)," *Ny Jord.* 62:47–59 (1975).
107. Låg, J. "Arsenic Pollution of Soils at Old Industrial Sites," *Acta Agric. Scand.* 28(1):97–100 (1978).
108. Låg, J. and E. Steinnes. "Regional Distribution of Halogens in Norwegian Forest Soils," *Geoderma* 16:317–325 (1976).
109. Låg, J. and E. Steinnes. "Regional Distribution of Se and As in Humus Layers of Norwegian Forest Soils," *Geoderma* 20:3–14 (1978).
110. Simesen, M. G. "Selenium and Se-Vitamin Deficiency Considered from a Geomedical Point of View," in *Geomedical Aspects in Present and Future Research*, J. Låg, Ed. (Oslo: Univeritatsforlagets, 1980) pp. 73–79.
111. Schalin, G. "Multiple Schlerosis and Selenium," in *Geomedical Aspects in Present and Future Research*, J. Låg, Ed. (Oslo: Univeritatsforlagets, 1980) pp. 81–102.
112. Da Silva Teixeira, A. J., E. M. Sequiera, M. D. Lucas, A. Santos Coutinho, and L. F. Mendia De Castro. "Soil Micronutriens Map of Portugal," *Agron. Lusit.* 31:293–304 (1971).
113. Sequeira, E. M. "Toxicity and Movement of Heavy Metals in Serpentinitic Soils (Northern-Eastern Portugal)," *Agron. Lusit.* 30:115–154 (1968).
114. Santhos Couthinho, A., A. J. Da Silva Teixeira, E. M. Sequeira, and M. D. Lucas. "Ferro, Manganesio e Zinco, Totai e Extraiveis, em Aluviossolos Calcarios Antigos da Campina de Faro," *Agron. Lusit.* 33:275–298 (1972).
115. Lucas, M. D. "Um Caso de Deficiência de Molindénio em Couve-flor na Leziria do Ribatejo," *Agron. Lusit.* 35(2):169–180 (1974).
116. Lucas, M. D. "Deficiência de Molibdénio em Melao num Planosolo da Região de Tavira, *Agron. Lusit.* 37(2):151–162 (1976).
117. Vieria E Silva, J. M. and A. M. Ventura. "Distribuiçao e Localizaçao do Zinco e do Cobre num Perfil de Gabro na Regiao de Beja," *Memorias e Noticias*. Publ. Mus. Lab. Mineral. Geol., Univ. Coimbra, No. 98 (1984).
118. Castelo Branco, M. A., E. M. Sequiera, and H. Domingues. "Trace Elements Map of Portugal's Soil. Present Situation," Newsletter from the F.A.O. European Cooperative Network on Trace Elements. 5th Iss., State University Gent, Belgium. (1987).
119. Sequiera, E. M. "Solos Arenosos da Regiao Mio-pli-pleistocénica a Sul do Tejo. IV. O Manganésio," *Agron. Lusit.* 41(2):121–137 (1981).

120. Sequiera, E. M. "Solos Arenosos da Regiao Mio-pli-pleistocénica a Sul do Tejo. III. O Zinco," *Agron. Lusit.* 41(1):97–113 (1980).
121. Sequiera, E. M. "Solos Arenosos da Regiao Mio-pli-pleistocénica a Sul do Tejo. V- O Cobre," *Agron. Lusit.* 41(2):149–164 (1981).
122. Sequiera, E. M. "Solos Arenosos da Regiao Mio-pli-pleistocénica a Sul do Tejo. VI. O Boro, O Molibdénio e o Cobalto," *Agron. Lusit.* 41(3–4):193–212 (1982).
123. Vieria, E., J. M. Silva, and H. Domingues. "Distribuiçao de Microelementos em Solos Derivados de Granitos e de Xistos," *Geociencias Aveiro.* 3(1–2):235–248 (1988).
124. Da Silva Teixeira, A. J., E. M. Sequiera, M. D. Lucas, and M. J. Santos. "Solos Derivados de Granidos e Xistos da Regiao Noreste de Portugal. Micronutrientes Totais e Extraéveis," *Pedologia Oeiras* 16(1):1–99 (1981).
125. Da Silva Teixeira, A. J., E. M. Sequiera, M. D. Lucas, and M. J. Santos. "Solos Arenosos da Regiao Mio-plio-plistocenica a Sul do Tejo. I. Caracteristicas. Micronutrientes Totais e Extraéveis," *Agron. Lusit.* 40(1):41–78 (1980).
126. Magalhaes, M. J., E. M. Sequeira, and M. D. Lucas. "Copper and Zinc in Vineyard of Central Portugal," *Water Air Soil Pollut.* 26:1–17 (1985).
127. Guitian, F. and I. Lopez. "Suelos de la Zona Humida Espanola, Serpentinas. 1 Morfologia y Datos Generales," *An. Edafol. Agrobiol.* 39(3–4):403–415 (1980).
128. Lopez, I. and F. Guitian. "Suelos de la Zona Humida Espanola. Suelos Sobre Serpentinas. 2 Oligoelementos y Relacion Ca/Mg en Suelos y Vegatation," *An. Edafol. Agrobiol.* 40(1–2):1–10 (1981).
129. Carballas, T. "Niquel en los Suelos de la Provincia de la Coruna," *An. Edafol. Agrobiol.* 24:267–293 (1985).
130. Cala Rivero, V., J. Rodriguez, and A. Guerra. "Contaminacion por Metales Pesados en los Suelos de la Vega de Araniguez. I- Pb, Cd, Cu, Zn, Ni, Cr," *An. Edafol. Agrobiol.* 14(11–12):1595–1608 (1985).
131. Boluda Hernandez, R., V. Andreu Perez, V. Pons Marti, and J. Sanchez Diaz. "Contenido de Metales Pesados (Cd, Co, Cr, Cu, Ni, Pb, Zn) en Suelos de la Comarca La Plama de Requera-Utiel (Valencia)," *An. Edafol. Agrobiol.* 47(11–12):1485–1502 (1988).
132. Aller, A. J. and L. Deban. "Total and Extractable Contents of Trace Metals in Agricultural Soils of the Valderas Area, Spain," *Sci. Total Environ.* 79:253–270 (1989).
133. Calvo De Anta, R. and V. Tovar Caballero. "Limitaciones a la Fertilidad en Areas Serpentinizadas de Galicia," *An. Edafol. Agrobiol.* 46(3–4):433–448 (1987).
134. Calvo De Anta, R., M. L. Fernandez Marcos, and M. A. Veiga Vila. "Composicion de la Solucion del Suelo en Medios Naturales de Galicia," *An. Edafol. Agrobiol.* 46(5–6):621–641 (1987).
135. Perez, J. D. and F. Gallardo-Lara. "Influence of Applied Vegetation Water on Micronutrient Availability in Soil," in *Proc. 3th Symp. Int. sur le Role des Oligoéléments en Agricolture* (Brussels, 1988) pp. 231–237.
136. Gallardo Lara, F. and M. Torres-Martin. "Dynamics of Copper Fractions in the Soil-plant System Under Condition of Intensive Forage Cropping," *Z. Pflanzenernaehr. Bodenkd.* 153:1–2 (1990).
137. Aller, A. J., J. L. Bernal, and M. J. Nozal. *Geochemistry of Trace Elements* (Commun. INIA, Serie Tecnologia Agraria, 1989) pp. 5–38.

138. Gallardo-Lara, F. and R. Nogales. "Effect of the Application of Town Refuse Compost on the Soil-Plant System: A Review," *Biol. Waste* 19:35–61 (1987).
139. Nogales, R., M. Gomez, and F. Gallaro-Lara. "Town-Refuse Compost as a Potential Source of Zinc for Plants," in *Proc. 5th Int. Conf. Heavy Metals in the Environment* (Athens, 1985) pp. 487–489.
140. Nogales, R., J. Robles, and F. Gallardo-Lara. "Boron Release from Town Refuse Compost as Measured by Sequential Plant Uptake," *Waste Manage. Res.* 5:513–520 (1987).
141. Nogales, R., J. Robles, and F. Gallardo-Lara. "Interactive Effect of Applied Town Refuse Compost and Mineral Complements on the Sequential Copper Uptake by Ryegrass," *Agric. Mediter.* 117:3–7 (1987).
142. Nogales, R., A. Navarro, M. T. Baca, and F. Gallardo-Lara. "DTPA-Extractable Micronutrients in Soils of Contrasting pH Affected by Organic Wastes and Elemental Sulphur," in *Trend in Trace Elements, FAO-INRA* (Bordeaux, 1989) pp. 93–105.
143. Melkerud, P. A. "Clay Mineralogical Composition of Weathering Profiles Associated with Spruce and Birch Stands," *Geol. Foeren. Stockholm Foerh.* 107(4):301–309 (1986).
144. Tyler, G., D. Berggren, B. Bergvist, U. Falkengren-Greup, L. Folkenson, and A. Rümling. "Soil Acidification and Metal Solubility in Forest of South Sweden," in *Effect of Air Pollutants, Especially Acidic Deposition, on Forest, Agriculture and Wetlands*, N.A.T.O. Adv. Res. Int. Workshop (Toronto, May 12–17, 1985).
145. Tyler, G. "Uptake, Retention and Toxicity of Heavy Metals in Lichens," *Water Air Soil Pollut.* 47(3–4):321–333 (1989).
146. Majdi, H. and H. Persson. "Effect of Road-Traffic Pollutants (Pb and Cd) on Tree Fine-Roots Along a Motor Road," *Plant Soil* 119:1–5 (1989).
147. Bergkvist, B. "Leaching of Metals from a Spruce Forest Soils as Influenced by Experimental Acidification," *Water Air Soil Pollut.* 31:901–918 (1986).
148. Balsberg Påhlsson, A. M. "Toxicity of Heavy Metals (Zn, Cu, Cd, Pb) to Vascular Plants," *Water Air Soil Pollut.* 47(3–4):287–319 (1989).
149. Bergkvist, B. "Soil Solution Chemistry and Metal Budgets of Spruce Forest Ecosystems in Sweden," *Water Air Soil Pollut.* 33:131–154 (1987).
150. Bergkvist, B., L. Folkeson, and D. Berggren. "Fluxes of Cu, Zn, Pb, Cd, Cr and Ni in Temperate Forest Ecosystems," *Water Air Soil Pollut.* 47(3–4):217–286 (1989).
151. Bååth, E. "Effects of Heavy Metals in Soil on Microbial Processes and Pollution (A Review)," *Water Air Soil Pollut.* 47:335–379 (1989).
152. Nilsson, S. I. and B. Bergkvist. "Aluminium Chemistry and Acidification Processes in a Shallow Podsol on the Swedish West-Coast," *Water Air Soil Pollut.* 20:311–329 (1983).
153. Schmitt, H. W. and H. Sticher. "Prediction of Heavy Metal Contents and Displacement in Soils," *Z. Pflanzenernaehr. Bodenkd.* 149:157–171 (1976).
154. Wyttenbach, A., S. Bajo, and L. Toblee. "Major and Trace Element Concentration in Needles of *Picea Abies*: Levels, Distribution Functions, Correlations and Environmental Influences," *Plant Soil* 85:313–325 (1985).
154a. Muller, M. "Bodenbildung auf silikatunterlage in der alpinen stufe des oberengadins (zentralalpen, Schweiz),"*Catena* 14:419–437 (1987).
155. Keller, L. and P. H. Brunner. "Waste-Related Cadmium Cycle in Switzerland," *Ecotoxicol. Environ. Saf.* 7(1):41–50 (1983).

156. Keller, L. and W. Fluckinger. "Immissionsokologische Untersuchungen an Dauerbeobachtungsflachen im Wald des Kantons Zurich/Schweiz," *Erste Ergebniss der Beobachtungsperiode*. VDI-Berichte 560. p. 253. 1985.
157. Bradley, R. I. "Trace Elements in Soils Around Llechryd, Dyfed, Wales," *Geoderma* 24:17–23 (1980).
158. Davies, B. E. "A Graphical Estimation of the Normal Lead Content of Some British Soils," *Geoderma* 29:67–75 (1983).
159. Davies, B. E. "Baseline Survey of Metals in Welsh Soils," in *Environmental Geochemistry ed Health*, I. Thornton, Ed. (London, April 16–17, 1985).
160. Davies, B. E. and C. F. Paveley. "Background Metal Levels in the Welsh Environment," a report to the Welsh Office (June 1987).
161. Jones, K. C., P. J. Peterson, and B. E. Davies. "Silver Concentration in Welsh Soils and Their Dispersal from Derelict Mine Sites," *Miner. Environ.* 5:122–127 (1983).
162. Jones, K. C., B. E. Davies, and P. J. Peterson. "Silver in Welsh Soils: Physical and Chemical Distribution Studies," *Geoderma* 37:157–174 (1986).
163. McGrath, S. P., C. H. Cunliffe, and A. J. Pope. "Lead, Zinc, Cadmium, Copper, Nickel Concentrations in the Topsoils of England and Wales," in *Environmental Geochemistry ed Health*, I. Thornton, Ed. (London, April 16–17, 1985).
164. McGrath, S. P. "The Range of Metal Concentrations in Topsoils of England and Wales in Relation to Soil Protection Guidelines," in *Proc. 20th Annu. Conf. Trace Substances in Environmental Health* (University of Missouri, Columbia, 1986) pp. 242–252.
165. McGrath, S. P. "Computerized Quality Control, Statistics and Regional Mapping of the Concentrations of Trace and Major Elements in the Soil of England and Wales," *Soil Use Manage.* 1(3):31–38 (1987)
166. Archer, F. C. and J. H. Hodgson. "Total and Extractable Trace Element Contents of Soils in England and Wales," *J. Soil Sci.* 38:421–431 (1987).
167. Bradley, R., I. C. Rudeforth, and C. Wilkins. "Distribution of Some Chemical Elements in the Soils of North West Pembrokeshire," *J. Soil Sci.* 29:258–270 (1978).
168. Davies, B. E. "Mercury Content of Soils in Western Britain with Special Reference to Contamination from Base Metal Mining," *Geoderma* 16:183–192 (1976).
169. Musa, A. S. and E. M. Bridges. "Trace Element Anomalies Associated with Copper Deficiencies in the Towy Valley, Dyfed, Wales," in Soil-Related Pests and Diseases, Welsh Soils Discussion Group n°21, (1980) pp. 91–101.
170. Bridges, E. M. "Toxic Metals in Amenity Soils," *Soil Use Manage.* 3(5):91–100 (1989).
171. Davies, B. E. "Consequences of Environmental Contamination by Lead Mining in Wales," *Hydrobiologia* 149:213–220 (1987).
172. Colbourn, P. and I. Thornton. "Lead Pollution in Agricultural Soils," *J. Soil Sci.* 29:513–526 (1978).
173. Anderson, R. J. and B. E. Davies. "Dental Caries Prevalence and Trace Elements in Soil, with Special Reference to Lead," *J. Geol. Soc. London* 137:547–558 (1980).
174. Davies, B. E., D. Conway, and S. Holth. "Lead Pollution of London Soils: A Potential Restriction on Their Use for Growing Vegetables," *J. Agric. Sci. Camb.* 93:749–752 (1979).

175. Davies, B. E. and N. J. Houghton. "Distance-Decline Patterns in Heavy Metal Contamination of Soils and Plants in Birmingham, England," *Urban Ecol.* 8:285–294 (1984).
176. Tinker, P. B. "Trace Elements in Arable Agriculture," *J. Soil Sci.* 37:587–601 (1986).
177. Wilson, M. J. and M. L. Berrow. "The Mineralogy and Heavy Metal Content of Some Serpentinite Soils in North-East Scotland," *Chem. Erde Bd.* 37:181–205 (1978).
178. Ure, A. M., J. R. Bacon, M. L. Berrow, and J. J. Watt. "The Total Trace Element Content of Some Scottish Soil by Spark Source Mass Spectrometry," *Geoderma* 22:1–23 (1979).
179. Berrow, M. L., M. J. Wilson, and G. A. Reaves. "Origin of Extractable Titanium and Vanadium in the a Horizons of Scottish Podzols," *Geoderma* 21:89–103 (1978).
180. Berrow, M. L. and R. L. Mitchell. "Location of Trace Elements in Soil Profiles: Total and Extractable Contents of Individual Horizons," *Trans. R. Soc. Edinburgh: Earth Sci.* 71:103–121 (1980).
181. Berrow, M. L. and G. A. Reaves. "Trace Elements in Scottish Soils Developed on Greywackes and Shales: Variability in the Total Contents of Basal Horizon Samples," *Geoderma* 26:157–164 (1981).
182. Berrow, M. L. and G. A. Reaves. "Background Levels of Trace Elements in Soils," in *Proc. Int. Conf. Environmental Contamination* (London: CEP Consultants Ltd., July 1984) pp. 333–340.
183. Berrow, M. L. and G. A. Reaves. "Extractable Copper Concentrations in Scottish Soils," *J. Soil Sci.* 36:31–43 (1985).
184. Reaves, G. A. and M. L. Berrow. "Total Lead Concentrations in Scottish Soils," *Geoderma* 32:1–8 (1984).
185. Reaves, G. A. and M. L. Berrow. "Extractable Lead Concentrations in Scottish Soils," *Geoderma* 32:117–129 (1984).
186. Reaves, G. A. and M. L. Berrow. "Total Copper of Scottish Soils," *J. Soil Sci.* 35:583–592 (1984).
187. McKeague J. A. and M. S. Wolynetz. "Background Levels of Minor Elements in Some Canadian Soils," *Geoderma* 24:299–307 (1980).
188. Vinogradov, A. P. *The Geochemistry of Rare and Dispersed Chemical Elements in Soils*, 2nd ed. (New York: Consultants Bureau, 1959).
189. Kabata-Pendias, A. and H. Pendias. *Trace Elements in Soils and Plants* (Boca Raton, FL: CRC Press, 1984).
190. Adriano, D. C. *Trace Elements in the Terrestrial Environment* (Springer-Verlag, 1986).

3

Background Levels and Environmental Influences on Trace Metals in Soils of the Temperate Humid Zone of Europe

Alina Kabata-Pendias, Stanislaw Dudka, Anna Chlopecka, and Teresa Gawinowska

Institute of Soil Science and Plant Cultivation, Osada Palacowa, 24-100, Pulawy, Poland

ABSTRACT

Soils of central-eastern Europe have developed under the predominating influence of podzolization and acidification processes. Soils of Europe have been virtually exposed for a long period of time to large emissions of trace pollutants. Consequently, the net mass balance of trace metals indicates an enrichment in surface soils of some regions. Depending upon the source of metal loading, the enrichment factor for metals vary from less than 10 to over 100 (or more) times.

An estimation of background levels of trace metals in European soils is a most complex problem because truly pristine ecosystems no longer exist. The method adopted for interpretation of background metal concentrations in surface soils of Poland was based on baseline values. The data obtained for soils from uncontaminated areas indicate that trace metal levels in these soils have not been changed significantly by pollution. The concentrations of trace metals in soils from uncontaminated areas of Poland are fairly similar to the values reported for soils of neighboring countries as well as to mean values estimated as an average content of soils of the world.

The overall mean concentrations of some trace metals in uncontaminated soils of the temperate humid zone in Europe are (in mg/kg, dry-weight basis) Mn — 500, Cr — 75, Pb — 35, Ni —20, Cu — 15, Cd — 0.5. These values are not necessarily background values, but they appear not to be significantly affected by pollution.

INTRODUCTION

The natural concentrations of trace elements in soils depend upon the amounts present in the parent rocks from which the soils form and upon soil-forming processes closely related to climatic conditions. Thus, the natural budget of trace elements in the soil should be an effect of those factors as well as of geochemical properties of an element. Depending upon these factors, either processes of downward leaching in the soil profile or processes of accumulation in the upper soil horizons would predominate.

These generalized rules allow soil scientists to predict the natural (baseline, background) contents of many elements based on their typical occurrence in parent rocks. Trace element contents commonly reported for main types of rocks (Table 1) show that argillaceous sediments (siltstones, shales, and clay-loams) almost always contain more trace elements than sands, sandstones, and limestones. The distribution of trace elements among various magmatic rocks is more complex, although a general trend can be observed that levels of trace metals in mafic magmatic rocks is higher than these in acid rocks (Table 1). Therefore, recent soils, especially those enriched either in clay or in silt fractions, contain greater amounts of several trace elements than those derived from other parent rocks, in particular, sands and acid igneous rocks.

Soils are the most significant sink for all trace elements released into the environment by man's activities, including industrial, municipal, and agricultural sources. Thus, present status of trace elements of soils either in natural or in agricultural ecosystems reflects their sources both from parent material and from anthropogenic addition. During recent decades trace inorganic pollutants have been distributed so widely that even in the remote regions soils show anthropogenically increased levels of certain trace metals. Trace metals enter the soils by a number of pathways, and their behavior and fate in soils differ based on the source and species of pollution.

Soils of Europe have been exposed for long periods of time, and to relatively large emissions of trace pollutants compared to soils of other continents. Therefore, studies on trace element status in soils and attempts to estimate their baselines are an important contribution to defining the magnitude of environmental pollution.

Studies on trace elements in soils have been carried out to varying extents in central-eastern Europe. This chapter presents some data available from several countries of this region.

Table 1. Trace Elements in Main Soil-Parent Rocks (Values Commonly Found, mg/kg, Dry-Weight Basis)

	Magmatic Rocks			Sedimentary Rocks		
Element	Mafic rocks	Intermediate rocks	Acid rocks	Argillaceous sediments and shales	Sandstones	Limestones, dolomites
Cd	0.13–0.22	0.13	0.09–0.20	0.22–0.30	0.05	0.035
Co	35–50	1.0–10	1–7	11–20	0.3–10	0.1–3.0
Cr	170–200	15–50	4–25	60–100	20–40	5–16
Cu	60–120	15–80	10–30	40	5–30	2–10
Hg	0.0X	0.0X	0.08	0.18–0.40	0.04–0.10	0.04–0.05
Mn	1200–2000	500–1200	350–600	500–850	100–500	200–1000
Mo	1.0–1.5	0.6–1.0	1–2	0.7–2.6	0.2–0.8	0.16–0.40
Ni	130–160	5–55	5–15	50–70	5–20	7–20
Pb	3–8	12–15	15–24	18–25	5–10	3–10
V	200–250	30–100	40–90	100–130	10–60	10–45
Zn	80–120	40–100	40–60	80–120	15–30	10–25

From: Kabata-Pendias and Pendias.[1]

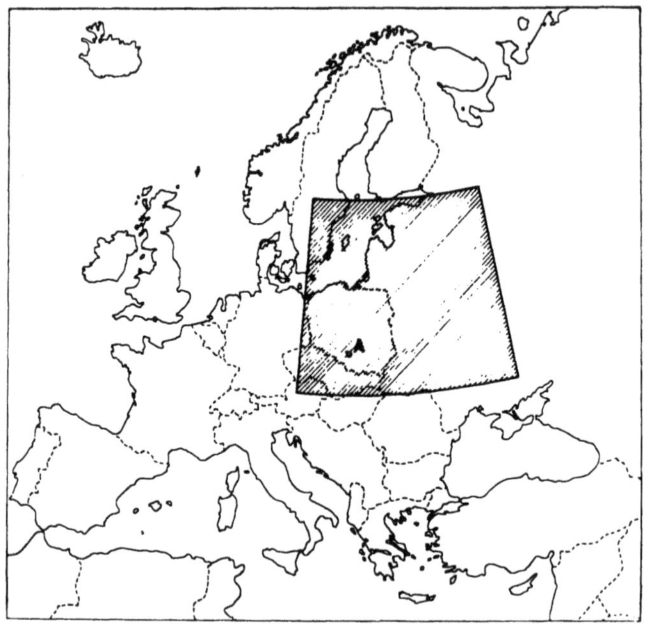

FIGURE 1. Study area in Europe (shadowed plot) is located between the longitude 14°–36° and latitude 48°–60°. Area of soil pollution classes is indicated as point A.

STUDY AREA

Central-eastern part of Europe (Figure 1) is within the region of marine cool temperate climate, with some southern parts under a variable steppe climate. Predominating climatic features are annual isotherms +4°C, January and July isotherms –4°C and +16 C, respectively, and annual rainfall 500 to 750 mm. Surficial quaternary sediments covering the major part of that area are composed of glacial till and outwash. The southern regions are covered mainly by aeolian sediments (loess). Although Hungary and Romania are not included in the area under consideration, available data for soils of these countries are also cited to provide information on the metal concentrations in these soils as well. In this zone two dominating vegetation regions are distinguished: (1) mixed coniferous/deciduous boreal forest and (2) sub-boreal oak and pine forest. Soils developed under the described conditions belong mainly to the following major soil groups as defined by FAO/UNESCO,[3] with approximate U.S. taxonomy equivalents in parentheses:[1] (1) Podzols (Spodosols), (2) Luvisols (Alfisols), (3) Podzoluvisols (Spodosols-Alfisols), and (4) Cambisols (Inceptisols) (Figure 2). Among the listed major soil types, other soils are frequently distinguished: Rendzinas (Rendosols), Fluvisols (Fluvents), Chernozems (Borolls), Gleysols (Aquic, part), and Histosols.

Podzols and podzoluvisols predominate in forests, grasslands, and arable cropping areas. The main crops grown at these soils are cereals and potatoes.

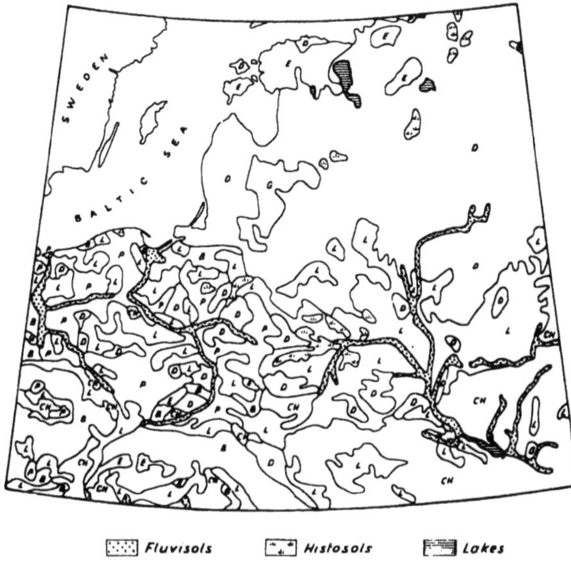

FIGURE 2. Major soil units in the study area (based on data of FAO-UNESCO[3]). Major soil units: B — Cambisols, D — Podzoluvisols, L — Luvisols, P — Podzols, E — Rendzinas, CH — Chernozems, G — Gleysols.

Luvisols and cambisols are most often used for cropping of sugar beets, cereals, and potatoes. Rendzinas occur under a broad climate range. In northern regions (e.g., Estonia) and in the hilly and mountain zones, growing of agricultural crops is highly important, while in southern regions (e.g., Czechoslovakia), fruit and vine production are the principal enterprises. Fluvisols are widely distributed along the major rivers but occupy limited areas. Intensive arable cropping is carried out on these soils, with crops varying upon climatic conditions. Chernozems are of wide agricultural use, from arable cropping to grazing and meadows. In most regions, however, they are most productive for wheat, maize grain, sugar beets, and alfalfa. Gleysols are commonly associated with cool marine climatic conditions and with local high water table. Common farming practices conducted on these soils are cereal grain and livestock production. Histosols are largely confined to cool and cold temperature climate. They are used for both arable and grassland farming.

Due to great regional variability in both soil properties and climatic conditions, as well as the impact of anthropogenic factors, any general description gives only an approximation of the possible development of ecosystems. Recent climatic trends in this part of Europe suggest the possible significant effects of global thermal and hydrological changes. These changes can cause aridization and salinization of the soils. There is simultaneously observed the strong impact of acid precipitation which is known to bring about soil acidification as well as other kinds of soil degradation.[2]

HISTORICAL BACKGROUND

The first information on trace elements in soils appeared in the 19th and 20th centuries in relation to micronutrient deficiencies. Thus, the first regional studies were conducted for commonly deficient micronutrients (mainly Cu and B, with some emphasis on Co and Mo). Other investigations were conducted to produce geochemical maps for ore prospecting. Results of both kinds of studies did not contribute much to an overall picture of trace element distributions among different soils and various regions.

The development of soil mapping and environmental geochemistry led to studies on trace element occurrence in soils. However, the stage of research varies for each country and, therefore, it is difficult to compare data of different authors for different soils and countries. The trace metals that have been most commonly studied are Cd, Cu, Mn, Pb, and Zn. Other elements such as As, Cr, Hg, Ni, and V have been investigated to a lesser extent. Data for other trace elements, including rare-earth elements (REE), are very seldom available.[1]

Collection and evaluation of data from many sources is made difficult by the unconfirmed reliability of analytical data, inadequacy in information on sampling and analytical techniques used, and inconsistent description of soils. If not otherwise indicated, the total metal contents are assumed to be expressed on a dry-weight basis.

Trace metals extracted from soils with different solutions have been most commonly analyzed for agricultural purposes. Whether these forms of trace metals are readily available to crop plants is still a matter of investigation and discussion.[1,4] Various extractants are also broadly used for analyses of soils polluted with trace metals. In most cases, soluble pools of trace metals are a function of their total contents of soils, and, therefore, these data are also informative on total contents. However, if such data are to be used in environmental and/or agricultural studies, a detailed description of the extraction methods should be included, and care should be taken to avoid any comparison of data based on different extraction methods.

DEFICIENCY OF TRACE METALS

In the formation of soils in the range from padzols to luvisols, the processes of leaching have predominated over other processes. Thus, these soils are very likely to be deficient in most nutrients, including trace metals, unless the soils have received micronutrient fertilizers or have become contaminated. In addition, low pH and Al toxicity are closely associated with these soils. Deficiency of other metals, such as Mn, Co, Mo, and Fe, are not strictly connected with soil taxonomic units. Rather, it is related to soil of different texture classes in which physical and chemical variables differ, and there are the significant factors contributing to micronutrient deficiencies.[1]

Numerous studies have been carried out on soils deficient in trace metals to crop plants. In the described area, up to 30 to 40% of the farmland has been observed to be deficient in easily soluble micronutrients.[5] Deficiencies of trace metals to agricultural crops are associated mainly with podzols, podzoluvisols, and histosols. These soil units are widely distributed in the temperate humid zone of Europe (Figure 2). A common agricultural practice is based on soil tests to establish micronutrient phytoavailability and on supplying micronutrient fertilizers to correct potential deficiencies. These practices, however, have to be carefully used in the areas under the impact of acidification and aerial pollution.

BACKGROUND METAL CONTENTS

Background metal concentrations in surface soils of the described area of Europe are rather difficult to establish. Trace metal loading in soils from various sources has already been appreciable in this area and, in many cases, it has altered the natural metal status and cycling in soils. One of the methods of interpretation of trace metal contents of soils is the calculation of baseline values.[6–8] Geochemical baselines represent a concentration specific to one area and time and are not always true backgrounds. They are calculated as ranges of concentrations which typically include about 95% of all observations. In contrast, geochemical backgrounds should represent natural concentrations, which ideally exclude human influence.[6] Baseline values for trace metals in soils from uncontaminated areas similar to observed ranges of concentrations from the some geographic area of interest indicate no significant change to their contents resulting from pollution.

Data presented for main soil units in Poland (Table 2) can be accepted as good estimates of background levels because they correspond to both baseline values and common concentrations found in parent materials (Table 1). Although the variation among soil units is not very pronounced, cambisols and fluvisols contain more metals than podzols and luvisols. This is especially well observed for Zn and partly for Cu occurrence in soil units of both Poland and the entire studied area (Figures 3 and 4).

The distribution of trace metals among soil textural classes (e.g., sandy, silty, loamy soils) is differentiated more strongly than among taxonomic units (Table 3). In general, trace metal contents in sands are significantly lower than in loams and loesses. This is specially true for Mn, Cr, Cu, Ni, and Zn, because these metals are mainly dependent on pedogenic sources. In case of Cd and Pb, for which anthropogenic sources are believed to be more significant, the association of their levels with soil textural classes is weaker. Also, distribution of Cd and Pb among soil units is only slightly differentiated, especially for Poland (Figures 5 and 6).

Data for the total contents of trace metals in soils are sparse for the studied area. Typically, only soluble fractions of these metals are determined, and the

Table 2. Metal Contents of Main Soil Units in Poland (mg/kg and, Unless Noted, Dry-Weight Basis)

	Podzols (n = 31)		Luvisols (n = 34)		Cambisols (n = 51)		Fluvisols (n = 8)	
Element	Mean	Range	Mean	Range	Mean	Range	Mean	Range
Fe %	0.48	0.1–1.1	0.53	0.1–1.7	1.07	0.4–2.5	1.76	0.7–2.8
Mn	219.1	36.5–775	233.90	58.5–1005	378.4	135.0–780	492.50	290.0–1415
Zn	36.87	3.1–762	34.36	11.7–120	58.02	24.0–725	83.58	46.2–110
Cu	5.33	1.0–13.5	5.37	1.0–19.5	9.71	4.5–24.0	15.92	7.0–31.0
Cr	11.79	4.0–22.0	14.06	4.0–26.0	28.60	12.0–68.0	47.50	27.0–67.0
Ni	5.31	1.3–17.5	6.00	3.0–19.0	12.30	3.5–57.0	21.56	2.4–67.5
Cd	0.39	0.1–5.0	0.35	0.1–1.8	0.45	0.1–6.5	0.48	0.4–1.0
Pb	17.52	4.5–286	15.29	8.0–54.5	20.43	9.4–225	22.28	15.0–27.8

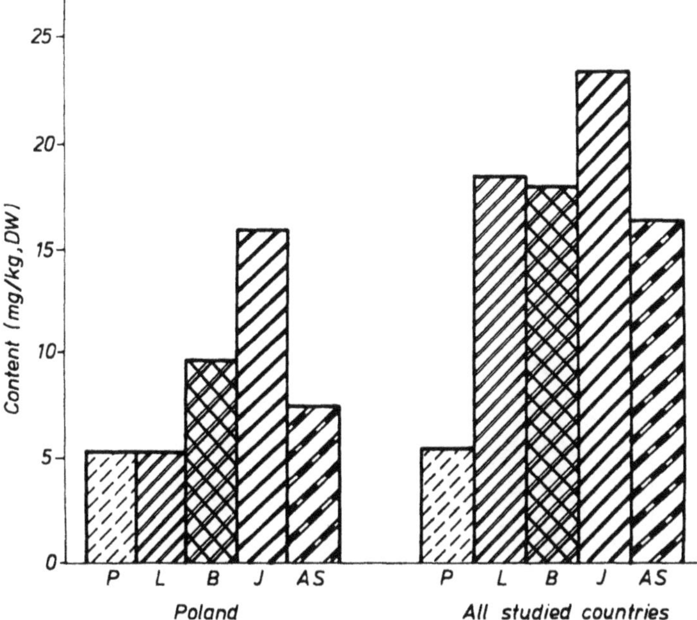

FIGURE 3. Cu content of soil units of Poland and all studied countries (mean values). Soil units: P — Podzols, L — Luvisols, B — Cambisols, J — Fluvisols, AS — all examined soils.

data from different countries are hardly comparable.[9–11] Also, the occurrence of metals in particular soil units is difficult to classify due to poor soil description. For example, podzoluvisols distinguished by the FAO-UNESCO as a dominant soil unit in some regions of the studied area (Figure 2) are not cited by most authors; therefore, data for these soils are apparently included in podzols, luvisols, or in some other soil units.[3]

Manganese

Manganese distribution among soils of various countries (Table 4) shows a pattern similar to that observed in the soils of Poland (Table 2). The lowest mean content is in podzols (220 mg/kg) and the highest in fluvisols (490 mg/kg). Average Mn contents for the different soils range from 150 to 1090 mg/kg, and its overall mean* concentration in soils of major units of the temperate humid zone in Europe is 505 mg/kg. The grand mean** of Mn content of unpolluted soils worldwide is estimated to be 440 mg/kg.[1]

---

* Arithmetical mean calculated for mean values given by various authors for different soil units of the studied area.

** Mean value estimates an average element content of soils of the world.

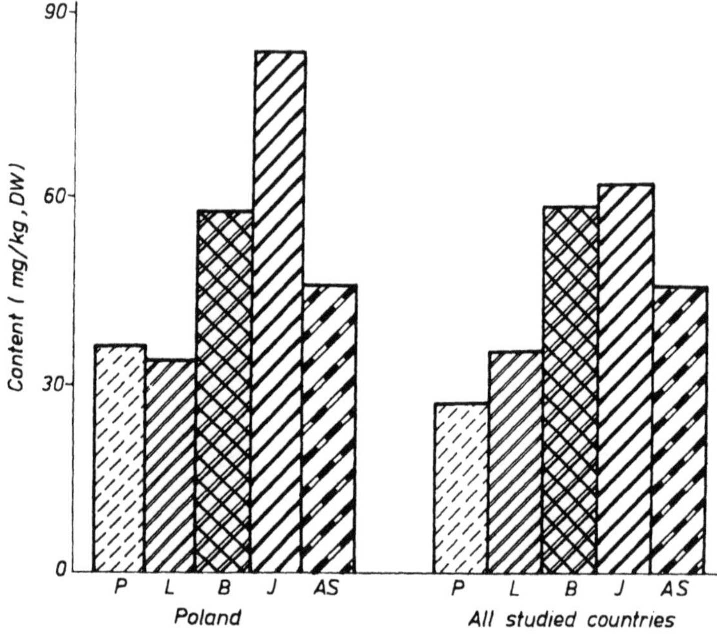

FIGURE 4. Zn content of soil units of Poland and all studied countries (mean values). Soil units: P — Podzols, L — Luvisols, B — Cambisols, J — Fluvisols, AS — all examined soils.

Table 3. Metal Contents of Main Soil Kinds in Poland (mg/kg and, Unless Noted, Dry-Weight Basis)

	Sands (n = 85)		Loesses (n = 17)		Loams (n = 30)	
Element	**Mean**	**Range**	**Mean**	**Range**	**Mean**	**Range**
Fe, %	0.56	0.05–1.7	1.26	0.6–1.9	1.50	0.7–2.8
Mn	240.9	37–1005	435.7	260–740	438.1	186–1415
Zn	37.4	3–762	60.0	28–116	74.8	37–725
Cu	5.8	1–20	10.4	5–16	13.6	6–31
Cr	14.1	4–43	37.2	15–64	38.3	16–68
Ni	6.0	1–39	17.9	7–33	18.3	2–68
Cd	0.3	0.1–5	0.46	0.1–1	0.53	0.2–6.4
Pb	16.8	5–286	19.5	9–33	22.5	11–137

Zinc

Zinc status shows a relatively uniform distribution among soil units (Table 5). However, its lower concentrations are always found in podzols (28 mg/kg) and luvisols (35 mg/kg), and higher levels are present in fluvisols (60 mg/kg) and histosols (58 mg/kg). Similar patterns of variation of Zn levels were observed also for soils of Poland (Table 2; Figure 4). Mean concentrations of Zn for all

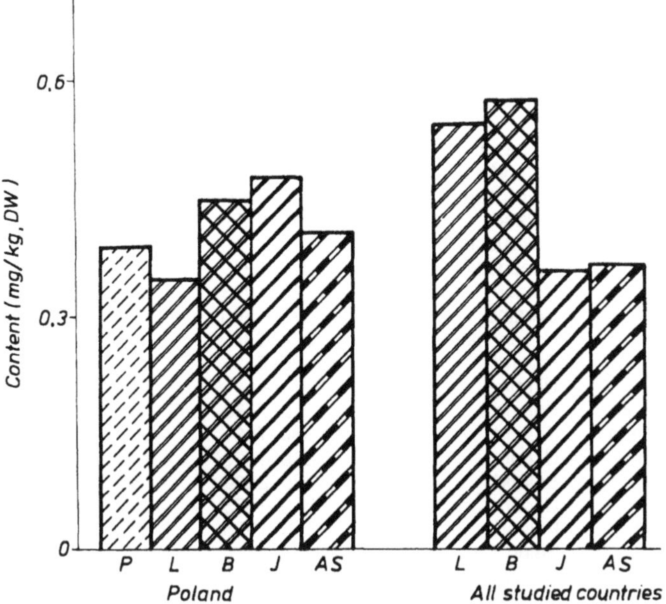

FIGURE 5. Cd content of soil units of Poland and all studied countries (mean values). Soil units: P — Podzols, L — Luvisols, B — Cambisols, J — Fluvisols, AS — all examined soils.

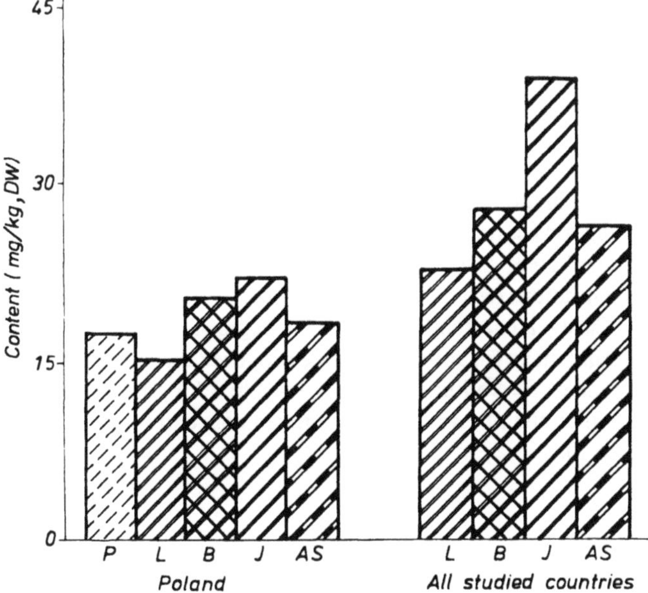

FIGURE 6. Pb content of soil units of Poland and all studied countries (mean values). Soil units: P — Podzols, L — Luvisols, B — Cambisols, J — Fluvisols, AS — all examined soils.

Table 4. Manganese Content (mg/kg, Dry-Weight Basis) of Surface Layers of Major Soil Units in Central-Eastern Europe

Soil Unit	Country	Range	Mean	Ref.
Podzols	Bulgaria	451–883	—	12
	Poland	15–1535	—	13–15
	Belorussia and Ukraine	135–310	217	16, 17
Luvisols	Poland	110–1060	470	14, 15
	Belorussia	—	370	17
Cambisols	Czechoslovakia[a]	230–321	273	10
	Poland	45–1065	420	13–15
	Ukraine	270–1300	475	16
Gleysols	Czechoslovakia[a]	249–284	266	10
	Poland	85–890	495	13, 14
	Ukraine	—	190	16
Rendzinas	Poland	50–7750	440	14, 18
Chernozems	Poland	380–700	560	13, 14
	Ukraine	340–1100	745	16, 19, 20
Histosols	Poland	20–2200	150	15
	Ukraine	510–1465	1005	16
Fluvisols	Poland	150–1965	1085	14
	Belorussia	—	240	17
Various soils	Hungary[b]	120–600	—	21
	Romania	194–1870	755	22
	Ukraine	690–1250	950	16

[a] Soluble in 1 *M* HCl.
[b] Data for whole soil profiles of forest soils.

soils of the studied area are 50 mg/kg, with the range of average values for the soil units from 10 to 105 mg/kg. The grand mean for Zn content of soils worldwide is calculated for 64 mg/kg.[1]

Copper

Copper levels in the investigated soils range widely for soil units (Table 6, Figure 3), being lowest for podzols (6 mg/kg) and highest for fluvisols (24 mg/kg) of the studied area. Also, soils of Poland (Table 2) contain Cu in a similar range of concentrations (from 5 mg/kg in podzols to 16 mg/kg in fluvisols). The overall mean Cu level for all soils of the studied region is 16 mg/kg, which is quite close to the grand mean for soils worldwide proposed recently to be 20 mg/kg.[1] The average values for soil units range from 3 to 30 mg/kg, which is a much broader range than the observed range in mean concentrations of soil units in Poland (Figure 3).

Chromium

Chromium in soils of all major units (Table 7) ranges within the average values, from 20 to 220 mg/kg. It is present in the lowest mean content in podzols (36 mg/kg) and highest in fluvisols (70 mg/kg) and chernozems (135 mg/kg), whereas mean concentrations of Cr in soils of Poland range from 10 mg/kg

Table 5. Zinc Content (mg/kg, Dry-Weight Basis) of Surface Layers of Major Soil Units in Central-Eastern Europe

Soil Unit	Country	Range	Mean	Ref.
Podzols	Poland	5–220	24	14, 15
	Russia and Ukraine	4–57	31	23, 24
Luvisols	Czechoslovakia[a]	8–12	10	10
	Poland	17–127	47	14, 15
	Romania	25–118	61	25
	Ukraine	40–55	48	23
Cambisols	Czechoslovakia[a]	11–16	13	10
	Poland	13–362	68	14, 15
	Romania	37–101	75	25
	Russia and Ukraine	9–77	35	23, 24
Gleysols	Czechoslovakia[a]	9–17	14	10
	Poland	13–98	52	14
	Russia	27–79	53	24
Rendzinas	Czechoslovakia[a]	—	11	10
	Poland	58–150	77	14, 15
	Russia and Ukraine	23–71	47	23, 24
Chernozems	Bulgaria	63–97	—	26
	Poland	33–82	62	14
	Russia	39–82	57	24
Histosols	Bulgaria	—	80	26
	Poland	13–250	60	14, 15
	Russia and Ukraine	8–74	34	23, 24
Fluvisols	Bulgaria	—	62	26
	Poland	55–124	85	14
	Russia	34–49	42	24
Various	Bulgaria	39–99	65	26, 27
	Romania	27–113	57	25
	Russia and Ukraine	31–192	105	28, 23, 24

[a] Soluble in 1 *M* HCl.

(podzols) to 45 mg/kg (fluvisols). The overall mean calculated for soil units of the studied area is 75 mg/kg, while the grand mean of Cr content of soils worldwide is given as 54 mg/kg.[1]

Nickel

Average nickel content of soils of the studied area (Table 8) ranges from 5 to 44 mg/kg. The overall mean for Ni in studied soils (22 mg/kg) is the same as the grand mean concentration in soils worldwide.[1] The effect of soil units on Ni concentration is clearly observed for the soils of Poland (Table 2), where the lowest mean Ni concentration was found in Podzols (5 mg/kg) and the highest in fluvisols (20 mg/kg).

Cadmium

Cadmium occurs in surface soils (Table 9) with a range of mean values from 0.06 to 0.9 mg/kg. The overall mean of 0.5 mg/kg for Cd in studied soils is the same as the worldwide grand mean.[1] Cadmium concentrations in soils, both of

Table 6. Copper Content (mg/kg, Dry-Weight Basis) of Surface Layers of Major Soil Units in Central-Eastern Europe

Soil Unit	Country	Range	Mean	Ref.
Podzols	Poland	1–26	8	13–15
	Russia	2–29	11	24
Luvisols	Czechoslovakia[a]	4–5	5	10
	Poland	8–54	19	13–15
	Romania	3–34	18	25
Cambisols	Czechoslovakia[a]	4–9	6	10
	Poland	4–36	16	13–15, 29
	Romania	9–44	27	25
	Russia	4–21	12	24
Gleysols	Czechoslovakia[a]	4–9	6	10
	Poland	3–53	13	14, 35
Rendzinas	Czechoslovakia[a]	—	8	10
	Poland	7–54	16	14, 15
	Russia	7–23	15	24
Chernozems	Bulgaria	26–38	29	31
	Czechoslovakia[a]	8–20	14	10
	Poland	7–53	19	14, 35
	Russia and Ukraine	16–70	28	20, 24
Histosols	Czechoslovakia[a]	—	3	10
	Poland	1–113	6	30, 32, 35
	Russia	5–23	13	24
Fluvisols	Poland	16–29	22	14
	Russia	12–36	25	24
Various	Russia and Ukraine	2–70	36	16, 24

[a] Soluble in 1 *M* HCl.

Poland (Table 2) and of the entire studied area, do not show a pronounced differentiation for soil units. However, some accumulation of the element in luvisols and cambisols of the temperate humid zone of Europe is observed (Figure 5).

Lead

Mean lead contents of surface soils of central-eastern Europe range from 15 to 65 mg/kg (Table 10). The overall mean for Pb in studied soils is 35 mg/kg; this value is somewhat higher than the grand mean calculated for soils worldwide (30 mg/kg).[1] Lead levels in soils do not correlate with the soil units either separately for soils of Poland or for all soils of the studied area (Figure 6). Ranges among soil units of Poland are lower than in soil units of the other countries of the studied area (Figure 6).

TRACE METAL POLLUTION

Trace metal loading from various sources is already high in certain areas of Europe.[1] As a result of atmospheric deposition of metallic compounds and/or particles, the content of surface soils has been building up in both industrial areas

Table 7. Chromium Content (mg/kg, Dry-Weight Basis) of Surface Layers of Major Soil Units in Central-Eastern Europe

Soil Unit	Country	Range	Mean	Ref.
Podzols	Poland	30–91	51	29
	Belorussia	18–25	21	17
Luvisols	Bulgaria	77–128	—	33
	Poland	21–38	29	34
	Romania	12–86	29	25
	Belorussia	—	84	17
Cambisols	Bulgaria	107–122	115	35
	Poland	35–81	58	29
	Romania	19–73	40	25
	Belorussia and Ukraine	—	51	17, 36
Gleysols	Poland	27–100	57	29
	Ukraine	—	85	36
Chernozems	Bulgaria	116–173	153	35
	Ukraine	71–195	121	19, 20
Fluvisols	Bulgaria	—	91	35
	Belorussia and Ukraine	—	55	19, 17
Various	Bulgaria	71–1085	221	35
	Poland	28–107	60	37
	Ukraine[a]	30–110	54	36

[a] Forest soils.

Table 8. Nickel Content (mg/kg, Dry-Weight Basis) of Surface Layers of Major Soil Units in Central-Eastern Europe

Soil Unit	Country	Range	Mean	Ref.
Podzols	Poland	1–52	7	15
	Belorussia and Ukraine	5–15	11	17, 38
Luvisols	Poland	7–70	19	15
	Belorussia	—	11	17
Cambisols	Poland	10–104	25	15, 39
	Romania	9–62	33	25
	Belorussia and Ukraine	—	24	17, 36
Gleysols	Ukraine	—	36	36
Rendzinas	Poland	7–41	21	18
Chernozems	Russia and Ukraine	14–40	30	19, 20, 38
Histosols	Poland	1–50	9	13, 40
	Belorussia	—	5	17
Fluvisols	Belorussia	—	10	17
Various	Romania	24–60	44	25
	Ukraine	27–75	42	36
	Ukraine[a]	22–55	33	19, 36

[a] Forest soils.

and in remote regions of Europe, depending on a long-range transport of emissions. Point sources of pollution have contributed also to elevated metal levels in soils.

Spatial distribution of metal emissions in Europe show a widely different pattern.[59] Emissions calculated (in t/yr) for the studied area vary with location; for Zn <1 to 1580, Cu 0.2 to 346, Cd 0.1 to 48, and Pb 15 to 1030 (Figure 7 and 8).

Table 9. Cadmium Content (mg/kg, Dry-Weight Basis) of Surface Layers of Major Soil Units in Central-Eastern Europe

Soil Unit	Country	Range	Mean	Ref.
Luvisols	Poland	0.18–0.25	0.20	14
	Romania	0.20–2.70	0.90	25
Cambisols	Poland	0.08–0.58	0.26	14
	Romania	0.50–1.60	0.90	25
Gleysols	Poland	0.14–0.96	0.50	14
Rendzinas	Poland	0.38–0.84	0.62	14, 15
Chernozems	Bulgaria	0.55–0.71	0.61	41
	Poland	0.18–0.58	0.38	14
Fluvisols	Bulgaria	—	0.42	41
	Poland	0.24–0.36	0.30	14
Various	Bulgaria	0.24–0.35	0.29	41
	Poland	0.09–1.80	0.44	14, 42

Table 10. Lead Content (mg/kg, Dry-Weight Basis) of Surface Layers of Major Soil Units in Central-Eastern Europe.

Soil Unit	Country	Range	Mean	Ref.
Luvisols	Poland	14–32	26	14
	Romania	5–41	19	25
Cambisols	Poland	13–52	25	14
	Romania	14–33	21	25
	Ukraine	—	40	36
Gleysols	Poland	20–49	30	14
	Ukraine	—	67	36
Rendzinas	Poland	17–46	29	14, 18
Chernozems	Poland	19–29	25	14
Histosols	Poland	18–85	—	43
Fluvisols	Poland	13–49	39	14
Various	Romania	8–20	15	25
	Ukraine	—	61	36
	Russia and Ukraine[a]	10–56	37	36, 44

[a] Forest soils.

Trace metal loading are first accumulated in surface soil layer. Consequently, as metals are mobilized and as acidification of soils increases, the metals partially leach downward to groundwater. Thus, the measurements of trace metal status of polluted soils represents values for the net mass balance of metals in surface soils.[1] In addition to their loadings, the budgets of metals in soils are affected by several other factors. Among those factors, soil properties, vegetation, and climate play the most important roles.

An estimation of background levels of trace metals in European soils is a most complex problem because a pristine state no longer exists. Background data collected according to unbiased and representative sampling designs, avoiding the contaminated areas, are probably only available for Poland.[7,8] In most cases, data given for polluted soils are not related to a particular soil unit and/or soil textural class. Nevertheless, from the comparison of the concentrations of metals in polluted soils with their ranges in major soil units, it is quite obvious that

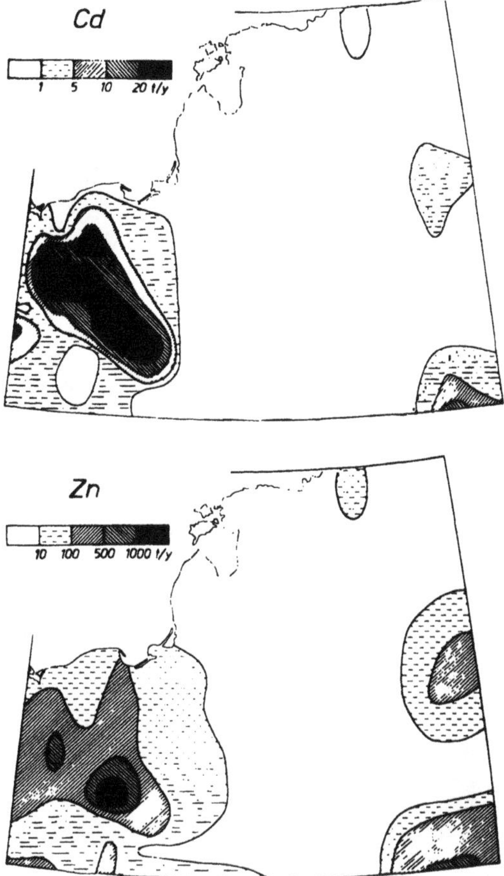

FIGURE 7. Spatial distribution of Zn and Cd emissions in the temperate humid zone of Europe. (Based on data calculated by Pacyna, J. M., *Spatial Distribution of the As, Cd, Cu, Pb, V, and Zn Emissions in Europe Within a 1.5° Grid Net* (NILU, Lillestrom, 1985) p. 22. With permission.)

so called "natural" levels have been exceeded frequently by several times. Depending upon the source of metal loading, the enrichment factors (EF) for trace metals may vary from less than 10 to more than 100 (Table 11).

The highest pollution, expressed as EF, was observed for Cd and Cu in soils of a vicinity of metal smelters. The accumulation of each metal (Cu, Zn, Cd, and Pb) was lowest among polluted soils in areas of gardens and orchards (Table 11).

SOIL POLLUTION INDEXES

Concern towards trace metal pollution of agricultural land is related to both human health (crop quality) and soil degradation (crop quantity). Both aspects are interrelated. However, the evaluation of soil degradation is of vital

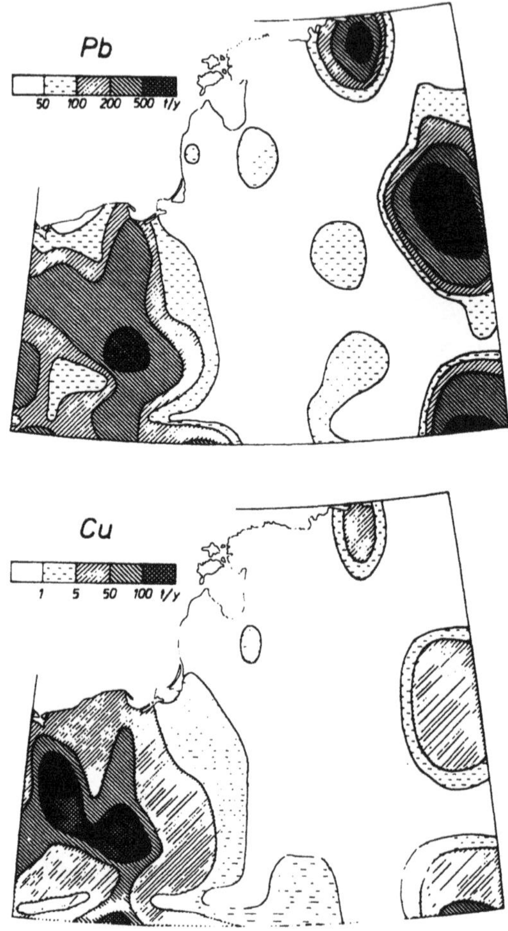

FIGURE 8. Spatial distribution of Pb and Cu emissions in the temperate humid zone of Europe. (Based on data calculated by Pacyna, J. M., *Spatial Distribution of the As, Cd, Cu, Pb, V, and Zn Emissions in Europe Within a 1.5° Grid Net* (NILU, Lillestrom, 1985) p. 22. With permission.)

importance for establishing guidelines for the control of agricultural environment pollution. An attempt was made to meet this demand, and a field study was conducted for the selected region that is heavily industrialized and polluted but still agriculturally important (Figures 1 and 9). The contents of Cd, Zn, and Pb and related pH values were transformed to unified units, which were mapped by isoline method. Critical levels of these metals were proposed based on general environmental, regional, and agricultural conditions. Five classes of soil pollution are related to proposed various land use. Results of this study showed that considerable polluted soils (III and IV classes of degradation) predominate, while slightly polluted (class I degradation) and very heavily polluted soils (class

Table 11. Metal Contamination of Surface Soils of Central-Eastern Europe (Concentrations in mg/kg, Dry-Weight Basis)

Metal	Source of Pollution	Country	Range	Ref.
Cu	Metal-processing industry	Bulgaria	24–2015	31
		Poland	72–125	45
	Urban gardens and orchards	Poland	12–240	46
		Ukraine	50–83	47
	Vineyards	Ukraine	50–83	47
	Fertilized farmland	Poland	80–1600	32
Zn	Metal-processing industry	Poland	1665–5567	48
	Nonferric metal mining	Russia	400–4245	49
Cd	Metal-processing industry	Bulgaria	2–5	50
		Poland	6–270	48, 51, 52
		Romania	12[a]	53
	Urban garden	Poland	0.4–5	46
	Sludge-treated or fertilized farmland	Hungary	3–6	54
		Poland	0.4–107	55
Pb	Metal-processing industry	Poland	72–1350	51, 56
		Kirghizia (Asia)	3000[a]	57
	Nonferric metal[b] mining	Kirghizia (Azia)	21–3044	58
	Urban garden	Poland	17–165	46

[a] Mean value.
[b] Non-European part of C.I.S.

V degradation) occur within small plots (Figure 9). A special guideline for the land use was established for this region.[60]

SUMMARY AND CONCLUSION

Pollution of soils with trace metals in the temperate humid zone of Europe is of great ecological concern. In several regions of Europe anthropogenic inputs have been observed to influence the status of trace metals in soils as well as in other environmental components.

Anthropogenic global changes in soils are reflected mainly by:

- metal enrichment in surface soil layers;
- acidification, penetrating deep in the soil profile;
- decrease in soil buffer capacity;
- mobilization of most metals (including aluminum);
- migration of trace metals to groundwaters;
- general losses in soil fertility;
- degradation in quality and quantity of crop plants.

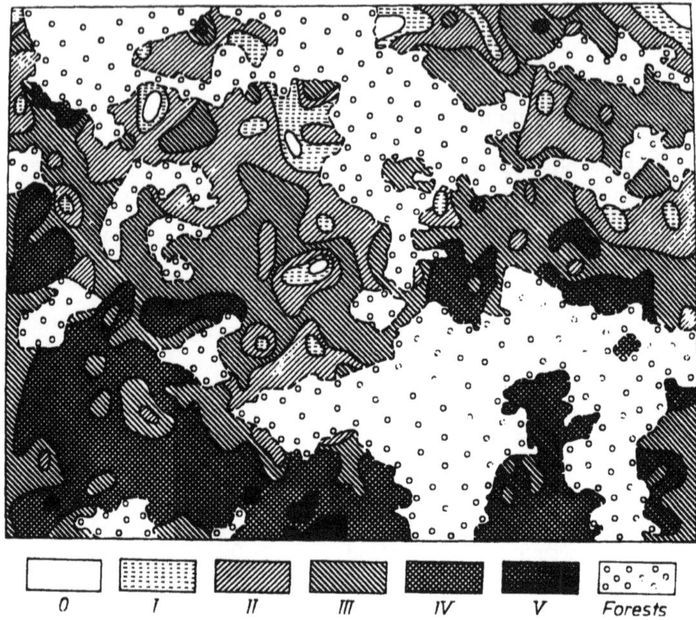

FIGURE 9. Soil pollution classes in a small area located as indicated on Figure 1 (based on initial scale 1:50,000). Numbers denote classes of soil pollution: 0 — unpolluted, I — slightly polluted, II — medium polluted, III — considerably polluted, IV — heavily polluted, V — very heavily polluted.

Mean values and ranges of trace metal contents in soil units calculated from existing databases seem to provide acceptable background values and can serve for the assessment of changes at the local and regional scales. Due to complex interacting factors involved in long-term changes in availability of metals to crops and in pollution of groundwaters, the monitoring of soil pollution is needed for the general prediction of anthropogenically altered metal cycling.

Guidelines to be established for controlling metal loading (from long-distance and local sources) to soils should take into account all environmental and health aspects and should be obligatory in the soil management at the temperate humid zone of Europe.

REFERENCES*

1. Kabata-Pendias, A. and H. Pendias. *Trace Elements in Soils and Plants*, 2nd ed., (Boca Raton, FL: Lewis Publ., Inc., 1992) p. 365.
2. Paces, T. "Natural and Anthropogenic Flux of Major Elements from Central Europe," *Ambio* 11:206–208 (1982).
3. *FAO-UNESCO Soil Maps of the World.* (UNESCO, Paris, 1974).
4. Wild, A. and L. H. P. Jones. "Mineral Nutrition of Crop Plants," in *Russell's Soil Conditions and Plant Growth*, Wild, A., Ed. (Longman Sci. Tech. Publ., Harlow, Essex, 1988) p. 69.
5. Kabata-Pendias, A. and M. Piortowska. "Present Status of Trace Elements in Soils and Plants of Poland," *Newsletter from FAO European Network on Trace Elements*, Gent, 5:45–55 (1987).
6. Tidball, R. R. and R. J. Ebens. "Regional Geochemical Baselines in Soils of the Powder River Basin, Montana-Wyoming," in *Geology and Energy Resources of the Powder River Basin*, Laudon, R. B., Ed., 28th Ann. Field Conf. Casper, WY (1976) pp. 299–310.
7. Kabata-Pendias, A., B. Gałczyńska, and S. Dudka. "Baseline Zinc Content of Soils and Plants in Poland," *Environ. Geochem. Health* 11:19–24 (1989).
8. Kabata-Pendias, A. and S. Dudka. "Evaluating Baseline Data for Cadmium in Soils and Plants in Poland," in *Element Concentrations Cadasters in Ecosystems (ECCE)*, Lieth, H. and B. Markert, Eds. (Weinheim: VCH Verlagsgesellschaft, 1990) pp. 265–280.
9. Zyrin, N. G. and G. D. Belicyna. *Trace Elements in Soils of USSR* (Moscow: Izd. Moskowskogo Universiteta, 1981) p. 250 (Ru).
10. Benes, S. and J. Pabianova. *Background Contents and Classification of Trace Elements in Soils* (VS Z, Praha, 1986) p. 203 (Cz).
11. Szukalski, H. *Microelements in Agricultural Production* (Warsaw: PWRL, 1979) p. 320 (Po).
12. Mitev, Kh. and Gyurov, G. "Manganese in Bulgarian Pseudopodsols," *Nauchni Tr.* 22:63–67 (1973) (Bu).
13. Boratyński, K., E. Roszyk, and M. Zięcka. "Review on Research on Microelements in Poland (B, Cu and Mn)," *Rocz. Glebozn.* 22: 205–264, (1971) (Po).
14. Kabata-Pendias, A. "Heavy Metal Concentrations in Arable Soils of Poland," *Pamiet. Pulawski* 74:101–111 (1981) (Po).
15. Kabata-Pendias, A. and M. Piotrowska. "Total Contents of Trace Elements in Soils of Poland," Materiały IUNG, S8, Pulawy, Poland, (1971) p. 47 (Po).
16. Krupskiy, N. K., L. P. Golovina, A. M. Aleksandrova, and T. I. Kisiel. "Distribution of Manganese in Soils of Ukrainian Forest-Steppe Zone," *Pochvovedenie* 11:41–46 (1978) (Ru).
17. Lukashev, K. J. and N. N. Pietukhova. "Minor Elements in Landscapes of Byelorussian S.S.R.," *Pochvovedenie* 8:47–60 (1974) (Ru).

* Publications Written in Languages Other Than English and German are Indicated by Letters in Parentheses: (Bu), Bulgarian; (Cz), Czechoslovakian; (Hu), Hungarian; (Po), Polish; (Ro), Romanian; and (Ru), Russian.

18. Sapek, A. and P. Sklodowski. "Concentration of Mn, Cu, Pb, Ni, and Co in Rendzinas of Poland," *Rocz. Glebozn.* 27:137–144 (1976) (Po).
19. Adierikhin, P. G., N. A. Protasova, and D. J. Shcheglov. "Microelements in Soil-Plant System in Central-Chernozem Region," *Agrokhimia* 6:102–112 (1978) (Ru).
20. Jakushevskaya, I. V. and A. G. Martynienko. "Minor Elements in the Landscapes of Separated Forest Stand Steppe," *Pochvovedenie* 4:44–53 (1972) (Ru).
21. Elek, E. "Investigation of the Manganese Supply in the Drainage Basin of the Lókos Brook," *Agrokem. Talajtan* 15:277–285 (1966) (Hu).
22. Bajescu, J. and A. Chiriac. "Trace Element Distribution in Brown Lessive Soils and Lessive Soils," *Stiinta Solului* 6:45–53 (1968) (Ro).
23. Golovina, L. P., M. N. Lysenko, and T. J. Kisiel. "Content and Distribution of Zinc in Soils of Ukrainian Woodland (Polesie)," *Pochvovedenie* 2:72–79 (1980) (Ru).
24. Zborishchuk, J. N. and N. G. Zyrin. "Copper and Zinc in the Ploughed Layer of Soils of the European USSR," *Pochvovedenie* 1:31–36 (1978) (Ru).
25. Rauta, C., S. Carstea, A. Mihailescu, and R. Lacatusu. "Some Aspects of Pedochemical and Biogeochemical Research in Romania," in *Proc. Int. Symp. Geochemistry and Health*, Thornton, I., Ed., (London: Imperial College, 1985) pp. 236–244.
26. Stanchev, L., G. Gyurav, and N. Mashev. "Cobalt as a Trace Element in Bulgarian Soils," *Izv. Centr. Nauch. Inst. Pochvozn. Agrotekh. "Pushkarov,"* 4:145–152 (1962) (Bu).
27. Mirchev, S. "Zinc and Copper Compounding Forms in Chernozems and Forest Soils," *Pochvozn. Agrokhim.* 13:84–31 (1978) (Bu).
28. Bondarenko, G. P. "Seasonal Dynamics of Mobile Forms of Trace Elements and Iron in Bottomland Soils of the Ramenskoe Widening of the Moskow River," *Nauchn. Dokl. Vyssh. Shk. Biol. Nauki* 4:202–218 (1962) (Ru).
29. Roszyk, E. "Contents of Vanadium, Chromium, Manganese, Cobalt, Nickel, and Copper in Lower Silesian Soils Derived from Loamy Silts and Silts," *Rocz. Glebozn.* 19:223–247 (1968) (Po).
30. Sapek, B. and H. Okruszko. "Copper Content of Hay and Organic Soils of Northeastern Poland," *Z. Probl. Post. Nauk Roln.* 179:225–236 (1976) (Po).
31. Tchuldziyan, H. and G. Khinov. "On the Chemistry of Copper Pollution of Certain Soils," *Pochvozn. Agrokhim.* 11:41–46 (1976) (Bu).
32. Sapek, B. "Copper Behavior in Reclaimed Peat Soil of Grassland," *Rocz. Nauk Roln.* 80F:13–39 (1980) (Po).
33. Garbanov, S. "Content and Distribution of Chromium in Main Soil Types of Bulgaria," *Pochvozn. Agrokhim.* 4:98–106 (1975) (Bu).
34. Glinski, J., J. Melke, and S. Uziak. "Trace Elements Content in Soils of Polish Carpathian Footland Region," *Rocz. Glebozn.* 19D: 73–82 (1968) (Po).
35. Naidenov, M. and A. Travesi. "Nondestructive Neutron Activation Analysis of Bulgarian Soils," *Soil Sci.* 124:152–160 (1977).
36. Prikhodko N. N. "Vanadium, Chromium, Nickel and Lead in Soils of Pritissenskaya Lowland and Piedmonts of Zakarpatie," *Agrokhimiya* 4:95–102 (1977) (Ru).
37. Dobrzański, B., J. Gliński, and S. Uziak. "Occurrence of Some Elements in Soils of Rzeszów Voyevodship as Influenced by Rocks and Soil Types," *Ann. UMCS* 24e:1–26 (1970) (Po).

38. Ogoleva, V. P. and L. N. Tcherdakoa. "Nickel in Soils of the Volgograd Region," *Agrokhimiya* 9:105–111 (1980) (Ru).
39. Czarnowska, K. and B. Gworek. "Heavy Metals in Some Soils of the Central and Northern Regions of Poland," *Roczn. Glebozn.* 5:41–57 (1987) (Po).
40. Sapek, A. and B. Sapek. "Nickel Content in the Grassland Vegetation," in *Proc. Nickel Symp.*, Anke, M., H. J. Schneider, and Chr. Brackner, Eds. (Jena, E. Germany: Fridrisch-Schiller University, 1980) pp. 215–220.
41. Petrov, I. I., D. L. Tsalev, and I. S. Lyotchev. "Investigation on Arsenic and Cadmium Content in Soils," *Higiena Zdravyeopazv.* 22:574–580 (1979) (Bu).
42. Czarnowska, K. "The Accumulation of Heavy Metals in Soils and Plants in Warsaw Area," *Pol. J. Soil Sci.* 7:117–122 (1974)
43. Sapek, A. "The Role of the Humus Substances in Podzol Soil Development," *Stud. Soc. Sci. Torun.* 7:1–95 (1971) (Po).
44. Akhtirtsev, B. P. "Content of Trace Elements in Grey Forest Soils of the Central Chernozem Belt," *Agrokhimiya* 9:72–80 (1965) (Ru).
45. Kabata-Pendias, A., E. Bolibrzuch, and P. Tarłowski. "Impact of a Copper Smelter on Agricultural Environments," *Rocz. Glebozn.* 32:207–214 (1981).
46. Czarnowska, K. "Heavy Metal Contents of Surface Soils and Plants in Urban Gardens," paper presented at Symp. on Environmental Pollution, Płock, November 12, 1982 (Po).
47. Kiriluk, V. P. "Accumulation of Copper and Silver in Chernozems of Vineyards," in *Microelements in Environment*, Vlasyuk, P. A., Ed. (Kiyev: Naukova Dumka, 1980) p. 76 (Ru).
48. Faber, A. and J. Niezgoda. "Contamination of Soils and Plants in a Vicinity of the Zinc and Lead Smelter. I. Soils," *Rocz. Glebozn.* 33:93–107 (1982) (Po).
49. Letunova, S. V. and V. A. Krivitskiy. "Concentration of Zinc in Biomass of Soil Microflora in South-Urals Copper-Zinc Subregion of Biosphere," *Agrokhimiya* 6:104–111 (1979) (Ru).
50. Petrov, I. I. and D. L. Tsalev. "Atomic Absorption Methods for Determination of Soil Arsenic Based on Arsine Generation," *Pochvozn. Agrokhim.* 14:20–25 (1979) (Bu).
51. Widera, S. "Contamination of the Soil and Assimilative Organs of the Pine Tree at Various Distances from the Source of Emission," *Arch. Ochrony Srodowiska* 3/4:141–146 (1980) (Po).
52. Greszta, J., S. Braniewski, and E. Chrzanowska. "Heavy Metals in Soils and Plants Around a Zinc Smelter," in *3rd Natl. Conf. Effects of Trace Pollut. on Agricul. Environ. Quality*, Vol. 2, Kabata-Pendias, A., Ed. (Puławy: IUNG, 1985) pp. 58–61 (Po).
53. Rauta, C., S. Carstea, and A. Mihailescu. "Influence of Some Pollutants on Agricultural Soils in Romania," *Arch. Ochrony Srodowiska* 1/2:33–37 (1987).
54. Vermes, L. "Results of Research Work and Status of Regulation of Heavy Metal Contamination Concerning Sewage Sludge Land Application in Hungary," *Arch. Ochrony Srodowiska* 1/2:21–32 (1987).
55. Umińska, R. *Assessment of Hazardous Levels of Trace Elements to Health in Contaminated Soils of Poland* (Warszawa: Inst. Medycyny Wsi, 1988) p. 188 (Po).
56. Manecki, A., Z. Klapyta, M. Schejbal-Chwastek, A. Skowroński, J. Tarkowski, and M. Tokarz. "The Effect of Industrial Pollutants of the Atmosphere on the Geochemistry of Natural Environmental of the Niepolomice Forest," *PAN Miner. Trans.* 71:58 (1981) (Po).

57. Wazhenin, I. G. and W. A. Bolshakov. "All-Union Conference of the Joint Departments on Methodical Principles of Mapping Soil Contamination with Heavy Metals and Methods of Their Determination," *Pochvovedenie* 2:151–159 (1978) (Ru).
58. Niyazova, G. A. and S. V. Letunova. "Microelements Accumulation by Soil Microflora at the Conditions of the Sumsaraky Lead-Zinc Biogeochemical Province in Kirghizya," *Ekologiya* 5:89–100 (1981) (Ru).
59. Pacyna, J. M. *Spatial Distribution of the As, Cd, Cu, Pb, V, and Zn Emissions in Europe Within a 1.5° Grid Net* (Lillestrom: NILU, 1985) p. 22.
60. Witek, T., A. Kabata-Pendias, and M. Piotrowska. "Primary Results of Evaluation of an Impact of Chemical Degradation of Soils on Agricultural Land Quality," *Surficial Recultivation of Degraded Land* (Wyd. Geod. Gosp. Grunt. U. Woj. Katowice, 1989) pp. 168–176 (Po).

4

Metal Contamination of Flooded Soils, Rice Plants, and Surface Waters in Asia

Zueng-Sang Chen

Department of Agricultural Chemistry, National Taiwan University, Taipei, Taiwan 10764, Republic of China

ABSTRACT

This chapter summarizes the causes of the soil pollution in Asian countries, which is mainly due to the metals discharged from various mine and chemical plants into the irrigation system or rivers used for flooded agricultural soils in the developed and developing countries, particularly Japan and the Republic of China.

Metals concentration data of natural (unpolluted) and polluted flooded soils, surface waters, and rice plants collected from Asian countries, particularly from Japan, Mainland China, and Taiwan, were used to study the microdistribution of metals in different rice plant parts, uptake and transport in the rice plants, and study the relationship between metals concentration in the rice grains, roots, leaf and paddy soils in different countries.

Finally, present criteria were summarized for allowable concentrations of Cd and Hg in the rice grain, and the critical concentration of As, Cd, Cr, Cu, Hg, Ni, Pb, and Zn allowed for paddy soils in Japan and the Republic of China.

INTRODUCTION

Industrial waste water, sewage sludge, and solid waste materials have been discharged into the environment. Due to improper control or illegal discharges, these materials, including toxic metals and toxic organic substances, have been discharged into irrigation systems, ground water, and even drinking water sources. Irrigation water is required for rice production, especially in Mainland China, Taiwan, Japan, Korea, and other Asian countries. When polluted water is used for irrigation on paddy soils, toxic substances may be incorporated into plant parts. These toxic substances may thus enter the food chain and possibly affect the health of animals and humans.

The objectives of this chapter are to: (l) present data on the metal concentration of uncontaminated and polluted flooded soils, rice plants, and surface waters in Asia and (2) compare the relationships between metal concentration in flooded soils and in rice plants.

METALS DISCHARGED FROM VARIOUS MINES AND CHEMICAL PLANTS INTO IRRIGATION SYSTEMS FOR FLOODED SOILS

Metal contamination of flooded soils was found to be due primarily to irrigation with waste waters discharged from various mines and chemical plants. For example, cadmium pollution was found in paddy soils irrigated with discharged water from the Kamioka Mine in Japan;[1] cadmium, zinc, lead, and copper pollution have been found in paddy soils irrigated with waste water discharged from the Ikuno Mine in Japan;[2] cadmium and zinc pollution has been reported in paddy soils irrigated with waste water from various mines in the Yoneshiro River Basin in Japan;[3] arsenic pollution was found in arable lands near the Sasagadani Mine in Shimane Prefecture;[4] cadmium and lead pollution occurred in agricultural soils of northern Taiwan from the discharge of two chemical plants producing the stabilizing components of plastics;[5] mercury was discharged from the Wangshan Mine of the Keitzou Prefecture, China; copper was discharged from the Yangping Mine of the Changshi Prefecture, China; and cadmium was discharged Shangyang Mine of Laioling Prefecture, China.[6]

METALS CONTAMINATION OF SURFACE WATER

The metals concentration of selected surface waters in the polluted area of Asia is shown in Table 1. The concentrations of Cd in the surface water of irrigation canals in the polluted area in Japan and Taiwan are from 0.14 to 0.22 and from 0.01 to 50.00 mg/L, respectively. The concentrations of lead in the

Table 1. Metal Contents (mg/kg) of Surface Waters in Irrigation Canals of Some Polluted Areas in Asia

Country	Element	Range	Pollution Sources	Ref.
Japan	Cd	0.14–0.22	Various mines	Homma[3]
	Cu	1.59–3.39		
	Zn	3.40–6.20		
	Pb	2.10–2.40		
Taiwan	Cd	0.01–50.00	Chemical plant	Chen[5]
	Pb	0.12–3.59		
China	Hg	8.8–23.3	Mercury mine	Yang & Kuboi[6]

surface water of the irrigation canals of the polluted areas in Japan and Taiwan are from 2.10 to 2.40 and from 0.12 to 3.59 mg/L, respectively.[3,5] The variation in concentration of Cd or Pb in the surface water is possibly due to the different extent of pollution and distribution of the sampling sites in different countries. The concentration of mercury in the surface water of the pollution area of China range from 8.8 to 23.3 mg/L.[6]

Since there are no data currently available about Cr, Ni, As, and other elements, they are not shown in Table 1.

The concentrations of Cd, Cu, Zn, Pb, and Hg in the surface water of the polluted area are much higher than those of the allowable concentration of natural irrigation water. The allowable concentrations of As, Cd, Cu, Hg, Zn, and Pb in the irrigation system of Japan are less than 0.05, 0.005, 0.02, 0.01, 0.5, and 0.1 mg/L, respectively[6] (see Table 1).

METHODS OF METALS ANALYSIS IN FLOODED SOILS AND RICE PLANTS

The metals in the paddy soils are generally extracted with (1) chelating agents, such as DTPA, EDTA, HEDTA, NTA, and EGTA;[7–9] (2) ammonium bicarbonate-0.005 *M* DTPA (pH 7.6);[10] (3) dilute acid, such as 0.1N HC1;[11] 0.05 *N* HCl and 0.025 *N* H_2SO_4;[12] and (4) ammonium acetate. These extractants are all used for estimating plant-available metals rather than total or potentially available metals. Total metal contents of rice plants were determined after digestion with concentrated nitric and perchloric acid.

In Taiwan, the Republic of China, in order to quickly estimate the plant-available concentration of metals in polluted soils, the recommended extractant is 0.1 *N* HCl. The ratio of soil to extraction solution is 1:10 (10 g soil to 100 mL 0.1 *N* HCl) with 1 h shaking at 180 oscillations per minute. For example, in Japan, 1 *N* HCl extractable As of soil is defined as available As to the rice plant. In other countries, DTPA- and EDTA-type extractants are the more popular reagents to extract the bioavailable metals in soils and to predict their plant uptake.

Table 2. Concentrations of Metals Extracted with 0.1 *N* HCl in the Unpolluted Flooded Soils in Asia

Country	Elements	Sample Number	Range (mg/kg)	Mean (mg/kg)	Ref.
China (Mainland)	As		0.12–9.61		Yang & Kuboi[6]
	Cd		0.10–0.19		
	Cu		20.6–32.2		
	Cr		23.8–65.9		
	Hg		0.03–0.38		
	Pb		17.4–24.8		
	Zn		54.9–85.5		
Japan	As	97	1.2–38.2	9	Iimura[14]
	Cd	147	0.12–41	0.45	
	Cu	408	11–120	32	
	Co	169	2.4–23.5	9	
	Cr	190	16–337	64	
	Hg	292	ND–2.90	0.32	
	Ni	379	9–412	39	
	Pb	407	6–189	29	
	Zn	408	13–258	99	
Taiwan	As	1240	ND–23.97	4.94	Wang et al.[13]
	Cd	1240	ND–0.22	0.06	
	Cu	1240	ND–11.0	4.77	
	Cr	1240	ND–1.80	0.25	
	Hg	1240	ND–0.24	0.09	
	Ni	1240	ND–4.50	1.77	
	Pb	1240	ND–13.0	7.02	
	Zn	1240	ND–16.0	8.09	

METALS CONCENTRATION OF NATURAL- AND POLLUTED-FLOODED SOILS

Metals Concentration of Natural Flooded Soils in Asia

Paddy soils are distributed mainly in Southeast Asia, especially in Mainland China, Taiwan, Japan, Korea, Thailand, and India. In undeveloped countries, flooded soils are less likely to be polluted by waste water discharged from chemical plants.

The concentration of metals in uncontaminated flooded soils in Mainland China, Taiwan, and Japan is shown in Table 2. The concentration of metals of paddy soils is extracted with dilute HCl to estimate plant availability. These results indicate that the natural concentrations of heavy metals in Mainland China and Japan are higher than those of elements in Taiwan. It is suggested that concentration of heavy metals in the natural agricultural soils is significantly affected by the properties of the parent materials in soils in the different countries. In Table 2 it is also indicated that concentrations of As, Cu, Pb, and Zn in the soils of Taiwan are less than 10 mg/kg.[13] However, the concentrations of Cu, Ni, and Pb in Japan are nearly 30 mg/kg, and the concentrations of the Cr and Zn are 64 and 99 mg/kg, respectively.[14] In Mainland China, the natural flooded soils contain high amounts of Cr and Zn. The concentration ranges are from 23.8 to 65.9 mg/kg for Cr and from 54.9 to 85.5 mg/kg for Zn.[6]

Metals Contamination of Flooded Soils in Asia

In developed and developing countries, such as Japan and the Republic of China, paddy soils were polluted by the waste water discharged from various mines, metal refineries, metal plating factories, or chemical plants. The metals concentration in the contaminated flooded soils is shown in Table 3.

There are considerable data collected on this topic in *Heavy Metal Pollution in Soils of Japan*.[15] The most well known pollution problem is Itai-Itai disease in Jinzu River of Japan. The average concentration of estimated plant-available Cd in flooded soils of this area is just greater than 2.0 mg/kg.[1] The contamination by Cd in northern Taiwan paddy soils is more serious than in Japan paddy soils. The concentration of Cd (4.72 to 180 mg/kg) in the polluted soils of Taiwan is two to ten times higher than that in the Japanese polluted soils.[17–22]

The copper contamination of agricultural soils of Japan and Mainland China is also very serious. The concentration range of copper in Japanese polluted soils is from 612 to 976 mg/kg[1] and in Mainland China polluted soils from 200 to 2124 mg/kg.[6]

The other serious metal contamination problem in paddy soils is mercury contamination in the area surrounding the Nifu Mine of the Seiwa village in Mie Prefecture, Japan[23] and in the area surrounding the Wagshan Mine in Kweichou Prefecture, China.[6] The mean concentration of mercury in the polluted soils of Japan is about 24.2 mg/kg,[23] but its natural concentration is only 0.65 mg/kg. The concentration of mercury in polluted soils of China ranges from 2.3 to 420 mg/kg and is much higher than the natural concentration (0.3 to 1.3 mg/kg).[6]

There are also arsenic, lead, and zinc pollution problems Asia. The arsenic concentration of polluted soils in the Sasagadani Mine of Japan ranges from 7.6 to 137.9 mg/kg, and the concentration of As is from 23 to 36 mg/kg.[4] High lead contamination of paddy soils was found in Taiwan. The concentration of Pb in the polluted paddy soils ranges from 352 to 3145 mg/kg[18,20] (see Table 3).

Vertical Distribution of Metals in the Polluted Flooded Soils

Generally, metals transported by air or waste water accumulate in the plow layer which is about 20 cm in depth from the soil surface, as shown in Figures 1, 2, and 3. The distribution of metals in the soil profile is significantly influenced by soil properties (such as texture and hydraulic conductivity) and the chemical nature of particular metals.[3,15]

A more detailed study of the distribution of metals in the sandy soils of a polluted area in northern Taiwan revealed Cd and Zn had moved to depths of 30 to 40 cm, while Pb was only moved to about 20 cm. The results, shown in Figure 3, suggest that the movement of zinc and cadmium in the paddy soil profiles is faster than lead.[15]

Yamane[24] also investigated the vertical distribution of total arsenic in two polluted soils by the Sasagadani Mine, shown in Figure 4; one is in gray soils where adequate drainage occurs, and the other in heavy gray soils where drainage

Table 3. Metal Concentrations in Polluted Flooded Soils in Asia

Country	Location or River	Element	n[b]	Concentration[a] Range (mg/kg)	Mean (mg/kg)	Ref.
Japan	Jinzu River	Cd	6	1.25–6.40	2.23	Morishita[1]
	Maruyama	Cd	1	5.35–22.2		Asami [2]
		Zn	1	1310–1780		
	Yoneshiro River	Cd	228	0.74–3.59	2.02	Homma[3]
	Nakajuku	Cd	10	0.48–1.18	0.80	Hirata[22]
	Sasagadani Mine	As	74	7.6–137.9	36.7	Tsutsumi[4]
		As	74	2.8–101.6[c]	23.3[c]	Tsutsumi[4]
	Gunma	Cu	78	510–2020	976	Morishita[1]
	Morita	Cu	2	456–768	612	Morishita[1]
		As	2	26.8–37.9	32.4	Morishita[1]
	Seiwa Villag	Hg	13	2.04–98.6[d]	24.2[d]	Morishita[23]
Taiwan	Kuan-In	Cd	20	45–1319	378	Lee et al.[18]
		Pb	20	67–12740	3145	Lee et al.[18]
	Lu-Tzu	Cd	36	6.2–1486	180	Lyuu et al.[20]
			118	1.12–148.15	16.35	Chen[17]
			45	0.22–21.73	4.72	Li & Ling[19]
		Pb	37	22.3–973	352	Lyuu et al.[20]
			154	6.3–126.7	25.8	Chen[17]
China (Mainland)	Wangshan	Hg		2.3–420[a]		Yang & Kuboi[6]
	Yungping	Cu		200–2124		Yang & Kuboi[6]

[a] The metals were extracted with the 0.1 *N* HCl method (10 g soils per 100 ml 0.1 *N* HCl, shaking 1 h) except where noted otherwise.
[b] n: sample numbers
[c] Extracted with 1 *N* HCl.
[d] Total content.

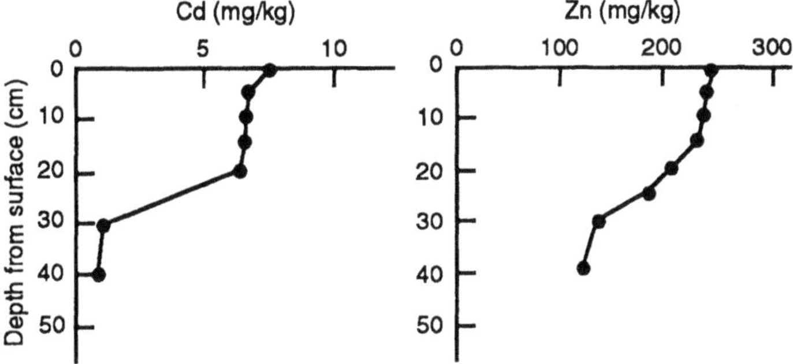

FIGURE 1. Vertical distribution of Cd and Zn in polluted paddy soils in Hosogoe area in Japan by the method of nitric acid-pechloric acid digestion. (From Homma S., in Heavy Metal Pollution in Soils of Japan. Kitagishi, K. and I. Yamane, Eds. [Japan Scientific Societies Press, 1981] pp. 139. With permission.)

is very poor. It seems that in gray soils As is settled upon the soils, then gradually dissolved into the soil solution as arsenite and transported downward along with water movement. But at the subsurface layer, where the redox potential is rather high, As is retained by the soils, probably as insoluble arsenate after the conversion from arsenite (see Figures 1 to 4).

Distribution of Metals in Different Soil Fractions of the Polluted Soils

Metals in different soil fractions or chemical forms of polluted soils have been widely investigated.[16,25–30] However, metal distribution in the different soil fractions of polluted, flooded soils in Asia was studied only by Chang[16] and Chen and Chang.[31]

Generally, soil fractions were divided into (1) exchangeable, (2) organic matter bound, (3) Mn oxide bound, (4) amorphous Fe oxide bound, (5) crystalline Fe oxide bound, and (6) residual fractions, including sand, silt, and clay. An example of the sequential extraction methods for metals proposed by Shuman[29] is shown in Table 4.

According to the analytical procedures suggested by Shuman,[29] Chen and Chang[31] divided the heavy metals of the polluted soil in Northern Taiwan into eight fractions: water soluble, exchangeable, organic matter, free manganese oxide, free iron oxide, sand, silt, and clay fractions. The sum of the sand, silt, and clay fractions was considered the residual fraction. Three soil pedons were selected from Paleudults (clayey soils) and Udipsamments (sandy soils)[32] to study the distribution of Cd and Pb in the different soil fractions. As shown in Table 5, Cd in the Taiwan polluted soils predominantly exists in the exchangeable fraction (47% of total content) and organic matter-bound fraction (32% of

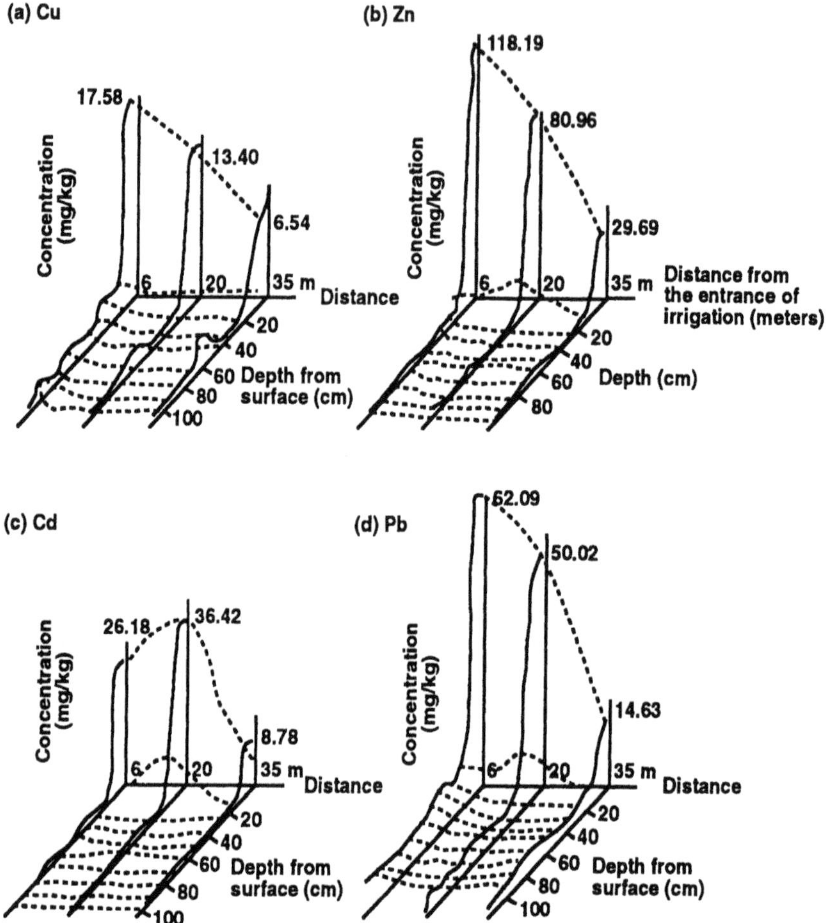

FIGURE 2. The three dimensional distribution of HCl-extractable (a) Cu, (b) Zn, (c) Cd, and (d) Pb in northern Taiwan paddy soils. (From Chang, R. Y. "Movement and Distribution in the Soil Profile of Cu, Zn, Cd, and Pb in the Polluted Soils," Master's thesis, National Taiwan University, Taipei, ROC [1989]. With permission.)

total content) in the clayey soils of Paleudults, and only a small amount of Cd exists in the other six soil fractions. Similar results were found in the sandy soils of Udipsamments. These results also indicate that the exchangeable and organic matter bound fractions can be regarded as the bioavailable or plant-available form and can be easily taken up by the growing rice in paddy soils.[16]

Pb exists mainly in the residual fraction (48% of total content) and free Mn and Fe oxide fraction (10 and 20% of total content, respectively) in the clayey soils of Paleudults. In the Udipsamments soils, there are different distribution patterns of heavy metals in the soil fractions. The Pb in the polluted sandy soils is predominantly distributed in the residual, exchangeable, and free oxide

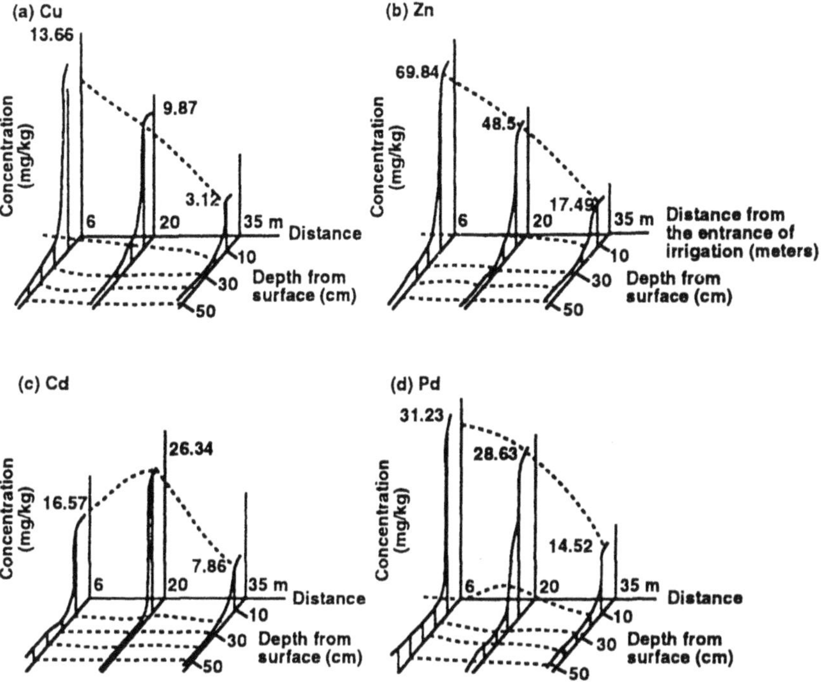

FIGURE 3. The three dimensional distribution of DTPA-extractable (a) Cu, (b) Zn, (c) Cd, and (d) Pb in northern Taiwan paddy soils. (From Chang, R. Y. "Movement and Distribution in the Soil Profile of Cu, Zn, Cd, and Pb in the Polluted Soils," Master's thesis, National Taiwan University, Taipei, ROC [1989]. With permission.)

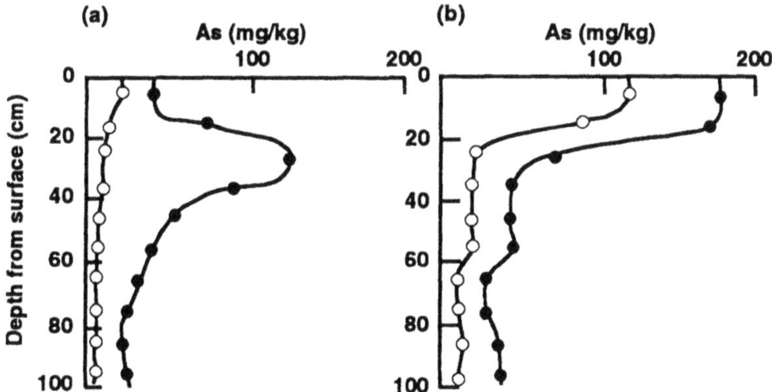

FIGURE 4. Two basic patterns in vertical distribution of As in soil polluted by the Sasagadani Mine, Japan. (a) Gray soil, (b) heavy soil. O: 1 *N* HCl-sol. As, ●: total As. (From Yamane, T., in *Mechanism and Analysis of Soil Pollution*, Shibuya, M. and M. Sangyotosho, Eds. [Industrial Publ. Co., Tokyo, 1979] pp. 38–71. With permission.)

Table 4. A Sequential Extraction Method for Metals in Soils

Step	Fraction	Solution	Soil (g)	Solution (ml)	Conditions
1.	Exchangeable	1 M $Mg(NO_3)_2$ (pH7)	10	40	Shake 2 h
2.	Organic matter	0.7 *M* NaOCl (pH8.5)	10	20	Boiling water bath, 30 min; stir occasionally (repeat)
3.	Mn oxides	0.1 *M* NH_2 OH.HCl (pH2)	1[a]	50	Shake 30 min
4.	Amorphous Fe oxides	0.2 *M* $(NH_4)_2$ C_2O_4 H_2O– 0.2 *M* H_2C_2 O_4 (pH3)	1	50	Shake 4 h in the dark
5.	Crystalline Fe oxides	Solution as for step 4 plus ascorbic acid	1	50	Boiling water bath, 30 min; stir occasionally (repeat)
6.	Sand	0.11 *M* $Na_4P_2O_7 \cdot 10H_2O$	8[b]	10	Shake end-over-end 16 h, wash silt and clay through 270-mesh screen with a jet of water
7.	Silt	HF, HCl, and HNO_3			
8.	Clay	HF, HCl, and HNO_3			

[a] From step 2, 1 g that is dried, ground, and passed through a 35-mesh screen.
[b] From step 2 after putting sample through step 5 twice. (From Shuman, L.M., *Soil Sci.* 140:11–21 [1985].)

Table 5. Distribution of Cd and Pb Among Different Soil Fractions of Udipsamments and Paleudults Polluted Soils in Taiwan

			Soil Fractions (mg/kg)						Soil Fractions (% of Total Concentration)					
	Soils Region[a]	Depth (cm)	Water Soluble	Exchangeable	Organic Matter	Free Oxides Mn	Free Oxides Fe	Residual[b]	Water Soluble	Exchangeable	Organic Matter	Free Oxides Mn	Free Oxides Fe	Residual[b]
Cd	A	0–10	0.37	11.06	7.27	0.52	1.09	2.01	2	47	32	3	6	11
		10–20	0.18	3.30	3.35	0.62	1.55	2.65	2	27	28	5	14	25
	B	0–10	0.15	5.93	4.93	0.12	0.59	1.90	1	44	36	1	4	14
		10–20	0.57	6.41	6.58	0.14	0.63	2.05	4	39	40	1	4	13
Pb	A	0–10	1.10	18.91	6.20	12.82	24.78	56.97	1	16	5	10	20	48
		10–20	0.65	16.62	4.73	6.60	20.64	44.68	1	18	5	7	22	47
	B	0–10	2.39	34.22	17.72	14.17	28.93	40.62	2	25	13	10	21	28
		10–20	0.69	32.67	16.95	9.63	22.43	32.85	1	28	14	8	20	29

[a] A: Paleudults (3 soil pedons selected); B: Udipsamments (3 soil pedons selected).
[b] Residual fraction includes sand, silt, and clay fractions.

fractions (each soil fraction contains 25 to 30% of the total Pb) (Table 5). The difference of the Pb distribution in the two soils may result from the influence of the soil texture. In other words, a higher percentage of Pb can be found in the residual fraction in high clay content soils.[16,31]

METALS CONTAMINATION OF RICE PLANTS

Factors Affecting the Accumulation of Metals in Rice Grain

Soil conditions and soil management, including pH, applied fertilizers, organic matter content, liming, plowing, and water control in the paddy fields, significantly affect the metal content in the different parts of rice plants through the changes in the availability of metals in soils.

Ito and Iimura[33] examined the effect of heavy metal concentration in cultural solution on brown rice. The results are shown in Figure 5. Brown rice cultured in solution containing more than 1 mg/L of Cd has more than 5 mg/kg of Cd in the grain. However, brown rice cultured in a solution containing 0.03 mg/L of Cd has 1 mg/kg of Cd in the grain without any decrease in yield, whereas cultured in solution containing 0.1 mg/L of Cd has 3 mg/kg of Cd in the grain with a slight decrease in yield.

Chino[34] examined the effect of the time of heavy metal supply on the accumulation rate of heavy metals in rice grains using isotopes of Zn or Cd. It was found that either metal, absorbed on the tenth day after heading, was translocated to the grains at the highest rate (Figure 6).

Microdistribution of Metals in Rice Grains

Kitagishi et al.[35] determined the distribution of heavy metals in polluted rice grain from Cd polluted areas. The rice grain was divided into the embryo, the endosperm, and the bran layer (pericarp plus aleurone layer). The distribution of Cd in rice grains affected by varying degrees of Cd pollution is given in Figure 7, expressed in different terms. Although Cd concentration (nanograms per milligram of dry matter) is remarkably higher in the embryo and in the bran layer than that in the endosperm, more than 80% of the total Cd in the rice grain is distributed in the endosperm.

Yoshikawa et al.[36] determined the distribution of heavy metals in the outer layer of rice grains from Cd-polluted areas. The data show that Cd, Cu, and Zn are localized in the surface layer of the grain, especially in the aleurone layer (Figure 8). Similar results are also reported by Tanaka et al.[37] for Fe and Mn, and by Tatekawa[38] for Cd.

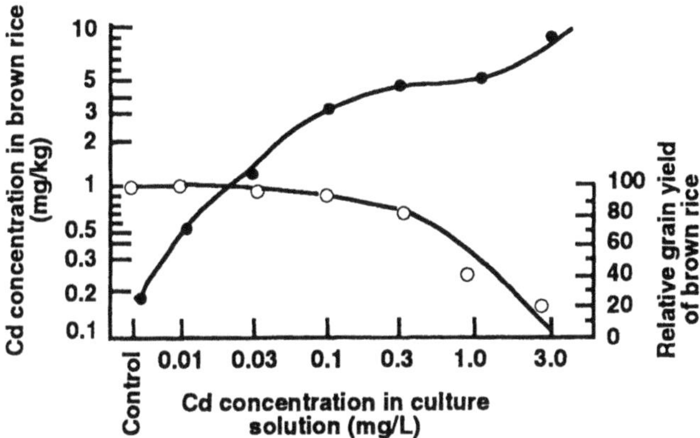

FIGURE 5. Effect of Cd concentration in culture solution on grain yield and Cd content of brown rice. ●: Cd concentration in brown rice; O: relative yield of brown rice. (From Ito, H. and K. Iimura. *J. Sci. Soil Manure Jpn.* 47:44–48 [1976]. With permission.)

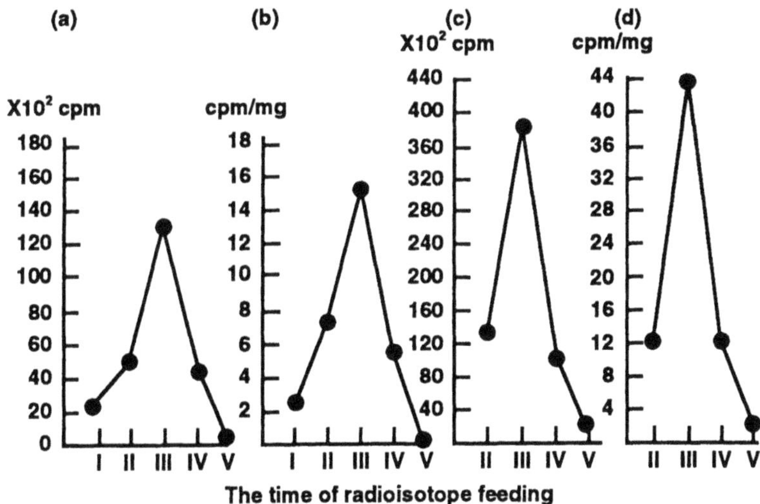

FIGURE 6. Accumulation of ^{65}Zn and ^{109}Cd as affected by the time of radioisotope feeding. (a) Abundance of ^{65}Zn per grain, (b) ^{65}Zn concentration, (c) abundance of ^{109}Cd per grain, (d) ^{109}Cd concentration. I — 20 days before heading; II — at the heading date; III — 10 days after heading; IV — 20 days after heading; V — 30 days after heading. (From Chino, M. *J. Sci. Soil Manure Jpn.* 44:204–210 [1973]. With permission.)

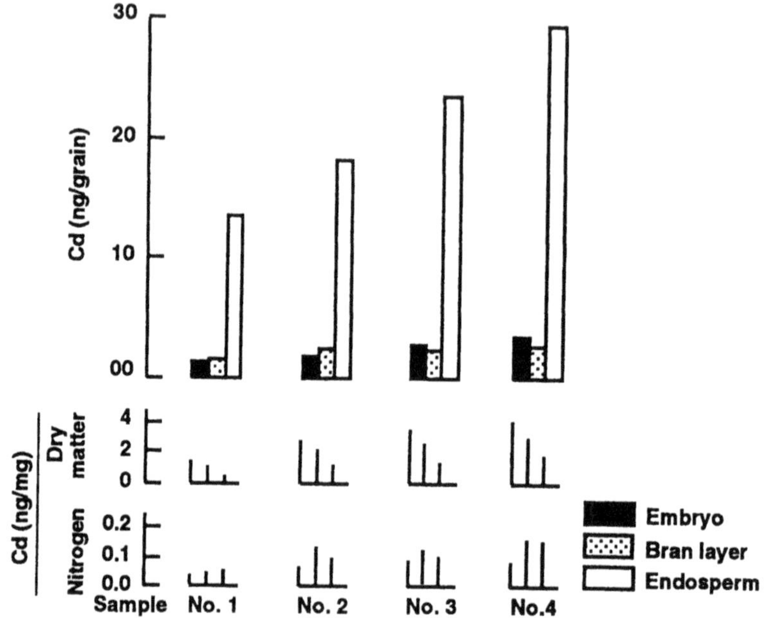

FIGURE 7. Distribution of Cd in polluted rice grain affected by varying degrees of cadmium pollution. (From Kitagishi, K., et al., *Rep. Environ. Sci. Mie Univ.* 1:129–141 [1976]. With permission).

Metals Concentration of Unpolluted and Highly Polluted Rice Plants in Asia

Metals concentration of polluted rice plants in Japan is shown in Table 6. The maximum concentrations of As, Cu, Cd, Hg, Pb, and Zn in the rice grain of Japan are 0.2, 6, 5.2, 4.9, 1.0, and 60 mg/kg, respectively.[14,39,40] The maximum concentrations of As, Cd, Cu, and Zn in the rice roots are 1182, 97, 560, and 4510 mg/kg, respectively.[5,16,41] The concentration of the heavy metals in the rice grain, roots, or other parts of the plants is significantly affected by the extent of pollution and different locations or countries.

Morishita[39] collected and compared the difference and ratio (MPC/NPC) of maximum polluted concentration (MPC) to nonpolluted concentration (NPC) as shown in Table 7. The results indicate that rice grown in contaminated soils transported and accumulated higher quantities of Cd and Hg among the heavy metals studied. The ratio of maximum polluted concentration to nonpolluted concentration is 60 and 120 for Cd and Hg, respectively, representing very high ratios. But the ratio is very low (=2) for Zn, Pb, and Cu. The ratio for As is 20 (Table 7).

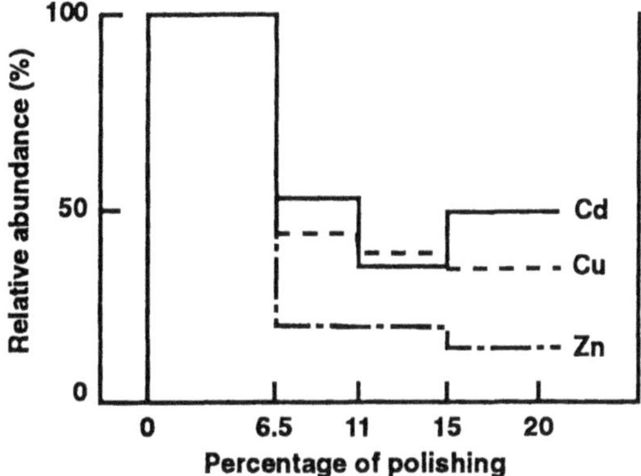

FIGURE 8. Distribution of heavy metals in the outer layer of the rice grain. (From Yoshikawa, T., S. Kusaka, T. Zikihara, and T. Yoshida. *J. Sci. Soil Manure Jpn.* 48:523–528 [1977]. With permission.)

The concentration of Hg in the polluted rice growing in Wangshan Mercury Mine of Keitzo Prefecture of Mainland China is between 0.08 and 0.34 mg/kg. It is higher than that in the unpolluted rice (0.02 mg/kg) in the surrounding area.[6]

Uptake-Transport of Metals in Different Parts of Polluted Rice Plants

Tatekawa[38] and the Water Pollution Control Lab[42] had showed that metals were distributed differently among the organs. The highest concentration of Cd was found in the root (90 to 95% of the total uptake in the whole rice plants) followed by the culm, leaf sheath, leaf blade, and ear in decreasing order (Table 8).[38] The older the leaf blades, the higher the Cd concentration. Among tissues in the ear, the highest amount was found in the rachi branch, the next highest in the husk, and the lowest in brown rice (Table 8).

In order to study the distribution of Cd and Pb in different parts of the rice plant of northern Taiwan, each plant was divided into four parts: (1) roots, (2) leaf and stem, (3) hull, and (4) brown rice. The results of 33 rice samples selected from clayey polluted soils indicated about 88% of total Cd (75.07 mg/kg) and 70% of total Pb (14.15 mg/kg) taken up by the roots and adsorbed on the exterior surfaces of the roots; a small amount of Cd and Pb occurred in the leaf, stem, and hull, and only 1.5% of Cd uptake (0.85 to 1.71 mg/kg) and trace amount of Pb occurred in brown rice (Table 9).[5] Although the content of Cd in the brown rice is low, it still exceeded the pollution standard (>0.50 mg/kg in the brown rice) adopted by the Government of ROC. The concentration of Pb in the brown rice of polluted Paleudults soil is nondetectable. It indicates that it is very difficult for the rice plants to transport Pb from the roots to the grains.

Table 6. Metal Concentrations of Rice Growing in Polluted Soils in Asia

Element	Part	Range (mg/kg)	Mean (mg/kg)	Ref.
Cu	Grain	—	2.8	Kitagishi & Yamane[15]
	Rice-grain	—	4	
	Rice-roots	—	560	
	Grain	5–6		Morishita[39]
Zn	Brown rice	19–18	23	Kitagishi & Yamane[15]
	Grain		21	
	Root of rice		4510	
	Brown rice	50–60		Morishita[39]
Cd	Brown rice	0.5–5.2		Morishita[39]
	Brown rice	0.85–1.71	1.22	Chen[5]
	Roots (rice)	61.8–97.2	75.1	Chen[5]
Pb	Brown rice	0.8–1.0		Morishita[39]
Hg	Rice		4.90	Iimura et al.[40]
	Brown rice	0.03–0.26	0.16	Morishita[39]
As	Rice	—	0.19	Kitagishi & Yamane[15]
	Brown rice	0.15–0.20	0.18	Miyagi Pref. Agric. Res. Ctr.[41]
	Root	936–1182		Kitagishi & Yamane[15]
	Straw	8.2–10.0		Miyagi Pref. Agric. Res. Ctr.[41]
	Grains	0.11–0.20	—	Iimura et al.[40]
Ni	Rice (Leaves)	11–20	—	Kitagishi & Yamane[15]
Mn	Unpolished rice		26	Kitagishi & Yamane[15]

Table 7. Ratio of Metal Contents in Highly Polluted Rice to Nonpolluted Rice in Japan

Elements	Nonpolluted (mg/kg)	Maximum polluted (mg/kg)	Ratio
Cd	0.08	5.2	65
Zn	30–50	50–60	1.5
Pb	0.4–0.6	0.8–1.0	2
Cu	3–5	5–6	1.5
As	0.06	1.2	20
Hg	0.04	4.93	123

From Morishita, T. *Survival of Human Being and Natural Environment* (Toyko University Publication Press, Tokyo, 1977) pp. 287–295.

Morishita and Anayama[43] also indicated that, including Cd content of the roots, 4.67 to 8.45% of the soil Cd was absorbed by rice plants, and that 3.56 to 8.09% of the Cd absorbed by rice plants was transferred to and accumulated in the unpolished rice grains.

COMPARISONS OF RELATIONSHIP BETWEEN METAL CONCENTRATION IN FLOODED SOILS AND RICE PLANTS

Cd in the soil is soluble under oxidized conditions, while it is precipitated as CdS under reducing conditions. When rice plants are grown under water-

Table 8. Distribution of Metals Among Various Parts of the Rice Plants Grown in Soils with Normal and Polluted Levels of Metals in Japan

		Part of the Rice Plant						
Metals	Level of Concentration[a]	Brown Rice	Husk	Rachi Branch	Leaf Blade	Leaf Sheath	Culm	In the Root
Cd	N	0.05	0.28	0.15	0.4	0.2	0.22	3.13
	P			29.8	18.1	90.2	66.2	3650
		(ND)		(1%)[b]	(0.5%)	(2.3%)	(1.7%)	(95%)
Co	N	0.1	0.27	0.28	1.0	0.67	0.87	15.5
	P	0.4	6.2	2.5	3.3	14.5	45.9	980
		(0.04%)	(0.6%)			(1.4%)	(4.4%)	(93%)
Cu	N	3.3	2.2	2.6	4.2	3.7	4.7	105
	P	3.7	3.4	3.8	8.9	13.3	5.3	560
		(0.6%)			(1.5%)	(2.2%)		(94%)
Pb	N	ND	ND	ND	9.5	4.0	4.0	57.1
	P	ND	ND	ND	63.5	98.0	31.0	9070
		(ND)			(0.7%)	(1%)	(0.3%)	(98%)
Ni	N	0.46	0.86	1.23	2.79	1.64	1.7	7.1
	P	5.6	15.9	23.9	20.4	20.1	10.9	807
		(0.6%)	(1.8%)	(2.6%)	(2.3%)	(2.2%)	(1.2%)	(89%)
Zn	N	15.5	22.0	26.1	44.0	26.2	35.6	68.3
	P	20.5	37.4	72.2	77.2	100.0	96.5	4510
		(0.4%)	(0.8%)	(1.5%)	(1.6%)	(2.0%)	(2.0%)	(92%)

[a] N: normal; P: polluted or excess.
[b] Percentage of total concentration in the whole rice plant.

From Water Pollution Control Laboratory. 1973 Annual Report, pp. 70–93.

Table 9. Comparison of Cd and Pb in the Various Parts of the Rice Plant Selected from Different Sites in Taiwan

	Region	Sample Numbers	Polluted Soils (mg/kg)	Part of the Rice Plant (mg/kg)				Ref.
				Roots	Leaf & Stem	Hull	Brown Rice	
Cd	A2 (polluted)	33	2.35—150.23 (18.24)[a]	61.83—97.23 (75.07)[a] (88.2%)[b]	3.54—9.76 (6.95) (8.2%)	1.46—2.69 (1.79) (2.1%)	0.85—1.71 (1.22) (1.5%)	Chen (1989)
	Control area[c]	101	ND–0.23 (0.05)[a]	—	—	—	ND—0.10 (ND)[a]	Li & Ling[19]
Pb	A2 (polluted)	33	4.65—120.34 (28.45)[a]	9.03—22.94 (14.15)[a] (70.4%)[b]	3.30—5.86 (4.37) (21.4%)	0.61—3.17 (1.58) (7.9%)	ND (ND) (0%)	Chen (1989)
	Control area[c]	101	0.96—18.13 (8.02)[a]	—	—	—	ND (ND)[a]	Li & Ling[19]

Note: ND: Nondetectable.

[a] Mean.
[b] Percentage of total uptake concentration.
[c] The same soils outside the polluted area.

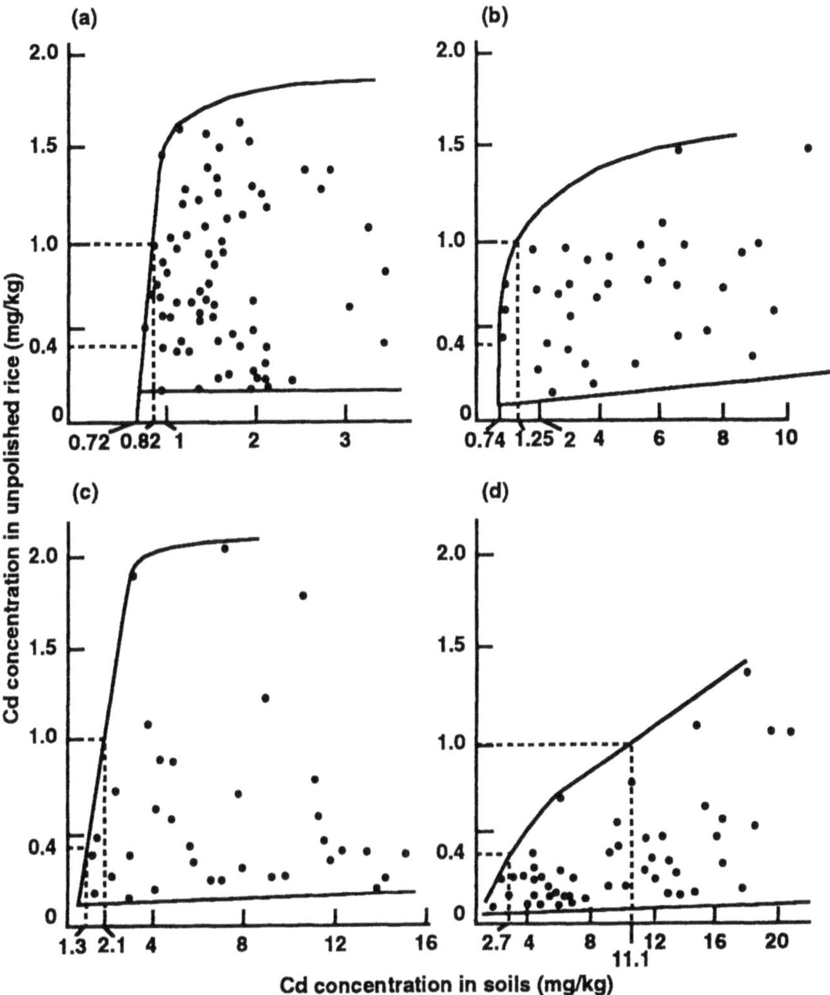

FIGURE 9. Comparison of relationship between cadmium concentration in soils and unpolished rice grains among Annaka, Bandai, Kurobe, and Fuchu districts in Japan. (a) Fuchu, (b) Kurobe, (c) Annaka, (d) Bandai. (Morishita, T. "A Study on the Mechanism of Heavy Metal Pollution of Soil-Crop Plants System," Ph.D. thesis, The University of Tokyo, Japan [1975]. With permission.)

deficient conditions, rice grain contains more Cd than when grown under flooded condition. So, the relationship between Cd concentration in soil and rice grain is poor at normal rice paddy conditions.[40]

Morishita[44] reported the relationship between Cd concentration in soils and rice plants in four districts of Japan: Annaka in Gunma Prefecture, Bandai in Fukushima Prefecture, Kurobe in Toyama Prefecture, and Fuchu in the basin in Toyama Prefecture (Figure 9). From these relationships, the empirically critical values of Cd in the soils corresponding to 0.4 or 1.0 mg/kg Cd of rice unpolished grain (level allowed in Japan) could be potentially determined for soils of each district.

In the case of Fuchu district in the Jinzu River basin, the critical values corresponding to 0.4 and 1.0 mg/kg Cd in the unpolished rice were 0.72 and 0.82 mg/kg Cd in soils, respectively, which is the lowest among these four districts.[44] But in the case of Bandai district, the critical values corresponding to 0.4 and 1.0 mg/kg Cd in unpolished rice were 2.7 and 11.1 mg/kg Cd in soils, respectively, which is the highest among the districts. These facts indicate that the Cd in the soils of Fuchu can be absorbed more easily by the rice plants than any other three districts in the study area. The differences likely are caused by differences in soil properties.

Although some soils have a high Cd concentration, it has been found that the areas of Japan with a heavy incidence of Itai-Itai disease most frequently have soils with a Cd concentration from 2 to 3 mg/kg. Similar results from northern Taiwan[5] indicate that the Cd concentration in the polluted soils ranges from 2 to 4 mg/kg and that the distribution of Cd in soils depends on the distribution of the irrigation river system.

High concentration of Cd has been detected in rice of Japan and Taiwan. The authorities of Toyama Prefecture detected 5.20 mg/kg Cd in unpolished rice produced in the Fuchu district in 1973. Morishita[1] detected 3.57 mg/kg Cd in the unpolished rice produced in soil with 4.2 mg/kg Cd in the Fuchu in 1974. Chen[17] and Li[19] also detected 4.30 and 4.19 mg/kg Cd in the polished rice produced in soils with 16.35 and 4.72 mg/kg Cd in the Taoyuan Prefecture of northern Taiwan, respectively (see Figure 9).

THE ALLOWABLE LEVEL FOR RICE PLANTS AND FLOODED AGRICULTURAL SOILS IN DIFFERENT COUNTRIES

Rice is the most important food in Asia, especially in Mainland China, Japan, Taiwan, and Korea. Therefore, the governments of these countries establish the critical concentration of metals allowed in the rice grain to protect the consumers.

The proposed critical allowable level of Cd and Hg in the rice grain in Japan, Taiwan, and Mainland China is shown in Table 10. The proposed critical allowable levels of Cd in the rice grain are 0.5 and 1.0 mg/kg by the governments of Republic of China and Japan, respectively.[2,5]

The proposed critical allowable level of metals in flooded agricultural soils in Japan and Taiwan is also shown in Table 10. The proposed allowable level of Cu, Ni, Pb, and Zn in Japanese flooded soils are 125, 100, 400, and 150 mg/kg, respectively.[2] The proposed allowable level of Cu, Ni, Pb, and Zn in Taiwan flooded soils are 100, 100, 120, and 80 mg/kg, respectively.[5] The proposed critical allowable concentration of As in the flooded soils in Japan and Taiwan are 15 and 60 mg/kg, respectively (see Table 10).

In summary, the extent and sources of metal contamination in paddy soils and rice plants were discussed in view of the potential role of rice grain as a vector

Table 10. The Allowable Levels for Rice Grain and Flooded Agricultural Soils in Japan and China

Country	Element	Rice Grain (mg/kg)	Flooded Soil (mg/kg)	Ref.
Japan	As	—	15[a]	Asami[2]
	Cd	1.0 (unpolished)	—	
	Cr	—	—	
	Cu	—	125[b]	
	Hg	—	0.5[c]	
	Ni	—	100[b]	
	Pb	—	400[b]	
	Zn	—	150[b]	
China (Taiwan)	As	—	60[c]	Chen[5]
	Cd	0.5 (polished)	10[b]	
	Cr	—	16[b]	
	Cu	—	100[b]	
	Hg	0.5	20[c]	
	Ni	—	100[b]	
	Pb	—	120[b]	
	Zn	—	80[b]	
China (Mainland)	Cd	0.4 (polished)	—	Yang & Kuboi[6]

[a] 1 *N* HCl extracted.
[b] 0.1 *N* HCl extracted.
[c] Total content.

of concern in transmitting metals to humans. By far, the rice growing areas in Asia are basically noncontaminated with metals from geologic and fertilizer sources; however, isolated instances have occurred where severe soil contamination of metals from anthropogenic sources, particularly Cd, have enriched the rice grain metal contents, resulting in human afflictions (the Itai-Itai disease in Japan is a good example). It is for this reason that several Asian countries, including Taiwan and Japan, have considered allowable levels of metals in paddy soils that potentially could elevate the metal contents of the rice grain. The metals of concern include Cd and Hg as well as As, Cr, Cu, Ni, Pb and Zn.

ACKNOWLEDGMENTS

The author wishes to thank Dr. Domy C. Adriano, Professor and Head of Biogeochemical Division, Savannah River Ecology Laboratory, The University of Georgia, U.S., and Dr. D. Y. Lee, Associate Professor of Department of Agricultural Chemistry, National Taiwan University, Taiwan, Republic of China, for their helpful discussions and comments on this chapter.

REFERENCES

1. Morishita, T. "The Jinzu River Basin: Contamination of Soil and Paddy Rice with Cadmium Discharged from Kamioka Mine," in *Heavy Metal Pollution in Soils of Japan*. Kitagishi, K. and I. Yamane, Eds. (Japan Scientific Societies Press, 1981) pp. 107–124.
2. Asami, T. "The Ichi and Maruyama River Basins: Soil Pollution by Cadmium, Zinc, Lead, and Copper Discharged from Ikuno Mine," in *Heavy Metal Pollution in Soils of Japan*. Kitagishi, K. and I. Yamane, Eds. (Japan Scientific Societies Press, 1981) pp. 149–163.
3. Homma S. "The Yoneshiro River Basin: Soil Pollution by Heavy Metals Discharged from Various Mines," in *Heavy Metal Pollution in Soils of Japan*. Kitagishi, K. and I. Yamane, Eds. (Japan Scientific Societies Press, 1981) pp. 137–148.
4. Tsutsumi, M. "Arsenic Pollution in Arable Land," in *Heavy Metal Pollution in Soils of Japan*. Kitagishi, K. and I. Yamane, Eds. (Japan Scientific Societies Press, 1981) pp. 181–192.
5. Chen, Z. S. "Cadmium and Lead Contamination of Soils, Rice Plants, and Surface Water in the Northern Taiwan," in *Symposium Abstracts of International Conference on Metals in Soils, Waters, Plants, and Animals*. (University of Georgia, Savannah River Ecology Laboratory, U.S.A., 1990) No. 37. p. 19.
6. Yang, J. R. and T. Kuboi. "Status and Policy of Soil Pollution in China," *J. Environ. Pollut. Control* 25(8):750–755 (1989).
7. Norvell, W. A. "Comparison of Chelating Agents as Extractants for Metals in Diverse Soil Materials," *Soil Sci. Soc. Am. J.* 48:1285–1292 (1984).
8. Norvell, W. A. and W. L. Lindsay. "Reactions of EDTA Complex of Fe, Zn, Mn, and Cu with Soils," *Soil Sci. Soc. Am. Proc.* 33:86–91 (1969).
9. Norvell, W. A. and W. L. Lindsay. "Reactions of DTPA Chelates of Iron, Zinc, Copper, and Manganese with Soils," *Soil Sci. Soc. Am. Proc.* 36:778–783 (1972).
10. Soltanpour, P. N. and A. P. Schwab. "A New Soil Test for Simultaneous Extraction of Macro- and Micro-Nutrients in Alkaline Soils," *Comm. Soil Sci. Plant Anal.* 8(3):195–207(1977).
11. Wear, J. I. and C. E. Evans. "Relationships of Zinc Uptake of Corn and Sorghum to Soil Zinc Measured by Three Extractants," *Soil Sci. Soc. Am. Proc.* 32:543–546 (1968).
12. Mehlich, A. "Uniformity of Expressing Soil Test Results. A Case for Calculating Results on a Volume Basis," *Comm. Soil Sci. Plant Anal.* 3:417–424 (1972).
13. Wang, Y. P., G. C. Li, and Z. S. Chen. "Data Base and Distribution of Heavy Metals in Taiwan Soils," in *Proc. 1st Workshop of Soil Pollution Prevention in Taiwan* (Taiwan, National Chung-Hsiung University, 1989) pp. 137–152.
14. Iimura, K. "Background Contents of Heavy Metals in Japan Soils," in *Heavy Metal Pollution in Soils of Japan*, Kitagishi, K. and I. Yamane, Eds. (Japan Scientific Societies Press, 1981) pp. 19–35.
15. Kitagishi, K. and I. Yamane. *Heavy Metals Pollution in Soils of Japan* (Toyko: Japan Science Society Press, 1981) p. 302.
16. Chang, R. Y. "Movement and Distribution in the Soil Profile of Cu, Zn, Cd, and Pb in the Polluted Soils," Master's thesis, National Taiwan University, Taipei, (1989).

17. Chen, Z. S. "Heavy Metals Contents of Rice in the Northern Area of Chi-Lee Chemical Plant, Taoyuan, Taiwan," in Report of Council of Agriculture, Taiwan, (1988).
18. Lee, C. D., S. L. Chang, S. P. Wang, C. C. Houng, Y. P. Hsu, and K. C. Yu. Unpublished results (1983).
19. Li, G. C. and H. T. Ling. Unpublished results (1988).
20. Lyuu, S. T., C. D. Lee, and S. L. Chang. Unpublished results (1984).
21. Asami, T. "Maximum Allowable Limits of Heavy Metals in Rice and Soils," in *Heavy Metal Pollution in Soils of Japan*. Kitagishi, K. and I. Yamane, Eds. (Japan Scientific Societies Press, 1981) pp. 257–274.
22. Hirata, H. "Annaka: Land Polluted Mainly by Fumes and Dust from Zinc Smelter," in *Heavy Metal Pollution in Soils of Japan*. Kitagishi, K. and I. Yamane, Eds. (Japan Scientific Societies Press, 1981) pp. 149–163.
23. Morishita, T., K. Kishino, and S. Idaka. "Mercury Contamination of Soils, Rice Plants, and Human Hair in the Vicinity of a Mercury Mine in Mie Prefecture, Japan," *Soil Sci. Plant Nutr.* 28(4):523–534 (1982).
24. Yamane, T. "Actual State and Countermeasure Against Arsenic Pollution in Shimane Prefecture," in *Mechanism and Analysis of Soil Pollution*, Shibuya, M. and M. Sangyotosho, Eds. (1979) pp. 38–71.
25. Tessler, A., P. G. C. Campbell, and M. Bisson. "Sequential Extraction Procedure for the Specification of Particular Trace Metals," *Anal. Chem.* 51:844–850 (1979).
26. Kuo, S. and B. L. McNeal. "Effects of pH and Phosphate on Cadmium Sorption by a Hydrous Ferric Oxide," *Soil Sci. Soc. Am. J.* 48:1040–1044 (1984).
27. Shuman, L. M. "Zinc, Manganese, and Copper in Soil Fractions," *Soil Sci.* 127:10–17 (1979).
28. Shuman, L. M. "Sodium Hypochlorite Methods for Extracting Microelements Associated with Soil Organic Matter," *Soil Sci. Soc. Am. J.* 47:656–660 (1983).
29. Shuman, L. M. "Fraction Methods for Microelements," *Soil Sci.* 140:11–21 (1985).
30. Miller, W. P., D. C. Martens, and L. W. Zelazny. "Effect of Sequence in Extraction of Trace Metals from Soils," *Soil Sci. Soc. Am. J.* 50:598–601 (1986).
31. Chen, Z. S. and R. Y. Chang. "Form and Distribution of Cd, Pb, and Zn in the Pollution Soils of Taiwan," in *Proc. 2nd Workshop of Soil Pollution Prevention. Taipei, Taiwan* (Taiwan, National Taiwan University, 1990) pp. 131–145.
32. "Soil Taxonomy: A Basic System of Soil Classification for Mapping and Interpreting Soil Surveys," Soil Survey Staff, U.S. Dept. Agriculture Handb. No. 436, U.S. Govt. Printing Office, Washington, D.C. (1975).
33. Ito, H. and K. Iimura. "Absorption of Zn and Cd by Rice Plants. II: Effect of Cd," *J. Sci. Soil Manure Jpn.* 47:44–48 (1976).
34. Chino, M. "The Distribution of Heavy Metals in Rice Plants Influenced by the Time and Path of Supply," *J. Sci. Soil Manure Jpn.* 44:204–210 (1973).
35. Kitagishi, K., M. Ohashi, Y. Tokai, and M. Umebayashi. "Distribution and Localization of Heavy Metals Within Rice Grain, Produced on Paddy Fields Contaminated with Cd, and Chemical Forms of Cd in Rice Endosperms," *Rep. Environ. Sci. Mie Univ.* 1:129–141 (1976).
36. Yoshikawa, T., S. Kusaka, T. Zikihara, and T. Yoshida. "Distribution and Formation of Heavy Metal Elements in Rice Plants. I. Distribution of Heavy Metal Elements in Rice Grains Using an Electron Probe X-Ray Microanalyzer (EPMA)," *J. Sci. Soil Manure Jpn.* 48:523–528 (1977).

37. Tanaka, K., T. Yoshida, and Z. Kasai. "Distribution of Mineral Elements in the Outer Layer of Rice and Wheat Grains, Using Electron Microprobe X-Ray Analysis," *Soil Sci. Plant Nutr.* 20:87–91 (1974).
38. Tatekawa, H. "Studies on the Analysis and the Countermeasure for Improvement of Paddy Fields Polluted by Heavy Metals Particularly Cd, in Fukushima Prefecture," *Spec. Bull. Kushima Pref. Agric. Exp. Stn.* 1:1–64 (1978).
39. Morishita, T. "Pollution Dynamics of Heavy Metals in the Soil-Crop System," in *Survival of Human Being and Natural Environment*, T., Morishita, Ed. (Tokyo, Tokyo University Publication Press, 1977) pp. 287–295.
40. Iimura, K., H. Ito, M. Chino, T. Morishita, and H. Hirata. "Behavior of Contaminant Heavy Metals in Soil-Plant System," in *Proc. Int. Sem. Soil Environmental Fertilizer Management and Intensive Agriculture*. Tokyo, pp. 357–368 (1977).
41. "Annual Report on the Countermeasures Against Soil Pollution," Miyaggi Pref. Agric. Research Center, Japan, pp. 22–28 (1977).
42. "1973 Annual Report Water Pollution Control Laboratory," Water Pollution Control Lab. Agricultural Central Experiment Station (Japan), pp. 70–93 (1973).
43. Morishita, T. and H. Anayama. "A Field Examination Study on the Reclamation Methods of the Cadmium Polluted Soil in the Jinzu River Basin," in *Basic Researches on the Mechanism of Contamination of Soil-Plant System by Heavy Metals and Decontamination*, Kumazawa, K., Ed. Japan, pp. 51–59 (1974).
44. Morishita, T. "A Study on the Mechanism of Heavy Metal Pollution of Soil-Crop Plants System," Ph.D. thesis, The University of Tokyo, Japan (1975).

5

The Transfer of Cadmium from Agricultural Soils to the Human Food Chain

Andrew P. Jackson* and Brian J. Alloway

Environmental Science Unit, Department of Geography, Queen Mary and Westfield College, University of London, Mile End Road, London E1 4NS, UK

ABSTRACT

Cadmium has no known essential biological function and is potentially toxic to both plants and animals. This paper reviews the behavior of Cd in the human food chain. The transfer of Cd along the soil-plant-human pathway is considered for soils that have been treated with sewage sludge. Only the soil-plant human pathway is considered because food crops dominate the human Cd exposure profile. The importance of this mode of soil contamination is placed in both a regional and global context. Consideration is given to the key processes determining the dynamics of Cd in agricultural systems. Cadmium concentrations for two key food groups, potatoes and cereal crops are reviewed. Soil factors affecting the bioavailability of Cd to crop plants are reviewed. The contribution of Cd from the atmosphere to the overall Cd burden of the plant is examined. The prediction of Cd concentrations in food crops is examined with reference to the use of predictive models and diagnostic soil testing procedures.

---

* Present address: Environmental Resources Limited, Eaton House, Wallbrook Court, North Hinksey Lane, Oxford OX2 0Q5, U.K.

Dietary exposure to Cd is assessed. The exposure profiles of a number of countries are compared and contrasted. Individual food groups may be classified in order of their sensitivity to changes in the soil Cd concentration. Techniques to determine human exposure to Cd are briefly reviewed; these range from total diet studies to risk assessment methodologies. The speciation of Cd in foods may alter the bioavailability of Cd and its partitioning within the body. A number of papers indicate that Cd in plants is primarily associated with a group of Class III metallothioneins called phytochelatins. The available data referring to the speciation of Cd in plants are presented.

INTRODUCTION AND OBJECTIVES

This paper presents a review of the behavior of Cd in the human food chain in which Cd arises from soils treated with sewage sludge. This particular mode of soil contamination was chosen due to the considerable body of work published on the subject and because of its relevance to the wider issues associated with the human health implications of the disposal and utilization of sewage sludges. The key processes operating in the transfer of Cd from sewage sludge treated soils to people will tend to be the same for all metals, and in this respect Cd is used as a model element. Although the paper is by no means exhaustive, its aim is to outline some of the key issues and processes and, where possible, to present areas of uncertainty.

CADMIUM IN THE ENVIRONMENT

Concentrations of Cd in the Environment

Cadmium is a relatively rare element and, as such, is usually present in environmental media at low concentrations. Bowen[1] has estimated a mean Cd concentration of 0.1 $\mu g\ g^{-1}$ in the earth's crust; Cd concentrations in the geosphere vary from 0.005 $\mu g\ g^{-1}$ in schists, to 219 $\mu g\ g^{-1}$ in black shales.[2] The Cd burdens and residence times of the principal global reservoirs are shown in Table 1.

Mean Cd concentrations range from 0.005 $\mu g\ kg^{-1}$ in glaciers to 4.5 $\mu g\ g^{-1}$ in the particulate organic matter of the marine compartment of the hydrosphere. The soil residence time is estimated to be 3000 years; this implies that contemporary practices that tend to elevate the concentration of soil Cd will have a cumulative effect.

Concentrations of Cd in uncontaminated soils are closely correlated with those of their parent materials. Typical concentrations for uncontaminated soils are <1 $\mu g\ g^{-1}$; however, concentrations may be significantly elevated by some human activities or by the weathering of parent materials with high Cd concentrations, e.g., black shales. Some typical soil Cd concentrations are given in Table 2.

Table 1. Cadmium Burdens and Residence Times in the Principal Global Reservoirs

Reservoir	Pool Mass (g)	Cadmium Concentration	Total Cadmium in Pool (g)	Residence Time
Atmosphere	$5.1 \times 10^{18} m^3$	$0.03\ ngm^{-3}$	1.5×10^8	7 d
Hydrosphere				
Oceans				
Dissolved	1.4×10^{24}	$0.06\ \mu g\ kg^{-1}$	8.4×10^{13}	2.1×10^4 yrs
Suspended particulates	1.4×10^{18}	$1.00\ \mu g\ g^{-1}$	1.4×10^{12}	—
Particulate organic matter	7×10^{16}	$4.5\ \mu g\ g^{-1}$	3.2×10^{11}	1.3 yrs
Fresh waters				
Dissolved	0.32×10^{20}	$0.05\ \mu g\ kg^{-1}$	1.6×10^9	—
Sediments	6.5×10^{17}	$0.16\ \mu g\ g^{-1}$	1.0×10^{11}	3.6 yrs
Glaciers	1.65×10^{22}	$0.005\ \mu g\ kg^{-1}$	8.2×10^{20}	—
Groundwater	4×10^{18}	$0.1\ \mu g\ kg^{-1}$	4×10^8	—
Sediment pore waters	3.2×10^{23}	$0.2\ \mu g\ kg^{-1}$	6.4×10^{13}	—
Swamps & marshes biomass	6×10^{15}	$0.6\ \mu g\ g^{-1}$	3.6×10^9	—
Biosphere				
Marine plants	2×10^{14}	$2.0\ \mu g\ g^{-1}$	4×10^8	18 d
Marine animals	3×10^{15}	$0.6\ \mu g\ g^{-1}$	3.6×10^9	—
Land plants	2.4×10^{18}	$0.3\ \mu g\ g^{-1}$	7.2×10^{11}	20 d
Land animals	2×10^{16}	$0.3\ \mu g\ g^{-1}$	6×10^9	—
Freshwater biota	2×10^{15}	$3.5\ \mu g\ g^{-1}$	7×10^9	3.5 yrs
Human biomass	4×10^9 persons	50 mg $person^{-1}$	2×10^8	1 to 40 yrs
Terrestrial litter	2.2×10^{18}	$0.6\ \mu g\ g^{-1}$	1.3×10^{12}	42 yrs
Lithosphere	5.7×10^{25}	$0.5\ \mu g\ g^{-1}$	2.8×10^{19}	109 yrs
Sedimentary rocks	2.5×10^{24}	$1.0\ \mu g\ g^{-1}$	2.5×10^{18}	—
Shale and clay	1.9×10^{24}	$1.3\ \mu g\ g^{-1}$	2.47×10^{18}	—
Limestone	0.35×10^{24}	$0.08\ \mu g\ g^{-1}$	2.1×10^{16}	—
Sandstone	0.3×10^{24}	$0.07\ \mu g\ g^{-1}$	2.1×10^{16}	—
Soils (to 100 cm)	3.3×10^{20}	$0.2\ \mu g\ g^{-1}$	6.6×10^{13}	3000 yrs
Organic fraction	6.8×10^{18}	$0.9\ \mu g\ g^{-1}$	6.1×10^{12}	> 200 yrs

From Nriagu, J. O., in *Cadmium in the Environment. Part I: Ecological Cycling.* Nriagu, J. E., Ed. (New York: Wiley-Interscience, 1980) pp. 2–12.

Table 2. Soil Cd Concentrations

Description of Soil	Concentration ($\mu g\ g^{-1}$) Range	Mean	Ref.
Agricultural, Denmark	0.03–0.9	0.22	2
Agricultural, Sweden	0.03–2.3	0.22	2
Uncultivated, Canada	0.01–0.1	0.07	2
Agricultural, U.S.	0.005–2.4	0.27	2
Agricultural, U.K.	0.01–2.4	1.00	2
Sludge-treated, U.K.	0.27–158.7	21.7	7
Sludge-treated, U.K.	0.10–26.2	—	4
Experimental, Denmark	0.10–0.20	—	31
Arable, Denmark	—	0.26	30
Polluted paddy soils, Japan	2.19–7.47	—	29
Agricultural, U.S.			
Western	0.20–0.49	0.33	46
North central	0.20–0.94	0.37	46
Northeast	0.08–0.21	0.17	46
Southern	0.03–0.44	0.15	46
Allotments, Netherlands	0.15–3.17	0.72	172
Agricultural, Netherlands	0.04–14	0.50	173

Of the studies referred to in Table 2, soil Cd concentrations range from 0.005 to 158.7 $\mu g\ g^{-1}$. However, the mean Cd concentration of agricultural soils probably lies in the range 0.10 to 1.00 $\mu g\ g^{-1}$; the application of sewage sludges to such soils will tend to elevate the Cd concentration of the plough layer.[2] A survey of sewage sludge-treated soils in the U.K. showed Cd concentrations ranging from 0.10 to 26.2 $\mu g\ g^{-1}$.[4] The excessive application of sludge to a soil or the application of sludges heavily contaminated with Cd may lead to very high soil Cd concentrations. Peak Cd concentrations of 26, 61, 64, and 159 $\mu g\ g^{-1}$ have been reported by Chumbley and Unwin,[4] Pike et al.,[5] and Alloway et al.[6,7]

Behavior of Cd in the Environment

The environmental behavior of Cd has been described by a number of models. Nriagu[3] has reported a phenomenological model that describes the global biogeochemical cycling of Cd. This model comprises a number of sources, sinks, and pathways and enables the global fluxes of Cd to be assessed. The biogeochemical cycling of Cd has been described in a number of smaller systems, for example marshes[8] and forests.[9,10]

Yost[11,12] has applied systems theory to the study of the Cd cycle in the U.S. in order to assess the pathways of Cd exposure to humans. The main fluxes of Cd in this system are summarized in Figure 1.

Cadmium inputs to the sewerage system are an important component of the model, and the sludges so generated represent a significant sink for Cd. Models such as this may be used to assess the relative value of strategies to control human Cd exposure.

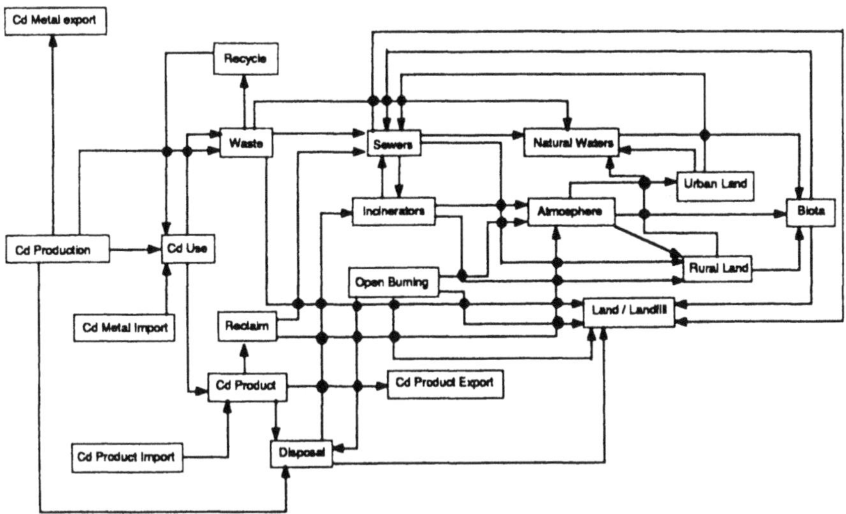

FIGURE 1. Primary elements of the U.S. Cd environmental flow system (From Yost, K. J. *Experientia* 40:157–164 [1984]. With permission.)

CADMIUM DYNAMICS IN AGRICULTURAL SYSTEMS

The aim of this section is to examine one approach to the determination of the behavior of Cd in agricultural soils. Systematic approaches to the assessment of the behavior of contaminants in the environment have been widely employed.[11–21] In many of these models, the environment is subdivided into several compartments and the transfers of a specific contaminant quantified between them. The scales of such models may vary from global inventories, such as those by Nriagu,[3,19] to national inventories.[22–24] Further refinement of this approach enables the transfer of a contaminant to and from a specific compartment of the environment to be studied in more detail; for example, Hovmand[25] has produced an analysis of the cycling of lead, Cd, Cu, Zn, and Ni in the plough layer of Danish agricultural soils.

A number of global estimations of Cd levels in soils and of soil contamination have been made by Nriagu.[3,19] Comparison of the anthropogenic and natural emissions of trace elements to the atmosphere has demonstrated human activity to be the key agent in the global cycling of Cd and a number of other elements.[26] The ratio of natural to anthropogenic emissions for Cd was calculated to be 0.15; only for the emission of lead was the anthropogenic component more significant than that for Cd.

Hutton[17] prepared an inventory for Cd in the Member States of the European Community, listing inputs to the environment from a variety of sources. The behavior of Cd in the soil-plant system was assessed using the exposure commitment approach.[15] The total Cd input to soils away from a point source was calculated to be 8 g ha^{-1} a^{-1}. Inputs from the application of sewage sludges were

Table 3. Mass Flows of Cd in Agricultural Soils

	Mass Flow (g ha^{-1} a^{-1})			
	UK[22]	FRG[174]	Denmark[40]	Netherlands[55]
A	3	3–27	2	3.5
B	0.94	10–25	0.12	5.5
C	4.30	3-7	3	
D	No data	~1	No data	No data
E	No data	0.5–10	1.3	2.5
F	No data	<0.6–>5	No data	No data
G	No data	0.3–>8	0.4	No data
H	No data	<1–>2	2	1.2

"considered to be too small on a national or regional basis to warrant inclusion." Projections for dietary exposures to Cd 100 years hence were made using four different scenarios, all of which indicated an increase in the Cd exposure of the human population.

The levels of Cd in soils are increasing due to Cd emissions from a variety of human activities; for example, the concentration of Cd in Danish soils is increasing by between 1 and 3 μg kg^{-1} a^{-1}.[27] Analyses of archived soil samples from the Rothamsted Experimental Station in the U.K. have indicated an increase of 27 to 55% in the soil Cd burden since the 1850s.[28] Other studies of soil Cd levels have reported levels greater than those anticipated for uncontaminated areas.[29–31]

In an attempt to examine the means by which levels of Cd have been increasing, the mass balance or soil budget approach has often been adopted. For example, Sauerbeck[32] has assessed the situation in the F.R.G. A summary of the data from the U.K., F.R.G, and Denmark is given in Table 3 and Figure 2. A problem with data from specific field trials is that the Cd budget does not balance and often gives a low recovery for the system as a whole.[33,34] Previous studies, observing this feature, have tried to explain it in terms of either changes in soil bulk density or by high losses due to leaching. McGrath[33] observed that this could be due the physical movement of soil from the experimental plot in which the study was being conducted; this process was subsequently modeled by McGrath and Lane.[35] By accounting for the movement of soil from the field trial area, an increase in the recovery of Cd and a number of other metals was observed. It is clear that work such as this will increase the estimated soil Cd retention time and enable the more accurate application of field trial data to the wider agricultural situation.

If data can be found to quantify the values A to H (g ha^{-1} a^{-1}), the mass balance for Cd in agricultural soils can be calculated; the rate of change in the mass of Cd in a given volume of soil is given by the following equation:

$$\Delta Cd = (A + B + C + D + G) - (H + F)$$

The key to these fluxes is presented in Figure 2.

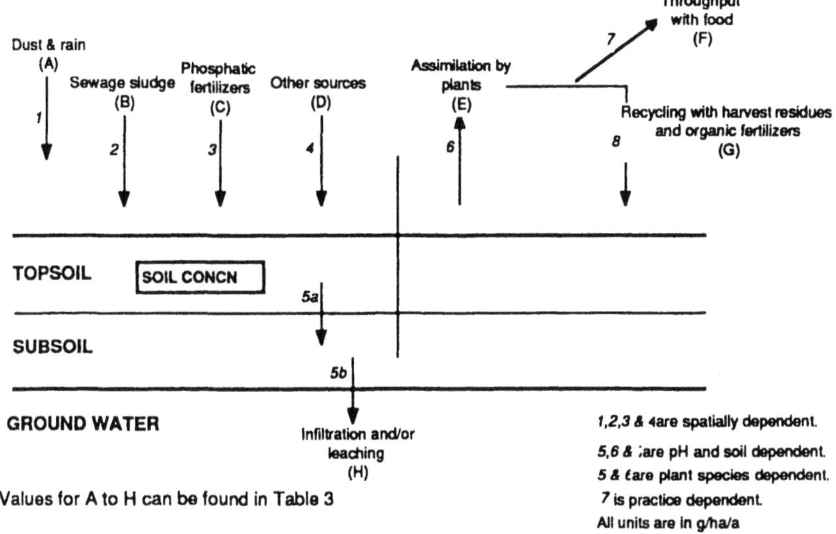

FIGURE 2. A Cd budget for agricultural soils. (From Kloke, A., D. R. Sauerbeck, and H. Vetter, in *Changing Metal Cycles and Human Health*, Nriagu, J. O., Ed. [Springer-Verlag, Berlin, 1984] pp. 113–141. With permission.)

Cadmium Inputs

In this section, published values for outputs from and inputs to agricultural soils are briefly reviewed. In some cases it is not possible to provide data for agricultural soils; therefore data for other systems will also be quoted. The aim of this section is to indicate the degree of variability between specific situations and to emphasize key data gaps. Three sources of Cd inputs will be considered:

- atmospheric deposition
- the application of phosphatic fertilizers
- the application of sewage sludges or other organic wastes.

Outputs from agricultural soils via plant uptake and leaching are discussed. There may well be other processes contributing to changes in the soil Cd burden, but these have yet to be quantified.

The key role of human industrial activities in soil contamination at the global level is shown in Table 4.[20] What these data do not convey is any information about spatial distribution, with the implication that although the global emissions to soil may be low, local emissions may be very significant. This is particularly true of the municipal sewage sludge source category, where local inputs may be as high as 80 g ha^{-1} a^{-1} but the global mean is 0.11 g ha^{-1} a^{-1}.[36]

Table 4. Global Estimates of Cd Inputs to Soils

Source Category	Cadmium Emission ($\times 10^3$ t a^{-1}) Median	Range
Agricultural and food wastes	1.50	0–3.0
Animal wastes, manure	0.70	0.2–1.2
Logging and other wood wastes	1.10	0–2.2
Urban refuse	4.20	0.88–7.5
Municipal sewage sludge	0.18	0.02–0.34
Coal fly ash and bottom ash	7.20	1.5–13
Wastage of commercial products	1.20	0.78–1.6
Atmospheric fallout	5.30	2.2–8.4
Fertilizer production	0.14	0.03–0.25
Total input	22.00	5.6–38

From Nriagu, J. O. and J. M. Pacyna. *Nature (London)* 333:134–139 (1988).

Atmospheric Deposition

The atmosphere is an important source of Cd inputs to agricultural soils. A wide range of values, often reflecting the degree of local industrial activity, exist. Monitoring in Greenland gives an annual deposition rate of 0.06 g ha^{-1} a^{-1}, compared with values of 44.4 g ha^{-1} a^{-1} in New York City and 135.6 g ha^{-1} a^{1} in an area adjacent to the Avonmouth smelter.[37] Annual atmospheric deposition of Cd in rural areas varies from 0.6 to 25 g ha^{-1} a^{-1} in the European Community.[38] Long-range atmospheric transport of Cd is an important process and has been successfully modeled.[39]

Estimates of Cd inputs to agricultural soils from the atmosphere may be made using a variety of means. Analysis of an archived soil collection by Jones et al.[28] has given some observed data that compare favorably with Nriagu's[19] predicted values of deposition from the atmosphere. Taking the cumulative net input of Cd between 1966 and 1980, a mean value of 19.9 g ha^{-1} a^{-1} can be calculated as being attributable to atmospheric sources. This value is considerably higher than that used in a Cd budget for the U.K. by Hutton and Symon,[22] a nationwide mean of 3 g ha^{-1} a^{-1} was used for this study. It should be stressed that Cd deposition rates are not uniform, and it may be that the semi-rural site (Rothamsted Experimental Station, Harpenden, U.K.) from which Jones et al. collected their samples was not representative of a typical rural site in the U.K. However, other similar studies of this site have also shown it not to be anomalous.

Soil Amendments

Two of the most important sources of soil contamination are phosphatic fertilizers and municipal sewage sludges.

Many phosphatic fertilizers applied to land contain high levels of Cd, as can be seen in Table 5. Fertilizers made from magmatic phosphates tend to contain

Table 5. Concentrations of Cd in Rock Phosphate Imported into the U.K.

Source Country	Concentration (mg kg^{-1})
U.S.	6.5
Senegal	71.0
Morocco	18.0
Tunisia	18.0

From Hutton, M. and C. Symon. *Sci. Total Environ.* 57:129–150 (1986).

only negligible concentrations of Cd, whereas those made from sedimentary phosphates will contain a range of concentrations, some of which may be very high.[40] Sewage sludges often contain elevated concentrations of Cd as a result of the disposal of waste from industrial activities.

In an experiment in California in which triple superphosphate was applied to a soil over a period of 36 years, the concentration of Cd in the top 15 cm of the experimental plot was 1.2 $\mu g\ g^{-1}$ as opposed to 0.07 $\mu g\ g^{-1}$ in the control plot.[41] An approximately fivefold increase in the Cd concentration of the upper 2.5 cm of the soil profile was observed in a similar experiment reported by Williams and David.[42]

Annual inputs of Cd to an agricultural soil from phosphatic fertilizers vary as widely as do the concentrations in the fertilizers themselves. A mean annual input of 3.5 g $ha^{-1}\ a^{-1}$ is reported by Kloke et al. for soils in the FRG.[24] Inputs to soils in the U.K. are estimated to be 22 t of Cd per annum, a mean of 4.3 g $ha^{-1}\ a^{-1}$ if the total area of arable land in the U.K. is assumed to be 5.12×10^6 ha.[22] Cadmium inputs to an experimental plot at Rothamsted receiving applications of superphosphate over a 96-year period were estimated to be 5 g $ha^{-1}\ a^{-1}$.[43]

The application of sewage sludges to agricultural soils is a potentially advantageous means by which to supply nutrients to the soil.[44–46] However, it has long been recognized that sludges may contain high levels of potentially toxic elements such as Cd.

Sludges are generally applied to land in the vicinity of their production, due to high transport costs. This will tend to give a highly nodal national distribution pattern and means that although the total input of Cd from sewage sludges may be low on a national basis,[47] it may well be very high on a local basis.[20] Most national estimates of the input of Cd to agricultural soils are probably in the region of two orders of magnitude lower than the inputs to those soils actually receiving sludges. Hutton and Symon[22] estimated an annual input of 0.96 g ha^{-1} to agricultural soils in the U.K. Hansen and Tjell[40] give a lower estimate for Denmark, 0.12 g $ha^{-1}\ a^{-1}$.

Inputs of Cd from the atmosphere, sewage sludges, phosphatic fertilizers, and other sources are spatially heterogeneous. However, contributions of Cd from the atmosphere and phosphatic fertilizers tend to display less spatial variation

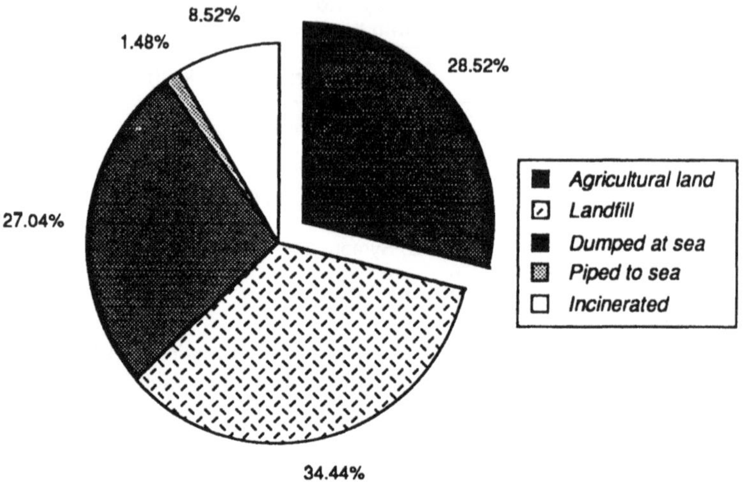

FIGURE 3. Inputs of Cd to the UK environment via sewage sludges. (From Hutton, M. and C. Symon. *Sci. Total Environ.* 57:129–150 [1986]. With permission.)

than those from sewage sludge. At the local level, controlling inputs of Cd from sewage sludges may be important. The disposal of sludges to agricultural land is an important part of the overall introduction of sludge-borne Cd to the environment, taking 28.5% of the total Cd content of sludges (Figure 3). The future trends in the production of sewage sludges point to an increase in the quantity of sludge generated but not to an increase in the concentration of Cd in these sludges.[17] The disposal routes are expected to change in emphasis, with the prediction that by 1990, 75% of the sludge produced in the European Community will be disposed of to land.[17] This may imply that the total quantity of Cd applied to the soil from sewage sludges will increase. However, sludge incineration has considerable potential for development and could prove to be an important component of future sludge management strategies, thereby reducing the quantity of sewage sludges applied to agricultural soils.

Cadmium Outputs

Leaching

Quantifying the losses of Cd from a soil due to leaching and associated processes is an area of some considerable uncertainty. The majority of studies examining this output have tended to imply that losses of Cd due to leaching are not a major component of the total losses from a soil. The influence of soil properties, especially soil pH, is another source of variation. Data from the examination of soil profiles indicates that the majority of the soil Cd burden will remain in the topsoil or plough layer.[33,41–43,48]

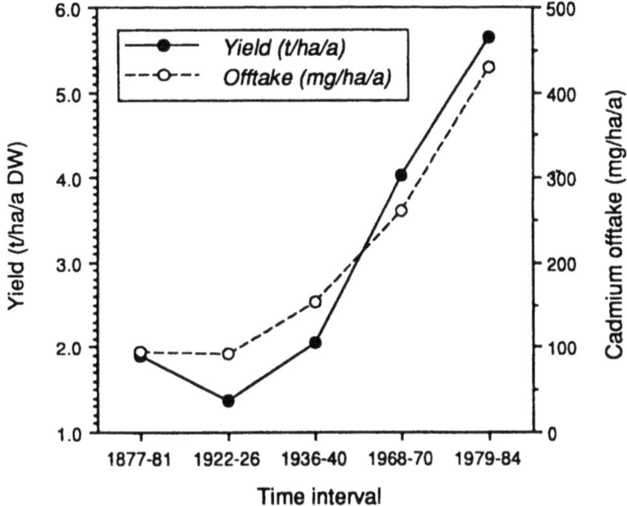

FIGURE 4. Cadmium offtakes via wheat grain and yields from an experimental plot at Rothamsted, U.K. (From Jones, K. C. and A. E. Johnston. *Environ. Pollut.* 57:199–216 [1989]. With permission.)

Estimates of annual losses by leaching and associated processes include 5 g ha^{-1} a^{-1} [49] and 1.5 g ha^{-1} a^{-1}.[27,40] A number of values are given by Kabata-Pendias and Pendias ranging from 0.3 g ha^{-1} a^{-1} for agricultural land in Denmark to 7 g ha^{-1} a^{-1} for a deciduous forest in Tennessee.[50] McGrath has compared the predicted concentrations of Cd in soil water necessary to account for the losses from an experimental plot with observed values; the observed values fell over an order of magnitude short of the predicted concentration.[33]

Plant Accumulation

Values for the offtake of Cd in the harvested crop tend to be low when compared with the total mass of Cd in the plough layer. In a soil to which sewage sludge had been applied, McGrath estimated that the total Cd offtake in the crops grown over a 20-year period was 180 g.[33] This represents only 0.28% of the total amount of metal added in sewage sludges. Data from Jones and Johnston give Cd offtakes in wheat grain varying from 0.09 to 0.43 g ha^{-1} a^{-1}, although there was "little evidence of long-term increases in crop Cd ...", the offtake did increase with time due to improved yields (Figure 4).[51]

Jackson showed that Cd offtakes from sludge-treated soils via lettuce and cabbage ranged from 0.25 to 21.67 g $ha^{-1}a^{-1}$ and from 0.09 to 42.95 g $ha^{-1}a^{-1}$, respectively.[52] Boo has collated a number of studies in which crop Cd offtakes were reported.[53] Two independent studies were compared in which offtakes via potatoes were approximately equal (1.20 and 1.40 g Cd ha^{-1} a^{-1}); however, winter oats (0.79 and 3.2 g ha^{-1} a^{-1}) and spring barley (0.79 and 4.50 g Cd ha^{-1} a^{-1}) did not show such uniformity. In addition to these crop-specific data, Boo also compared crop offtake data from studies by Henkens,[54] CCRX,[55] Breimer and Smilde,[56] and RIVM,[57] who reported values of 1.0, 2.5, 1.4, and 2.3 g Cd ha^{-1} a^{-1}, respectively.

CADMIUM IN FOOD CROPS GROWN ON SEWAGE SLUDGE-TREATED SOILS

The Accumulation of Cd by Crops Growing on Sewage Sludge-Treated Soils

Typical Crop Concentrations

The main source of Cd to a plant growing on a soil with elevated levels of Cd is the soil itself, although the deposition of atmospheric Cd will play a more important role at sites with background levels of Cd in the soil and at sites where the concentration of Cd in the atmosphere is high.[58] The accumulation of a trace element from the soil may be described by a number of means, one of which is the concentration factor (CF-value)[58,59] or accumulation ratio.[6] The CF value is determined by dividing the concentration of Cd in the plant by that in the soil; plants concentrations have been expressed in both the dry and fresh weights. However, at background or only slightly elevated soil concentrations, atmospheric deposition may make a significant contribution to the plant trace element burden; this is particularly pertinent in the case of Cd.[25,60–62] The atmospheric input of Cd implies that the CF value will tend to overestimate the actual output of Cd from the soil. The contribution of atmospheric Cd to the total Cd concentration of a number of food crops is shown in Table 6. In order to account for the contribution of Cd from the atmosphere, an additional coefficient, the CF_A value, was derived by Chamberlain.[59] This is determined by dividing the soil-derived Cd concentration of the plant by the concentration of Cd in the soil. Chamberlain's work was concerned mainly with Pb, for which the coefficient was first derived, but may be applied equally well to the study of Cd dynamics. For example, a recent study by Dollard and Davies has shown that lettuce plants grown in the vicinity of a nonferrous metal smelter have a CF value of 13.8 ± 4.5 and a CF_A value of 1.85 ± 0.41.[58] This implies that the output of Cd from a soil via the harvested crop could be overestimated by a factor of around seven times. It should be noted that the use of the CF_A value for Cd is usually less critical than it is for lead, due to the relatively greater bioavailability of soil Cd.

It has been observed that plants accumulate Cd at different rates and that the final concentration of Cd in plant tissues will differ between species growing concurrently on the same soil, as can be seen in Figure 5. The accumulation of Cd by plants growing on sewage sludge-amended soils also shows this effect.[63–65]

Table 6. The Contribution of Cd From the Atmosphere to the Total Plant Concentration

Food Crop	Percentage of Cd Burden From Atmospheric Sources	Ref.
Barley (grain)	41–58	
Kale (leaf)	36–60	
Carrot (root)	37–52	25
Wheat (grain)	21	
Rye (grain)	17–28	
Cabbage (leaves)	36–60	
Radish (tuber)	25–47	
Turnip (root)	5–6	
Spinach (leaves)	23	60
Carrot (root)	4–8	
Lettuce (leaves)	7–21	
Radish (tuber)	24–28	
Spinach (leaves)	8–10	61
Lettuce (leaves)	6–10	
Beans	99.5	
Lettuce	86–99	
Carrot	48–90	58[a]
Cabbage	86–92	

[a] It should be noted that this experiment was performed in the vicinity of a metal smelter.

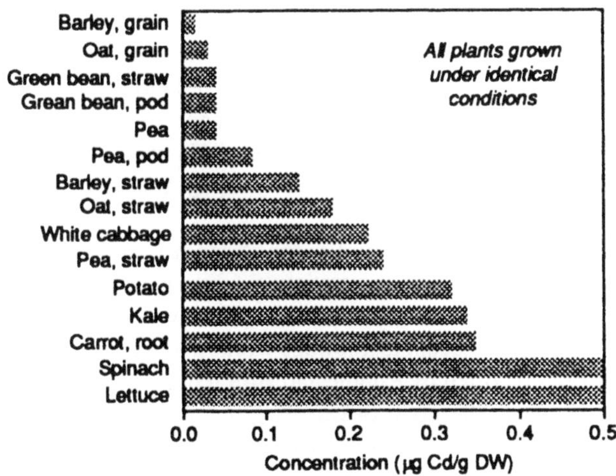

FIGURE 5. Concentrations of Cd in plants grown under identical soil conditions. (From Hansen, J. A. and J. C. Tjell, in *Environmental Effects of Organic and Inorganic Contaminants in Sewage Sludge*, Davis, R. D., G. Hucker, and P. L'Hermite, Eds. [Reidel, Dordrecht, 1983] pp. 91–113. With permission.)

Table 7. Concentrations of Cd in Potato Tubers

Description	Concentration ($\mu g\ g^{-1}$)		Ref.
	Mean	Range	
Sludged soils from southern U.S.	—	0.08–0.51 dwt	175
U.S. sludge soils	—	0.12–0.23 dwt	63
Experimental soils			
Control soil	0.11 dwt	—	46
Sludge soil	0.10 dwt	—	46
ADAS survey of U.K. sludged soils	0.60 dwt	—	4
Metal smelter site	—	0.03–0.20 dwt	66
Control site in FRG			
Peeled	0.06 dwt	—	67
Unpeeled	0.12 dwt	—	67
Survey of EC	—	0.02–0.09 dwt	74
Netherlands survey	0.03 fwt	0.01- 0.09 fwt	173
Spanish survey	0.013 fwt	0.01–0.02 fwt	176
MAFF survey			
Control soils	<0.03 fwt	<0.01–0.06 fwt	145
Sludge soils	0.14 fwt	0.06–0.21 fwt	145
Shipham	0.13 fwt	0.03–0.30f wt	145

From Jackson, A. P. "The Bioavailability of Cadmium From Sewage Sludge-Amended Soils," Ph.D. thesis, University of London (1990).

The reported Cd concentrations in potatoes and cereal crops are shown in Tables 7 and 8. These two crops are of particular importance as they represent a considerable proportion of the average mass of food consumed.

As discussed below, the interpretation of Cd concentrations in crops can be complicated by marked inter- and intra-specific variations. Data from Davies and Crews[66] and Kampe[67] indicate that the skin or peel of potato tubers contains a higher Cd concentration than that in the flesh. This makes direct comparison of the data in Table 7 difficult because only rarely are the details of sample preparation reported.

Mean Cd concentrations in grain range from 0.004 $\mu g\ g^{-1}$ dwt[68] to 0.75 $\mu g\ g^{-1}$ dwt.[69] The peak concentration (5.60 $\mu g\ g^{-1}$ dwt) was reported by Bingham and Page[70] for wheat grown on soil spiked with a relatively high level of Cd. A study of soils adjacent to a sewage works and receiving large volumes of sludge has been reported by Rundle and Holt reported mean concentrations 0.69 μg Cd g^{-1} dwt in the grain of cereals growing on heavily sludged land adjacent to a sewage works.[71] This value is approximately one order of magnitude higher than those found in the majority of samples of grain grown on control soils.

A problem with the interpretation of grain or cereal concentrations is that it is unclear whether or not they are for whole grain samples or flours, etc., as this will affect the significance of the concentration. Davis et al.[72] report that the concentration of Cd in flour is only 57% of that found in the whole grain. Therefore, when assessing the dietary exposure to Cd from grain products, care

Table 8. Concentrations of Cd in Cereal Grains

Description	Concentration (μg g⁻¹) Mean	Range	Ref.
Comparison of pot and field trials for wheat	0.10 dwt	—	177
Netherlands survey	0.07 fwt	0.02–0.35 fwt	173
Retrospective analysis of archived samples of wheat grain	0.055 dwt	0.02–0.11 dwt	75
Field trial with sludge soils	<0.01 dwt	—	178
Whole grain samples	0.03 dwt	—	73
Control soils, wheat grain samples	—	<LOD–0.024 dwt	175
Sludge-amended soils, U.S.	—	0.02–0.05 dwt	95
Spiked soils, wheat grain	—	0.10–5.60 dwt	70
Wheat grain from field trials	—	0.02–0.05 dwt	63
Wheat grain samples produced by organic farming methods in FRG	—	0.015–0.075 dwt	179
MAFF survey, cereals food group	—	<0.02–0.05 dwt	132
Wheat grain from Stoke Bardolph	0.69 dwt	—	71
Wheat grain from Royston trial	—	0.05–0.70 dwt	47
National survey, Sweden			
Winter wheat	0.06 fwt	—	180
Spring wheat	0.04 fwt	—	180
Barley from sludged soils	—	0.04–0.07 dwt	180
Sludged soils in lysimeters, wheat grain	0.75 dwt	—	69
Spiked soils			
Wheat	—	0.50–1.90 dwt	174
Barley	—	0.60–3.50 dwt	174
Wheat grain from pot trials	—	0.025–0.68 dwt	181
EC background concentration in wheat	—	0.03 dwt	17
Wheat grain from sludged soils	—	0.025–0.68 dwt	182
CEC survey			
Flour	—	0.02–0.15 dwt	74
Grain	—	0.01–0.04 dwt	74
Wheat from U.S. survey	0.047 dwt	0.014–0.21 dwt	46
Wheat grain from field trials			
Control soils	0.004 dwt	—	68
Sludged soils	0.016 dwt	—	68
Corn grains	—	0.01–0.02 dwt	183

From Jackson, A. P. "The Bioavailability of Cadmium From Sewage Sludge-Amended Soils," Ph.D. thesis, University of London (1990).

should be taken to accurately describe the nature of the food being consumed. Micco et al. analyzed a variety of grain products and found that bran had a higher Cd concentration than other components.[73] Similar results are reported by Kloke et al.[24] Cd accumulation by rice grains follows a comparable trend, with unpolished rice having a higher Cd concentration than polished rice.[29] The assumption that whole grain products will necessarily contain higher Cd concentrations is contradicted by some sources.[74] Analysis of archived samples has shown that there do not appear to be marked increases in the concentration of Cd in cereal grains through time.[51,75]

Table 9. Concentrations of Cd in the Edible Components of Food Crops

		Concentration (μg Cd g^{-1} dwt)		
Crop Class	**n**	**Minimum**	**Maximum**	**Mean**
Fruits	190	0.0043	0.012	0.005
Vegetables				
Seed	394	0.0160	0.130	0.028
Root/bulb	878	0.0290	0.710	0.208
Fruit	322	0.0210	0.540	0.237
Leaf	297	0.0930	0.880	0.560
Field crops				
Grain	1302	0.0140	0.210	0.047

From Page, A. L., A. C. Chang, and M. El-Amamy, in *Lead, Mercury, Cadmium and Arsenic in the Environment,* Hutchinson, T. C. and K. M. Meema, Eds. (John Wiley & Sons, New York, 1987) pp. 119–146.

Intra-Plant Variation

In addition to inter- and intra-specific variations in Cd concentrations in plants, marked differences also occur in the accumulation of the metal between the various plant organs, as shown in Table 9. These characteristics are important in the context of the dietary exposure profile, as they will determine, to an extent, the degree of exposure from the food when levels of Cd in the soil are elevated by the application of sewage sludge.

In an experiment in which potato plants were grown in sand cultures spiked with Cd salts, concentrations in plant tissues were reported to decline in the order:[52]

roots > stems and leaves > tubers

The response of tissue Cd concentrations to changes in Cd exposure from a Cd-spiked nutrient solution is shown in Figure 6.

Stem, leaf and tuber concentrations show an initial resistance to Cd accumulation that gradually decreases with increasing exposure. Root concentrations do not show this trend and respond linearly to Cd exposure.

Soil Factors Affecting Cd Bioavailability

In addition to plant factors affecting Cd concentrations in crops, a considerable number of soil variables have been demonstrated to influence Cd bioavailability.[2,40,76–78] The concentration of Cd in the soil is probably the most influential factor. A number of studies have shown statistically significant linear relationships between soil and plant Cd concentrations. For example, Chumbley and Unwin reported highly significant correlations for cabbage and lettuce.[4] Jackson found relationships between soil Cd concentrations and those of the

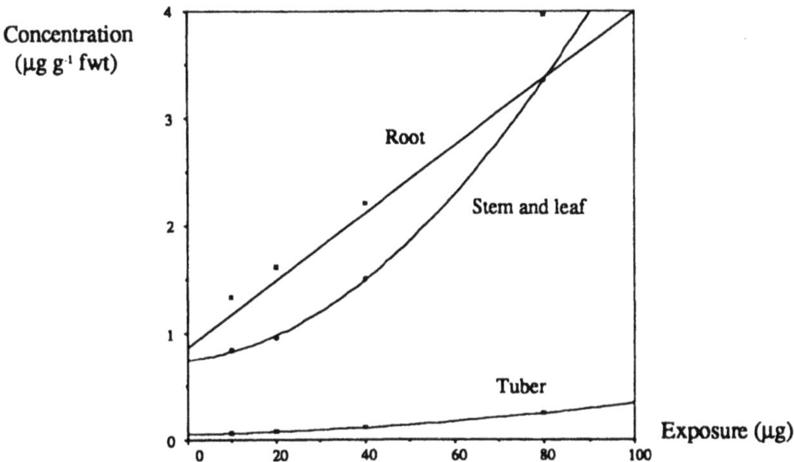

FIGURE 6. The change in potato plant tissue Cd concentrations with exposure from a nutrient solution. (From Jackson, A. P. "The Bioavailability of Cadmium from Sewage Sludge-Amended Soils," Ph.D. thesis, University of London, U.K. [1990].)

edible tissues of lettuce, cabbage, and potato.[52] An equation of the form shown below was derived:

$$\log \text{Plant Cd} = a + b \log \text{Soil Cd}$$

The values for the intercept (a) and the regression coefficient (b) are given in Table 10.

Soil pH has been demonstrated to be significantly negatively correlated with plant Cd concentrations, and it is generally through the manipulation of soil pH that Cd bioavailability may be managed. Andersson and Nilsson found that Cd uptake by fodder rape (*Brassica napus*) was inversely related to soil pH.[79] Christensen showed that the adsorptive capacity of soils for Cd increased by a factor of three for each increase of one pH unit between pH 4.0 to 7.7.[80] In an experiment in which three crops were grown on a number of sewage sludge-treated soils, Jackson and Alloway reported that the application of lime was shown to reduce the bioavailability of Cd to cabbage (by 43%) and lettuce (by 41%) but not to potato tubers.[7] Alloway et al. reported that crop Cd concentration ratios were highest for plants grown on acid soils.[81]

The sorptive properties of a soil determine the partitioning of Cd between the solid and liquid phases and therefore the potentially bioavailable pool of soil Cd. A number of soil variables have been found to influence the ratio of Cd sorbed per unit weight of soil to the solute Cd concentration; this ratio is referred to as the distribution coefficient or K_d value. Christensen reported that soil pH, Ca concentration, and competition by other elements (especially Zn) affected the K_d values of several soils with relatively low concentrations of Cd.[80,82–86]

Table 10. Regression Parameters Describing the Relationship Between Soil and Crop Cd Concentrations

Crop	a	b	r	R^2
Cabbage	−0.55	0.52[a]	0.001	0.55
Lettuce	0.44	0.43[b]	0.001	0.33
Potato	−1.26	0.54[a]	0.001	0.57

[a] $HF/HClO_4$ decomposition.
[b] HNO_3 extraction.

From Jackson, A. P. "The Bioavailability of Cadmium from Sewage Sludge-Amended Soils," Ph.D. thesis, University of London (1990).

Neal and Sposito investigated Cd sorption at soil solution concentrations ranging from 0.001 to 12 μmol Cd kg^{-1}.[87] S-curve adsorption isotherms were observed for sludged soils with a high content of organic matter indicating the complexation of Cd in a soluble form at the lowest concentrations. However, when the soluble organic matter content of these soils was removed, their adsorption isotherms became L-shaped, which is the normal shape found when cadmium is adsorbed on the solid phase of soils.

Cabrera et al. have investigated the effect of humic acid on Cd-uptake by barley plants.[88] Humic acid-Cd complexes were not shown to be readily absorbed by plants, implying that low concentrations of humic acids may reduce Cd bioavailability. Cadmium concentrations in plant shoots were reported to be inversely related to the concentration of humic acid in solution. In a series of solution culture experiments reported by Tyler and McBride, the addition of humic acid was shown to reduce Cd activity in the solution and uptake by corn roots.[89] Humic acid had no effect on Cd translocation in corn plants.

The effects of soil temperature on Cd bioavailability to vegetables grown on sewage sludge-treated soil has been reported by Giordano et al.[90] Heating the soil to 27°C significantly elevated Cd concentrations in broccoli and potato; concentrations in the other vegetables grown were also elevated but not to the point where they became statistically significant. An experiment in which plants were grown at temperatures ranging from 9 to 24°C has been reported by Siriratpuriya et al.[91] Lettuce Cd concentrations were shown to increase with temperature when grown on sewage sludge-treated soils.

Cadmium speciation in the soil solution of sludge-treated soils may well play a role in its bioavailability. Brummer has stated that element bioavailability is a function of at least three parameters:[92]

- the total amount of potentially available elements (the quantity factor),
- the concentration or activity and ionic ratios of elements in the soil solution (the intensity factor), and
- the rate elements transfer from solid to liquid phases and to plant roots (reaction kinetics).

The bioavailability of Cd is clearly affected by a number of soil variables simultaneously; therefore multivariate statistical methods may be used to provide empirical models.

Bingham has derived equations to predict the concentration of Cd in the leaf tissue of lettuce and Swiss chard grown on eight soils treated with sewage sludge; these were:[77]

$$\text{Lettuce Cd} = 21.97 + 20.43\ \text{SE Cd} + 3.92\ \text{CEC}\ (R^2 = 0.82)$$
$$\text{Chard Cd} = 629.02 + 31.79\ \text{SE Cd} - 93.46\ \text{pH}\ (R^2 = 0.89)$$

(where SE Cd = Cd concentration of saturation extract, CEC = cation exchange capacity)

Multivariate equations were used by Alloway et al. to describe the accumulation of Cd by cabbage, lettuce, radish, and carrot grown on soils contaminated from a variety of sources.[6] The equations derived for sewage sludge-treated soils are shown below:

$$\log \text{cabbage Cd CR} = 1.48 \log \text{CEC} - 0.82 \log \text{Cd}\ (HNO_3) - 0.27\ (R^2 = 0.71)$$
$$\log \text{lettuce Cd CR} = 4.04 - 4.25 \log \text{pH} - 0.61 \log \text{Cd}\ (HNO_3)\ (R^2 = 0.77)$$
$$\log \text{carrot Cd CR} = 3.16 \log \text{CEC} - 1.63 \log \text{Cd}\ (HNO_3) - 4.75 \log \text{pH}\ (R^2 = 0.92)$$
$$\log \text{radish Cd CR} = 0.85 \log \text{LOI} - 0.95 \log \text{Cd}\ (HNO_3) - 0.45\ (R^2 = 0.83)$$

(where Cd CR = Cd concentration ratio, Cd (HNO_3) = nitric acid extractable Cd and LOI = loss-on-ignition)

In a study of potato, lettuce, and cabbage grown on a number of sewage sludge-treated soils, Jackson and Alloway derived multivariate equations to describe both crop Cd concentrations and CRs.[7] The equations in which the crop Cd concentration is the dependent variable are shown below:

$$\log \text{cabbage Cd} = 0.50 - 0.44 \log T + 0.58 \log \text{Cd}\ (CaCl_2)\ (R^2 = 0.75)$$
$$\log \text{lettuce Cd} = 3.52 - 3.11 \log \text{pH} - 0.50 \log \text{LOI} + \log \text{Cd (DTPA)}\ (R^2 = 0.63)$$
$$\log \text{potato Cd} = 0.59 \log \text{Cd (DTPA)} - 1.06\ (R^2 = 0.76)$$

(where T = time elapsed since the last application of sewage sludge, Cd ($CaCl_2$) = 0.05 *M* $CaCl_2$ extractable Cd and Cd (DTPA) = 0.005 *M* DTPA extractable Cd)

In this section, much emphasis has been placed upon the role of soil variables in affecting Cd bioavailability to food crops. However, treating soils with sewage sludges may well change a number of soil variables; for example, sludge application may cause an initial drop in pH, increase the adsorptive capacity of metals through the addition of organic matter and hydrous oxides, and alter the porosity of soils through the physical conditioning effect. As argued by Corey et al., the sludge matrix itself plays a key role in Cd adsorption in the soil environment.[93]

Temporal Changes in Crop Cd Concentrations

A factor unique to soils amended with sewage sludge is the effect of the residual period, i.e., the time between the application of sludge and the plant being harvested. Given the long residence time of Cd in soils, any radical changes in its bioavailability over the residual period would be very important to the overall assessment of dietary exposure. A number of field trials have been conducted from which changes in the bioavailability of Cd during this period of time may be examined.

With applications of sludge to an acidic coarse sandy soil, at both 10 and 100 t ha^{-1} a^{-1}, Juste and Solda reported decreases in the Cd concentration of maize leaves.[94] The most marked decrease, 1980, corresponded with a considerable increase in Ni concentrations. It was suggested that the decrease in leaf Cd concentrations was caused by low temperatures in the rooting zone, Ni uptake was not reduced with the consequence of reducing the yield. Observed changes after the first year fall into two groups, one in which bioavailability remains relatively constant and another in which it gradually declines. Over the residual period, the uptake of Cd by corn was seen to fall in a field trial reported by Bidwell and Dowdy, as can be seen in Figure 7.[95]

The rate of decrease was largely dependent upon the initial Cd input. In a similar study by Hinesly et al., the bioavailability of Cd to corn (*Zea mays*) also fell, with the concentrations in grain falling from 0.44 to 0.07 μg g^{-1} dwt over the 4-year residual period.[96] Kelling et al. reported a fall in both the concentration of Cd in plant tissues and in the DTPA-extractable soil Cd over a 4-year residual period.[97] Analysis of sludge-amended soils from the Woburn Market Garden Experiment in the U.K. has shown that over a 22-year residual period, the proportion of Cd extractable from the soil by EDTA has remained constant.[33] Data from field trials conducted by Dowdy et al., which examined Cd uptake by *Phaseolus vulgaris*, showed no change over the 4-year residual period.[98] A similar observation for a wider variety of crops has been made by Larsen.[99] An isotope-aided study by Lonsjo demonstrated an initial rise in the ^{109}Cd transfer factor ($m^2 kg^{-1}$ dwt) for spring wheat after the application of ^{109}Cd spiked sludge; this fell to a near constant accumulation thereafter, as can be seen in Figure 8.[100]

Three applications of ^{109}Cd were made to two of the soils used in this study:

- ^{109}Cd (0)
- ^{109}Cd spiked sludge (Sl)
- ^{109}Cd spiked sludge and lime (Sl & L)

For soil 1 (pH6.0, 35% clay, 1.6% organic matter), the transfer factors in 1972 were highest for the soils receiving the ^{109}Cd without the sludge matrix, the addition of lime to the soil did not reduce the transfer factor in the first year. Mean transfer factors for 1973 to 1980 were 1.34, 1.34, and 0.67 for the 0, Sl, and Sl and L treated soils, respectively. This implies that the sludge matrix had no marked effect on ^{109}Cd bioavailability. Soil 6 (pH6.1, 6% clay, 4.5% organic

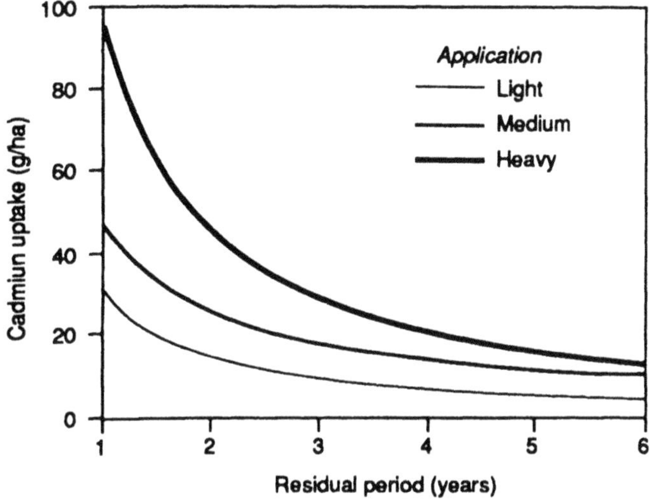

FIGURE 7. Modeled values for the accumulation of Cd by corn during the residual period. (From Bidwell, A. M. and R. H. Dowdy. *J. Environ. Qual.* 16:438–442 [1987]. With permission.)

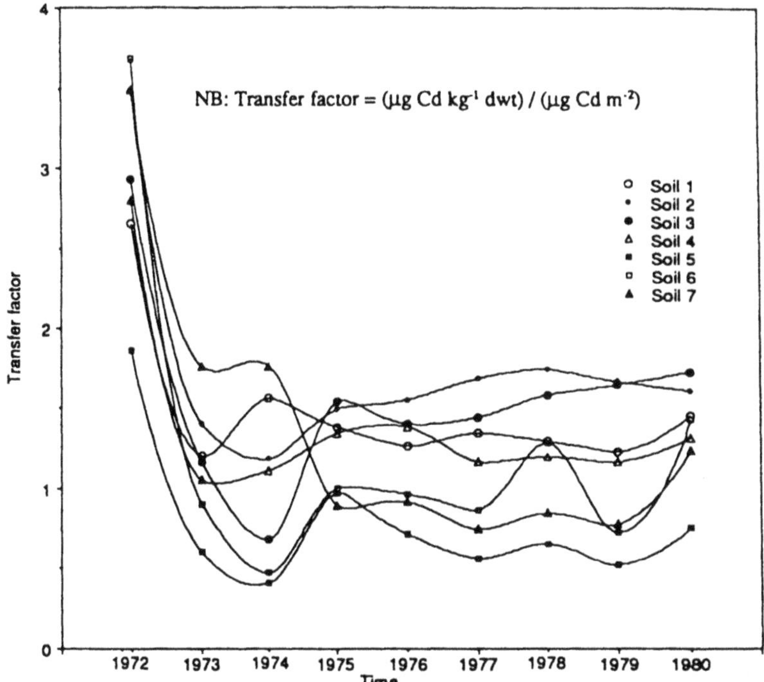

FIGURE 8. Uptake of ^{109}Cd during the residual period. (From Lonsjo, H., in *Utilization of Sewage Sludge on Land: Rates of Application and Long-Term Effects of Metals*, Berglund, S., R. D. Davis, and P. L'Hermite, Eds. [Reidel, Dordrecht,1984] pp. 135–145. With permission.)

matter) did show some evidence of the sludge matrix influencing ^{109}Cd transfer factors. Over the period 1973 to 1980, mean transfer factors for the three treatments declined in the order 0 > Sl > Sl+L.

Soil 1 had a relatively high sorptive capacity that was not significantly enhanced by the application of sludge, whereas the sorptive capacity of soil 6 was much lower and therefore enhanced by the application of sludge. As can be seen from Figure 8, the transfer factor fluctuates. The data from this experiment are particularly valuable because they were collected from seven markedly different soils that received Cd at levels below those normally found in the use of sewage sludge in agriculture (1.6 mg ha^{-1}). The longest residual period examined in the papers referred to above is that reported by McGrath (22 years), a relatively short period of time when one considers the average residence time for Cd in agricultural soils.[33] In none of the papers reviewed was there any evidence of a sudden increase in the bioavailability of Cd, as measured either directly by plant assay or by the use of soil extractants.

Predicting Concentrations in Crops

If the potential dietary exposure to Cd from soils treated with sewage sludge is to be predicted, a set of relationships has to be found that will enable the prediction of Cd concentrations in the edible components of crop plants. Such models should be applicable over a range of soil Cd concentrations, soil physicochemical characteristics, crops, and pathways of human exposure.

Mathematical Models

Modeling the uptake of heavy metals from the soil into the plant may be approached in two ways: (1) on the basis of current theories, a theoretical model may be developed and then tested against observed data, or (2) data is collected and then analyzed in order to develop an empirical model. Most of the models presented below may be said to be equilibrium models, with the implicit assumption that temporal fluctuations in rates of reaction/transfer do not occur. Jackson and Smith give a generalized form for equilibrium models to describe the transfer of radionuclides or stable elements as[101]

$$C_a = f_a \sum_g J_g C_g$$

where J_g = daily intake of food g; C_g = concentration of the element in food g; f_a = equilibrium diet-to-animal product transfer factor, and C_a = concentration of the element in animal product. Equilibrium models are relatively simple and do not allow the flexibility afforded by the more complex dynamic models. Dynamic models usually describe the rates of flux between the compartments of a system and take the form:

$$\frac{\partial}{\partial_t} q_i(t) = \sum_{i=1}^{n} \mu_{ij} q_j(t) + I_i(t) \qquad \text{for } i = 1 \text{ to } n$$

where $q_i(t)$ = content of compartment i at time t; $I_i(t)$ = rate of intake into compartment i at time t; μ_{ij} with $i \neq j$ is the constant rate of transfer from compartment i to j and n = number of compartments.

The first step in most empirical modeling techniques is to quantify the accumulation of heavy metals in the plant and to identify the variables that enable a prediction of this characteristic to be made. A wide variety of both soil and plant characteristics have been proposed. Models developed by Rappaport et al.[102] and Browne et al.,[103] based upon soil parameters, describe the plant uptake of Zn and Cd respectively. The model proposed by Browne is shown below:

$$\log P = a + b \log Cd_{DTPA}$$

Where P = plant Cd concentration, Cd_{DTPA} = DTPA extractable Cd, and a and b are linear regression coefficients. b was found to be principally a function of soil pH and cation exchange capacity, and a was primarily a function of the plant species and may be related to the selectivity coefficient.[104] The Zn accumulation model of Rappaport et al. is based upon DTPA extractable and organically bound Zn.[102]

A Cd and Zn accumulation model based upon a nutrient uptake model, soil characteristics, and kinetic parameters has been developed by Mullins et al. to describe plant uptake from sewage sludge-amended soils.[105] Sensitivity analyses showed that root growth constants, average root radius, water influx rate, and soil solution metal concentrations were the most influential variables in the model. This model was based upon a more general model developed by Barber, later refined by Cushman, and recognizes three modes of ion-influx kinetics:[106]

- passive ion movement independent of respiration energy,
- passive ion uptake along an electrochemical gradient, dependant on respiration energy, and
- active ion uptake against an electrochemical gradient requiring respiration energy.

In the comparison of Cd and Zn accumulation by Mullins et al., Cd proved to be the least accurately predicted.[105]

A deterministic model for Cd uptake from sewage sludge-amended soils has been developed by Christensen and Tjell.[107] It divides the total plant Cd concentration into three fractions, based upon their source:

- Cd from the topsoil,
- Cd from the subsoil, and
- Cd from the atmosphere.

Plant uptake from the topsoil is described by the following expression:

$$b = P.\ T_t\ .\ C_t = P.\ T_t\ .(S_t\ /K_{d,t})$$

where b = root uptake from the topsoil, P = plant factor (constant for a specific plant), T = transpired amount of water, C = solute Cd concentration, S = soil Cd concentration, t = index for topsoil, and K_d = Cd distribution coefficient. Plant uptake from the subsoil is described by:

$$c = P.\ T_s.\ (S_s./\ K_{d,s})$$

where c = root uptake of Cd from the subsoil and s = index for subsoil. Foliar uptake of atmospheric Cd is considered to be a constant for a given plant species in a given environment. The solute Cd concentration, S, is assumed to be governed by two main processes, adsorption onto the solid phase and precipitation. The model accurately predicted the uptake of Cd from sewage sludge treated soil; however, the tested data base was restricted in both the number of crops and soil types.

Models have the potential to predict the accumulation of Cd by plants grown on soils treated with sewage sludge and provide valuable indications of areas in which processes are not fully understood and where key parameters are lacking.

Soil Extractants

Diagnostic soil testing procedures enable empirical approximations of Cd bioavailability to be made. Such tests usually require only the measurement of one variable and allow comparisons to be made across a wide range of soils. The majority of soil tests for Cd bioavailability require the use of an extracting solution; such solutions can be divided into three groups: dilute acids, chelating agents, and neutral salts. Many of the soil tests using these extractants were originally developed for the diagnosis of micronutrient deficiency problems, e.g., the DTPA soil test. Care must be taken when applying such tests to element sufficiency studies.[108,109]

A great deal of research effort has been put into finding the ideal soil extractant for predicting the bioavailability of Cd from sludge-amended soils. The results indicate that the use of the neutral salts may be most suitable.[108,110–115] Hani and Gupta recommended the use of 0.1 *M* $NaNO_3$ as an extractant.[112,113] This test has the disadvantage of extracting less Cd than the more commonly used 0.05 *M* $CaCl_2$ test, but has the advantage of causing less vapor phase interference problems during analysis by electrothermal atomization atomic absorption spectrometry. Despite the analytical problems of the 0.05 *M* $CaCl_2$ test, it has been quite widely used and found to produce satisfactory results.[114–116]

DIETARY EXPOSURE TO CD FROM FOOD CROPS

Introduction

In this section the dietary exposure of humans to Cd arising from the ingestion of food crops is examined. Concentrations of Cd in a number of foods are reviewed and placed in the context of the total Cd exposure. A comparison of the methods used to assess Cd exposure is presented. A brief review of the speciation of Cd in foods is given together with an examination of the potential effects of speciation on the bioavailability of Cd in foods.

Cadmium and Health

The incidence of acute toxicity is rare and Cd exposure through the diet is predominantly a chronic problem. The principal feature of its toxic action is the very long body retention time, implying that the concentration of Cd in target organs will increase with time. Cadmium is therefore classed as a cumulative toxin. The kidney, or more specifically the renal cortex, is the main target organ in which Cd has a mean residence time of 30 or more years.[117] The ratio of the concentration of Cd in the renal cortex to that in the whole kidney is 1:1.25.[118] A wide variety of estimates for the body retention time have been made, these range from 16 to 30 years.[119]

The fate of Cd in the body is determined by two factors:

- the concentration of Cd binding ligands in an organ
- the stability of the resulting complex[120]

A renal cortex concentration of 200 $\mu g\ g^{-1}$ fwt has been shown to induce renal dysfunction which may be manifested in a variety of ways, the most common of which is proteinuria.[121] Metabolic models may be used to determine the daily exposure necessary to give this critical renal concentration of Cd. The PCC-10, or population critical concentration, is the Cd concentration in the renal cortex above which symptoms of renal dysfunction are expected to occur in 10% of the exposed population.[118] The rationale behind the PCC-10 has been questioned, as the toxicity of Cd is influenced by a wide variety of factors, both intrinsic and extrinsic. Cadmium toxicity is subject to some complication as it would seem that certain species are more bioavailable than others and may therefore result in different concentrations in the target organs.[122–124] Extrinsic factors, such as the trace element composition of the diet, have, in laboratory animals, been shown to decrease the effects of Cd toxicity; an extensive review of this aspect is given by Fox.[125] Cadmium toxicity is enhanced by deficiencies of Zn, Cu, Se, Fe, and Ca. The fiber content of the diet also has an antagonistic effect upon its action; Cd appears to be bound to a wide variety of dietary fibers.[123] The order of binding efficiencies is cellulose < glucomannan < pectin < sodium alginate < sodium carboxymethyl cellulose < lignin.[123] Despite all of these complicating factors, a PCC-10 of 180 to 220 μg g^{-1} fwt in the renal cortex is still used as the threshold concentration.[126]

The pathway via which the Cd enters the body is very important, as different routes will lead to differential rates of entry into metabolic pathways. Bennett defines the exposure to a substance as a combination of the concentration of that substance and the time for which it is at the target point.[15] The speciation of Cd may play a crucial role in the relative risks associated with a given exposure. The main sites of Cd exposure are the gastrointestinal tract and the pulmonary system. The main sources of exposure are the diet and the atmosphere. Concentrations of Cd in ambient air remain relatively low, ~3 ng kg^{-1} at rural sites in the U.K.,[127] and so the diet remains the primary source of exposure to those members of the population who do not smoke or who are not occupationally exposed.[128] A feature of exposure to the pulmonary system is that it absorbs Cd far more efficiently than the GI tract, 15 to 30% compared with 4 to 6%.[17] Absorption of Cd by the GI tract is highly variable, research on human subjects giving a mean of 4.6 ± 4.0%.[129]

A number of exposure limits have been proposed by several organizations, most of which are based upon that issued by the FAO/WHO in 1972. The FAO/WHO makes its recommendations on the basis of a decision-making structure, the product of which is the allocation of an unconditional, conditional, or temporary acceptable daily intake (ADI). The allocation of an ADI for Cd was considered to be unsuitable for a variety of reasons, one of which was the cumulative nature of the dose. Therefore the term provisional tolerable weekly intake (PTWI) was used, implying by the word "provisional" the tentative nature of the evaluation and by "tolerable" the inference that any exposure is not wholly acceptable.

A PTWI of 400 to 500 μg was proposed on the assumption that all of this exposure came from the diet, that 200 μg g^{-1} fwt in the renal cortex was the critical concentration, that absorption by the GI tract was 5% efficient, and that daily excretion was 0.005% of the total body burden. Care should be taken when applying this PTWI to models, as this is only an assessment of exposure and not of dose. The actual dose is dependent upon many factors, including metal speciation and the exposure route. Another feature of the PTWI is that it is based upon the body of the average Caucasian and as such may be too high for populations with a smaller mean body size. For example, the PTWI for the Japanese population is 325 μg.[121] This corresponds to a maximum exposure of 1 μg Cd kg^{-1} body weight per day.

Concentrations in Foods and Total Diet Studies

The diet generally exposes people in the European Community to between 18 and 48 μg Cd day^{-1}, of which only about 5% is from the consumption of water.[17] Given that the diet is the most important route of Cd exposure, an assessment of the sources of contamination to agroecosystems is necessary. Cadmium contamination of the soil is an important process as all crops will take up Cd from

Table 11. National Means for Dietary Exposure

Country	Intake (mg Cd week^{-1})
West Germany	0.40
West Germany*	0.20
Poland*	0.13
Japan	0.27
New Zealand	0.11
Australia	0.15
Belgium	0.35
Denmark	0.21
Italy	0.38
U.S.	0.23
Britain	<0.15

Note: Data for children signified by *.

From Sherlock, J. C. *Experientia* 40:152-156 (1984).

the soil; however, the degree to which Cd of plant origin impinges upon the diet of both people and livestock depends largely upon which component of the plant or animal is consumed. All of the data quoted in this section are from assessments which have employed standard diet techniques to quantify dietary exposure. The dietary exposure to Cd varies quite considerably between countries, as can be seen from Table 11.

Dietary exposure to Cd has three basic components:

- total mass of food consumed,
- composition of the diet, i.e., how much of the daily intake is derived from what food classes, and
- the Cd concentration of the food classes.

The qualitative and quantitative aspects of diet composition will therefore produce an exposure profile; the differences in this exposure profile create differentials in dietary exposure. As an example of this phenomena, it is useful to compare the dietary exposure of four countries, Finland, West Germany, Japan,[131] and Britain[132] (see Table 12).

The most striking feature of these data is the very low overall exposures from the Finnish and British diets when compared to those from the West German and Japanese diets. In terms of the quantities of food consumed, the West German and Finnish diets are very similar, yet the exposure from the Finnish diet is over three times lower. Consumption of cereal products is approximately the same but cereals consumed in West Germany contain a higher mean Cd concentration, which leads to an added exposure of 4.4 μg Cd day^{-1} from this source. The lower than average Cd concentration of foods consumed in Finland applies across a wide range of food classes.[133] Exposure from vegetables is very high in the West German profile due to a higher overall consumption and concentration. It is harder to make such direct comparisons between the data from Britain and the other three countries due to a different classification of the food groups.

Table 12. Cadmium Exposure Profiles from Finland, West Germany, Japan, and Britian

	Exposure (μg Cd day^{-1})			
	Finland	Germany	Japan	Britain
Cereals	6.6 (267)	11.0 (254)	17.6 (503)	5.0 (240)
Roots and tubers	2.4 (238)	10.4 (221)	2.1 (72)	<2.0 (159)
Sugars and honey	0.5 (109)	0.4 (122)	0.5 (73)	<1.0 (95)
Pulses	0.1	0.1	0.1	—
Nuts and oilseeds	0.0 (2)	0.1 (12)	2.3 (32)	—
Vegetables	1.4 (87)	15.4 (188)	10.3 (299)	2.0 (157)
Fruits	0.3 (219)	3.0 (287)	0.9 (178)	1.2 (84)
Meat and offals	2.0 (169)	4.3 (268)	1.0 (82)	2.3 (138)
Eggs	0.1 (29)	0.6 (47)	0.5 (45)	—
Fish and seafood	0.7 (78)	1.2 (27)	21.4 (239)	<0.2 (15)
Milk	1.1 (711)	1.2 (329)	0.7 (136)	<2 (360)
Oils and fats	0.7 (60)	1.2 (80)	0.4 (40)	<1.0 (90)
Spices	0.0 (1)	0.4 (2)	0.0 (2)	—
Stimulants	0.4 (41)	1.1 (28)	0.2 (9)	—
Alcoholic beverages	0.2 (180)	1.5 (489)	0.6 (160)	—
Total	15.0 (2191)	50.0 (2354)	60.0 (1870)	19.0 (1456)

Note: Figures in parentheses give the data for the mass of food consumed from each food class. For Britain all data for vegetables other than potatoes given in the vegetable line and data for potatoes are in the roots and tubers line.

From Louekari, K. and S. Salminen. *Food Add. Contam.* 3:355–362 (1986).

The average Japanese diet gives a Cd exposure higher than both the Finnish and the West German diets, even though the mass of food consumed is lower. The three main sources of Cd exposure are cereals, vegetables, and fish/ seafood, the latter giving an exposure of 21.4 μg. This enhanced exposure to Cd is due to both the high consumption of foods of this type and also to their higher concentration of Cd. These differences in exposure profiles clearly have implications for standard setting both intra- and internationally. For example, a higher Cd exposure could be indicative of a different and more susceptible diet composition and not of higher levels of environmental Cd contamination. The mean body size of the exposed population is a pertinent factor as it implies that the mass of the target organ may differ and the mean Cd concentration will not be the same for a given exposure. Therefore, the PTWI for Japan is 325 μg Cd, which is 75 μg lower than that for people in the U.S.

The foods that should receive the most attention are those that form the largest component of the total intake in terms of grams of food per week. In the "typical" western European diet, the two most important groups are cereal products and vegetables. The consumption of cereals accounts for 30 to 40% of the total dietary exposure to Cd of people in the European Community.[17]

Most of the data presented in this section have only referred to food from sources considered to be uncontaminated by Cd; however, soil contamination may arise from several sources, one of which is treatment with sewage sludges. It is necessary to be able to target those foods that will be affected by this land management/waste disposal practice and to assess the degree to which they will be impacted.

Different components of the exposure profile will respond in varying ways to an increase in the soil Cd concentrations; thus, the relative importance of the specific components of the diet will also change. If a component of the diet contributes to the overall exposure to a large extent but does not markedly respond to a change in the soil concentration, then it may be said to have a buffered response. The term "buffered" is used in preference to "insensitive," as the exposure from these components is ameliorated by the position of the component in the food chain[134,135] or by virtue of the part of the component consumed (e.g., fruits). However, if a component of the diet contributes significantly to the overall exposure and markedly responds to a change in the soil Cd concentration, then it may be said to have a sensitive response. Using this relationship between changes in the soil concentration and dietary exposure, the components of the diet may be divided into groups. In this example the components of the diet correspond to those used by the Food and Drug Administration in the U.S.[136]

Sensitive	Intermediate	Buffered
Grain & cereals	Dairy products	Meat & fish
Potatoes	Legumes	Garden fruits
Root vegetables		Fruits
Leafy vegetables		Oils & fats
		Beverages
		Sugars & adjuncts

Differential responses to changes in soil Cd concentrations exist within these groups, and their classification will vary depending upon the exposure profile of the country concerned. Leafy vegetables, although not responsible for a large component of the dietary exposure, are placed in the sensitive group as they respond very markedly to increases in soil Cd concentration and so have the potential to raise the total exposure. The inclusion of meats and fish in the buffered group is based upon the assumption that the consumption of offals and shellfish continue at their generally low levels. It should be noted that the above classification will only apply to the average consumer; people who have excessive consumption of certain foods may well have exposures well above the average.[137,138]

The Assessment of Exposure

In order to assess the effect of dietary Cd to the consumer, some measure of the degree of exposure is necessary. A number of different methods exist, but they do not always give the same exposure for a given situation. The four main methods of exposure assessment are

- the standard or total diet study,
- the duplicate meal,
- fecal analysis, and
- diary studies.

The standard method is used in the majority of national surveys, such as the Survey of Household Food Consumption and Expenditure in the U.K. and the Food and Drug Administration's Compliance Program in the U.S. This technique involves an assessment of the mean consumption of foods from a variety of food classes, followed by their chemical analysis to arrive at a mean Cd concentration.[139] By multiplying together the mean consumption and the mean Cd concentration of a certain class, a measure of the dietary exposure from this class is obtained. The main problem with this method is that although it gives an indication of the typical dietary exposure to Cd, it does not enable an assessment of the exposure of critical groups to be made. The interpretation of exposures from samples quoted to have Cd concentrations less than the limit of detection is also a potential source of error.

The duplicate diet and diary study methods can be used to assess the exposure to Cd of a more specific group of people. They are also useful if a particular component of the diet is under examination. The duplicate diet method requires the preparation of an extra meal or component thereof for each meal prepared in a given household; the meal/component is then analyzed. Although this technique may well provide a more targeted data set, it does require a considerable investment in the form of analytical effort. In an attempt to examine the dietary intake of Cd by a small population living in an area with a very high soil Cd concentration (Shipham, U.K.) the duplicate diet and dietary record studies of exposure were compared. The duplicate diet study was found to underestimate the exposure predicted by the dietary record studies.[140] This difference has often been observed and is thought to be related either to a problem with the use and interpretation of detection limits during the analytical phase of the assessment[128,141] or due to a slight change in the dietary habits of the participants.[142,143] This discrepancy, as found during studies of Shipham residents, is shown in Table 13.

The assessment of dietary exposure using fecal analysis is less frequently employed than either the duplicate meal or standard diet methods. The main problem with this method is the high variability in the absorptive capacity of the GI tract.[129]

A more cost effective means of obtaining data for groups of people who may be said to be most at risk has been proposed by Coomes et al.,[144] who found that a linear relationship exists between the arithmetic mean of consumption and specific measures of extreme consumption of a particular food. In this method, consumption greater than or equal to the 90th percentile of the consumption frequency distribution is used as a measure of extreme consumption. At the 95th percentile the mass of food consumed in any particular food is given by the equation:

Table 13. Comparison of Exposure Assessment Techniques used at Shipham, U.K.

	Mean	Range
Diary estimate	0.25	0.14 – 0.52
Duplicate diet study	0.20	0.04–1.08
National average estimate based on total diet study	0.14	0.09 – 0.18

Note: All units are mg Cd wk^{-1} $person^{-1}$.

From Sherlock, J. C. and B. Walters. *Chem. Ind.* 13:505–508 (1983).

$$Y = 2.0X + 111$$

where Y = consumption at the 95th percentile and X = mean consumption.

In this way data can be taken from total diet studies, such as the National Food Survey, and used to predict the enhancement in Cd exposure produced by extreme consumption of a particular food. The main disadvantages of this approach stem from the fact that it cannot be used to predict extreme exposure from combinations of foods.[144]

In light of the concern over dietary exposure, several attempts have been made to restrict the dietary exposure to Cd. Most of these aim to control the levels of Cd in the soil and sludges;[145,146] an alternative approach is to propose limits on the Cd concentrations in specific food groups.[24,53] In the FRG a set of proposals known as the Guidelines '79 were proposed to limit Cd, Pb and Hg concentrations in foods;[147] in the Netherlands the Dutch Food and Drugs Act has set maximum permissible Cd concentrations in foods. The Dutch recommendations were studied by the Landbouw-Advies-Commissie Milieukritische Stoffen (LAC) which attempted to relate them to soil Cd concentrations; it was concluded that their scientific basis was limited.[53] Maximum concentrations in a variety of food groups are shown in Table 14. The threshold concentrations of Cd in these food groups are well below those that would produce overt phytotoxic effects in even the most Cd-sensitive members of the groups.

The use of exposure commitment assessment for examining the pathways of contaminants to people is based upon the commitment method as applied by the United Nations Scientific Committee on the Effects of Atomic Radiation (UNSCEAR) to the study of radiation exposure.[14] The basic method has been developed to cover several contaminants,[18] of which Cd is one. The assumptions of the method include that pollutant transfers follow first order kinetics and that risk is directly proportional to concentration.

The assessment procedure begins with the identification of sources and pathways of the pollutant. Figure 9, taken from MARC Report number 26, shows the pathways at background levels, to which the majority of the population are exposed.[17] Increased exposure from the application of sewage sludge to arable

Table 14. Guidelines for Concentrations of Cd in Foods

Food Group	Maximum Permissible Concentration (mg kg^{-1} fwt)	
	FRG	Netherlands
Cereals	0.10	0.15
Potatoes	0.10	0.10
Green vegetables	0.10	—
Sprout vegetables	0.10	—
Fruits	0.10	—
Root vegetables	0.05	—
Pomaceous fruits	0.05	—
Stone fruits	0.05	—
Berries	0.05	—
Lettuce, endive, spinach	—	0.20
Leek, carrot	—	0.20
Curly kale	—	0.10
Mushrooms	—	0.10
Cucumbers, gherkins	—	0.03
Beetroot, sprouts, cauliflower, headed cabbage, tomatoes, paprika, onions, shallots, pulses	—	0.10

From de Boo, W. *Toxicol. Environ. Chem.* 27:53–63 (1990).

soils is considered to be so negligible as to be not worthy of inclusion, as it only represents a small input at the level of the nation state. The situation at the local level is considerably different, with sludges accounting for up to 90% of the total Cd input to a given soil.[14]

Although the exposure commitment assessment technique is a useful means by which to assess the behavior and possible risks associated with Cd in the large scale, it is not well suited to the examination of Cd exposure from sewage sludge treatment of arable soils. The heterogeneous distribution of Cd exposures from this source are better suited to a less general or large-scale scenario. A smaller unit of study, perhaps analogous to a catchment, may be more appropriate.

In addition to the problems of scale, there are other problems that are principally associated with the calculation of transfer factors between compartments. Bennett[14] identifies four sources of error:

- nonlinearity,
- synergistic, antagonistic, or additive reactions between the contaminant and other variables,
- speciation, and
- compartment heterogeneity.

Cadmium can be shown to exhibit all of these characteristics. It would appear that Cd speciation affects the risk associated with a given exposure to the human consumer. If proven, this would contradict the basic assumption that risk is directly proportional to concentration.

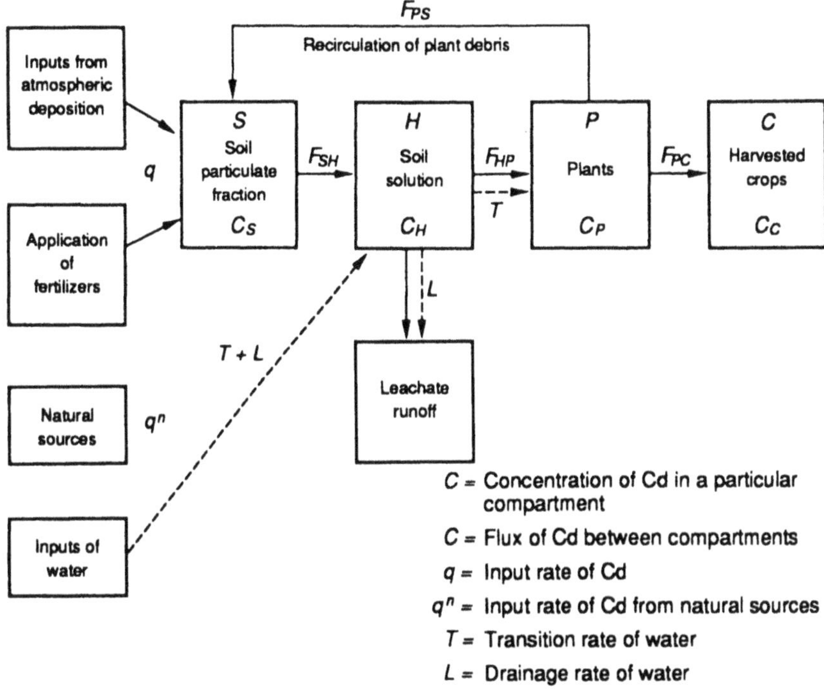

FIGURE 9. Cadmium transfers in agroecosystems. (From Hutton, M. "Cadmium in the European Community," MARC Report No. 26. [London: University of London, 1982]. With permission.)

The United States Environmental Protection Agency (USEPA) has used a model to relate sludge application rates and dietary exposure. An estimated PTWI of 525 μg was calculated independently of that of the WHO/FAO; in order to conform with the more widely used limit, the recommended maximum daily exposure was lowered to 70 μg Cd per day. The median dietary exposure in the U.S. in 1974 was 40 μg Cd per day; the model was therefore aimed at predicting the application rate that would increase the exposure by 30 μg d^{-1}.

To establish a dose response function for the soil-plant system, the model uses data describing the relationship between sludge application rates to soil and plant Cd concentrations. In order to assess the concentration of Cd in the impacted food classes necessary to increase the exposure to that which is the maximum permissible, a multiplication factor can be applied to existing concentrations; the factor is calculated using the equation below:

$$MF = (X_1 - Y)/(X - Y)$$

where MF = the multiplication factor, X = μg Cd per day in diet at the present time, X_1 = projected μg Cd per day in the diet, and Y = exposure from food classes not impacted by the application of Cd to the soil.[148] The contribution of the particular crop to the total exposure is then taken from the Compliance Programme of the Food and Drug Administration, multiplied by the factor given in the equation, and used to assess the application rate necessary to elevate the overall dietary exposure from this source. Not all crops respond in the same way to increasing Cd concentrations in the soil, and so the integrated response by all components of the diet has to be calculated. The model takes soil pH into account, an acid soil is one with a pH of between 5.5 and 5.7 and a neutral soil is one with a pH of between 6.1 and 6.4. Two types of diet, "normal" and lacto-ovo-vegetarian, were compared; the latter showed the most sensitive response to soil Cd loading. Using the basic model, a number of scenarios can be simulated in order to assess Cd exposure, given a variety of different factors. Naylor and Loehr give three possible scenarios:[149] Scenario 1 — a person gets his or her entire diet from soils to which sewage sludge has been applied. The soil has a pH of 7 and all crops respond to the increase in the Cd concentration as does lettuce; Scenario 2 — the soil pH is greater than 7 and the person consumes only grain products that were grown on sludge-amended soils; and Scenario 3 — sludge is only applied to soils used for the production of animal feed crops.

In the U.K., the risks associated with contaminated soils are assessed through a system of generic trigger concentrations.[150,151]

The Speciation of Cd in Foods

The bioavailability of Cd is influenced by a wide variety of intrinsic and extrinsic factors. Cadmium speciation is an intrinsic factor and as such may well play a key role in governing the degree of bioavailability and, therefore, the effect of a given exposure.[152]

In foods, the predominant Cd species are those bound to low molecular weight proteins called metallothioneins or phytochelatins. In meat and meat products Cd has been shown to be bound to Class I metallothioneins;[153] in plant-based foods Cd is bound to phytochelatins or phytometallothioneins (PC), Class III metallothioneins. An early definition of metallothionein is given in a report from the "First International Meeting on Metallothionein and Other Low Molecular Weight Metal-Binding Proteins" in which the following five distinguishing features were recognized:[154]

- molecular weight 6 to 7 kD,
- high metal content,
- characteristic amino acid composition (high cysteine content, no aromatic amino acid, or histidine)
- optical features characteristic of metal thiolates, and
- unique amino acid sequence.

Nomenclature has been expanded to include metal-binding proteins in plants; therefore, a metallothionein is any polypeptide that resembles equine metallothionein in several of its features. Within this definition there exist three classes:[155]

Class I: polypeptides with locations of cysteine closely related to those in equine metallothionein.
Class II: polypeptides with locations of cysteine only distantly related to those of equine renal metallothionein.
Class III: atypical nontranslationally synthesized metal thiolate polypeptides such as cadystin and phytometallotionein or phytochelatin.

Emphasis will only be placed upon the Class III metallothioneins in foods, which shall be referred to as phytochelatins.

The bioavailability and fate of Cd after ingestion may be determined, at least in part, by its speciation.[152,156,157] Cd bound to phytochelatins represents the dominant form of exposure, as this is likely to be the dominant species in vegetables, grain, etc. Although Cd in foods may be bound to other proteins, many of these will be denatured during the cooking process, unlike metallothioneins which are relatively heat stable.[153,157,158] Comparisons of the fate of Cd fed to mice as $CdCl_2$ or as Cd-thionein suggest that the final body distribution of Cd may differ with speciation.[122] A greater proportion of the dose administered as Cd-thionein accumulated in the most sensitive organs, the kidneys, whereas that administered as $CdCl_2$ accumulated in the liver; the concentration of Cd in the kidney was similar irrespective of the species administered.[159,160] A similar experiment to that conducted by Cherian, but using wheat grain as the source of Cd, has not repeated these observations.[161]

A number of studies of phytochelatins and other Cd-binding proteins in vascular plants have been made, including studies on wheat grain,[161] cabbage,[162] soybean,[163] *Agrostis gigantea*,[164] rice,[165] tomato,[166] and lettuce.[167] The available data are summarized in Table 15.

The phytochelatins isolated from tomato, maize, and cabbage are similar to animal or Class I MT in four respects:[168]

- synthesis is stimulated by Cd exposure
- high cysteine content
- do not have any aromatic amino acids
- ~10 kD characteristic produced by gel-filtration studies

The main differences between the phytochelatins from the plant tissues and the MT is that they differ in their elution characteristics at high ionic strength. Table 16 shows the differences in amino acid composition between the phytochelatins in three vascular plants and human MT.

The main differences between phytochelatins and human MT are the higher serine concentration of MT and the higher glutamic acid concentration in the phytochelatins. The differences between the phytochelatins are less pronounced. A further complicating factor is the changes in speciation that occur in the gastrointestinal tract upon ingestion.[170,171]

To conclude, it is widely recognized that the speciation of Cd in foods may well be a factor governing its bioavailability. The precise mechanism(s) in which Cd speciation has an influence is not clear; however, different species have been

Table 15. Cadium Binding Proteins Isolated From Plant Tissues

Sample	Molecular Weight of Cd-Binding Species (Da)	Ref.
Cabbage leaves	1×10^4	162
Lettuce leaves	3.2×10^3	167
Tomato roots	1×10^4	168
Zea mays roots	8 & 3.5×10^3	184
Wheat grain	1.1×10^4	161
Soybean plants	>50, 13.8 & 2.3×10^3	163
Agrostis gigantea roots	3.7×10^3	164
Rice plants	3.31×10^4	165
Tomato roots	1×10^4	166
Bean plants		
Roots	5 & 10×10^3	
Leaves	0.7 & 5×10^3	185

From Jackson, A. P. "The Bioavailability of Cadmium From Sewage Sludge-Amended Soils," Ph.D. thesis, University of London (1990).

Table 16. Amino Acid Composition of Cd-Binding Proteins From Different Sources[168,169]

Amino Acid	Human MT	Cabbage PC	Tomato PC	Maize PC
Cys	32.8	28.9	25.6	40.3
Asx	6.6	4.2	5.4	2.8
Glx	3.3	39.0	53.3	35.1
Gly	8.2	11.1	12.8	10.4
Ser	13.1	1.9	1.9	1.5
Thr	3.3	1.5	0.3	1.0
Pro	3.3	1.7	<LOD	1.6
Ala	11.5	2.1	0.9	1.5
Val	1.6	0.9	<LOD	0.9
Met	1.6	0.7	—	—
Ile	1.6	3.4	<LOD	0.6
Leu	0	1.9	0.2	1.0
Tyr	0	0.9	<LOD	0.5
Phe	0	0	<LOD	0.5
Lys	13.1	1.0	0.5	1.0
His	0	0.8	0.2	0.5
Arg	0	0	<LOD	0.7

Note: PC — phytochelatin; MT — metallothionein. <LOD — less than limit of detection, All values are expressed in mole percent. Amino acid abbreviations are as follows: Cys — cysteine; Asx — aspartic acid; Glx — glutamic acid; Gly — glycine; Ser — serine; Thr — threonine; Pro — proline; Ala — alanine; Val — valine; Met — methionine; Ile — Isoleucine; Leu — leucine; Tyr — tyrosine; Phe — phenylalinine; Lys — lysine; His — histidine; Arg — arginine.

shown to produce different degrees of accumulation in target organs. The speciation of Cd in plants differs from that in animals, and this may alter the metabolic dose derived from the same exposure from these two food sources.

CONCLUSIONS

Cadmium is a nonessential metal that poses a potential hazard to human health through excessive exposure. It has a relatively high soil-plant concentration ratio, and therefore the variables controlling its bioavailability are of paramount importance in determining the concentrations of Cd in food crops.

The residence time for Cd in soils is estimated to be up to hundreds, or possibly more than a thousand years; consequently, care must be taken to minimize the extent to which soils become contaminated. The application of sewage sludges to soils often results in an increase in Cd concentration.

Cadmium bioavailability is controlled by a number of soil variables, including Cd concentration, pH, Ca concentration, adsorptive properties of the soil constituents, and the antagonistic effects of other metals. Soil pH is negatively correlated with crop Cd uptake, and liming soils to raise their pH is generally the most effective management strategy for reducing the biavailability of Cd.

Sewage sludge can be a major source of Cd in the soils to which it is applied, although, on a global scale, it accounts for a much smaller contribution to soils than the inputs of Cd from the atmosphere and phosphatic fertilizers. Apart from increasing the metal concentrations, the addition of sewage sludges changes several soil properties. These include a beneficial physical soil conditioning effect, increased capacity to adsorb metals, and decreased pH. These effects have a major influence on the bioavailability of Cd. It is therefore important to understand the duration of effects in the residual period after the application of sewage sludge. The available evidence indicates either a decrease in the bioavailability of Cd or that it remains fairly constant. However, these conclusions are only based on comparatively short-term investigations, relative to the estimated residence time.

Although the soil-crop concentration ratios for Cd are usually higher than for most other trace metals, the amounts of the metal removed from the soil in crops are relatively small and do not significantly reduce the level of contamination in sludge-treated soils. Other losses of Cd from soils, such as leaching, are negligible.

Of the wide range of soil tests available for predicting the relative availability of trace metals in soils, it would appear that DTPA and dilute neutral reagents such as $CaCl_2$ and $NaNO_3$ are the most generally useful reagents for Cd.

The accumulation of Cd by plants varies between species and cultivars and also between organs within a plant. Cadmium has the characteristic of accumulating in the leaves.

Studies of the speciation of Cd in plant tissues have indicated that the predominant forms are bound to low molecular weight proteins called phytochelatins. There is a possibility that the distribution of the Cd in the body after assimilation may be affected by its speciation.

There is concern that median dietary exposures in the U.S. and several other countries are at levels of around half of the recommended daily exposure to Cd of 70 μg/d. Diets in the former West Germany and in Japan were significantly higher than those in Great Britain and Finland. Various models have been used to determine safe application rates for sewage sludges on agricultural land.

A number of areas of uncertainty exist in the assessment of Cd transfers to the human food chain; these may be addressed by research on the following areas.

- A clarification of the processes and rates by which Cd is lost from the plough layer.
- Atmospheric deposition of Cd. In particular, data are needed to characterize those processes that give rise to the majority of the atmospheric Cd in rural or semi-rural areas.
- A more accurate characterization of the interception of atmospheric Cd by food crops. In particular, data are required to allow such processes to be accurately modeled, e.g., deposition velocities, absorption coefficients, interception fractions, and half lives for the retention of particles on the surfaces of crop plants.
- The development of a comprehensive database of soil-plant concentration ratios. Such a database should cover as many crop and soil types as possible. The data from which such ratios are derived should have been subject to stringent quality assurance procedures. It is essential that the relative contributions of Cd from atmospheric sources and the soil be differentiated and that only Cd from the latter source be used in the calculation of the ratio.
- Such an undertaking would require the quantification of those pathways that contribute to the Cd burden of crops, i.e., resuspension of soil, interception of Cd from the atmosphere, and soil-root transfer.
- A program of work examining the speciation of cadmium in soil, paying particular attention to the soil solution. With the rapid development of on-line detectors for chromatography systems, many of the problems previously encountered in such studies may be resolvable.
- An examination of the relationship between Cd species in the soil solution and their availability for root uptake.
- Characterization of the decomposition of organic matter in sewage sludge-treated soils and the effect of its decay on the bioavailability of Cd through time.
- Further characterization of the speciation of Cd in foods, with particular emphasis on those foods that make a significant contribution to the overall exposure profile.
- An examination of the effect of speciation upon the fraction of Cd absorbed across the gastrointestinal tract.
- A study of the effects of food processing on the concentrations of Cd in food products relative to concentrations in the raw materials. Such a study would enable the concentrations of Cd determined in crops to be adjusted so that they more accurately reflect the concentration in the crop at the point of consumption.

REFERENCES

1. Bowen, H. J. M. *Environmental Chemistry of the Elements* (London: Academic Press, 1979).
2. Alloway, B. J. "Cadmium," in *Heavy Metals in Soils*. Alloway, B. J., Ed. (Glasgow: Blackie & Son, 1990) pp. 100–121.
3. Nriagu, J. O. "Human Influence on the Global Cadmium Cycle," in *Cadmium in the Environment. Part 1: Ecological Cycling*, Nriagu, J. O., Ed. (New York: Wiley-Interscience,1980) pp. 2–12.
4. Chumbley, C. J. and R. J. Unwin. "Cadmium and Lead Content of Vegetable Crops Grown on Land with a History of Sewage Sludge Application," *Environ. Pollut.* 4B:231–237 (1982).
5. Pike, E. R., L. C. Graham, and M. W. Foyden. "An Appraisal of Toxic Metal Residue in the Soils of a Disused Sewage Farm. II," *J. Assoc. Public Anal.* 13:48–63 (1975).
6. Alloway, B. J., A. P. Jackson, and H. Morgan. "The Accumulation of Cadmium by Vegetables Grown on Soils Contaminated from a Variety of Sources," *Sci. Total Environ.* 91:223–236 (1990).
7. Jackson, A. P. and B. J. Alloway. "The Transfer of Cadmium from Sewage Sludge Amended Soils into the Edible Components of Food Crops," *Water Air Soil Pollut.* 57–58:873–881 (1991).
8. Hazen, R. E. and T. J. Kneip. "Biogeochemical Cycling of Cadmum in a Marsh Ecosystem," in *Cadmium in the Environment. Part 1: Ecological Cycling*, Nriagu, J. O., Ed. (New York: Wiley-Interscience, 1980) pp. 399–424.
9. Martin, M. H. and P. J. Coughtrey. "Cycling and Fate of Heavy Metals in a Contaminated Woodland Ecosystem," in *Pollutant Transport and Fate in Ecosystems*, Coughtrey, P. J., M. H. Martin, and M. H. Unsworth, Eds. (Oxford: Blackwell Scientific, 1987) pp. 314–336.
10. Sopper, W. E. and S. N. Kerr. "Cadmium in Forest Ecosystems," in *Cadmium in the Environment. Part 1: Ecological Cycling*, Nriagu, J. O., Ed. (New York: Wiley-Interscience, 1980) pp. 655–667.
11. Yost, K. J. "Environmental Exposure to Cadmium in the United States," in *Cadmium 79*. Edited Proc. 2nd Int. Cadmium Conf. Cannes (London: Metal Bulletin Limited, 1980) pp. 11–20.
12. Yost, K. J. "Cadmium in the Environment and Human Health: An Overview," *Experientia* 40:157–164 (1984).
13. Yost, K. J. and L. J. Miles. "Environmental Health Assessment for Cadmium: A Systems Approach," *J. Environ. Health* 14:285–311 (1979).
14. Bennett, B. G. "Exposure Committment Assessments to Environmental Pollutants; Summary Exposure for Lead, Cadmium and Arsenic," MARC Report No. 23. (London: University of London, 1981).
15. Bennett, B. G. "The Exposure Commitment Method for Pollutant Exposure Evaluation," *Ecotoxicol. Environ. Saf.* 6:363–368 (1982).

16. Bennett, B. G. "Modeling Exposure Routes of Trace Metals from Sources to Man," in *Changing Metal Cycles and Human Health*, Nriagu, J. O., Ed. (Berlin: Springer-Verlag, 1984) pp. 345–356.
17. Hutton, M. "Cadmium in the European Community," MARC Report No. 26. (London: University of London, 1982).
18. Jones, K. C. and B. G. Bennett. "Human Exposure to Environmental Polychlorinated Dibenzo-p-dioxins and Dibenzofurans: An Exposure Commitment Assessment for 2,3,7,8–TCCD," *Sci. Total Environ.* 78:99–116 (1989).
19. Nriagu, J. O. "Global Inventory of Natural and Anthropogenic Sources of Metals to the Atmosphere," *Nature (London)* 279:409–411 (1979).
20. Nriagu, J. O. and J. M. Pacyna. "Quantitative Assessment of Worldwide Contamination of Air, Water and Soils by Trace Metals," *Nature (London)* 333:134–139 (1988).
21. Rauhut, A. "The Cadmium Balance and Cadmium Emissions in the European Community," in *Cadmium 79*. Edited Proc. 2nd Int. Cadmium Conf., Cannes (London: Metal Bulletin Limited, 1980) pp. 80–82.
22. Hutton, M. and C. Symon. "The Quantities of Cadmium, Lead and Mercury Entering the UK Environment from Human Activities," *Sci. Total Environ.* 57:129–150 (1986).
23. Hutton, M. and C. Symon. "Sources of Cadmium Discharge to the UK Environment," in *Pollutant Transport and Fate in Ecosystems*, Coughtrey, P. J., M. H. Martin, and M. H. Unsworth, Eds. (Oxford: Blackwell Scientific, 1987) pp. 223–239.
24. Kloke, A., D. R. Sauerbeck, and H. Vetter. "The Contamination of Plants and Soils with Heavy Metals and the Transport of Metals in Terrestrial Food Chains" in *Changing Metal Cycles and Human Health*, Nriagu, J. O., Ed. (Berlin: Springer-Verlag, 1984) pp. 113–141.
25. Hovmand, M. F., J. C. Tjell, and H. Mosbaek. "Plant Uptake of Airborne Cadmium," *Environ. Pollut.* 30A:27–38 (1983).
26. Nriagu, J. O. "A Global Estimate of Natural Sources of Atmospheric Trace Metals," *Nature (London)* 338:47–49 (1989).
27. Tjell, J. C., T. H. Christensen, and F. Bro-Rasmussen. "Cadmium in Soil and Terrestrial Biota, with Emphasis on the Danish Situation," *Ecotoxicol. Environ. Saf.* 7:122–140 (1983).
28. Jones, K. C., C. J. Symon, and A. E. Johnston. "Retrospective Analysis of an Archived Soil Collection. II. Cadmium," *Sci. Total Environ.* 67:75–89 (1987).
29. Asami, T. "Pollution of Soils by Cadmium," in *Changing Metal Cycles and Human Health*, Nriagu, J. O., Ed. (Berlin: Springer-Verlag, 1984) pp. 95–111.
30. Tjell, J. C. and M. F. Hovmand. "Metal Concentrations in Danish Arable Soils," *Acta Agric. Scand.* 28:81–89 (1978).
31. Tjell, J. C. and T. H. Christensen. "Evidence of Increasing Cadmium Contents of Agricultural Soils," in *Heavy Metals in the Environment* (Edinburgh: CEP, 1985).
32. Sauerbeck, D. "Zur Bedeutung des Cadmiums in Phosphatdungemitteln," *Landbrau* 32:192–197 (1982).
33. McGrath, S. P. "Long-Term Studies of Metal Transfers Following the Application of Sewage Sludge," in *Pollutant Transport and Fate in Ecosystems*, Coughtrey, P. J., M. H. Martin, and M. H. Unsworth, Eds. (Oxford: Blackwell Scientific, 1987) pp. 301–317.

34. Williams, D. E., J. Vlamis, A. H. Pukite, and J. E. Corey. "Metal Movement in Sludge-Amended Soils: A Nine Year Study," *Soil Sci.* 143:124–131 (1987).
35. McGrath, S. P. and P. W. Lane. "An Explanation for the Apparent Losses of Metals in a Long-Term Field Experiment with Sewage Sludge," *Environ. Pollut.* 60:235–256 1989.
36. Davis, R. D. and E. F. Coker. "Cadmium in Agriculture with Special Reference to the Utilization of Sewage Sludge on Land," Water Research Centre, Rep. TR139 Stevenage, Herts (1980).
37. Williams, C. R. and R. M. Harrison. "Cadmium in the Atmosphere," *Experientia* 40:29–36 (1984).
38. Lahmann, E., S. Munari, V. Amicarelli, P. Abbaticchio, and Gabellieri. "Heavy Metals: Identification of Air Quality and Environmental Problems in the European Community," Vol. 1 (Luxembourg: CEC, 1986).
39. Pacyna, J. M., B. Ottar, U. Fanza, and W. Maenhaut. "Long-Range Transport of Trace Elements to Ng Alesund, Spitsbergen," *Atmos. Environ.* 19:857–865 (1985).
40. Hansen, J. A. and J. C. Tjell. "Sludge Application to Land — Overview of the Cadmium Problem," in *Environmental Effects of Organic and Inorganic Contaminants in Sewage Sludge*, Davis, R. D., G. Hucker, and P. L'Hermite, Eds. (Dordrecht: Reidel, 1983) pp. 91–113.
41. Mulla, D. J., A. L. Page, and T. J. Ganje. "Cadmium Accumulations and Bioavailability in Soils from Long-Term Phosphorus Fertilization," *Environ. Qual.* 9:408–412 (1980).
42. Williams, C. H. and D. J. David. "The Accumulation in Soil of Cadmium Residues from Phosphate Fertilizers and Their Effect on the Cadmium Content of Plants," *Soil* 12:86–93 (1976).
43. Rothbaum, H. P., R. L. Goguel, A. E. Johnston, and G. E. G. Mattingly. "Cadmium Accumulations in Soils from Long-Continued Applications of Superphosphate," *J. Soil Sci.* 37:99–107 (1986).
44. Chaney, R. L. "Effective Utilization of Sewage Sludge on Cropland in the United States and Toxicological Considerations for Land Application," in *Proc. 2nd Int. Symp. Land Application of Sewage Sludge* (Tokyo: Association for the Utilization of Sewage Sludge, 1988) pp. 77–105.
45. Mays, D. A. and P. M. Giordano. "Benefits from the Land Application of Municipal Sewage Sludges," (TVA Circular, Tennessee Valley Authority, 1988).
46. Page, A. L., A. C. Chang, and M. El-Amamy. "Cadmium Levels in Soils and Crops in the United States," in *Lead, Mercury, Cadmium and Arsenic in the Environment*, Hutchinson, T. C. and K. M. Meema, Eds. (New York: John Wiley & Sons, 1987) pp. 119–146.
47. Davis, R. D. "Cadmium in Sludges Used as a Fertilizer," *Experientia* 40:117–126 (1984).
48. Davis, R. D., C. H. Carlton-Smith, J. H. Stark, and J. A. Campbell. "Distribution of Metals in Grassland Soils Following Surface Applications of Sewage Sludge," *Environ. Pollut.* 49:99–115 (1988).
49. Bowen, H. J. M. "Soil Pollution," *Educ. Chem.* 72–76 (1975).
50. Kabata-Pendias, A. and H. Pendias. *Trace Elements in Soils and Plants* (CRC Press, Boca Raton, FL, 1984).

51. Jones, K. C. and A. E. Johnston. "Cadmium in Cereal Grains and Herbage from Long-Term Experimental Plots at Rothamsted, UK," *Environ. Pollut.* 57:199–216 (1989).
52. Jackson, A. P. "The Bioavailability of Cadmium from Sewage Sludge-Amended Soils," Ph.D. thesis, University of London, U.K. (1990).
53. de Boo, W. "Cadmium in Agriculture," *Toxicol. Environ. Chem.* 27:53–63 (1990).
54. Henkens, C. H. "Policy Regarding Cadmium Supply in Arable Farming," (Report Consultentschap Bodemaangelegenheden in de landbouw, Wageningen, The Netherlands, 1983).
55. "Cadmium the Burden on the Dutch Environment," Coordinatie — Commissie voor de metingen van Radioactiviteit en Xenobiotische stoffen (CCRX), Ministry of Public Health, Planning and Environment, (The Hague, The Netherlands, 1985).
56. Breimer, T. and K. W. Smilde. "Effect of Organic Fertilizer Rates on the Heavy Metal Contents in the Top Soil of Arable Land," (Proefstation voor de Akkerbouw en de Groenteteelt in de Vollegrond, Lelystad, The Netherlands, 1986).
57. "Draft Basic Document on Cadmium," Ryksinstitut voor Volksgezondheid en Milieuhygiene (RIVM, Bilthoven, The Netherlands, 1987).
58. Dollard, G. J. and T. J. Davies. A Study of the Contribution of Airborne Cadmium to the Cadmium Burdens of Several Vegetable Species (United Kingdom Atomic Energy Authority, Didcot, 1989).
59. Chamberlain, A. C. "Fallout of Lead and Uptake by Crops," *Atmos. Environ.* 17:693–706 (1983).
60. Harrison, R. M. and M. B. Chirgawi. "The Assessment of Air and Soil as Contributors of Some Trace Metals to Vegetable Plants. I. Use of a Filtered Air Growth Cabinet," *Sci. Total Environ.* 83:13–34 (1989a).
61. Harrison, R. M. and M. B. Chirgawi. "The Assessment of Air and Soil as Contributors of Some Trace Metals to Vegetable Plants II. Translocation of Atmospheric and Laboratory-Generated Cadmium Aerosols to and Within Vegetable Plants," *Sci. Total Environ.* 83:35–45 (1989b).
62. Harrison, R. M. and M. B. Chirgawi. "The Assessment of Air and Soil as Contributors of Some Trace Metals to Vegetable Plants. III. Experiments with Field Grown Plants," *Sci. Total Environ.* 83:47–62 (1989c).
63. Dowdy, R. H. and W. E. Larson. "The Availability of Sludge-Bourne Metals to Various Vegetable Crops," *J. Environ. Qual.* 4:278–282 (1975).
64. Keefer, R. F., R. N. Singh, and D. J. Horvath. "Chemical Composition of Vegetables Grown on an Agricultural Soil Amended with Sewage Sludges," *J. Environ. Qual.* 15:146–152 (1986).
65. Kim, S. J., A. C., Chang, A. L. Page, and J. E. Warneke. "Relative Concentrations of Cadmium and Zinc in Tissue of Selected Food Plants Grown on Sludge-Treated Soils," *J. Environ. Qual.* 17:568–573 (1988).
66. Davies, B. E. and H. M. Crews. "The Contribution of Heavy Metals in Potato Peel to Dietary Intake," *Sci. Total Environ.* 30:261–264 (1983).
67. Kampe, W. "Cd and Pb in the Consumption of Foodstuffs Depending on Various Contents of Heavy Metals. Results of Total Diet Studies," in *Processing and Use of Sewage Sludge*, L'Hermite, P. and H. Ott, Eds. (Dordrecht: Reidel, 1984) pp. 334–349.

68. Campbell, W. T., R. W. Miller, J. H. Reynolds, and T. M. Schreeg. "Alfalfa, Sweetcorn and Wheat Responses to Long-Term Application of Municipal Waste Water to Cropland," *J. Environ. Qual.* 12:243–249 (1983).
69. Webber, M. D. and T. L. Monks. "Cadmium Concentrations in Field and Vegetable Crops. A Recommended Maximum Cadmium Loading to Agricultural Soils," in *Environmental Effects of Organic and Inorganic Contaminants in Sewage Sludge*, Davis, R. D., G. Hucker, and P. L'Hermite, Eds. (Dordrecht: Reidel, 1983) pp. 130–137.
70. Bingham, F. T. and A. L. Page. "Cadmium Accumulation by Economic Crops," in *International Conference on Heavy Metals in the Environment* (Edinburgh: CEP, 1975) pp. 433–442.
71. Rundle, H. L. and C. Holt. "Output of Heavy Metals in the Produce from a Historic Sewage Farm," in *International Conference on Heavy Metals in the Environment 1* (Edingburgh: CEP, 1983) pp. 354–357.
72. Davis, R. D., J. H. Stark, and C. H. Carlton-Smith. "Cadmium in Sludge Treated Soil in Relation to Potential Human Dietary Intake of Cadmium," in *Environmental Effects of Organic and Inorganic Contaminants in Sewage Sludge*, Davis, R. D., G. Hucker, and P. L'Hermite, Eds. (Dordrecht: Reidel, 1983) pp. 137–147.
73. Micco, C., R. Onori, M. Miraglia, L. Gambelli, and C. Brera. "Evaluation of Lead, Cadmium, Chromium, Copper and Zinc by Atomic Absorption Spectroscopy in Darum Wheat Milling Products in Relation to the Percentage of Extraction," *Food Add. Contam.* 4:429–435 (1987).
74. CEC (1978) "Criteria (Dose/effect Relationships) for Cadmium," Report of a working group of experts prepared for the Commission of the European Communities, Directorate-General for Social Affairs, Health and Safety Directorate. (Pergamon, Oxford, 1978).
75. Lorenz, H., H. D. Ocker, J. Bruggemann, P. Weigert, and M. Sonneborn. "Content of Cadmium in Cereals of the Past Compared with the Present," *Liebens-Unter. Uno-Forsch.* 183:402–405 (1986).
76. Adriano, D. C. *Trace Elements in the Terrestrial Environment* (Berlin: Springer-Verlag, 1986).
77. Bingham, F. T. "Bioavailability of Cd to Food Crops in Relation to Heavy Metal Content of Sludge-Amended Soil," *Environ. Health Perspect.* 28:39–43 (1979).
78. Sommers, L., V. Van Volk, P. M. Giordano, W. E. Sopper, and R. Bastian. "Effects of Soil Properties on Accumulation of Trace Elements by Crops," in *Land Application of Sludge. Food Chain Implications*, Page, A. L., T. J. Logan, and J. A. Ryan, Eds. (Chelsea, MI: Lewis Publishers, 1987) pp. 5–24.
79. Andersson, A. and K. O. Nilsson. "Influence of Lime and Soil pH on Cd Availability to Plants," *Ambio* 3 (5):198–200 (1974).
80. Christensen, T. H. "Cadmium Soil Sorption at Low Concentrations: I. Effect of Time, Cadmium Load, pH and Calcium," *Water Air Soil Pollut.* 21:105–114 (1984a).
81. Alloway, B J., I. Thornton, G. A. Smart, J. C. Sherlock, and M. J. Quinn. "Metal Availability," *Sci. Total Environ.* 75:41–69 (1988).
82. Christensen, T. H. "Cadmium Soil Sorption at Low Concentrations: II. Reversibility, Effect of Changes in Solute Composition and Effect of Soil Ageing," *Water Air Soil Pollut.* 21:115–125 (1984b).

83. Christensen, T. H. "Cadmium Soil Sorption at Low Concentrations: III. Prediction and Observation of Mobility," *Water Air Soil Pollut.* 26:255–264 (1985a).
84. Christensen, T. H. "Cadmium Soil Sorption at Low Concentrations: IV. Effect of Waste Leachates on Distribution Coefficients," *Water Air Soil Pollut.* 26:265–274 (1985b).
85. Christensen, T. H. "Cadmium Soil Sorption at Low Concentrations: V. Evidence of Competition by Other Heavy Metals," *Water Air Soil Pollut.* 34:293–303 (1987a).
86. Christensen, T. H. "Cadmium Soil Sorption at Low Concentrations: VI. A Model for Zinc Competition," *Water Air Soil Pollut.* 34:305–314 (1987b).
87. Neal, R. H. and G. Sposito. "Effects of Soluble Organic Matter and Sewage Sludge Amendments on Cadmium Sorption by Soils at Low Cadmium Concentrations," *Soil Sci.* 142:164–172 (1986).
88. Cabrera, D., S. D. Young, and D. L. Rowell. "The Toxicity of Cadmium to Barley Plants as Affected by Complex Formations with Humic Acid," *Plant Soil* 105:195–204 (1988).
89. Tyler, L. D. and M. B. McBride. "Influence of Ca, pH and Humic Acid on Cd Uptake," *Plant Soil* 64:259–262 (1982).
90. Giordano, P. M., D. A. Mays, and A. D. Behel. "Soil Temperature Effects on Uptake of Cadmium and Zinc by Vegetables Grown on Sludge-Amended Soil," *Environ. Qual.* 8:233–236 (1979).
91. Siriratpuriya, O., E. Vigerust, and A. R. Selmer-Olsen. "Effect of Temperature and Heavy Metal Application on Metal Content in Lettuce," *Meld. Nor. Landbrukshogsk.* 64(7) (1985).
92. Brummer, G. W. "Heavy Metal Species, Mobility and Availability" in *"The Importance of Chemical Speciation"* in *Environmental Processes,* Bernhard, M., F. E. Brinckman, and P. J. Sadler, Eds. (Berlin: Springer-Verlag, 1986) pp. 169–192.
93. Corey, R. B., L. D. King, C. Lue-Hing, D. S. Fanning, J. J. Street, and J. M. Walker. "Effects of Sludge Properties on Accumulation of Trace Elements by Crops," in *Land Application of Sludge. Food Chain Implications*, Page, A. L., T. J. Logan, and J. A. Ryan, Eds. (Chelsea, MI: Lewis Publishers, 1987) pp. 25–27.
94. Juste, C. and P. Solda. "Heavy Metal Availability in Long-Term Experiments," in *Factors Influencing Sludge Utilization Practices in Europe*, Davis, R. D. and P. L'Hermite, Eds. (Dordrecht: Reidel, 1986) pp. 13–23.
95. Bidwell, A. M. and R. H. Dowdy. "Cadmium and Zinc Availability to Corn Following Termination of Sewage Sludge Applications," *J. Environ. Qual.* 16:438–442 (1987).
96. Hinesly, T. D., Y. E. L. Ziegler, and G. L. Barrett. "Residual Effects of Irrigating Corn with Digested Sewage Sludge," *Environ. Qual.* 8:35–38 (1979).
97. Kelling, K., D. R. Keeney, L. M. Walsh, and J. A. Ryan. "A Field Study of the Agricultural Use of Sewage Sludge: III. Effect on Uptake and Extractability of Sludge-Borne Metals," *J. Environ. Qual.* 6(4):352–358 (1977).
98. Dowdy, R. H., W. E. Larson, J. M. Titrud, and J. J. Latterell. "Growth and Metal Uptake of Snap Beans Grown on a Sewage Sludge-Amended Soil: A Four Year Field Study," *J. Environ. Qual.* 7:252–257 (1978).
99. Larsen, K. E. "Cadmium Content in Soils and Crops After Use of Sewage Sludge," in *Utilization of Sewage Sludge on Land: Rates of Application and Long-Term Effects of Metals*, Berglund, S., R. D. Davis, and P. L'Hermite, Eds. (Dordrecht: Reidel, 1984) pp. 157–165.

100. Lonsjo, H. "Isotope-Aided Studies on Crop Uptake of Cadmium Under Swedish Field Conditions," in *Utilization of Sewage Sludge on Land: Rates of Application and Long-Term Effects of Metals*, Berglund, S., R. D. Davis, and P. L'Hermite, Eds. (Dordrecht: Reidel, 1984) pp. 135–145.
101. Jackson, D. and A. D. Smith. "Generalised Models for the Transfer and Distribution of Stable Elements and Radionuclides in Agricultural Systems," in *Pollution Transport and Fate in Ecosystems*, Coughtrey, P. J., M. H. Martin, and M. H. Unsworth, Eds. (Oxford: Blackwell Scientific, 1987) pp. 385–402.
102. Rappaport, B. D., D. C. Martens, T. W. Simpson, and B. D. Reneau, Jr. "Prediction of Available Zinc in Sewage Sludge-Amended Soils," *J. Environ. Qual.* 15:133–136 (1986).
103. Browne, C. L., Y. M. Wong, and D. R. Buhler. "A Predictive Model for the Accumulation of Cadmium by Container-Grown Plants," *J. Environ. Qual.* 13:184–188 (1984).
104. Poelstra, P., M. J. Frissel, and El-Bassam. "Transport and Accumulation of Cd Ions in Soils and Plants," *Z. Pflanzenernaehr. Bodenkd.* 142:848–864 (1979).
105. Mullins G. L., L. E. Sommers, and S. A. Barber. "Modelling the Plant Uptake of Cadmium and Zinc from Soils Treated with Sewage Sludge," *Soil Sci. Soc. Am. J.* 50:1245–1250 (1986).
106. Barber, S. A. *Soil Nutrient Bioavailability: A Mechanistic Approach* (New York: Wiley-Interscience, 1984).
107. Christensen, T. and J. Tjell. "Interpretation of Experimental Results on Cadmium Crop Uptake from Sewage Sludge Amended Soil," in *Processing and Use of Sewage Sludge*, L'Hermite, P. and H. Ott, Eds. (Dordrecht: Reidel, 1983) pp. 358–369.
108. Jackson, A. P. and B. J. Alloway. "The Bioavailability of Cadmium to Lettuce and Cabbage in Soils Previously Treated with Sewage Sludges," *Plant Soil* 132:179–186 (1991).
109. O'Connor, G. A. "Use and Misuse of the DTPA Soil Test," *J. Environ. Qual.* 17:715–718 (1988).
110. Alloway, B. J., A. R. Tills, and H. Morgan. "The Speciation and Availability of Cadmium and Lead in Polluted Soils," in *Trace Substances and Environmental Health XVIII*, Hemphill, D. D., Ed. (University of Missouri, Columbia, 1984), pp 187–201.
111. Alloway, B. J. and H. Morgan. "The Behaviour and Availability of Cd, Ni and Pb in Polluted Soils," in *Contaminated Soils*, Assink, J. W. and W. J. van den Brink, Eds. (Dordrecht : Martinus Nijhoff, 1986) pp. 101–113.
112. Hani, H. and S. Gupta. "Total and Biorelevant Heavy Metal Contents and Their Usefulness in Establishing Limiting Values in Soils," in *Environmental Effects of Organic and Inorganic Contaminants in Sewage Sludge*, Davis, R. D., G. Hucker, and P. L'Hermite, Eds. (Dordrecht: Reidel, 1983) pp. 121–129.
113. Hani, H. and S. Gupta. "Chemical Methods for the Biological Characterization of Metal in Sludge and Soil," in *Processing and Use of Organic Sludge and Liquid Agricultural Waste*, L'Hermite, P., Ed. (Dordrecht: Reidel, 1985) pp. 157–167.
114. Morgan, H. and B. J. Alloway. "The Value of Soil Chemical Extractants for Predicting the Availability of Cadmium to Vegetables," in *Trace Substances and Environmental Health XVIII*, Hemphill, D. D., Ed. (University of Missouri, Columbia, 1984) pp. 539–547.

115. Sanders, J. R., S. P. McGrath, and T. McM. Adams. "Zinc, Copper and Nickel Concentrations in Ryegrass Grown on Sewage Sludge-Contaminated Soil of Different pH," *J. Sci. Food Agric.* 37:961–968 (1986).
116. Sauerbeck, D. and P. Styperek. "Predicting the Cadmium Availability from Different Soils by $CaCl_2$ Extraction," in *Processing and Use of Sewage Sludge*, L'Hermite, P. and H. Ott, Eds. (Dordrecht: Reidel, 1984) pp. 431–435.
117. Elinder, C-G., L. Jousson, M. Piscator, and B. Rauster. "Histopathological Changes in Relation to Cadmium Concentration in Horse Kidneys," *Environ. Res.* 26:1–21 (1981).
118. Kjellstrom, T., C.-G. Elinder, and L. Friberg. "Conceptual Problems in Establishing the Critical Concentration of Cadmium in Human Kidney Cortex," *Environ. Res.* 33:284–295 (1984).
119. Joint FAO/WHO Expert Committee on Food Additives, "Evaluation of Certain Food Additives and the Contaminants Mercury, Lead and Cadmium," WHO Tech. Rep. Series No. 505. FAO Nutrition Meetings Report Series No. 51 (World Health Organization, Geneva, 1972).
120. Kagi, J. H. R. and H. J. Hopke. "Biochemical Interactions of Mercury, Cadmium and Lead," in *Changing Metal Cycles and Human Health,* Nriagu, J. O., Ed. (Berlin: Springer-Verlag, 1984) pp. 237–250.
121. Kjellstrom, T. and G. F. Nordberg. "A Kinetic Model of Cadmium Metabolism in the Human Being," *Environ. Res.* 16:246–269 (1978).
122. Fox, M. R. S. "Cadmium Bioavailability," *Fed. Proc.* 42:1726–1729 (1983).
123. Fox, M. R. S. "Nutritional Factors That May Influence the Bioavailability of Cadmium," *Environ. Qual.* 17:175–180 (1988).
124. McKenzie, J. M. "Bioavailability of Trace Elements in Foodstuffs and Beverages," in *Changing Metal Cycles and Human Health*, Nriagu, J. O., Ed. (Berlin: Springer-Verlag, 1984) pp. 187–198.
125. Fox, M. R. S. "Nutritional Influences on Metal Toxicity: Cadmium as a Model Toxic Element," *Environ. Health Perspect.* 29:95–104 (1979).
126. Foulkes, E. C. "The Critical Level of Cadmium in Renal Cortex: The Concept and its Limitations," *Environ. Geochem. Health* 8:91–94 (1986).
127. Cawse, P. A. "Trace and Major Elements in the Atmosphere at Rural Locations in Great Britain 1972–81," in *Pollutant Transport and Fate in Ecosystems*, Coughtrey, P. J., M. H. Martin, and M. H. Unsworth, Eds. (Oxford: Blackwell Scientific, 1987) pp. 89–113.
128. Louekari, K., U. Uusitalo, and P. Pietinen. "Variation and Modifying Factors of the Exposure to Lead and Cadmium Based on an Epidemiological Study," *Sci. Total Environ.* 84:1–12 (1989).
129. McLellan, J. S., P. R. Flanagan, M. J. Chamberlain, and L. S. Valberg. "Measurements of Dietary Cadmium Absorption in Humans," *Toxicol. Environ. Health* 4:131–138 (1978).
130. Sherlock, J. C. "Cadmium in Foods and the Diet," *Experientia* 40:152–156 (1984).
131. Louekari, K. and S. Salminen. "Intake of Heavy Metals from Foods in Finland, West Germany and Japan," *Food Add. Contam.* 3:355–362 (1986).
132. "Survey of Cadmium in Food: First Supplementary Report. The Twelfth Report of the Steering Group on Food Surveillance," The Working Party on the Monitoring of Food Stuffs for Heavy Metals, Ministry of Agriculture Fisheries and Food (London: HMSO, 1983).

133. Koivistoinen, P. "Mineral Element Composition of Finnish Foods: N, K, Ca, Mg, P, S, Fe, Cu, Mn, Zn, Mo, Co, Ni, Cr, F, Se, Si, Pb, Al, B, Br, Hg, As, Cd, Pb and Ash," *Acta Agric. Scand. Suppl.* 22 (1980).
134. Brams, E. and W. Anthony. "Cadmium and Lead Through an Agricultural Food Chain," *J. Environ. Qual.* 28:295–306 (1983).
135. Bache, C. A., W. H. Gutenmann, D. Kirtland, and D. J. Lisk. "Cadmium in Tissues of Swine Fed Barley Grown on Municipal Sludge-Amended Soil," *J. Food Saf.* 8:199–204 (1987).
136. Drury, J. S. and A. S. Hammons. "Cadmium in Foods. a Review of the Worlds Literature," EPA-560/2–78–007 (1979).
137. McKone, T. C. and P. B. Ryan. "Human Exposures to Chemical Contaminants Through Food Chains: An Uncertainty Analysis," *Environ. Sci. Technol.* 23:1154–1163 (1989).
138. Nriagu, J. O. "A Silent Epidemic of Environmental Metal Poisoning?" *Environ. Pollut.* 50:139–161 (1988).
139. Global Environmental Monitoring System "Guidelines for the Study of Dietary Intakes of Chemical Contaminants," WHO Offset Publication No. 87 (World Health Organization, Geneva, 1985).
140. Barltrop, D. "Evaluation of Cadmium Exposure from Contaminated Soil," in *Contaminated Soil,* Assink, J. W. and W. J. van den Brink, Eds. (Dordrecht: Martinus Nijhoff, 1986) pp. 169–179.
141. Dabeka, R. W., A. D. McKenzie, and G. M. A. Laeroix. "Dietary Intakes of Lead, Cadmium, Arsenic and Fluoride by Canadian Adults: A 24–Hour Duplicate Diet Study," *Food Addit. Contam.* 4:89–102 (1987).
142. Morgan, H., G. A. Smart, and J. C. Sherlock. "Intakes of Metal," *Sci. Total Environ.* 75:71–101 (1988).
143. Sherlock, J. C. and B. Walters. "Dietary Intake of Heavy Metals and its Estimation," *Chem. Ind.* 13:505–508 (1983).
144. Coomes, T. J., J. C. Sherlock, and B. Walters. "Studies in Dietary Intake and Extreme Food Consumption," *R. Soc. Health* 102:119–123 (1982).
145. Ministry of Agriculture, Fisheries and Food. "The Sludge (Use in Agriculture) Regulations 1989," No. 1263, (London: HMSO, 1989).
146. "Code of Practice for Agricultural Use of Sewage Sludge," U.K. Department of the Environment, (London: HMSO,1989).
147. Bundesgesundheitsamt "Richtwerte '79 fur Blei, Cadmium und Quecksilber in und auf Lebensmitte," *Bundesgesundheitsblatt* 22:282–283 (1979).
148. Ryan, J. A., H. R. Pahren, and J. B. Lucas. "Controlling Cadmium in the Human Food Chain: A Review and Rationale Based on Health Effects," *Environ. Res.* 27:251–302 (1982).
149. Naylor, L. M. and R. C. Loehr. "Increase in Dietary Cadmium Consumption as a Result of Application of Sewage Sludge to Agricultural Land," *Environ. Sci. Technol.* 5:81–886 (1981).
150. Morgan, H. and D. L. Simms. "Setting Trigger Concentrations for Contaminated Land," in *Contaminated Soil '88,* Wolf, K., W. J. van den Brink, and F. J. Colon, Eds. (Dordrecht: Kluwer, 1988) pp. 327–337.
151. Simms, D. L. "Towards a Scientific Basis for Regulating Lead Contamination," *Sci. Total Environ.* 58:209–224 (1986).
152. Chmielnicka, J. and M. G. Cherian. "Environmental Exposure to Cadmium and Factors Affecting Trace-Element Metabolism and Metal Toxicity," *Biol. Trace Element Res.* 10:243–262 (1986).

153. Crews, H. M., J. R. Dean, L. Ebdon, and R. C. Massey. "Application of High-Performance Liquid Chromatography — Inductively Coupled Plasma Mass Spectrometry to the Investigation of Cadmium Speciation in Pig Kidney Following Cooking and In Vitro Gastro-Intestinal Digestion," *Analyst* 114:895–899 (1989).
154. Nordberg, M. and Y. Kojima. "Metallothionein and Other Low Molecular Weight Metal-Binding Proteins," in *Metallothionein: Proceedings of the First International Conference on Metallothionein and Other Low Molecular Weight Metal-Binding Proteins,* Kagi, J. H. R. and M. Nordberg, Eds. (Basel: Birkhauser-Verlag, 1979) pp. 41–124.
155. Fowler, B. A., C. E. Hilderbrand, Y. Kojima, and M. Webb. "Nomenclature of Metallothionein," in *Metallothionein II: Proceedings of the Second International Conference on Metallothionein and Other Low Molecular Weight Metal-Binding Proteins*, Kagi, J. H. R. and Y. Kojima, Eds. (Basel: Birkhauser-Verlag, 1987).
156. Cherian, M. G., R. A. Gower, and L. S. Valberg. "Gastrointestinal Absorption and Organ Distribution of Oral Cadmium Chloride and Cadmium-Metallothionein in Mice," *J. Toxicol. Environ. Health* 4:861–868 (1978).
157. Klein, D., H. Greim, and K. H. Summer. "Stability of Metallothionein in Gastric Juice," *Toxicology* 41:121–129 (1986).
158. Maitani, T., M. P. Waalkes, and C. D. Klaassen. "Distribution of Cadmium After Oral Administration of Cadmium-Thionein to Mice," *Toxicol. Appl. Pharmacol.* 74:237–243 (1984).
159. Cherian, M. G. "Metabolism and Potential Toxic Effects of Metallothionein" in *Metallothionein: Proceedings of the First International Conference on Metallothionein and Other Low Molecular Weight Metal-Binding Proteins,* Kagi, J. H. R. and M. Nordberg, Eds. (Basel: Birkhauser-Verlag, 1979) pp. 337–345.
160. Cherian, M. G. "Absorption and Tissue Distribution of Cadmium in Mice After Chronic Feeding with Cadmium Chloride and Cadmium Metallothionein," *Bull. Environ. Contam. Toxicol.* 30:33–36 (1983).
161. Wagner, G. J., E. Nulty, and M. LeFevre. "Cadmium in Wheat Grain: Its Nature and Fate After Ingestion," *J. Toxicol. Environ. Health* 13:979–989 (1984).
162. Wagner, G. J. "Characterization of the Cadmium-Binding Complex of Cabbage Leaves," *Plant Physiol.* 76:797–805 (1984).
163. Casterline, J. L. and N. M. Barnett. "Cadmium-Binding Components in Soybean Plants," *Plant Physiol.* 69:1004–1007 (1982).
164. Rauser, W. E. "Isolation and Partial Purification of Cadmium-Binding Protein from the Roots of the Grass *Agrostis Gigantea*," *Plant Physiol.* 74:1025–1029 (1984).
165. Kaneta, M., H. Hikichi, S. Endo, and N. Sugiyama. "Isolation of a Cadmium-Binding Protein from Cadmium-Treated Rice Plants (*Oryza sativa* L.)," *Agric. Biol. Chem.* 47:417–418 (1983).
166. Bartolf, M., E. Brennan, and C. A. Price. "Partial Characterization of a Cadmium-Binding Protein from the Roots of Cadmium-Treated Tomato," *Plant Physiol.* 66:438–441 (1980).
167. Henze, W. and F. Umland. "Speciation of Cadmium and Copper in Lettuce Leaves. *Trace Element Anal. Chem,*" in *Med. Biol.* 4:501–507 (1987).

168. Rauser, W. E. "The Cd-Binding Protein from Tomato Compared to Those of Other Vascular Plants," in *Metallothionein II: Proceedings of the Second International Conference on Metallothionein and Other Low Molecular Weight Metal-Binding Proteins*, Kagi, J. H. R. and Y. Kojima, Eds. (Basel: Birkhauser-Verlag, 1987) pp. 301–308.
169. Grill, E. "Phytochelatins, the Heavy Metal Binding Peptides of Plants: Characterization and Sequence Determination," in *Metallothionein II: Proceedings of the Second International Conference on Metallothionein and Other Low Molecular Weight Metal-Binding Proteins*, Kagi, J. H. R. and Y. Kojima, Eds. (Basel: Birkhauser-Verlag, 1987) pp. 317–322.
170. Mills, C. F. "The Influence of Chemical Species on the Absorption and Physiological Utilization of Trace Elements from the Diet or Environment," in the *Importance of Chemical "Speciation" in Environmental Processes*, Bernhard, M., F. E. Brinckman, and P. J. Sadler, Eds. (Berlin: Springer-Verlag, 1986) pp. 71–83.
171. van Dokkum, W. "The Significance of Speciation for Predicting Mineral Bioavailability," in *Nutrient Availability: Chemical and Biological Aspects*, Southgate, D., I. Johnston, and G. R. Fenwick, Eds. (Cambridge: Royal Society of Chemistry, 1989) pp. 89–96.
172. van Lune, P. "Cadmium and Lead in Soils and Crops from Allotment Gardens from The Netherlands," *Neth. J. Agric. Sci.* 35:207–210 (1987).
173. Wiersma, D., B. J. van Goor, and N. van der Veen. "Cadmium, Lead, Mercury and Arsenic Concentrations in Crops and Corresponding Soils in The Netherlands," *J. Agric. Food Chem.* 34:1067–1074 (1986).
174. Kloke, A. "Tolerable Amounts of Heavy Metals in Soils and Their Accumulation in Plants," in *Environmental Effects of Organic and Inorganic Contaminants* in *Sewage Sludge*, Davis, R. D., G. Hucker, and P. L'Hermite, Eds. (Dordrecht: Reidel, The Netherlands, 1983) pp. 171–176.
175. King, L. D. "Agricultural Use of Municipal and Industrial Sludges in Southern United States," *Southern Cooperative Series Bull. 314*. (North Carolina State University, Raleigh, 1986) p. 52.
176. Zurera, G., B. Estrada, F. Rincon, and R. Pozo. "Lead and Cadmium Contamination Levels in Edible Vegetables," *Bull. Environ. Contam. Toxicol.* 38:805–812 (1987).
177. Naylor, L. M., M. Barmasse, and R. C. Loehr. "Uptake of Cadmium and Zinc by Corn on Sludge Treated Soil," *Bio Cycle* 37–41 (1987).
178. Rappaport, B. D., D. C. Martens, R. B. Reneau, Jr., and T. W. Simpson. "Metal Availability in Sludge Amended Soils with Elevated Metal Levels," *J. Environ. Qual.* 17:42–47 (1988).
179. Horner, E. and Kurfurst, U. "Cadmium in Wheat Grain from Biological Cultivation," *Fresenius Z. Anal. Chem.* 328:386–387 (1987).
180. Berglund, S. "Plant Uptake of Cadmium. Summary of Swedish Investigations Up To and Including 1981," in *Environmental Effects of Organic and Inorganic Contaminants in Sewage Sludge*, Davis, R. D., G. Hucker, and P. L'Hermite, Eds. (Dordrecht: Reidel, The Netherlands, 1983) pp. 176–200.
181. Vigerust, E. and A. R. Selmer-Olsen. "Basis for the Metal Limits Relevant to Sludge Utilisation," paper presented at the seminar Treatment and Use of Sewage Sludge, Bonn (1985).

182. Vlamis, J., D. E. Williams, J. E. Corey, A. L. Page, and T. J. Ganje. "Zinc and Cadmium Uptake by Barley in Field Plots Fertilized Seven Years with Urban and Suburban Sludge," *Soil Sci.* 139:81–87 (1985).
183. Hyde, H. C., A. L. Page, F. T. Bingham, and R. J. Mahler. "Effect of Heavy Metals in Sludge on Agricultural Crops," *Water Pollut. Cont. Fed.* 51:2475–2486 (1979).
184. Bernhard, W. R. and J. H. R. Kagi. "Purification and Characterization of Atypial Cadmium-Binding Polypeptides from *Zea mays*," in *Metallothionein II: Proceedings of the Second International Conference on Metallothionein and Other Low Molecular Weight Metal-Binding Proteins*, Kagi, J. H. R. and Y. Kojima, Eds. (Basel: Birkhauser-Verlag, 1987) pp. 309–315.
185. Weigel, H. J. and H. J. Jager. "Subcellular Distribution and Chemical Form of Cadmium in Bean Plants," *Plant Physiol.* 65:480–482 (1980).

6

Long-Term Application of Sewage Sludge and Its Effects on Metal Uptake by Crops

C. Juste and M. Mench

Institut National de la Recherche Agronomique, Station d'Agronomie, Centre de Recherches de Bordeaux, BP 81, F-33883 Villenave d'Ornon Cedex, France

ABSTRACT

The effects of long-term application of sewage sludge (i.e., at least 10 years in duration) on metal distribution in the soil profile, the response of crop yields, and the bioavailability of metals were reviewed using results from field trials mostly located in the European Community and the U.S. In almost all of the studies, sludge-borne metals appeared to remain in the zone of sludge incorporation (e.g., 0–15 cm depth). The metal recoveries resulting from mass balances of metals added into the soil ranged from 30 to 90%. Lateral soil movement was the main explanation of the progressive "disappearence" of metals from experimental plots.

Phytotoxicity due to sludge-borne metals was rarely observed on grain crops. Harmful effects on some legume plants could be explained by the detrimental influence of metals on the microbial activity of soils, especially nitrogen fixation. The sewage sludge application exhibited a positive effect on plant growth in 65% of cases. Generally, investigators have focused on metals that have high concentrations in sludge (e.g., Zn, Mn, Cu) or known to be hazardous (e.g., Cd, Pb, Ni). Cadmium, Ni, and Zn were the most bioavailable, whereas Cr and Pb plant uptake was insignificant. Among these metals, the concentration of a given metal species in plants grown on sludge-treated soils was related to the sludge application rate, to the annual or cumulative application of sludge, or to the time following cessation of sludge application. Conflicting results were due to the various kinds of sludges, soils, climatic conditions, and cultural practices in the reviewed field trials. However, the total metal input to the soil was the major factor influencing metal concentration in plant tissues. A trend of a progressive decrease in metal uptake was observed with time following sludge application.

Further information is needed to evaluate (a) long-term changes in factors controlling metal bioavailability of the soils, (b) long-term changes of soil properties as a result of metal accumulation, and (c) the behavior of other trace elements such as As, Be, Co, Hg, Mo, Se, and Tl, whose concentrations in sludge could be significant relative to plant and animal nutrition.

INTRODUCTION

The interest in applying municipal sewage sludges on cropland has grown considerably in past decades. The sludge is a valuable source of N, P, minor elements (e.g., Cu, Fe, Mn, Zn), and organic matter that would benefit growing crops. In addition, sewage sludge disposal to agricultural soils is an alternative available to the municipal agencies that treat waste waters. However, besides the beneficial effects, concern over the potential hazards associated with the considerable heavy metal content of sludges (e.g., Cd, Cr, Ni, Pb, Zn) has received attention by many researchers. Indeed, soil contamination by sludge-borne metals can result in a decrease of crop yields due to metal phytotoxicity or in an increase of metal transfer into the food chain.

Two kinds of metal effects can be considered when sludges are applied to soils:

- a short-term effect due to the easily available forms of metal originating from sludges;
- a long-term effect resulting from progressive changes in metal speciation or soil properties occuring year to year in the soil-plant system following sludge application or after applications cease.

Short-term effects of metals are now quite well understood, and numerous papers have been published on this subject. On the other hand, long-term field experiments are essential for a better understanding and prediction of the long-term behavior of metals in soil when sludges are utilized on a large scale. Results from laboratory and greenhouse studies have a limited value for this purpose. Therefore, field experiments are essential because all environmental factors must be taken into account, i.e., climatic factors, physical properties of soil, the heterogeneous distribution of metals and roots in the soil profile, agricultural practices, and cropping history. Moreover, long-term experiments represent the only valuable approach to accurately assess the cumulative and residual effects of metals as sludge progressively decays in soil. Unfortunately, several constraints are attached to the long-term experiments, such as the need of perennial sites and changes in cultural practices or cultivars with time. The biggest uncertainty lies in the lateral movement of soil particles across plot boundaries, inducing a possible contamination of the surrounding check plots during tillage operations.

Despite the above constraints, long-term field experiments must be considered as an irreplaceable means of increasing the knowledge of the potential hazards associated with the use of sewage sludges in cropland.

The aim of this paper is to illustrate the cumulative and residual effects of heavy metals in sludge-treated soils using data drawn from a number of field trials conducted in several countries. The following three main aspects are successively reviewed:

- Long-term accumulation and distribution of metals in sludge-treated soils
- Crop yields response to metal inputs into the soils through sludge application
- Plant metal uptake and long-term changes in the bioavailability of metals in sludge-treated soils.

A SURVEY OF LONG-TERM EXPERIMENTS WITH SLUDGE-BORNE METALS

The terms heavy metals, metals, and trace elements are used interchangeably in this writing, although trace elements is now preferred by most scientists. This term includes various chemical elements studied in regard to their bioavailability to plants. The objectives of this survey were to identify the long-term field experiments covering both the use of sewage sludge and the behavior of metals in soils and plants, and to produce a summary of the results obtained to highlight any fundamental gaps that might justify further researches.

A literature survey yielded 195 references. However, the number of experiments that can be qualified as "long term" (i.e., at least 10 years in duration) are very limited. Of these experiments, 23 were selected according to their accessibility (e.g., international publications or proceedings) and their applicability.

Most of the important experiments were in European Community (EC) member states (i.e., Germany, France, U.K., and Denmark) and the U.S. (Table 1). Results from Germany were conviently summarized by Sauerbeck and Styperek.[1,2] Indeed, this compilation does not represent an exhaustive analysis of the experiments conducted; several other locations may be included. We restricted the entry on the basis of duration of experiments, methods of cultivation, waste material tested, soil properties, and treatment levels. All data were from field trials. The test plant species experimented are outlined in the other tables devoted to bioavailability.

These long-term experiments may be divided into two categories: (a) experiments with continuing application of sewage sludge or (b) observations on residual effect following termination of sewage sludge application.

Table 1. Long-Term Field Experiments with Sewage Sludge

Country	Location	Experiment	Start	End or Residual	Material Used	Appl. Rate Mg DM/ha	Frequency	Soil Type	pH	%C	CEC	Ref.
France	Bordeaux	Bx1	1974		Digest & dehydr	10	Yearly	S	5.5	1.5		
		Bx1	1974		Digest & dehydr	100	2 Years	S	5.5	1.5		35–37
		Bx2	1976	1980 residual	Digest heat & dehyd	10	Yearly	S	5.5	2.2	5.6	
		Bx2	1976	1980 residual	Digest heat & dehyd	100	2 Years	S	5.5	2.2	5.6	
	Nancy	N1	1974	1981	Digest, lime + FeCl3 & dehydr	30–340	4 times	cL	7.1	1.8	16	12, 38
U.S.	Berkeley	Bel	1973		Sludge cake	45–180	Yearly	L	5.5	1.1	20	4, 5, 39
		Be2	1973	residual	Air-dry	45–225	Yearly	L	5.5	1.1	20	
	Fairland	F	1972	residual	Digest & dehydr	56–224	Once	sL	5.9			21, 30, 31
	Beltsville	B11	1976	residual	Heat Treated	56–224	Once	lS	5			
		B12	1976	residual	Limed digest	56–224	Once	lS	6.3			
		B14	1978	residual	Digested	50–100	Once	lS	5.5			
	Joliet	J1	1969	1973 residual	Liquid	max 61	Yearly	sL		0.85	14	8, 9
		J2	1968	1974	Liquid	max 130	Yearly	sL				
	Riverside	R1	1976		Composted	22.5–90	Yearly	lS	6	0.5	8.7	3, 6, 23, 24
		R2	1976		Composted	22.5–90	Yearly	L	7.1	2.5	14	
	St. Paul	SP		3 yrs + 6 residual	Limed sludge cake & unlimited	30–90 30–45	Yearly	sS	6.2			25, 26
U.K.	Stevenage	Ste	1976		Air dry	35–280	Once	cL	8	1.1		1, 2
	Woburn	W1	1942	1961 residual	Digested & dried	max 383	Yearly	lS	6.5			7, 10, 17
		W2	1942	1961 residual	Digested & dried	max 766	Yearly	lS	6.5			18–20
DK	Vejen	Ve1	1974		Dehydr	11–22	Yearly	lS	6.1	2.6		1, 2
		Ve2	1974		Dehydr	11–22	Yearly	sL	5.8	2.3		
		Ve3	1974		Dehydr	11–22	Yearly	S	6.3	2.7		
Germany	Bonn	Bn1	1958		Liq. sludge	5–10	2 years	uL	6.5	0.9	14	
		Bn2	1958		Sl. compost	2–12	2 years	uL	5.6	1.2	12	
		Bn3	1972		Liq. sludge	4.5–9	2 years	lS	6.6	1.1	14	
	Braunschweig	Bs1	1971		Liq. filt. sludge	5–15	Yearly	ulS	5.5	1.6	9.3	
		Bs2	1971		Liq. sludge	5–15	Yearly	ulS	6.7	0.9	8.4	

Bremen	Br1	1968	Liq. sludge	20	Yearly	peat	4.2	28.9	234	
	Br2	1972	Liq. sludge	12–20	Yearly	peat	4	16	190	1, 2
Giessen	Gi1	1972	Sl compost	120–200	3 years	L	6.6	1.1	11.6	
	Gi2	1979	Dehydr sl	4.5–9	Yearly	L	6.2	0.9	10.5	
	Gi3	1969	Liq. sludge	2.5–5	Yearly	L	5.9	0.8	12.4	
	Gi4	1969	Liq. sludge	2.5–5	Yearly	L	5.4	5.9	17.8	
München	Mü1	1980	Sludge cake	100–2000	Once	sL	6.6	7.9		
	Mü2	1976	Liq. filt sludge	100–690	Once	lS	5.8	2.3	16	
	Mü3	1978	Liq. sludge	5–10	Yearly	sL	6.3	1.4		
	Mü4	1979	Liq. sludge	1.5	Yearly	lS	6.4	1.7		
Speyer	Sp1	1981	Sl compost	5–30	Yearly	S	6.2			
	Sp2	1958	Diff. sl comp.	1.1–2.1	Yearly	S	5.5	0.7		
	Sp3	1972	Liq. sludge	15–30	3 years	S	6.8	0.8		
	Sp4	1974	Liq. sludge	30	2 years	S	6.8			

Table 2. Maximum Metal Inputs in Several Long-Term Experiments (kg/ha)

			Metal						
Sites	**Country**	**Experiment**	**Cd**	**Cu**	**Cr**	**Mn**	**Ni**	**Pb**	**Zn**
Bordeaux	France	Bx1	26	158	65	5679		659	4882
		Bx2	641	170	64	234	1337	231	976
Nancy	France		9	132		122	37	197	746
Woburn	U.K.	Market Garden	70	864	704		135	694	2158
Maryland	U.S.	Fairland	3	128			10	63	290
		Beltsville	1.3	116		323	7	97	286
St. Paul	U.S.		25						348
Riverside	U.S.		44	730	1180		301	936	2888
Joliet	U.S.	J1	58						1290
		J2	101						2358
Berkeley	U.S.	Be1	60						4937
		Be2	7.5						546

The waste material used generally consists of sewage sludge of various kinds such as liquid sludge, dehydrated sludge, and anaerobically digested sludge. In some cases composted sludge was used.

The metal concentrations in the sludges were always reported, but other parameters (pH, organic matter content, type of sludge process, etc.) were less frequently mentioned. Among the metal elements, concentrations of Cd and Zn were virtually always measured due to their large content in sludge, their high mobility, and because they represent the greatest risks of bioaccumulation in food chain. Copper, Cr, Ni, Pb were less frequently included for different reasons: generally their bioavailability is lower, and sometimes laboratories do not have access to analytical facilities for some of these elements. Iron and manganese are often present in large amounts in sludges but were less frequently measured because they are considered to be nontoxic. Other elements that are important to plant or animal nutrition like Co, Mo, Hg, Tl, V, As, B, Al, Be, and Se were rarely accounted for.

The maximum input of metals in various field experiments through sludge application is quantified in Table 2. The summary provides useful information for better understanding the experimental data and interpretation. As clearly evidenced in this table, metal inputs in the soil are greatly different from one site to another. For example, the maximum input of Cd and Zn ranged from 1.3 to 641 kg/ha and from 290 to 4937 kg/ha, respectively. Obviously, such differences in metal inputs induce marked differences in the metal behavior in soil and plants. As a consequence, a great deal of caution must be exercised to draw conclusions from data obtained in field trials.

In other respects, the generally high rate of Zn addition through land application of sludge calls the attention on the potentially important role of Zn on the behavior of other metals that were incorporated into soils at the same time to a lesser extent.

LONG-TERM ACCUMULATION AND DISTRIBUTION OF METALS IN SLUDGE-TREATED SOILS

Pattern of Metal Distribution in the Soil Profile

As expected, almost all researchers reported that incorporation of the sludges into the soil resulted in a marked increase of the soil metal content. There was also agreement that metals accumulated in the topsoil with no evidence of rapid downward movement, even many years after the termination of sludge application. Metals appeared to remain in the zone of sludge incorporation (e.g., 0 to 15 cm depth) as the result of their adsorption on hydrous oxides, clays and organic matter, the formation of insoluble salts, or the presence of residual sludge particles. However, in some cases, a slight but significant increase in the heavy metal content in soils below the zone of sludge incorporation was detected.[3] Such a limited metal movement was measured by Williams et al.[4,5] for Zn and by Chang et al.[6] for Cd. However, this increase in metal content beyond the topsoil layer must be viewed with caution because it could indicate a possible mixing with the underlying layers during the cultivation practices or an inconsistent soil sampling procedure.

The distribution of metal with depth in the soil profile is dependent on the cumulative metal inputs through sludge deposition. If the soil can be considered as a cation exchanger, the extent of downward movement of metals in soil is expected to increase with the metal loading rate. Therefore, it may be easier to detect metal enrichment in subsurface soil if a significant amount of metal is leached or metal loading is high.

The long-term experiments of Bordeaux (France) provided an exemple. In the fall of 1989, the metal distribution in the soil profile of each treatment plot was evaluated. Horizontal soil layers, 2 m^2 surface and 20 cm thick, were taken successively to 100 cm. Each collected layer was weighed, air dried, and sieved to pass a screen with 2 mm opening. A sample of this fine soil was digested using the aqua regia procedure, (i.e., 25% (v/v) 14 N HNO_3, 75% 12 N HCl), and the contents were brought to volume with distilled-deionized water. Metal concentrations were determined by a graphite furnace atomic absorption spectrometer with deuterium or Zeeman effect background correction. Despite its drastic and destructive character, the procedure used to evaluate the metal distribution in the soil profile was judged to be most likely accurate because possible artifacts, chiefly due to variations of bulk density, were so eliminated. The results show that for a given metal, the downward movement of an element increased with its loading rate (Table 3). A consistent increase in metal concentration occured only for Mn and Zn to a depth of 60 cm. Slight but significant increases of Cd, Ni, and Cu concentrations were found in the subsurface soil layers at the highest metal loading rates for the duration of the experiment (1974 to 1989 or 1976 to 1989).

The sewage sludge with higher Cd and Ni content (i.e., Bx2) resulted in

Table 3. Pattern of Heavy Metal Distribution in the Soil Profile from the Bordeaux Experiments (France) in Relation to Metal Loading Rate (Mean Values in 1989)

Metal	Cumulative Metal Inputs (kg ha^{-1})	Treatment[a]	Metal Distribution in the Soil Profile (% of the Metal Recovery)				
			0–20	20–40	40–60	60–80	80–100[b]
Cu	182	Bx 2-100	71	26	3	0	0
	36	Bx 2-10	100	0	0	0	0
	641	Bx 2-100	74	25	1	0	0
Cd	107	Bx 2-10	60	38	1	1	0
	26	Bx 1-100	73	27	0	0	0
	6083	Bx 1-100	60	16	15	6	3
Mn	1217	Bx 1-10	61	23	14	2	0
	234	Bx 2-100	84	16	0	0	0
Ni	1337	Bx 2-100	71	27	1	1	0
	223	Bx 2-10	59	41	0	0	0
Pb	659	Bx 1-100	75	24	1	0	0
	231	Bx 2-100	87	13	0	0	0
Zn	5134	Bx 1-100	71	27	1	1	0
	1027	Bx 1-10	59	41	0	0	0

[a] Bx 1-10 = Bx1 10 Mg/ha/y; Bx 1-100 = Bx1 100 Mg/ha/2 years; Bx 2-10 = Bx2 10 Mg/ha/y; Bx 2-100 = Bx2 100 Mg/ha/2 years.
[b] Soil layers expressed as centimeters.

Table 4. Total Metal Concentration (mg/kg) and pH in the Topsoil (0–25 cm Depth) of Plots from the Bordeaux (France) Experiments as a Result of Sewage Sludge Type and Loading Rate (Mean Values in 1989)

Cumulative Sludge Inputs (Mg DM/ha)	Experiment[a]	Metal							
		Cd	Cu	Cr	Mn	Ni	Pb	Zn	pH
0	Bx 2	1.30	4.5	3.7	23.0	3.6	10.6	8.1	5.8
0	Bx 1	0.33	14.2	6.8	33.0	2.4	17.9	19.2	5.4
50	Bx 2	27.90	15.6	7.2	33.0	73.8	22.1	45.7	6.8
160	Bx 1	1.00	18.9	8.2	312.0	6.4	44.7	199.0	6.2
300	Bx 2	96.00	45.5	21.0	69.0	247.0	44.3	155.3	7.1
800	Bx 1	5.71	66.7	23.1	1789.0	30.8	189.0	1074.0	5.8

[a] See Table 1.

greater increases of the Cd and Ni concentrations in the soil samples (Table 4). On the other hand, the municipal sludge from Bordeaux (i.e., Bx1) induced a marked raise in the soil Zn and Mn contents, due to its higher Zn and Mn concentrations.

Heavy Metal Balance in the Soil

Several articles have been published concerning sludge-borne metal balance in long-term experiments.[5,7–9,41] The metal recoveries resulting from mass balances of metals added into the soil ranged from 30 to 90%. Several hypotheses have been suggested to account for the incomplete metal recovery. McGrath,[7]

Table 5. Metal Recoveries in the High Rate Sludge Plots in the Bordeaux Experiments (France) (Mean Values in 1989)

Metal	Percent Recovery[a] Cumulative[b]	Residual[c]
Cd	56	45
Cu	80	70
Mn	100	49
Pb	68	36
Zn	72	43

[a] Metal recovered in the whole profile (kg ha^{-1}, 100 cm depth) vs. total metal amounts of metals applied in sewage (kg ha^{-1}), expressed in percent.
[b] Bx1.
[c] Bx2.

using data from the Market Garden Experiment at Woburn (U.K.), demonstrated that the losses of metal in sludge-treated plots were attributable neither to the small amount of metal removed by offtake in crops nor to the insignificant leaching of metal from the soil profile. Consequently, the most plausible explanation of the apparent lack of a metal balance was due to soil displacement during cultivation practices or water runoff. McGrath and Lane,[10] using a two-dimensional "dispersion model," found that the soil surface metal content from the Woburn experiment plots agreed with the predicted values derived from the model. Dispersion coefficients characterizing metal movements in the direction parallel or perpendicular to ploughing were 0.24 and 0.13 m^2 per tillage operation, respectively. Williams et al.[5] have evaluated the extent of lateral soil displacements by analyzing samples of soil collected inside and outside of sludged-plots. Their results agreed with those of McGrath[7] and confirmed that lateral soil movement was the most likely explanation for a lack of metal balance in the soil. Therefore, metals that are not taken up by crop in large quantities or leached appreciably through the soil profile (i.e., Pb or Cr) might be tracers to quantify soil movements.[10]

In long-term experiments where sewage sludge application has been terminated, a decrease of metal recovery often resulted because of soil displacement due to cultivation transports metals outside of the experimental area. McGrath,[7] in the Woburn experiment, demonstrated that in 1961, after termination of sewage sludge application, metal recoveries ranged from 55 to 80% inside sludge-treated plots, whereas in 1980, for the same metals, the amounts accounted for represented only 32 to 42% of the metal inputs.

Similar results were found in the Bordeaux experiments (Table 5). In experimental plots where Bx1 sludge was applied annually since 1974, metal recoveries in 1989 ranged from 54 to 100%. In contrast, in the Bx2 experiment plots where sludge application was terminated in 1980, metal recoveries in 1989 ranged from 36 to 70%. The progressive "disappearence" of metals from the experimental plot of the Bordeaux experiments was supported by the cadmium

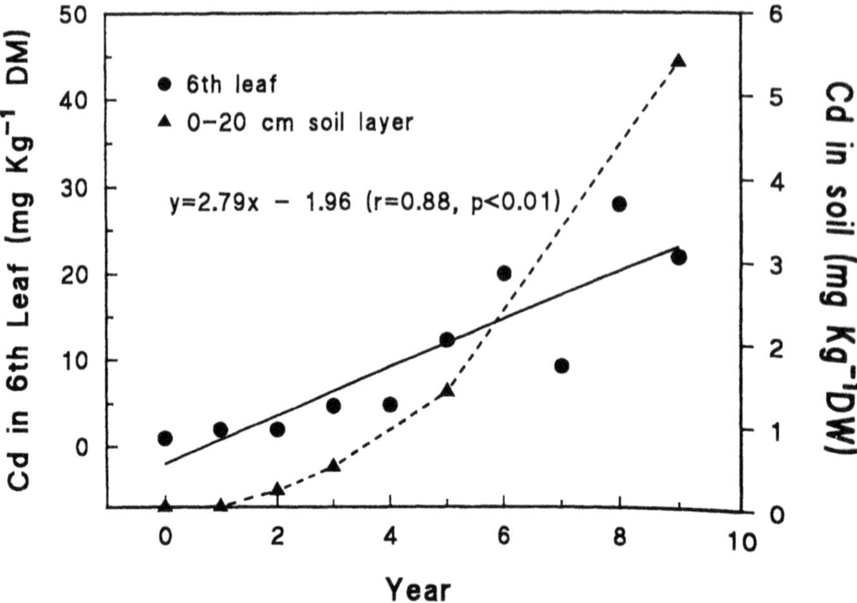

FIGURE 1. Increase in the concentration of Cd in the 6th leaf of corn plants and in the 0 to 20 cm soil layer from control plots in the Bx2 experiment at Bordeaux.

contamination of control plots in the Bx2 experiment (Table 4) and Cd uptake of corn plants growing on these plots increased with time (Figure 1). The uptake of other metals, however, was unchanged, and the concentrations of other metals did not increase as appreciably in comparison with their background levels as was the case for Cd. It appears that the importance of the metal's lateral movement may differ from one metal to another. This finding is an important matter of concern and requires further investigations to elucidate its cause.

Even if the lateral movement of metals bound to soil particles is the most probable cause of metal imbalance in the soil, other causes of metal loss should also be examined. Possible errors introduced by constraints in soil and sludge sampling, metal extraction procedures, and the analytical method must be considered, especially if small quantities of metals are involved in the balance calculation. In particular, inconsistent metal recovery of various extraction procedures is undoubtedly a major cause of underestimation. For example, 4 *N* HNO_3 is unable to completely extract metals held in certain compartments of the soil matrix. Another cause of inability to account for metals in the soil is the inaccurate measurements of the soil's bulk density and thickness.

Heavy Metal Extractability by Mild Chemical Reagents

For measurement of total metal content in the soil, extractions with a strong acid are necessary. This procedure, however, is not suitable to predict the mobility of metals in the soil profile and the bioavailability of metals to plants, or to provide a measurement of the long-term changes in these parameters following termination of sludge application.

For plant nutrition and soil fertility assessment, various chemical reagents have been used to estimate the fraction of the soil metal that is potentially available to plants. The most frequently employed reagents are dilute HCl, EDTA, acetic acid, NTA (nitrilotriacetic acid), DTPA, and $CaCl_2$. For some elements (e.g., Cd, Zn), sufficient amounts can be extracted by chelating agents such as EDTA or DTPA for routine analysis, and the correlations between the metal concentration in soil extracts and the metal uptake by crops are reasonable.[11,34] These reagents are still the most common metal extractants used in the majority of soil analysis laboratories. Recent results, however, indicate that dilute salt solutions (e.g., 0.1 *M* $Ca(NO_3)_2$, NH_4NO_3, $CaCl_2$) or water would be the most direct extractants to assess the mobility and the bioavailability of metals, as marked increase in the sensitivity of analytical procedures are now available.[42–45]

Generally, the EDTA- or DTPA-extractable metal concentration increases with the amounts of metal added to the soil through sewage sludge deposition. It has been observed, however, EDTA- or DTPA-extractable metals of the soil decrease with time, especially after the sewage sludge disposal is discontinued.[12] This phenomenon suggests that metals revert with time to less available forms in the soil. An additional process for the apparent decrease of metal is the lateral displacement of soil particles, above mentioned. As a consequence, the long-term changes in extractable metal content must be interpreted with caution. To avoid a possible misinterpretation, perhaps "available" metals of the soil is better expressed as the ratio of the extractable metal vs. total metal.[4] The relative metal availability calculated in this manner presents a consistent trend of the metal behavior with time. Nevertheless, the absolute quantity of metal extracted by appropriate reagents, such as DTPA, is still very useful to evaluate plant-available metal of the soil in the short-term. Morel and Guckert[12] attributed some seasonal changes in DTPA-extractable metals to the action of both microbial activity and plant root systems (i.e., acidification, release of chelating compounds).

THE RESPONSE OF CROP YIELDS TO HEAVY METALS IN LONG-TERM SEWAGE SLUDGE EXPERIMENTS

In field experiments, the most obvious way to quickly assess the beneficial, detrimental, or lack of sludge effect on crops is to measure yield response. Sludge is a very complex mixture of organic and mineral compounds. Therefore, plant response to waste addition cannot be related with assurance to a single factor. Besides, factors improving plant growth such as nutrient supply may mask the negative effects due to the presence of toxic compounds. Also, during a number of years when sludges are being applied, the concentration of metals in the soil only gradually increases to critical levels. Using yield alone would not give warning of approaching problems.

Table 6. Corn Grain Yield in the Bx1 Experiment (Bordeaux, France) with Cumulative Sludge Application (Average Values over the 1974 to 1986 Period)

Cumulative Sludge Inputs (Mg DM ha^{-1})	Treatment[a]	Grain Yield kg DM ha^{-1}
0	Control	9160 a
130	10 Mg DM/ha/y	9330 b
700	100 Mg DM/ha/2 y	9860 c

Note: Within a column, values followed by the same letter are not significantly different at the 5% level (Newman Keuls test).

[a] See Table 1.

From Juste, C. and P. Solda, in *Abstractsof the International Conference on Metals in Soils, Waters, Plants and Animals*, Orlando, Florida, April 30–May 3, 1990. University of Georgia, U.S. Department of Energy.

Based on the results from selected long-term experiments (Table 1), the application of sewage sludge exhibited a positive effect on plant growth in 65% of cases, regardless of the type of experiment. Most other results showed that yields from the sludge-treated plots were not statistically different from the control treated with mineral fertilizers. With sludge applications, yield enhancement as high as 30% was sometimes reported, especially under adverse growing conditions such as a drought.[14,15] Delivery of nitrogen, phosphorus, micronutrients, and organic matter during the decay of waste is likely the chief cause of the observed beneficial effect on crops.

Grain Crop Yields

In the Bordeaux experiment (i.e., Bx1), benefits of sludge application was noted every year from 1974 to 1986 in plots continuously cropped with maize (Table 6). Over the 12-year period, the average grain yield was 8% (p <0.01) above the control, despite a cumulative sludge loading as high as 700 Mg DM/ha. In this case, the yield enhancement probably resulted from an annual incorporation of 1000 to 1500 kg/ha of sludge-related nitrogen. The improvement of the nitrogen nutrition was confirmed by plant analysis. Moreover, phytotoxic symptoms were never observed on plants.

By contrast, phytotoxic effects due to an excess of metals were observed in the Bx2 Bordeaux experiment which received, during the 1976 to 1980 period, Cd- and Ni-contaminated sludges. These sludges contained unusually large concentrations of these metals, because the wastewater treatment plant received waste discharges from a Ni-Cd battery factory (Table 2). The cumulative Cd and Ni inputs into the soil were up to 641 and 1337 kg/ha, respectively, representing an extreme case. In 1980, the last year sludges were applied, yield from the high sludge-loading plots decreased sharply, down to 50% of the control (Figure 2). In the same season, at the early stage of growth, maize exhibited phytotoxic

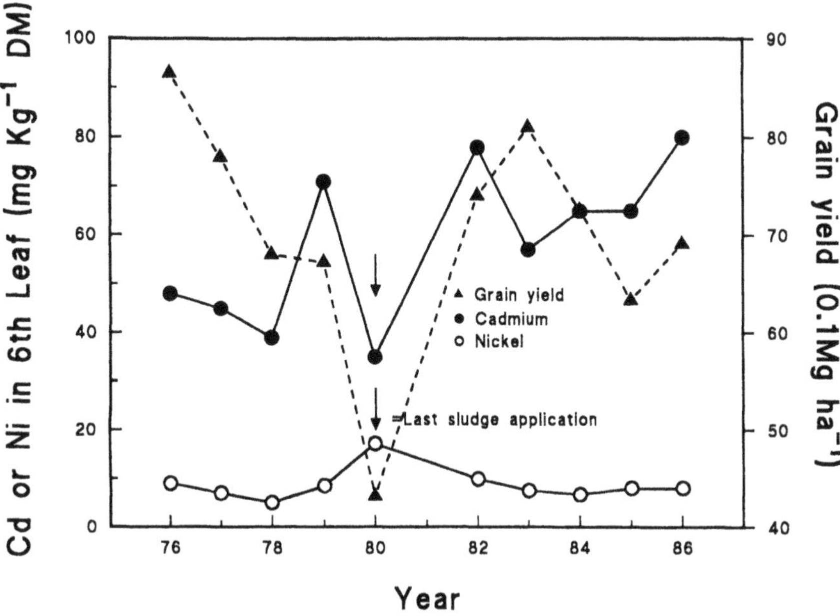

FIGURE 2. Relationship between the grain yield and the concentrations of Cd or Ni in the 6th leaf of corn plants in the Bx2 experiment at Bordeaux.

symptoms such as intervenal chlorosis and purple on stems. Owing to the sharp drop of yield, sludge deposition was discontinued. However, the residual effect of sludge application on corn yields and metal composition of plants was monitored in subsequent years. Grain yield of the high sludge-loading plots immediately started to recover and gradually leveled off to a yield level similar to that before 1980 (Figure 2). Maize yields from sludge-treated plots, especially from the high rate, were always smaller than that from control plots in subsequent years (Figure 3, Table 7). Additionally, signs of intervenal chlorosis up to the time of tasseling were apparent for some years following termination of sludge application.

The harmful effects observed on crops grown in 1980 and the consistently poor yields in the following years may be explained by Cd toxicity, Ni toxicity, or combined effects of Cd and Ni toxicity. Leaf compositions monitoring over the 1976 to 1986 period demonstrated that Ni toxicity was the most probable cause. The Ni concentration of the 6th leaf reached a peak in 1980, whereas Cd concentration dropped significantly (Figure 2), indicating a possible antagonistic effect of Ni on Cd uptake or translocation in the plants.

Legume Plant Yields

Legumes (e.g., beans, white clover) grown on sludge-treated soils often experience marked reductions in yields.[16,17] Giordano et al.[16] correlated the detrimental influence of sludge application on bean pod yield to the Zn

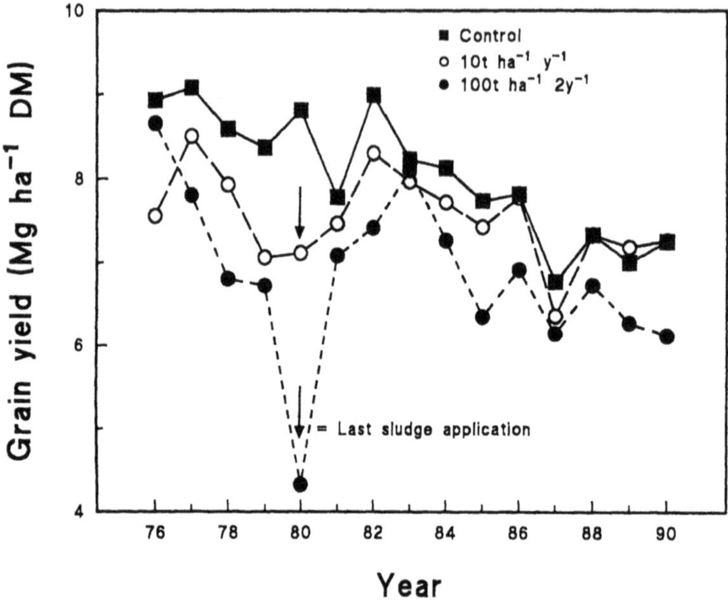

FIGURE 3. Corn grain yield during and after the period of sludge application (1976 to 1980) in the Bx2 experiment at Bordeaux.

Table 7. Corn Grain Yield in the Bx2 Experiment (Bordeaux, France) devoted to the Residual Effect of the Sludge with High Cd and Ni Content (Average Values over the 1976–1986 Period)

Treatment	Cumulative Sludge Inputs (Mg DM ha^{-1})	Grain Yield (kg DM ha^{-1})
Control	0	8406 a
10 Mg DM/ha/y	50	7682 b
100 Mg DM/ha/2 y	300	7032 c

Note: Within a column, values followed by the same letter are not significantly different at the 5% level (Newman Keuls test).

accumulation in plant tissue. Using a data from sludge- or manure-treated soils at the Woburn Market Garden Experiment, McGrath et al.,[17] Giller et al.,[18] and Koomen et al.[19] showed that heavy metal toxicity to white clover was not directly responsible for the decrease in plant dry matter yield. Instead, large concentrations of metals in the soil resulting from 20 years of sludge application were unfavorable to nitrogen fixation by *Trifolium repens* because in this type of soil only ineffective strains of *Rhizobium* survived.[18] Hence, metal accumulated in the soils, as a result of sludge applications, could be more harmful to the microbial activity of soils than to the development of higher plants. The decrease of growth and nitrogen fixation by blue-green algae observed by Brookes et al.[20] in the metal-contaminated soil of the Woburn Market Garden Experiment further substantiated that belief.

Phytotoxicity of Sludge-Borne Metals

There is much literature on primarily phytotoxic elements such as Zn and also Cu if concentrations are large enough (e.g., Zn at 300 mg/kg and above[2,9]) or at acid pH. However, sludge-borne metals do not represent a serious limitation for plant growth. Phytotoxicity is rarely observed even though large amounts of metals are added into the soil, resulting from sludge applications.[6,8,9,21,24,25,39] An exception is the Bx2 Bordeaux experiment in which Cd and Ni loadings were both high. Heckman et al.[21] reported that soybean plants grown on $CdCl_2$ amended soils rates exhibited Cd phytotoxicity symptoms, whereas plants grown in sludge-treated plots at equivalent Cd inputs in the form of sludge were not affected.

Several hypotheses have been put forward to account for the relatively low phytotoxic effect of sludge-borne metals. These include increased soil pH, formation of insoluble salts (e.g., phosphate, sulfate, and silicate salts), metal sorption by iron and manganese oxides, or organic matter.[46,47] The antagonistic effects among the sludge-metals may be another reason for their low phytotoxic effect.

HEAVY METAL ACCUMULATION BY PLANTS GROWN IN LONG-TERM SEWAGE SLUDGE EXPERIMENTS

Crops are links of the food chain, and metals entering the human diet are harmful to health. As a consequence, uptake of sludge-borne metals by crops was investigated by many researchers in long-term experiments that allow more realistic generalizations than glasshouse experiments.

Only limited number of metals were studied in majority of long-term experiments. Generally, investigators have focused on metals that have high concentrations in sludge (e.g., Zn, Mn, Cu) or on potentially hazardous metals (e.g., Cd, Pb, Ni). The technical difficulty, including the inaccuracy associated with analytical procedures of some elements, is an important but sometimes unmentioned reason for so few a number of metals being studied. Therefore, trace elements such as As, Be, Co, Hg, Mo, Se, and Tl, which may have significant concentrations in sludge and may play an important part in environment, are rarely investigated.

Heavy metal accumulation in plants as a result of sludge deposition can be approached by examining the influence of cumulative sludge application on metal accumulation in crops, long-term changes in metal accumulation in crops, and heavy metal accumulation in plants as influenced by metal species, plant species, and plant parts.

Influence of Cumulative Sludge Application on Metal Accumulation in Plants

Generally, metal levels of plants grown on sludge-treated soils are functions of the annual sludge loading rate. This trend was observed in the Bordeaux experiments, especially for Zn, Cd, and Ni (Figures 4, 5, and 6). However, long-term changes in other soil parameters could have also occured to affect chemical behavior of the metals. For example, an elevation of soil pH can reduce metal bioavailability, and consequently, the metal concentration in plant tissue would decline in spite of a continuous rise in the total metal content of the soil. In the Bx1 experiment at Bordeaux, the sludge applications resulted in a decrease of corn leaf Mn concentration compared with control (Figure 7). The marked decline of Mn bioavailability in the soil treated with the medium rate of sludge, despite yearly metal inputs averaging 75 kg Mn/ha, was due to Mn fixation induced by an increase of soil pH from 5.8 to 6.4. Similar results were reported by Chang et al.[23] and Soon et al.[22]

According to Chang et al.,[24] as the metal input increases, metal concentration of the plant tissue approaches an ultimate level, indicating a threshold of the metal uptake by plant. Using this concept, Chang et al.[24] established, for swiss chard, radish leaves, and tuber grown in a long-term experiment starting in 1976, a nonlinear regression mode:

$$y = a + b\,(1 - e^{-cx})$$

where y = concentration in plant tissue, x = total cumulative sludge loading, and a, b, and c are parameters whose values are determined by an adjustment.

Long-Term Changes in Heavy Metal Accumulation in Crops

For given amounts of a pollutant added to soil, the risk of food chain contamination is higher if the pollutant remains in the soil for a long period of time. The long-term effect of sludge addition on metal uptake by crops could be approached by comparing the effects of the annual application of sludge with that of the cumulative amounts in previous years and monitoring metal accumulation in plants in the years successive to termination of sludge deposition.

Metal Accumulation in Crops Resulting from Annual or Cumulative Addition of Sludges

Some researchers, notably Hinesly et al.,[9] stated that the annual input of sludge was a better indicator of metal levels in plant tissues than the cumulative total of input. A similar result was observed in the Bordeaux experiment, where sharp increases of the Mn and Zn concentrations in the 6th leaf were found to be related to the annual sludge inputs (Figures 4 and 7).

Some investigators attempted to quantify the influence of annual or cumulative sludge deposition on plant absorption of metals by defining the relationship

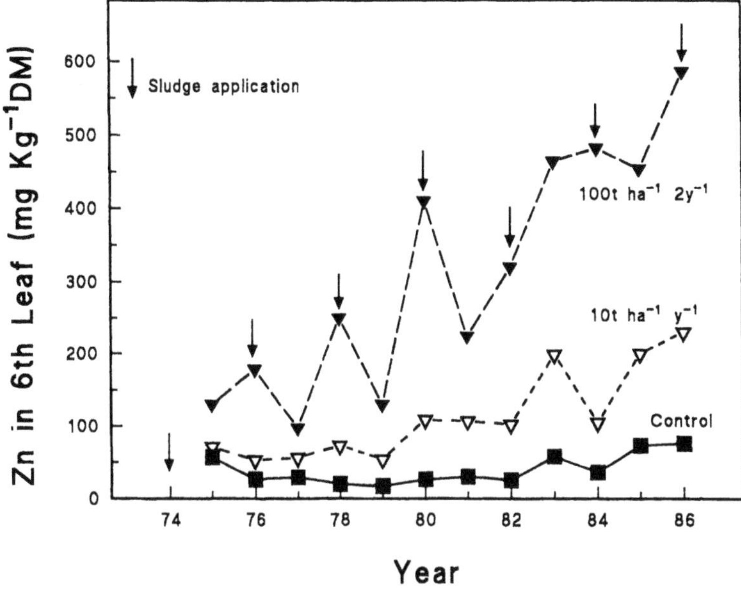

FIGURE 4. Effect of increasing sludge application on the concentration of Zn in the 6th leaf of corn plants in the Bx1 experiment at Bordeaux.

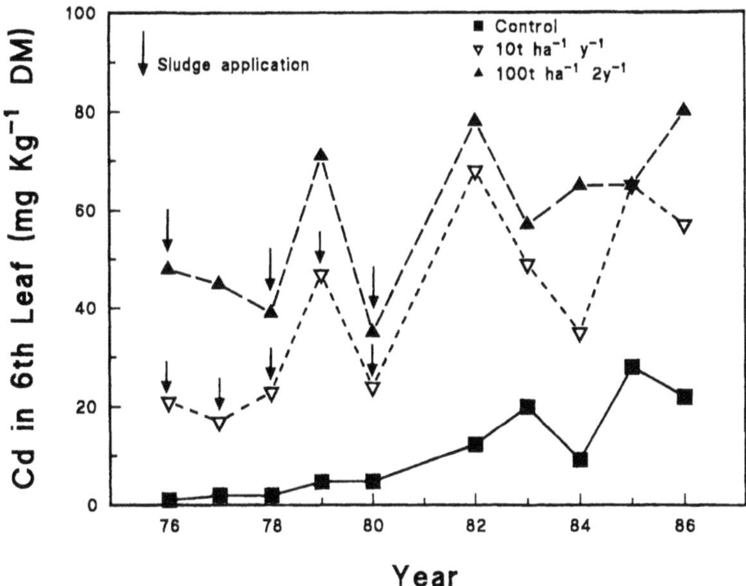

FIGURE 5. Effect of increasing sludge application on the concentration of Cd in the 6th of corn plants in the Bx2 experiment at Bordeaux.

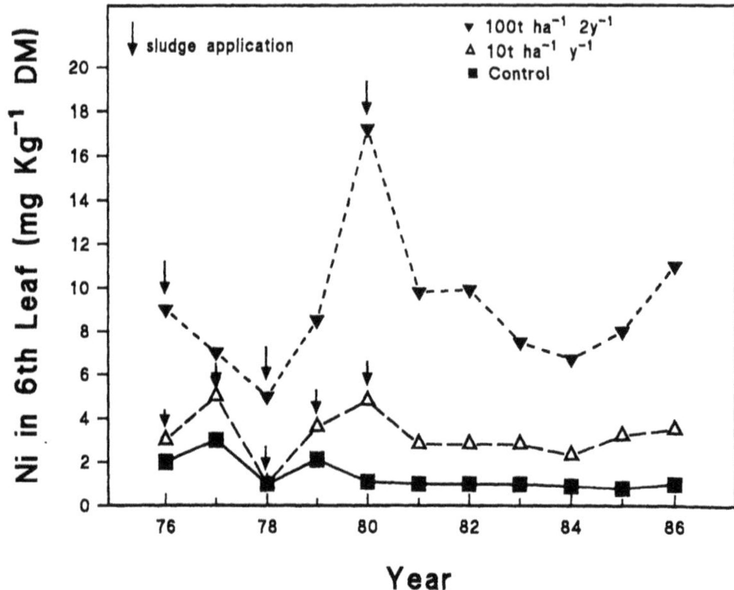

FIGURE 6. Effect of increasing sludge application on the concentration of Ni in the 6th leaf of corn plants in the Bx2 experiment at Bordeaux.

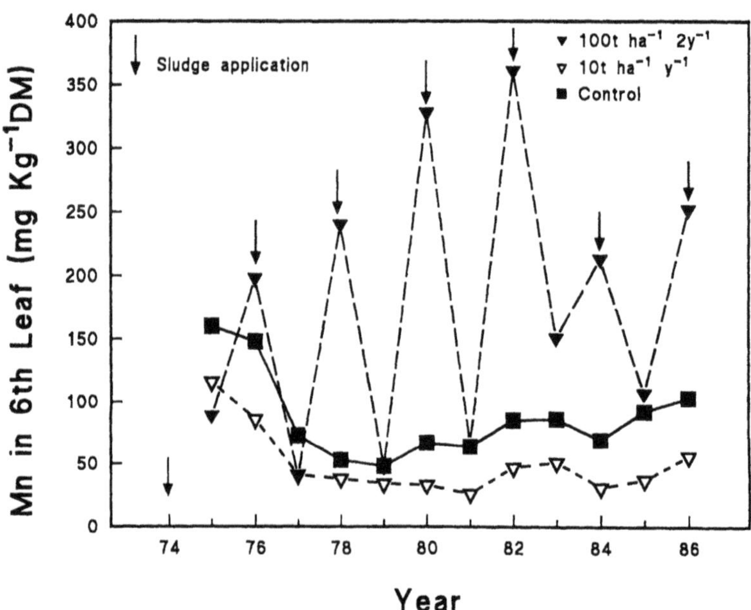

FIGURE 7. Effect of increasing sludge application on the concentration of Mn in the 6th leaf of corn plants in the Bx1 experiment at Bordeaux.

of metal levels in plants vs. either annual or cumulative metal input to the soil using regression analysis (Table 8).

Chang et al.,[24] Soon et al.,[22] and data from the Bordeaux experiments demonstrated that the cumulative metal input to the soil was the major factor determining the metal concentrations in plant tissues (Figures 8 and 9).

Long-Term Changes in Metal Accumulation by Crops After Termination of Sludge Deposition

Sauerbeck and Styperek[1,2] have summarized data from 15 European sewage sludge-field experiments that showed conflicting results on the metal bioavailability in years subsequent to termination of sludge application. In the St. Paul experiment, Cd uptakes by corn decreased significantly with time following terminal sludge application.[25] The data of Hinesly et al.[8] and Kelling et al.[33] showed similar trends. Results of several investigations, however, led to an opposite conclusion.[1,26–29,34] In the Bx2 experiment at Bordeaux, the Cd and Ni concentrations in the 6th leaf of corn did not change significantly for 6 years after sludge deposition was terminated (Figures 5 and 6). The Cd content of corn grain from sludge-treated plots remained significantly higher than that of plants from control plots (Table 9).

Heavy Metal Accumulation in Plants in Relation to Metal, Plant Species, and Plant Parts

The ability of plants to absorb metals from soil is controlled by the chemical species present and the physiological stage of plants at the time of metal translocation.

McGrath[7] estimated the metal uptake of plants grown on the sludge-treated plots of the Woburn experiment with a 20-year cropping sequence and illustrated plant ability to absorb different metals. Although only a small fraction of the added metals might be removed by crops, the results showed that Zn, Ni, and Cd were the most bioavailable metals, whereas Pb and Cr uptake was characterized by a very insignificant rate (Table 10).

In the Bordeaux experiments, the relationship between heavy metal concentrations in the 6th leaf of corn and those in the grain was evaluated in 1986, and the ratio of metal concentration in grain vs. metal concentration in the 6th leaf of plants grown in the control plots was calculated (Table 11). Zinc and Ni have the highest values for this ratio, and the ability of metals to accumulate in the grains is ranked in the following order: Zn > Ni > Fe, Cr > Cd > Cu > Mn > Pb. However, this ratio is an indicator and does not imply that the 6th leaf supplies metal in the grain.

Cadmium

In experiments with low Cd inputs, (i.e., J2 and SP, Tables 1 and 2), the Cd concentration in corn grain decreased to the background level in 4 to 6 years after

Table 8. Influence of Annual or Cumulative Metal Inputs to Soil on Plant Metal Uptake

Metal	Plant	Soil	Regression Equation	Ref.
Cd	Barley grain	Sandy loam	$Y = 0.29 - 0.038\ pH + 0.009\ A + 0.004\ B$	Chang et al.[24]
Cd	Barley grain	Loam	$Y = 0.35 - 0.054\ pH + 0.003\ A - 9 \times 10^{-6}\ B$	Chang et al.[24]
Zn	Barley leaf	Sandy loam	$Y = 126.53 - 15.67\ pH + 0.02\ A + 0.02\ B$	Chang et al.[24]
Zn	Barley leaf	Loam	$Y = 69.12 - 6.43\ pH + 0.0003\ A + 0.002\ B$	Chang et al.[24]
Zn	Corn stover	Loam	$Y = 17.9 - 0.00041\ B + 0.893\ B^2 - 0.079\ B\ pH$	Soon et al.[22]
Cd	Corn stover	Loam	$Y = 0.443 + 0.293\ B - 0.51\ B^2 - 0.0032\ pH$	Soon et al.[22]

Note: Y = Concentration in plant tissue; A = Metal input from the sludge application immediately before the crop; B = cumulative total metal input from previous sludge applications; pH = soil paste pH at the end of the growing season.

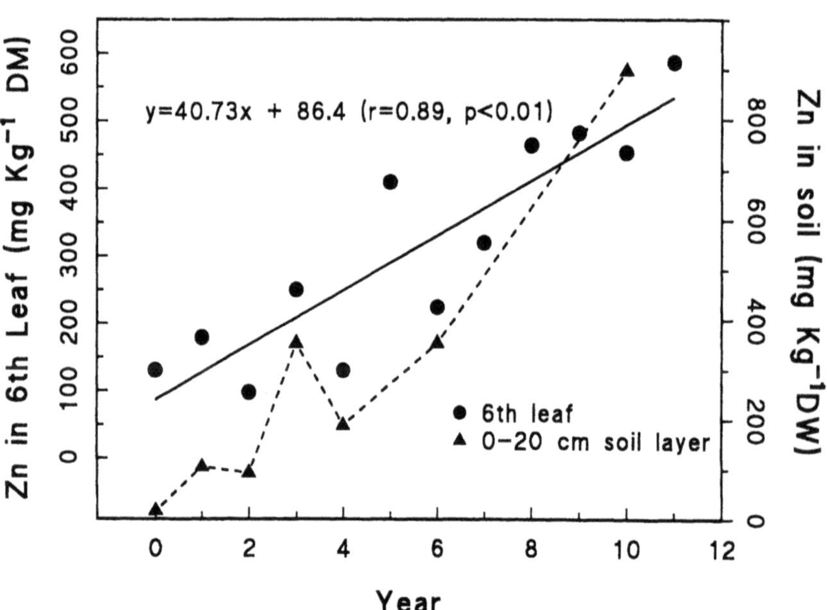

FIGURE 8. Increase in the concentration of Zn in the 6th leaf of corn plants and in the 0 to 20 cm soil layer from the high rate sludge plots (100 Mg DM/ha/2 y) during the period 1975 to 1986 in the Bx1 experiment at Bordeaux.

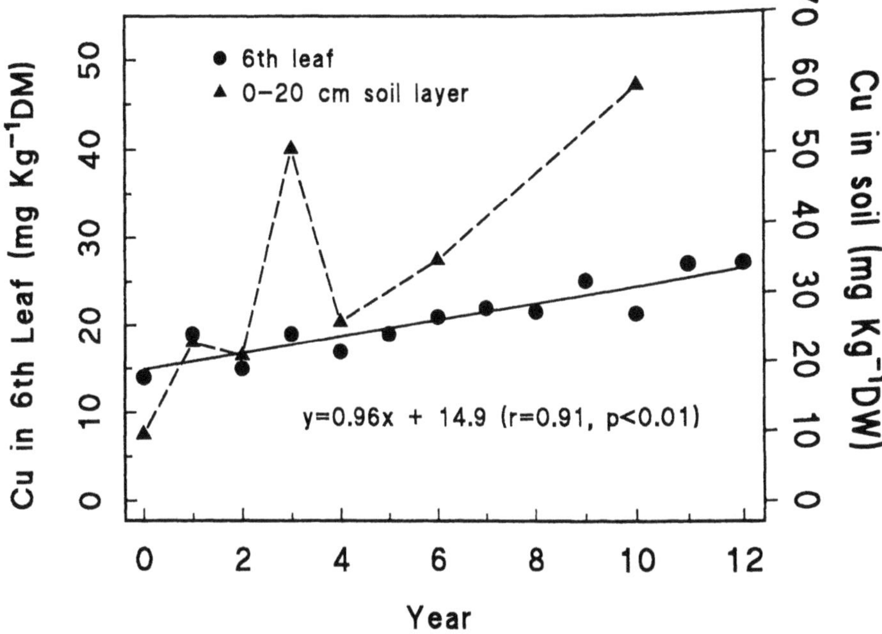

FIGURE 9. Increase in the concentration of Cu in the 6th leaf of corn plants and in the 0 to 20 cm soil layer from the high rate sludge plots (100 Mg DM/ha/2 y) during the period 1975 to 1986 in the Bx1 experiment at Bordeaux.

Table 9. Cd Concentrations in Corn Grain from the Bx2 Experiment (Bordeaux, France) in 1986

Treatment	Cumulative[a] Sludge Inputs (Mg DM ha^{-1})	Cadmium Concentration (µg kg^{-1} DM)
Control	0	166 a
10 Mg DM/ha/y	50	377 b
100 Mg DM/ha/every 2 y	300	435 c

[a] Sludge applications over the 1976–1980 period. Within a column, values followed by the same letter are not significantly different at the 5% level (Newman Keuls test).

terminating sludge application. In the Bx2 experiment at Bordeaux, the Cd concentration in corn grain from high sludge treated plots remained significantly higher than that of control plots at the 6th year after sludge deposition was terminated (Table 9). The amount of Cd accumulated in the grain was moderate compared to those in the leaf (Table 11), but it surely depends on the concentrations of each metal in the soil. In 1984 and 1986, Cd concentration of the 6th leaf of corn at the Bx2 experiment was approaching 80 mg Cd/kg and was among the highest observed value of Cd in the vegetative parts of a grain crop (Figure 2).

Table 10. Metal Removal by Harvested Crops as a Result of Metal Inputs from Sewage Sludge. Data from the Market Garden Experiment at Woburn (U.K.)

Metal	Metal Offtake 1960–1980 (kg ha^{-1})	Offtake (% of Total Inputs)
Zinc	11.74	0.57
Copper	1.33	0.16
Nickel	0.45	0.37
Cadmium	0.18	0.28
Chromium	0.22	0.03
Lead	0.41	0.06

From McGrath, S. P., in *Pollutant Transport and Fate in Ecosystems*, Coughtrey, P. J., J. H. Martin, and M. H. Unsworth, Eds. Special Publ. No. 6 of the British Ecological Society (Oxford: Blackwell Scientific, 1987) pp. 301–317.

Table 11. Ratio of Metal Concentration in Grain vs. Metal Concentration in the 6th Leaf Calculated for Corn Plants from the Control Plots in the Bx1 Experiment (Bordeaux, France)

Metal	R[a]
Cadmium	8.7
Copper	7.8
Chromium	9.7
Manganese	5.8
Nickel	30.8
Lead	0.8
Zinc	38.2

[a] $R = \dfrac{\text{Metal concentration in grain } (\mu g\ kg^{-1}) \times 100}{\text{Metal concentration in the 6th leaf } (\mu g\ kg^{-1})}$

Cereal grains always contained less Cd than their vegetative aerial parts (Tables 12 and 13). According to studies in Germany, a grain Cd concentration of about 1 mg/kg was frequently found in winter wheat grown on soil where the total Cd level is near or above 2 mg/kg. Maximum values of Cd concentration in grain of wheat rarely exceeded 3 mg/kg. Corn grain accumulated less Cd than other cereal crops, probably because it had an efficient barrier to restrict the metal translocation into the grain.

Total soil Cd of 5 to 45 mg/kg resulted in 3 to 18 mg Cd/kg in sugar beet leaves (Table 14). Roots of the same plants contained up to 2.5 mg Cd/kg. In experiments summarized in Sauerbeck and Stypereck,[2] the maximum observed Cd in sugar beets did not exceed 0.6 mg/kg. Sugar beets grown at the Woburn experiments, however, contained up to 11 mg/kg of Cd.

Mean values for Cd concentrations in soybean and tobacco leaves from the experiments on residual effects of sludge at Maryland are summarized in Table 15. The tobacco leaf is widely recognized as a Cd accumulator. Results of these long-term experiments supported the observations.[30,31]

Table 12. Metal Concentrations (µg/kg DM) in Grains from Long-Term Field Trials with Applications of Sewage Sludge

Location	Rate	Plant Species	Metal Inputs	Cd	Cu	Cr	Mn	Ni	Pb	Zn	Fe
Bordeaux (France)	0	Corn	Low	73	1,422	120	6,000		16	29,000	26,000
	Bx 1-10	Corn		82	1,348	103	5,000		22	35,000	26,000
	Bx 1-100	Corn	Max	85	1,807	70	7,000		16	58,000	30,000
	0	Corn	Low	260	1,100		5,000	1,200		28,000	31,000
	Bx 2-10	Corn		690	1,700	94	2,000	3,400	9	29,000	29,000
	Bx 2-100	Corn	Max	830	3,100	94	800	4,900	130	29,000	27,000
Nancy (France)	0	Corn			1,900		7,200			26,000	37,000
	340	Corn		140	2,000		6,000			26,000	49,000
Joliet (U.S.)	J1	Corn	Max	70/100						25,000/27,000	
	J2	Corn	Max	90/810						28,000/56,000	
St. Paul (U.S.)	0	Corn	Low	20	<100	<100		<100	<100	29,600	
	30	Corn		30						37,300	
	45	Corn		40						38,800	
	90	Corn	Max	60	1,500			800	600	39,400	
Woburn (U.K.)	W2	Barley		1,100	78,000	3,300		3,300	12,000	329,000	
Berkeley (U.S.)	Be1	Barley	Max	650						250,000	
	Be2	Barley	Max	200						75,000	
Riverside (U.S.)	0	Barley	Low	40							
	1			50							
	2	(Loam/sandy)		40/60							
	3	(Loam/sandy)	Max	70/130							
Germany[a]		Corn	Max	700						175,000	
		Winter wheat	Low							20,000	
			Max	3,000	<12,500			5,000		250,000	
				40							

[a] Sauerbeck and Styperek.[1,2]

Table 13. Metal Concentrations (Mg/kg DM) in Cereal Leaves from Long-Term Field Trials with Applications of Sewage Sludge

Location	Rate	Plant Species	Metal Inputs	Cd	Cu	Cr	Mn	Ni	Pb	Zn	Fe
Bordeaux (France)	0	Corn	Low	0.84	17	1.2	100	1.2	2.1	50	
	Bx1-10	Corn		0.96	17	0.8	50	0.8	2	200	
	Bx1-100	Corn	Max	3	27	5.7	250	0.8	1.8	600	
	Bx2-10	Corn		57	14	0.9	40	3.5	1.7	63	196
	Bx2-100	Corn	Max	80	18	0.9	31	9	1.7	78	204
Nancy (France)	0	Corn	Low	0.35	9		106			30	420
	340	Corn	Max	1.4	20		116			115	270
Joliet (U.S.)	J1	Corn	Low	0.3						23	
	J1	Corn	Max	2.1						45	
	J2	Corn	Low	0.2						59	
	J2	Corn	Max	10.9						293	
St. Paul (U.S.)	0	Corn	Low	0.08						18.5	
	30	Corn		0.88						66	
	45	Corn		1.26						79	
	90	Corn	Max	1.93						105	
Woburn (U.K.)	W2	Barley	Max	1.6	12	1.2		2.3	8.2	179	
Berkeley (U.S.)	Be1	Barley	Low	0.06						68	
	Be1	Barley	Max	3.4						820	
	Be2	Barley	Low	0.04						57	
	Be2	Barley	Max	0.5						76	
Riverside (U.S.)	0	Barley	Low							22.2	
	1										
	2	(Loam/sandy)									
	3	(Loam/sandy)	Max							57.3	
Germany[a]		Corn	Max	30				5		350	
		Winter wheat	Max	10	20			5	<5	450	

[a] Sauerbeck and Styperek[1,2] and Sauerbeck.[40]

Table 14. Metal Concentrations (mg/kg DM) in Row Crops from Long-Term Field Trials with Applications of Sewage Sludge

Location	Rate	Plant Species	Plant Part	Cd	Cu	Cr	Ni	Pb	Zn
Riverside (U.S.)	Highest	Radish	roots	1.5					100
			leaves	9					400
		Swiss chard		9					375
Woburn (U.K.)	Highest	Sugar beet	roots	11	303	20	12	22	784
			leaves	35	90	46	42	66	1506
		Carrots	roots		11	0.2	5	0.2	101
			leaves		46	4	14	7.8	454
		Potatoes	tubers	4.3	106	5.7	8.6	30	242
Germany[a]		Sugar beet	roots	2.5	25		15	25	1400
			leaves	17	20		40	30	600

[a] Sauerbeck and Styperek.[1,2]

Table 15. Metal Concentrations in Shoots (mg/kg DM) of Dicotyledonous Plants from Long-Term Field Trials with Applications of Sewage Sludge

Location	Rate	Treatment	Plant Species	Cd	Cu	Mn	Ni	Pb	Zn	Fe
Fairland (U.S.)	0		Tobacco	3.9	15	123	2.9	2.9	53	408
	1		Tobacco	10	31	187	3.2	3.2	162	388
	2		Tobacco	12.7	32	217	3.6	3.2	255	368
	3		Tobacco	16.4	36	307	5.4	3.1	389	368
Fairland (U.S.)	0		Soybean	0.06	8.7	49	2.5		24	
	56		Soybean	0.17	8.9	36	4.6		103	
	112		Soybean	0.25	9.5	32	6.1		165	
Beltsville (U.S.)		Nu-Earth	Soybean	1.75	7.5	27	4.3		59	
		CdC_{12}	Soybean	9.03	9.6	45	3		36	
		Lime-digested	Soybean	0.05	8.9	21	0.9		30	
Beltsville (U.S.)		Nu-Earth	Tobacco	69	30	136	13	3.6	335	200
		Lime-digested	Tobacco	3.8	26	32	1	5	48	252
München (FRG)			Spinach	15	16		3		400	
Germany[a]			Spinach	4	12		6	3	400	
			Peas	1	4		3	2	200	
			Beans	0.5	7		23	2	170	

[a] Sauerbeck.[40]

Table 16. Cr, Mn, and Pb Concentrations in Corn Tissues collected in 1986 from the Bx1 Experiment (Bordeaux, France)

Cumulative Sludge Inputs ($Mg\ ha^{-1}$)	6th Leaf			Grain		
	Cr[a]	Pb[a]	Mn[b]	Cr[a]	Pb[a]	Mn[b]
0	1236 a	2110 a	103 a	120 a	16 a	8 a
130	806 b	2010 a	56 b	103 a	22 a	7 a
700	568 c	1858 a	252 c	70 a	16 a	8 a

Note: Within a column, values followed by the same letter are not significantly different at the 5% level (Newman-Keuls test).

[a] $\mu g\ kg^{-1}$ DM.
[b] $mg\ kg^{-1}$ DM.

From Juste, C. and P. Solda, in *Abstractsof the International Conference on Metals in Soils, Waters, Plants and Animals*, Orlando, Florida, April 30–May 3, 1990. University of Georgia, U.S. Department of Energy.

Chromium

Chromium translocation from soil to crops is generally insignificant, and the lowest Cr levels were frequently found in grain, especially in corn grain (70 to 1700 μg/kg) (Table 12). In the Bx1 experiment at Bordeaux, Cr concentration in the 6th leaf of corn decreased with the cumulative sludge application, due possibly to a marked increase of the soil pH (Table 16).

Some field data were available for Cr accumulation in dicotyledonous species (Table 14). The data further confirmed that the translocation of this metal from soil to plant is insignificant. Sauerbeck and Styperek[1] concluded that Cr in the sludge-treated soils does not pose any serious problem for plant growth or food chain transfer of Cr.

Copper

Field trials at Bordeaux showed that Cu concentration in corn leaves was similar regardless of the treatment during the 6 years of the experiment. Afterwards, the Cu concentration of the plant tissues harvested from the high-sludge treatment plots progressively increased. Such a behavior could be an indication that Cu availability was enhanced as the organic matter in sewage sludge decayed or simply due to the increasing concentration of Cu in the soil (Figure 10).

Results from the Woburn experiment demonstrated that Cu concentration in barley grain harvested from soil with more than 200 mg Cu/kg could be very high (Table 12). In some cases, oats and corn grain concentrations might reach values as high as 20 mg/kg.[2] It would be interesting to test whether the Cu concentration in grain is related to its nitrogen content. Sauerbeck and Styperek[1] stated that Cu concentration in plant tissue would remain at a level similar to that of the control,

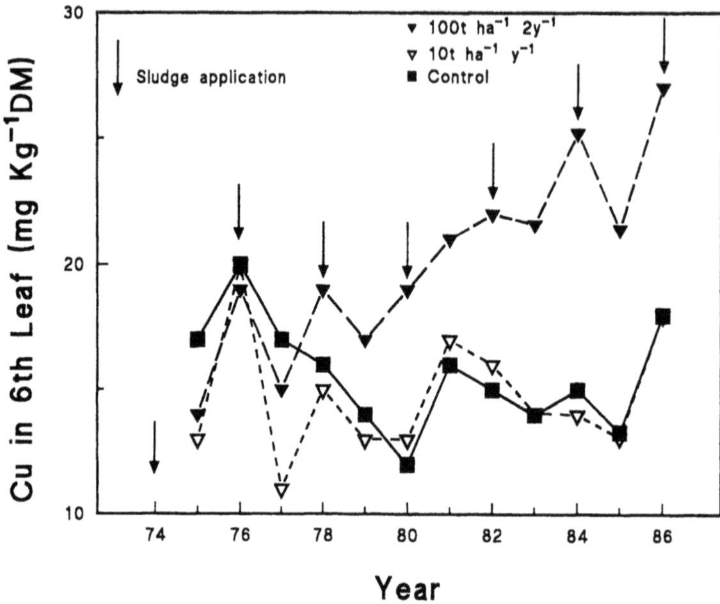

FIGURE 10. Effect of increasing sludge application on the concentration of Cu in the 6th leaf of corn plants in the Bx1 experiment at Bordeaux.

if the Cu concentration of the sludge-treated soil was <100 mg/kg, and they suggested that Cu in soil should not exceed 60 mg/kg instead of the 100 mg/kg.

Sugar beet roots may contain equal or higher amounts of Cu than the corresponding leaves (Table 14), although it is not known to what extent contamination of roots with soil contributes to this. The values observed in the German experiments varied from 5 to 30 mg/kg. Results from the Woburn experiment showed a significant transfer from soil to sugar beet roots, red beet, and potato tubers. Similar to radish, carrots accumulate larger amounts of Cu in leaves than in roots.

Results from the Fairland experiment showed that Cu concentration in tobacco is higher than that in soybean (Table 15), and Cu concentration of tobacco leaves increased with the Cu inputs to soils and leveled off at approximatively 36 mg/kg.

Lead

Caution must be exercised in evaluating Pb accumulated in plants because a significant part of the metal found in the leaves could be from aerial depositions. Pb uptake by crops was generally slight and was not influenced by the metal level or pH of the soil (Tables 12, 13, 14). Sauerbeck and Styperek[1,2] reported unexpected high levels of lead in potato leaves from plants grown on the sludge-treated soils of an experiment in Germany. The Pb concentration of tubers, however, was not affected. The result of Bordeaux experiments demonstrated that Pb uptake by corn grown on sludge-treated soils was insignificant (Table 12) and also poorly translocated into the grain (Table 11). Lead appears to be one of the least mobile and bioavailable metal in the sludge-soil-plant system.

Manganese

Accumulation of Mn in grain and dicotyledonous crops grown on sludge-treated soils is influenced by the soil pH. This correlation may be illustrated by results of the Bordeaux experiments (Figure 7), Mulchi et al.[31] (Table 15), and Kirkham.[34] Comparable observations were made by Soon et al.,[22] who concluded that Mn deficiency may be a potential problem in some soils when sludge is applied. It was also evident from the results of the Bordeaux experiments that Mn concentration in the corn leaves is higher than that in the corn grain, leading to a small ratio (Table 11).

As with corn, the increasing sludge application rates did not significantly increase Mn concentration in dicotyledonous plants (Table 15). On the contrary, Mn concentration of soybean grown on sludged-soil was lower than that on the control. Generally, the addition of sludge produces a significant reduction in Mn content of crops.[32] The chemical behavior of soil Mn is pH dependant, and the application of limed-sludges greatly reduced the Mn concentration in tobacco leaves.[31]

Nickel

Nickel is a relatively mobile element in sludge-treated soils. Results of an experiment in Germany show that Ni concentration in grains (i.e., summer and winter wheat) could reach 0.8 up to 5 mg/kg (Table 12). Ni also accumulated in barley and oats grown in soil that was high in organic carbon. Results from the Bordeaux experiments, however, indicated that Ni uptake by corn grown in sludge-treated soils was not substantial. But the ratio of Ni in vegetative parts vs. Ni in grain was lower than, for example, that of Cd (Table 11). The ability of Ni to be translocated into the grain of wheat or barley casts a question on the possible mechanisms of the enhanced transfer.

Sauerbeck and Styperek[2] reported that Ni accumulated in both sugar beet roots and leaves at concentrations ranging from 5 to 30 mg/kg (Table 14). The Ni concentrations in potato tubers and carrots grown on the same soils were lower than in beets (Table 14). They recommended that anthropogenic Ni inputs to soils be <50 mg Ni/kg.

The dicotyledonous plants accumulated Ni in a lesser extent (Table 15), and the Ni concentrations of soybean and tobacco were low (<10 mg/kg). The levels of Ni in tobacco leaves, however, exhibited a general increasing trend with increasing sludge loadings when unlimed sludges were used, but the reverse was true if lime-treated sludges were applied. Increase of soil pH resulting from application of lime-treated sludge reduced the Ni concentration in soybean and tobacco leaves (Table 15).

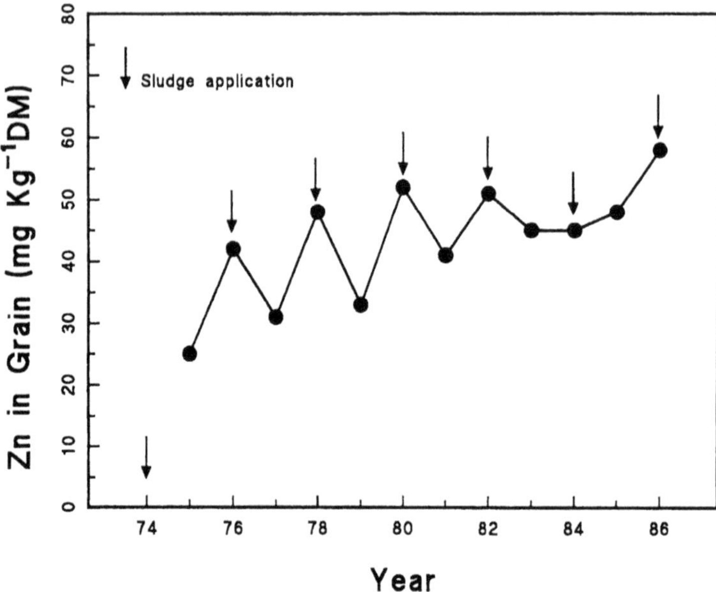

FIGURE 11. Effect of increasing sludge application on the concentration of Zn in the grain of corn plants in the Bx1 experiment at Bordeaux.

Zinc

Long-term field experiments demonstrated that Zn is the most bioavailable metal in sludge-treated soils. In the Bx1 experiment at Bordeaux, the Zn concentration in the corn's 6th leaf obtained from the high rate sludge plots was sharply higher than that in the control, from the beginning (Figure 4). Except for 1976 and 1977, the mean Zn concentrations of leaves progressively increased year after year and reached 600 mg/kg in 1986. But corn containing this level of Zn did not exhibit any phytotoxicity symptom. The phytotoxicity threshold of 300 mg/kg, quoted by Hinesly et al.,[9] perhaps underestimated plant tolerance of Zn. The Zn concentration of grain was significantly higher than the control and fluctuated yearly up to 1982 (Figure 11). It leveled off thereafter at approximately 60 mg/kg, indicating a higher mobility in plants than other metals (Table 11). Barley grown on sludge-treated soils also accumulated Zn, and metal concentration in barley leaves and grains reached as high as 820 and 250 mg/kg, respectively (Tables 12 and 13).

Results from experiments conducted at Woburn and in Germany showed that Zn concentration of sugar beet leaves could be as high or higher than Zn concentrations of the soils they were grown (Table 14). Sauerbeck and Styperek[1] reported that signs of Zn toxicity were apparent on the sugar beet crops grown on soils with high levels of Zn. The Zn accumulation in potato tubers and carrot roots, however, was significantly less than in beets. The Zn concentration in tobacco leaves was larger than that in soybean, although the crops were grown on different sites with different sludges (Table 15). Again, soils receiving lime-treated sludges had higher soil pH and produced plants that accumulated less Zn (Table 15).

CONCLUSIONS

Data from long-term field investigations were used to evaluate transfers of metals from sludges to soils and plants. Results may be used as the basis for developing guidelines to prevent metal induced toxicity to plants, animals, and humans when sewage sludges are applied to agricultural soils.

1. The sludge-borne metals in soils are accumulated in the surface layer. The lack of downward movement and the lateral displacement of metals caused by cultivation contributed to the varying metal recovery percentages with respect to time and metal species. The potential for metals to be transported outside of the experimental plots requires the attention of researchers and casts an uncertainty on evaluating data of the long-term experiments.

2. The sludge-borne metals do not cause serious toxicity to the majority of crops, even if large amounts of metal are added to soil.

3. Zinc is the most bioavailable sludge-borne metal. Cadmium and nickel are also bioavailable, but the ability for the absorbed Cd and Ni to move from vegetative parts into grains is substantially less than the Zn.

4. Despite results provided by the long-term field experiments, information is needed (1) to evaluate long-term changes in factors controlling metal bioavailability of the soils and (2) to determine long-term changes of soil properties as a result of metal accumulation.

In other respects, as stated by Sauerbeck and Styperek,[1] possible changes in the bioavailability of metallic contaminants after long-term residence in soils is a question that remains to be answered.

Finally, it is obvious that only few works have been devoted to evaluate plant absorption of other trace elements such as Be, Co, Hg, Mo, As, Se, and Tl during and following termination of sludge applications.

ACKNOWLEDGMENTS

We are grateful to P. Solda for valuable assistance in the management of long-term experiments at Bordeaux. We would like to thank Dr. S. P. McGrath, Rothamsted Exp. Station, AFRC, Harpenden (U.K.), and the Referees (Prof. A. C. Chang and Ms. Linda Candelaria of the University of California, Riverside; Dr. D. Sauerbeck, Inst. of Plant Nutrition and Soil, Braunschweig, Germany) for suggestions and comments. We wish also to express our sincere appreciation to additional institutions cooperating in this paper and to Dr. D. C. Adriano, University of Georgia, Savannah River Ecology Laboratory, Aiken SC (U.S.) for this fruitful cooperation.

REFERENCES

1. Sauerbeck D. R. and P. Styperek. "Long-term Effects of Contaminants," in *Processing and Use of Organic Sludge and Liquid Agricultural Wastes*, L'Hermite, P., Ed. Proc. 4th Int. Symp., Rome, October 8–11, 1985 (Dordrecht: Reidel, 1986) pp. 318–325.
2. Sauerbeck, D. R. and P. Styperek. "Heavy Metals in Soils and Plants of 25 Long-term Field Experiments Treated with Sewage Sludge," in *Agricultural Waste Management and Environmental Protection*, Welte, E. and I. Szabolcs, Eds. 4th Int. Symp. CIEC, Braunschweig (FRG), May 11–14, 1987 (Belgrade: International Scientific Centre of Fertilizers, CIEC; Braunschweig: Federal Agric. Res. Ctr., FAL, 1988) pp. 439–451.
3. Chang, A. C., J. E. Warneke, A. L. Page, and L. J. Lund. "Accumulation of Heavy Metals in Sewage Sludge-treated Soils," *J. Environ. Qual.* 13(1):87–91 (1984).
4. Williams, D. E., J. Vlamis, A. H. Pukite, and J. E. Corey. "Metal Movement in Sludge-treated Soils After Six Years of Sludge Addition. 1. Cadmium, Copper, Lead and Zinc," *Soil Sci.* 137(5):351–398 (1975).
5. Williams, D. E., J. Vlamis, A. H. Pukite, and J. E. Corey. "Metal Movement in Sludge-amended Soils: A Nine-year Study," *Soil Sci.* 143(2):124–130 (1987).
6. Chang, A. C., A. L. Page, and F. T. Bingham. "Heavy Metal Absorption by Winter Wheat Following Termination of Cropland Sludge Applications," *J. Environ. Qual.* 11(4):705–708 (1982).
7. McGrath, S. P. "Long-term Studies of Metal Transfers Following Application of Sewage Sludge," in *Pollutant Transport and Fate in Ecosystems*, Coughtrey, P. J., J. H. Martin, and M. H. Unsworth, Eds. Special Publ. No. 6 of the British Ecological Society (Oxford: Blackwell Scientific, 1987) pp. 301–317.
8. Hinesly, T. D., E. L. Ziegler, and G. L. Barrett. "Residual Effects of Irrigating Corn with Digested Sewage Sludge," *J. Environ. Qual.* 8(1):35–38 (1979).
9. Hinesly, T. D., R. L. Jones, E. L. Ziegler, and J. J. Tyler. "Effects of Annual and Accumulative Applications of Sewage Sludge on Assimilation of Zinc and Cadmium by Corn (*Zea mays* L.)," *Environ. Sci. Technol.* 11(2):182–188 (1977).
10. McGrath, S. P. and P. W. Lane. "An Explanation for the Apparent Losses of Metals in a Long-Term Field Experiment with Sewage Sludge," *Environ. Pollut.* 60:235–256 (1989).
11. Lindsay, W. L. and W. A. Norwell. "Development of a DTPA Soil Test for Zinc, Iron, Manganese and Copper," *Soil Sci. Soc. Am. J.* 42:421–428 (1978).
12. Morel, J. L. and A. Guckert. "Evolution en Plein Champ de la Solubilité dans DTPA des Métaux Lourds du Sol Introduits par des Epandages de Boues Urbaines Chaulées," *Agronomie* 4(4):377–386 (1984).
13. Grant, R. O. and S. E. Olesen. "Sludge Utilization in Spruce Plantations on Sandy Soils," in *Utilisation of Sewage Sludge on Land: Rates of Application and Long Term Effects of Metals*. Berglund, S., R. D. Davis, and P. L'Hermite, Eds., Proc. of a Seminar held at Uppsala, June 7–9, 1983 (Dordrecht: Reidel, 1984) pp. 77–90.
14. Piets, R. I., J. R. Peterson, T. D. Hinesly, E. L. Ziegler, K. E. Redborg, and C. Lue-Hing. "Sewage Sludge Application to Calcareous Strip-mine Spoil: I. Effect on Corn Yields and N, P, K, Ca, and Mg Compositions," *J. Environ. Qual.* 11 (4):685–689 (1982).

15. Piets, R. I., J. R. Peterson, T. D. Hinesly, E. L. Ziegler, K. E. Redborg, and C. Lue-Hing. "Sewage Sludge Application to Calcareous Strip-mine Spoil: II. Effect on Spoil and Corn Cadmium, Copper, Nickel and Zinc," *J. Environ. Qual.* 12 (4):463–467 (1983).
16. Giordano, P. M., J. J. Mortvedt, and D. A. Mays. "Effect of Municipal Wastes on Crop Yields and Uptake of Heavy Metals," *J. Environ. Qual.* 4(3):394–398 (1975).
17. McGrath, S. P., P. C. Brookes, and K. E. Giller. "Effects of Potentially Toxic Metals in Soil Derived from Past Applications of Sewage Sludge on Nitrogen Fixation by *Trifolium repens* L., " *Soil Biol. Biochem.* 20(4):415–424 (1988).
18. Giller, K. E., S. P. McGrath, and P. R. Hirsch. "Absence of Nitrogen Fixation in Clover Grown on Soil Subject to Long-term Contamination with Heavy Metals is Due to Survival of Only Ineffective *Rhizobium*," *Soil Biol. Biochem.* 21(6):841–848 (1989).
19. Koomen, I., S. P. McGrath, and K. E. Giller. "Mycorrhizal Infection of Clover is Delayed in Soils Contaminated with Heavy Metals from Past Sewage Sludge Applications," *Soil Biol. Biochem.* 22(6):871–873 (1990).
20. Brookes, P. C., S. P. McGrath, and C. Heijnen. "Metal Residues in Soils Previously Treated with Sewage-Sludge and Their Effects on Growth and Nitrogen Fixation by Blue-Green Algae," *Soil Biol. Biochem.* 18(4):345–353 (1986).
21. Heckman, J. R., J. S. Angle, and R. L. Chaney. "Residual Effects of Sewage Sludge on Soybean: I. Accumulation of Heavy Metals," *J. Environ. Qual.* 16(2):113–117 (1987).
22. Soon, Y. K., T. E. Bates, and J. R. Moyer. "Land Application of Chemically Treated Sewage Sludge: III. Effects on Soil and Plant Heavy Metal Content," *J. Environ. Qual.* 9(3):497–504 (1980).
23. Chang, A. C., A. L. Page, J. E. Warneke, M. R. Resketo, and T. E. Jones. "Accumulation of Cadmium and Zinc in Barley Grown on Sludge-Treated Soils: A Long-Term Field Study," *J. Environ. Qual.* 12(3):391–397 (1983).
24. Chang, A. C., A. L. Page, and J. E. Warneke. "Long-term Sludge Applications on Cadmium and Zinc Accumulation in Swiss Chard and Radish," *J. Environ. Qual.* 16(3):217–221 (1987).
25. Bidwell, A. M. and R. H. Dowdy. "Cadmium and Zinc Availability to Corn Following Termination of Sewage Sludge Applications," *J. Environ. Qual.* 16(4):438–442 (1987).
26. Dowdy, R. H., W. E. Larson, J. M. Titrud, and J. J. Latterell. "Growth and Metal Uptake of Snap Beans Grown on Sewage Sludge-amended Soil: A Four-year Field Study," *J. Environ. Qual.* 7(2):252–257 (1978).
27. Giordano, P. M., J. J. Mortvedt, and D. A. Lays. "Residual Effects of Municipal Wastes on Yield and Heavy Metal Content of Sweet Corn and Bush Beans," in *Agronomy Abstracts, 1979 Annual Meetings* (Fort Collins, CO: American Society of Agronomy, Crop Science Society of America and Soil Science Society of America, August 5–10, 1979) p. 29.
28. Larsen, K. E. "Cadmium Content in Soil and Crops after Use of Sewage Sludge," in *Utilisation of Sewage Sludge on Land: Rates of Application and Long-term Effects of Metals*. Berglund, S., R. D. Davis and P. L'Hermite, Eds., Proc. of a Seminar held at Uppsala, June 7–9, 1983 (Dordrecht: Reidel, 1984) pp. 157–165.

29. Berrow, M. M. and J. C. Burridge. "Persistence of Metals in Available Form in Sewage Sludge Treated Soils under Field Conditions," in *Proc. 3rd Int. Conf. on Heavy Metals in the Environment* (Amsterdam: Commission of the European Communities and World Health Organization, CEP Consultants Ltd. Sept. 1981) pp. 202–205.
30. Bell, P. F., C. A. Adamu, C. L. Mulchi, M. McIntosh, and R. L. Chaney. "Residual Effects of Land Applied Municipal Sludge on Tobacco. I. Effects on Heavy Metal Concentrations in Soils and Plants," *Tobacco Sci.* 33–37:46–50 (1988).
31. Mulchi, C. L., P. F. Bell, C. Adamu, and R. Chaney. "Long-term Availability of Metals in Sludge Amended Acid Soils," *J. Plant Nutr.* 10(9–16):1149–1161 (1987).
32. De Haan, S. "Results of Large-scale Field Experiments with Sewage Sludge as an Organic Fertilizer for Arable Soils in Different Regions of the Netherlands," *Long-term Effects of Sewage Sludge and Farm Slurries Applications.* Williams, J. H., G. Guidi and P. L'Hermite, Eds., Proc. of a Round-Table Seminar held at Pisa, Italy, September 25–27, 1984 (Dordrecht: Reidel, 1985) pp. 57–72.
33. Kelling, K. A., D. R. Keeny, L. M. Walsh, and J. A. Tyan. "A Field Study on the Agricultural Use of Sewage Sludge: III. Effect on Uptake and Extractability of Sludge-Borne Metals," *J. Environ. Qual.* 6(4):352–358 (1977).
34. Kirkham, M. B. "Trace Elements in Corn Grown on Long-term Sludge Disposal Site," *Environ. Sci. Technol.* 9(8):765–768 (1975).
35. Juste, C. and P. Solda. "Effect of a Long Term Application of Sewage Sludge on Heavy Metal Uptake by a Continuous Maize Crop in a Sandy Soil," in *Abstr. Int. Conf. on Metals in Soils, Waters, Plants and Animals*, Orlando, Florida, April 30–May 3, 1990. University of Georgia, U.S. Department of Energy.
36. Juste, C. and P. Solda. "Effect on a Long Term Sludge Disposal on Cadmium and Nickel Toxicity to a Continuous Maize Crop," in *Processing and Use of Organic Sludge and Liquid Agricultural Wastes.* L'Hermite, P., Ed. Proc. 4th Int. Symp. held in Rome, October 8–11, 1985 (Dordrecht: Reidel) pp. 336–347.
37. Juste, C. and P. Solda. "Heavy Metal Availability in Long-term Experiments," in *Factors Influencing Sludge Utilization Practices in Europe.* Davis, R. D., H. Haeni, and P. L'Hermite, Eds., Proc. of the Round-Table Seminar held in Liebefeld (CH) May 8–10, 1985 (London: Elsevier, 1986) pp. 13–23.
38. Morel, J. L., J. C. Pierrat, and A. Guckert. "Effet et Arrière-effet de l'Epandage de Boues Urbaines Conditionnées à la Chaux et au Chlorure Ferrique sur la Teneur en Métaux Lourds d'un Maïs," *Agronomie* 8:107–113 (1988).
39. Vlamis, J., D. E. Williams, J. E. Corey, A. L. Page, and T. J. Ganje. "Zinc and Cadmium Uptake by Barley in Field Plots Fertilized Seven Years with Urban and Suburban Sludge," *Soil Sci.* 139:81–87 (1985).
40. Sauerbeck, D. R. "Plant, Element and Soil Properties Governing Uptake and Availability of Heavy Metals Derived from Sewage Sludge," in *Water, Air, Soil Pollut.* 57–58: 227–237 (1991).
41. Luebben, S., E. Rietz, and D. Sauerbeck. "Heavy Metals in the Subsoil of Field Experiments Treated with Sewage Sludge," in *Abstr. Int. Conf. on Metals in Soils, Waters, Plants and Animals*, Orlando, FL, April 30–May 3, 1990. University of Georgia, U.S. Department of Energy.

42. Werner, W., T. Delschen, and C. Birke. "Soil-Plant Relationships of Heavy Metals After Long-term Application of Sewage Sludge," in *Proc. Int. Conf. Heavy Metals in the Environment*, Geneva (CH), September 1989 (Edinburgh: CEP Consultants Ltd., 1989) pp. 107–110.
43. Sauerbeck, D. and P. Styperek. "Predicting the Cadmium Availability from Different Soils by $CaCl_2$-Extraction," in *Processing and Use of Sewage Sludge*. L'Hermite, P. and H. Ott, Eds., (Dordrecht: Reidel, 1984) pp. 431–434.
44. Birke, C., G. Hasselbach, S. Lübben, and A. Schaller. "Suitability of Chemical Extraction Procedures for Prediction of Heavy Metal Transfer from Sewage Sludge-treated Soils into Plants," in *Abstr. Int. Conf. on Metals in Soils, Waters, Plants and Animals*, Orlando, FL, April 30–May 3, 1990. University of Georgia, U.S. Department of Energy.
45. Station Fédérale de Recherches en Chimie Agricole et sur l'Hygiène de l'Environnement, Ed. "Méthode pour Déterminer dans les Sols les Concentrations en Métaux Lourds Disponibles pour les Plantes et les Microorganismes et Vérification dans des Régions Contaminées," Rapport final COST 681 (Les Cahiers de la FAC Liebefeld, Bern, 2, 1990).
46. Emmerich, W. E., L. J. Lund, A. L. Page, and A. C. Chang. "Solid Phase Forms of Heavy Metals in Sewage Sludge-treated Soils," *J. Environ. Qual.* 11:178–181 (1982).
47. Bell, P. F., B. C. James, and R. L. Chaney. "Heavy Metal Extractability in Long-term Sewage Sludge and Metal Salt-amended Soils," *J. Environ. Qual.* 20:481–486 (1991).

7

Fate of Trace Metals in Sewage Sludge Compost

C. L. Henry and R. B. Harrison

University of Washington, College of Forest Resources, AR-10, Seattle, WA 98195

ABSTRACT

The nature of Cd, Cr, Cu, Ni, Pb, and Zn in sewage sludge compost was investigated, including studies on the solubility, retention, and uptake of metals by plants. Compost used in the studies was produced from a mixture of 3:1 (v/v) wood chips and 25% solids sewage sludge composted in large static piles for a period of 1 year. A laboratory study was conducted using sequential extractions with distilled water, $MgCl_2$, and pyrophosphate to establish the strength of trace metal bonding within the mature and two-year-old field-aged compost. After 9 months, compost was sampled by depth and analyzed for change in metal concentration. PVC tubes installed in field sites were removed and analyzed after 2, 6, 12 and 24 month incubations. Measurements of weight loss were used to assess decomposition, and, together with change of metals concentration, total loss of metals from the compost. Additionally, grass, lettuce, carrots, and tomatoes were field grown in 100% compost, 50% compost, soil fertilized with agronomic levels of NPK, and control soil. Vegetation samples were collected and analyzed for trace metal concentrations for two consecutive growing seasons. The bulk of trace metals in composted sewage sludge were not available either for plant uptake or leaching through the soil. Distilled water extracts accounted for <1% of the total for all metals in newly finished compost and less than 1% of total for Cd, Cr, Cu, and Pb in two-year-old compost. Exchangeable trace metals varied considerably, measured in the following order: Cu (0.8–0.9%) < Cr (3.5–7.7%) < Pb (6.8–7.4%) < Ni (12–16%) < Zn (15–52%) < Cd (32–48%). Residual metals (not leached by distilled water, $MgCl_2$, or $Na_4P_2O_7$) is nearly at or well over 50% for all metals, and for Cd, Cu, and Pb greater than 80%. Those metals that showed little water or $MgCl_2$ leaching (Cr and Cu) were similarly conservative after 9 months in the field. Metal concentrations increased in the compost due to reduction of compost mass by decomposition. Small loss of Cd and Zn and relatively large loss of Ni from the compost occurred after 9 months in the field. Uptake slopes of plants grown in sludge compost are in the order of Cd > Zn > Ni > Cu > Cr. In general, Cd, Cr, and Cu are similarly available for plant uptake in compost (at a significantly lower pH) as compared to the soil, while Ni and Zn are slightly more available. Uptake rates of trace metals by the plants in this study are generally in the order of lettuce > grass > carrots > tomatoes.

INTRODUCTION

Composting is a promising method of managing sewage sludge since sludge compost is often a more desirable material to handle than sludge in the original form. The primary objective of composting is to biologically stabilize organic matter and create a humus-like material. Advantages of sludge compost over sewage sludge include more acceptable appearance, significant pathogen and odor reduction, and potentially the dilution of sludge contaminants by the addition of bulking material.

Despite these potential improvements in quality, products manufactured from municipal sewage sludge still contain varying amounts of trace metals (Cd, Cr, Cu, Ni, Pb, Zn). Such metals are of environmental concern due to their potential effects on plant uptake and growth, and the further potential for increasing exposure of animals and humans to metals. The composting process, which is an aerobic process, may result in chemical forms of metals in sludge compost that act differently from those in anaerobically digested sludge when placed in the environment.

Trace Metal Extraction

The importance of the chemical form of metal on plant uptake or movement in soil is best illustrated in the Peer Review of Standards for the Disposal of Sewage Sludge.[1] Trace metals in a salt form (i.e., soluble in the soil solution) are much more available for plant uptake and potential toxicity than those held in the sludge organic matrix. The chemistry of sludge compost has not been studied nearly as exhaustively as uncomposted sludge; however, the scientific community has produced literature indicating the effect/fate of trace metals after composting and use as a soil amendment. Studies of chemistry of metals in sludge compost and soils receiving sludge compost applications have included both simple one-extractant and sequential extraction procedures in order to determine the chemical forms of metals.

As would be expected, Barbera found the application of compost to agricultural soils increased both the total and the available concentrations of some metals in the soil.[2] Significant linear relationships held among the different analyses performed on Zn and Pb including total, EDTA, and DTPA extractable. EDTA- and DTPA-extracted Zn were found to be linear with slopes of about 75 and 33%, respectively, of total Zn. Pb was shown to have similar slopes, but other metals did not develop linear relationships.

It has been suggested that metals become less available when sludge is added to soils.[3,4] A study was conducted on sandy loam and loam soils to which 0, 22.5, 45, and 90 dry metric tons (dt) ha^{-1} yr^{-1} composted sewage sludge was applied annually for 7 years. Soils were analyzed for relative metal solubility by a sequential extraction procedure to determine the distribution of solid phase chemical forms of the sludge-borne trace metals. In the untreated soils, essentially all the trace metals were present in either the solid-residue form (for Cr, Cu, Ni, and Zn) or the carbonate form (for Cd and Pb). The authors found that, although every extracted fraction showed some increase in the amount of trace metals with higher compost application rates, the most significant increases occurred in the carbonate fraction, and sometimes in the organically bound fraction. Three years after application the distribution of metals was essentially the same as the first year of application.

In another study, DTPA-extractable Zn, Cu, Cd, and Ni were monitored throughout a 2.4-year experiment where compost was added to a sandy soil.[5] Extractable Cd, Zn, and Cu decreased with time, indicating increased retention with time. Soil Ni extractability tended to increase with time.

Organo-mineral interactions were characterized by Vedy et al. by a granulometric fractionation and by chemical extractions for a compost enriched with trace metals.[6] The granulometric fractionation indicated that, in general, higher concentrations of metals were found in finer particle sizes (<50 μm). Chemical extraction suggested that both Cu and Zn are strongly associated with organic matter (67 to 88% and 66 to 96%, respectively), while organic chelates are scarce. Little redistribution of metals were found following a 21-day incubation period. The authors found that increasing Cd concentration up to 15 $\mu g\ g^{-1}$ had little effect on microbial respiration, while 1500 μg Cu g^{-1} caused a slightly higher reduction and 3500 μg Zn g^{-1} reduced respiration by over 50%.

Trace Metal Retention

Movement of trace metals through the soil profile has long been a concern, yet seldom is significant movement found following sludge application.[7] The lack of significant metal movement also appears to be the rule for sludge compost, largely due to the forms of metals suggested by chemical extractions.

Darmody et al. added composted municipal sewage sludge to a silt loam soil in a tree nursery at rates of 0, 150, and 300 dt ha^{-1}.[8] Soils were sampled to a depth of 150 cm in 25-cm increments each year for 3 years after compost addition, and pH and extractable Cu and Zn were analyzed.

Surface soil levels of Cu and Zn increased with increased application rate, as would be expected, and high levels of Cu and Zn remained stable in surface soil layers over time. However, some of the extractable elements added by compost were found to be mobile. In the 300 dt ha^{-1} plots, extractable Cu and Zn were significantly increased to a depth of 75 cm. Depth of significant increase was less in the 150 dt ha^{-1} plots. Zinc (but not Cu) showed a gradual decline in concentration at the surface. In the 300 dt ha^{-1} plots, Cu concentration reached a maximum in the B horizon.

Tackett et al. determined that pH can affect leaching rates of metals in some cases.[9] Short-term leaching rates of the metals Cd, Fe, Pb, and Zn from composted sewage sludge were determined over the pH range of 2.5 to 7.0 at unit intervals of 0.5 pH. Only Zn and Cd leached significantly faster as the pH was lowered, with both showing the greatest relative solubility increase over the pH range of 5.5 to 6.0.

Plant Uptake of Metals

The bulk of the research regarding trace metal addition to soils from compost applications has concentrated on the effects of metals on plants. Generally, increases in soil concentrations have caused increases in plant tissues. This appears to be true for both crops grown for direct human food consumption and for other plant species such as trees.

Direct Human Food Crops

Bledsoe found that vegetables (beans, beets, tomatoes, lettuce, carrots, spinach, and radishes) grew well in composted sludge in greenhouse and nursery trials, but that trace metal uptake was significantly increased.[10] Foliar metal levels were 15 to 50 times greater than the soil-grown plants. Normally, the levels of metals were not excessive for either plant nutrition or human consumption as found by Falahi-Ardakani et al.[11] They grew six vegetable species (broccoli, cabbage, lettuce, eggplant, pepper, and tomato) for 8 weeks in a medium of composted sewage sludge compost, perlite, and peat (equal parts by volume). Zinc and Cd were accumulated at rates of 4 to 10 mg wk^{-1} and 0.13 to 2.4 mg wk^{-1}, respectively, by the plants. Digested sewage sludge compost was applied to a sandy soil at rates of 0, 56, 112, and 448 dt ha^{-1}.[5] The pearl millet grown as a summer crop had Cd levels significantly increased with increasing rates of compost from about 0.40 to 1.62 μg g^{-1}. Tunison et al. composted sewage sludge with bark shredder screenings mixed with acid minespoil and tested to determine the effect on blueberries.[12] Berries grown in compost media contained <0.6 μg Cd g^{-1}, 4 to 5 μg Cr g^{-1}, 2 to 3 μg Cu g^{-1}, 1.2 μg Ni g^{-1}, <1.2 μg Pb g^{-1}, and 12 to 13 μg Zn g^{-1}. None of these were significantly different from the controls.

Some trace metal toxicity has been suggested from pot studies. Sterrett et al. grew tomato and cabbage transplants in a media containing sewage sludge compost from either a residential, low metal (LM) or industrially contaminated, high metal (HM) source, with 0.4, 6, and 60 mg kg^{-1} Cd, respectively.[13] The transplant quality (stem diameter, fresh and dry weight) of plants grown in LM were similar to controls; however, transplants grown in HM media were small, less developed, and exhibited symptoms of trace metal phytotoxicity. The concentrations of Zn, Fe, Mn, Cu, Pb, Cd, and Ni in tomato and of Zn, Cu, Cd, and Ni in cabbage plants grown in HM media were significantly higher than those in transplants grown in LM media. Lowering the pH of the HM media by addition of peat moss further increased the concentrations of Zn, Mn, Cu, and Ni in tomato and Zn and Mn in cabbage transplants, in some cases to phytotoxic levels. Later, Sterrett et al. found compost source had no significant effect on transplant quality of muskmelon seedlings.[14] Also, following planting, marketable yield of cabbage, tomato, and muskmelon was similar for controls, HM and LM transplants. Transplant media had little influence on the trace metal concentrations found in either foliar samples or the edible portions of the crops studied. Trace metal concentrations in foliar samples of all three crops were lower than those found in transplants. The lowest concentrations were found in tomato and muskmelon fruit and in cabbage heads.

Falahi-Ardakani et al. studied the effect of pH on Cd uptake.[15] They grew parthenocarpic cucumber plants in containers filled with 0, 25, and 50% by volume compost and potting medium (made from peat moss and vermiculite). Compost was made from ferric chloride precipitated, lime-stabilized, digested sewage sludge composted with wood chips. The Cd concentration of leaf and fruit samples from plants grown in media amended with 25 or 50% compost was unaffected by changes in pH from 7.2 to 3.4.

Tree Species

A number of studies have evaluated metal uptake by various tree species. Coleman et al. found as much as six times higher root concentration of Cd and Zn in nursery beds amended with 0, 513, 1000, and 1530 m^3 ha^{-1} of composted sewage sludge (3:1 fir-hemlock sawdust:municipal sewage sludge from Seattle) and planted with Douglas fir, noble fir, and ponderosa pine.[16] A field study was conducted to determine the benefits of land application of composted sewage sludge to white pine and hybrid poplar.[17] Composted municipal sewage sludge was disked in a silt loam soil at rates of 0, 150, and 300 dt ha^{-1}. Hybrid poplars grown on the compost-amended soil had a lower concentration of Zn in the leaves than the control during the 3-year study. No consistent differences in elemental composition of the white pine needles were apparent.

Korcak et al. applied digested sewage sludge compost to sandy soils at rates of 0, 56, 112, and 448 dt ha^{-1}.[5] Metal contents of leaf samples of red oak and black walnut seedlings taken the summer of the first growing season were low and within normal levels recorded in the literature. In another study, composted sewage sludge was used as a soil amendment for apple seedlings grown in a greenhouse. Compost was applied at rates equivalent to 0, 25, and 50 dt ha^{-1}. While plant growth was significantly increased by addition of the sludge compost, root Cd level was increased even though soil pH was maintained above 6.3. Tissue Zn, Cu and Ni were not consistently affected by waste additions.[18]

In another study, open pollinated apple seeds were germinated and grown for a period of 7 months in soil amended with two different compost sources at rates of 25, 50 or 100 dry kg ha^{-1}.[19] The composts were a high metal, limed and unlimed composted sludge and a low metal composted, lime stabilized sludge. Germination was unaffected by treatments, suggesting no initial toxicity. The soils treated with high-metal compost produced the lowest growth, particularly when unlimed. Elevated tissue metal levels indicated that Mn, Zn, Cu, and Ni were the probable causes of the reduced growth. In addition, Korcak used a lime stabilized sewage sludge compost as a surface amendment to improve the soil and nutritional status of a number of established pear cultivars grown on an acidic, low fertility site.[20] Foliage trace metals were not elevated and in most cases were decreased by addition of sludge compost over the course of the study. Fresh fruit or peel Cd were not significantly affected by the compost addition.

Plant Uptake of Metals in Relation to Extractable Metals

Attempts have been made to relate extractable metals to increases in plant uptake rates. Results have varied greatly. Korcak et al. found poor correlations between DPTA extractable metals and metal concentrations in pearl millet, red oak, or black walnut.[5] Barbera noticed significant uptakes of metals in corn silage and beet leaf.[2] Generally, variations in uptake of metals in plant tissues were related to variations of their available fraction in the soil. In particular, a strong linear relationship was found for Zn between its available fraction in the soil and the metal uptake in corn silage. In contrast, mineral fertilizers were found to have no effect on the levels of total and available metals in the soil or on the uptake of metals in plant tissues.

A strip-mined soil was amended with both raw and digested sewage sludge compost at rates of 56, 112, and 224 dt ha^{-1}. Soil pH was initially 2.9, while the addition of compost raised the pH to 4.0 for the two lower application rates and 5.0 for the high rate. Yield was improved dramatically for all compost rates; the highest yields occurred at higher rates. Both total and DTPA-extractable metals levels increased with increasing application rates. Plant Zn and Cd also increased; however, plant Cu and Ni decreased.[21]

Two soils that were amended annually with composted sewage sludge and cropped to barley were sampled at 4-week intervals following planting in 1981. The Cd and Zn concentrations in barley grown in the sludge-treated were consistently higher than those in the nontreated control. The percentage of Cd and Zn in each extracted fraction did not change appreciably throughout the growing season, yet the rate of uptake by barley progressively decreased with plant development. As a result, the part of the plant that developed at the later stages (i.e., grain) always contained less Cd and Zn than those that developed earlier.[4]

Purpose of Study

Three related studies were conducted in order to determine if laboratory characterization of the compost could predict the retention and release of metals after field application, and also predict the uptake of metals by plants grown in the sludge compost. Specific studies on the compost included: (1) the forms and solubility of metals in compost by use of a sequential extraction procedure, (2) the retention of metals in compost over time following field application, and (3) yearly determination of the uptake of metals by turfgrass, tomatoes, lettuce, and carrots.

METHODS AND MATERIALS

Compost used in these studies was produced by GroCo, Inc., from a mixture of 3:1 (v/v) wood chips and 25% (based on dry solids) anaerobically digested sewage sludge from the Municipality of Metropolitan Seattle (Metro). After thorough mixing, the material was placed in unaerated static piles approximately 10 m high. Three turnings took place during a 1-year composting period. The finished pile was sampled (with a grain sampler below the surface) at three depths (surface, 1 m, and 2 m) in four locations to provide material for sequential extractions. Compost from the finished pile was then transported to the University of Washington's Charles Lathrop Pack Research Facility (approximately 100 km south of Seattle, WA) for plant uptake and decomposition studies.

Sequential Extraction

Twelve samples of each compost were taken as described above, air-dried, and ground in preparation for sequential analyses as illustrated in Figure 1. In addition, compost that had been incubating in the field for 2 years was also analyzed in the same way as the freshly finished compost. Leachates from compost samples were collected by use of a constant-rate vacuum extractor using an influent rate of 1.4 cm h^{-1}. Solutions were analyzed by an inductively coupled argon-plasma spectrophotometer (ICP; Thermo Jarrell Ash ICAP 61E, Thermo Jarrel Ash, Franklin, MA) optimized for the trace metals of interest in this study.[22]

Trace Metal Retention

Elemental analyses on initial samples were performed by digestion using a wet oxidation procedure and analyzed by ICP.[23] Studies of metal retention in the compost involved two components: (1) change in metals concentration and (2) reduction of mass of the compost. Compost was applied to a depth of 25 cm in raised beds. After 9 months samples were taken with depth from the field and analyzed for elemental composition.

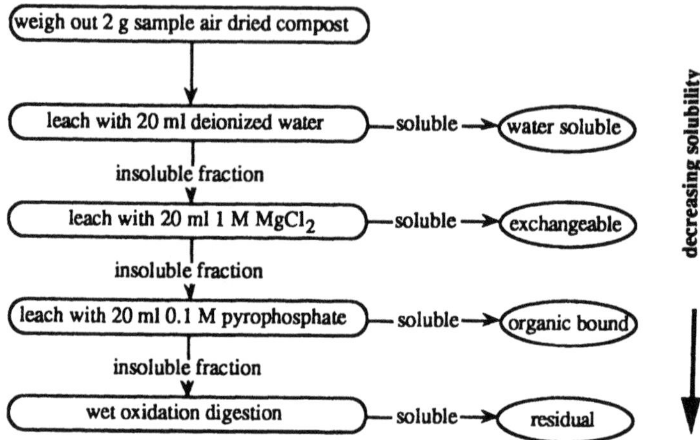

FIGURE 1. Sequential extraction procedure for determining solubility and forms of trace metals in the sludge compost.

Decomposition tubes were constructed of 10 cm high by 7 cm ID PVC tubes with fine polyethylene mesh glued to the bottom. Approximately 2 cm of soil was added followed by about 4 cm of either pure compost or a 1:1 mixture with soil. Half of the tubes received supplemental irrigation during the dry season. The reduction of mass (decomposition) was measured by comparing initial calculated dry weight of the material with measured dry weight (at 103°C) of samples that had been placed in the field 2, 6, and 12 months earlier.

Plant Uptake

Plots 1.2 m × 1.2 m in 0.3 m-deep raised nursery beds were constructed and filled with the following growth mediums: (1) control soil, (2) soil amended with commercial fertilizer, (3) compost only, and (4) compost/soil mixture at 1:1 with subsoil. Compost/soil mixtures were both layered (compost on top of soil) and mixed to determine if there were significant plant uptake and decomposition differences between the two application methods. Loading rates of metals (kg ha^{-1}) in the top 15 cm are given in Table 1.

The total metals in the compost were quite similar or less than those in the soil, even though the concentration in the compost was higher. This was due to the bulk density of the compost being considerably less than that of the soil. Soil was taken from the Indianola series, a well-drained sandy glacial outwash soil (mixed mesic Dystric Xeropsamments). The following plant species were grown in the summer of 1988 and 1989: (1) tomatoes, (2) lettuce, (3) carrots (skins intact), and (4) lawn grass. Information on the source of seed is given in Table 2. The beds were maintained free of weeds (by hand) and watered as necessary to avoid moisture deficit.

Samples of plants were taken at the end of each growing season. The total above-ground parts of grass and lettuce and the carrot root and the tomato fruit

Table 1. Loading Rates of Metals (kg ha^{-1}) in Control Soil, Compost Mixture, and Compost

	Cadmium	Chromium	Copper	Nickel	Zinc
Control soil	4.0	108	76	56	232
1:1 compost:soil					
Compost	2.7	30	55	11	123
Soil	2.0	54	38	28	116
Total	4.7	84	93	39	239
100% compost	5.4	60	110	21	245

Table 2. Plant Cultivars Grown in Compost

Plant	Years 1 and 2		Year 3	
Type	Seed Source	Cultivar	Seed Source	Cultivar
Lettuce	Ed Hume, Inc.	Parris Island Romaine	Burpee No. B-55228	Parris Island Cos
Carrots	Ed Hume, Inc.	Scarlet Mentes	Burpee No. B-59444	Toudo hybrid
Tomato	Ernst, Inc.	Early Growth	Burpee No. B-51755	Burpee VF Hybrid
Grass	Ernst, Inc.		Ernst, Inc.	

were sampled. Samples were collected by hand, pre-weighed for fresh weight, cut into smaller sizes for drying, oven dried at 70°C until at a constant weight, and ground to 100 mesh in a stainless mill with carbide blades. Samples were digested with HNO_3-$HClO_4$ wet oxidaton and analyzed by inductively-coupled plasma emission spectroscopy.

RESULTS AND DISCUSSION

Sequential Extraction

Data for the extraction of Cd, Cr, Cu, Ni, Pb, and Zn are presented Table 3 according to Figure 1 as a percentage of the total concentration of each metal. Newly finished compost has been compared with compost that has been out in the field for 2 years. Also included in Table 3 are data acquired from utilizing a similar extraction procedure on sludge originating about the same time from the Metro WWTP.

The process of sequential extraction helps establish the strength with which trace metals are bound to the compost/sludge matrix. The metal fraction removed by water can be assumed to be readily available for plant uptake and leaching. Water-soluble metals ranged from insignificant amounts for Pb, Cd, and Cr in most samples to a maximum of >9% for Ni in anaerobically digested sludge. Further extraction with $MgCl_2$ leaching removes metals that are considered weakly bound and exchangeable. This quantity of metals released ranged from insignificant amounts for Cu to over 52% for Zn in the finished compost.

Table 3. Extractable Trace Metals of Finished and 2-Year-Old Sewage Sludge Compost Compared to Sewage Sludge (% of total metals)

	Distilled	$MgCl_2$	Pyrophosphate	Residual
			(% of total)	
Cadmium				
Sludge	0.0	11.1a	1.9a	87.0c
Finished compost	0.0	48.1c	3.8b	48.1a
Field compost	0.3	32.1b	6.2b	61.4b
ANOVA	ns	**	*	**
Chromium				
Sludge	0.1	4.5	3.1a	92.3
Finished compost	0.0	3.5	9.6b	86.9
Field compost	0.5	7.7	11.5b	80.3
ANOVA	ns	ns	*	ns
Copper				
Sludge	0.2a	0.2a	2.7a	96.9
Finished compost	0.3ab	0.9b	4.8b	94.0
Field compost	0.7b	0.8c	2.4a	96.1
ANOVA	*	**	*	ns
Lead				
Sludge	0.0	4.7	2.9	92.3a
Finished compost	0.0	6.8	9.5	83.7b
Field compost	0.0	7.4	9.3	83.3ab
ANOVA	ns	ns	ns	*
Nickel				
Sludge	9.7c	11.9	9.2	69.2
Finished compost	0.2a	15.8	22.2	61.8
Field compost	4.0b	11.7	21.9	62.4
ANOVA	**	ns	ns	ns
Zinc				
Sludge	0.2a	1.0a	0.1	98.6c
Finished compost	0.5b	52.4c	0.1	47.0a
Field compost	2.6b	15.3b	0.0	82.2b
ANOVA	*	**	ns	**

Note: * and ** indicate analysis of variance is significant at $p = 0.05$ and $p = 0.01$ levels, respectively. ns indicates not significantly different at the $p = 0.05$ level.

Pyrophosphate ($Na_4P_2O_7$) extraction removes metals held by organic complexes and ranged from less than 0.2% for Zn up to 22% for Ni. Sequential analysis showed that the vast majority of metals are not water soluble and available for uptake initially. Most sludge and compost metals are also not exchangeable (except for Zn), nor are they in an organic form. This suggests that they would have much less long-term effect when added to soils than would be estimated from their total content.

Comparison to Sludge

Water solubility of metals in the sludge is generally very similar compared to composts. However, sludge had more soluble Ni and less soluble Zn. There was much less exchangeable Cd, Cu, and Zn in the sludge than in the composts. This may suggest that the composting process (i.e., further decomposition under aerobic conditions) releases to exchange sites the metals that are otherwise held during the anaerobic sludge digestion process. Composts also had a higher percentage of Cd, Ni, and Pb leached in the organic form than sludge. In all cases sludge had a higher residual fraction than the composts (differences statistically significant for Cd and Zn only).

Cadmium

Distilled water released insignificant amounts of Cd from compost, suggesting little is immediately availability for either leaching or plant uptake. In comparison, a larger quantity of Cd is "exchangeable" (leached with $MgCl_2$); 32% in the field compost compared to over 48% in the finished compost. The difference between the two may be a result of conversion to a more stable form with time in the field. Extractions with $Na_4P_2O_7$ indicated that 4 to 6% of the total Cd in the composts was organically bound.

Chromium

Less than 1% of the Cr was water soluble and only 3 to 7% by $MgCl_2$ extraction, suggesting Cr is essentially unavailable for plant uptake or leaching. This result is in keeping with the high retention of Cr observed in the literature. Approximately 9 to 12% of the Cr was organically bound. Though not statistically different, there appear to be slight changes of Cr with time from residual forms to organic and exchangeable forms.

Copper

Less than 2% of Cu was leached with either water or $MgCl_2$, and little was organically bound (2 to 5%), leaving the majority of Cu in a stable residual form (94 to 96%). Based on this extreme lack of solubility, it appears that Cu applied in compost will not be soluble or plant available even in the long-term.

Nickel

Ni is recognized as a relatively mobile trace metal and was the most water-soluble metal observed in this study (4% in the field compost). Significant amounts of Ni were also present in the exchangeable form (12 to 16%). In addition, considerable Ni was leached by $Na_4P_2O_7$ (22%). Yet over 60% of Ni remains in residual forms. Differences between composts were observed in the amounts of water extracted vs. $MgCl_2$. Whereas the field compost lost 4% of the Ni with water leaching, only 0.2% was lost in the finished compost. That 4% difference was reversed by leaching with $MgCl_2$.

Table 4. Trace Metal Concentrations with Depth ($\mu g\ g^{-1}$) in Sludge Compost after 9 Months Compared to Original Composition and Soil

	Soil	Initial Compost	2 cm	10 cm	18 cm	25 cm	Ave. % Change
Cd	2.0	7.3	7.6	6.9	6.4	7.7	−2
Cr	54	80.6	114	119	95.9	105	+35
Cu	38	148	177	167	115	181	+15
Ni	28	28.3	23.4	24.1	22.3	22.0	−19
Zn	116	330	398	368	313	351	+8
pH	6.0	4.4	4.7	4.7	4.6	4.6	

Lead

Pb was essentially not water soluble, but about 7% was removed by $MgCl_2$ extraction. Another 9 to 10% was organically bound. The two composts released Pb almost identically.

Zinc

As with Ni, significant amounts of Zn were water soluble. Also like Ni, the field compost had much more water soluble Zn than the finished compost. The exchangeable Zn was somewhat similar to exchangeable Cd, very large in the finished compost (>50%). Almost no Zn was held in the organic fraction (<0.2%).

Trace Metal Retention

Change in Metals Concentration

The change in metal concentration in a given soil horizon varies depending upon the susceptibility of the metal to leaching into lower soil profiles (Table 4). For instance, Ni (averaging a loss of 19%) decreased in concentration the greatest amount in compost during the 9-month period. On the other hand, Cr was strongly held and increased in concentration by 35%. Increased concentrations were expected for those elements that do not readily leach, as mass loss during decomposition will concentrate residual elements. These field observations follow data from the sequential extractions; Ni was easily leached with distilled water, while Cr was quite insoluble.

Decomposition Rates

The average decomposition (loss of mass) that occurred during the first 2 months of exposure was 33%. There was no significant difference in decomposition where the compost was incorporated or surface applied. Supplemental irrigation also had no significant effect. At 6 months, apparently no additional decomposition had occurred, and the decomposition measured actually appeared to be slightly lower than at 2 months, though this difference was not

Table 5. Cadmium Concentrations and Uptake Slopes in Lettuce, Carrots, Tomatoes, and Grass During Three Growing Seasons

	1988	1989	1990	Average Uptake Slope
		($\mu g\ g^{-1}$)		($\mu g\ g^{-1}$ (kg $ha^{-1})^{-1}$)
Lettuce				
Control	4.9	n.d.	3.5	1.05
NPK	1.9	n.d.	3.0	0.61
100% compost	8.4	7.0	6.5	1.35
50% compost	3.8	7.5	6.0	1.22
Carrots				
Control	1.6	2.7	1.1	0.45
NPK	1.9	1.8	0.9	0.38
100% compost	2.2	4.9	1.5	0.53
50% compost	1.5	4.1	1.3	0.49
Tomatoes				
Control	n.d.	n.d.	1.9	0.48
NPK	1.4	n.d.	2.3	0.46
100% compost	1.1	1.5	1.9	0.28
50% compost	1.0	1.3	1.0	0.23
Grass				
Control	1.6	2.6	2.0	0.52
NPK	1.2	1.9	1.6	0.39
100% compost	1.5	3.7	2.3	0.46
50% compost	1.1	0.9	2.1	0.29

statistically different (average 31%). Such a result is not wholly unexpected. Decomposition rates usually peak soon after application, and then decrease with additional time. Furthermore, the higher temperatures of the first 2 months of the experiment (summer months) enhanced the decomposition process during the peak period, while temperatures are decidedly cooler in the following 4 months. By the twelfth month an average of 38% decomposition had occurred, or about 6% additional from the 2- and 6-month decomposition. During the second year 3.5% additional decomposition occurred for a total of 41.5%.

A comparison of the decomposition measurements with the change in concentration of metals shows the following: (1) It can be seen that increases in the concentration of strongly retained metals are explained by loss of mass, (2) little or no change in concentration of metals still suggests loss (such as Cd or Zn) equivalent to the loss of mass, and (3) relatively large decreases in concentration (Ni = –19%) result in even larger losses of metals when combined with mass loss (total loss = 40 to 50%).

Plant Uptake

Cadmium

The concentrations and uptake rates of Cd in lettuce, carrots, tomatoes, and grass for the three growing seasons are presented in Table 5.

Lettuce. Lettuce and similar broad-leaf species are commonly regarded as relatively high assimilators of trace metals, and this was seen in the results of this study. Cadmium uptake by lettuce in the 100% compost bed was 8.4 $\mu g\ g^{-1}$ during the first year. During the second and third years there was a decided decrease in Cd (7.0 and 6.5 $\mu g\ g^{-1}$, respectively). This compares to fertilized and control lettuce average Cd concentrations of 1.9 to 4.9 $\mu g\ g^{-1}$. The 50% compost bed had lettuce Cd concentrations somewhat between the control and 100% compost (3.8 $\mu g\ g^{-1}$) during the first year, but this had nearly doubled by the second and third years (7.5 and 6.0 $\mu g\ g^{-1}$, respectively). The calculated average 3-year uptake slope for lettuce grown in the composts was 1.2 to 1.4 $\mu g\ g^{-1}$ (kg $ha^{-1})^{-1}$ compared to 0.6 to 1.1 $\mu g\ g^{-1}$ (kg $ha^{-1})^{-1}$ for control soil. Thus, even though the compost pH was about 1.4 units lower than that of the soil, uptake rates were not greatly different when compared to soil only.

Carrots. The concentration of Cd in carrots was considerably less than in lettuce. Average Cd concentrations in carrots grown in compost were 1.1 to 4.9 $\mu g\ g^{-1}$ compared to control and NPK concentrations of 1.2 to 2.7 $\mu g\ g^{-1}$. Unexplained is the high Cd concentration in carrots grown in compost during the second year. As with lettuce, uptake slopes were similar in the compost compared to the soil; however, in both cases uptake slopes were about half of those for lettuce.

Tomatoes. Fruits generally are regarded as relatively low assimilators of trace metals, and had the lowest average concentrations and uptake slopes. Due to poor growing conditions, some treatments during the first 2 years did not produce fruit. Thus it is difficult to compare treatments. It appears, however, that uptake slopes were considerably less in the compost compared to the soil (about half).

Grass. Concentrations of Cd in grass were 0.9 to 3.7 $\mu g\ g^{-1}$ for compost treatments, and 1.6 to 2.6 $\mu g\ g^{-1}$ for control and NPK treatments. Uptake slopes for grass grown in compost also appeared slightly less than those of control and NPK treatment.

Chromium

The concentrations and uptake rates of Cr in lettuce, carrots, tomatoes, and grass for the three growing seasons are presented in Table 6.

Lettuce. Lettuce was not the highest accumulator of Cr in this study; however, lettuce did contain about two times the Cr concentration of tomatoes and carrots. Chromium concentrations were essentially unaffected by treatment (1.5 to 6.6 and 2.1 to 4.1 $\mu g\ g^{-1}$ for lettuce in compost and control and NPK treatment, respectively). In the first year the control and NPK treatment had higher concentrations than the compost treated lettuce. During the third year, this was reversed. The high concentrations in year 3 for the lettuce grown in compost is unexplained. Uptake slopes are only slightly higher for the Cr in the composts than in the soils (0.05 to 0.07 and 0.04 to 0.05 $\mu g\ g^{-1}$ (kg $ha^{-1})^{-1}$, respectively), once again suggesting that, even with a lower pH, the cadmium in the compost is only slightly more available than in the soil.

Table 6. Chromium Concentrations and Uptake Slopes in Lettuce, Carrots, Tomatoes, and Grass During Three Growing Seasons

	1988	1989	1990	Average Uptake Slope
		(μg g^{-1})		(μg g^{-1} (kg ha^{-1})$^{-1}$)
Lettuce				
Control	2.1	n.d.	2.6	0.022
NPK	4.1	n.d.	2.9	0.032
100% compost	1.9	1.5	6.6	0.056
50% compost	1.5	2.1	3.3	0.027
Carrots				
Control	0.7	0.7	1.3	0.008
NPK	0.8	2.1	1.2	0.013
100% compost	0.7	3.7	1.6	0.033
50% compost	1.2	3.3	1.1	0.022
Tomatoes				
Control	n.d.	n.d.	1.9	0.018
NPK	0.5	n.d.	3.8	0.020
100% compost	1.8	0.3	2.0	0.023
50% compost	0.5	0.4	1.3	0.009
Grass				
Control	4.7	5.9	11.4	0.068
NPK	4.1	8.7	3.4	0.050
100% compost	2.9	3.1	2.2	0.046
50% compost	3.1	3.5	2.5	0.036

Carrots. In years 1 and 3 Cr concentrations in carrots in all treatments were similar (0.7 to 1.6 μg g^{1}). In year 2 the compost-treated lettuce had significantly greater Cr concentrations than the control and NPK treatment. Average uptake slopes for the compost-treated carrots were more than double those of carrots grown in soil only (0.022 to 0.033 compared to 0.008 to 0.013 μg g^{-1} (kg ha^{-1})$^{-1}$).

Tomatoes. Concentrations of Cr in tomatoes generally were below 2.0 μg g^{-1}. Uptake slopes for tomatoes in the compost treatments averaged about the same as in the soil.

Grass. Grass accumulated higher amounts of Cr than any other plant in this study. This was especially true for the control and NPK treatment, which reached average peak concentrations of 11.4 and 8.7 μg g^{-1}, respectively, over the 3 years. Concentrations of Cr in the grass grown in compost were always lower than in the control and NPK treatment, resulting in slightly lower uptake slopes.

Copper

The concentrations and uptake rates of Cu in lettuce, carrots, tomatoes, and grass for the three growing seasons are presented in Table 7.

Lettuce. Little difference was found in lettuce Cu concentrations within the treatments, with the exception of plants grown in 100% compost during year 3.

Table 7. Copper Concentrations and Uptake Slopes in Lettuce, Carrots, Tomatoes, and Grass During Three Growing Seasons

	1988	1989	1990	Average Uptake Slope
		($\mu g\ g^{-1}$)		($\mu g\ g^{-1}$ (kg $ha^{-1})^{-1}$)
Lettuce				
Control	8.3	n.d.	4.0	0.08
NPK	9.2	n.d.	8.6	0.12
100% compost	10.1	6.9	19.8	0.11
50% compost	8.4	5.5	6.7	0.07
Carrots				
Control	4.9	3.0	0.1	0.04
NPK	7.9	3.9	0.6	0.05
100% compost	6.9	6.3	4.4	0.05
50% compost	7.0	7.6	0.3	0.05
Tomatoes				
Control	n.d.	n.d.	5.0	0.07
NPK	5.6	n.d.	11.8	0.11
100% compost	4.2	5.7	6.5	0.05
50% compost	4.5	6.3	6.1	0.06
Grass				
Control	6.9	6.6	2.6	0.07
NPK	9.4	5.3	2.1	0.07
100% compost	10.5	8.7	10.9	0.07
50% compost	11.2	4.0	2.3	0.06

In the control and NPK plots Cu concentrations ranged from 4.0 to 9.2 $\mu g\ g^{-1}$, while in the compost plots the range was 5.5 to 10.1 $\mu g\ g^{-1}$ and the noted high concentration of 19.8 $\mu g\ g^{-1}$. Because soil and compost Cu concentrations are similar, uptake slopes are also similar; 0.07 to 0.11 for the compost compared to 0.08 to 0.12 $\mu g\ g^{-1}$ (kg $ha^{-1})^{-1}$ for the soil.

Carrots. Carrots grown in compost had only slightly higher Cu concentrations than the control and NPK treatment. All carrots showed a dramatic lowering of Cu concentration during year 3, suggesting that the Cu became less available over time. This is not consistent with the results found with either lettuce or tomatoes, but generally agrees with Cu concentrations in grass. Uptake slopes for carrots grown in compost again are essentially the same as those for the control and NPK treatment.

Tomatoes. Tomato Cu concentrations were similar for all treatments. Due to missing data, uptake slopes are hard to calculate and compare, but would also be similar.

Grass. Grass Cu concentrations followed the same trend as in lettuce, with similar concentrations and uptake slopes by treatment.

Nickel

Ni concentrations and uptake rates in lettuce, carrots, tomatoes, and grass for three growing seasons are presented in Table 8.

Lettuce. Nickel concentrations in lettuce were not significantly different between treatments with the exception of year 3 in the 100% compost. Ni loading

Table 8. Nickel Concentrations and Uptake Slopes in Lettuce, Carrots, Tomatoes, and Grass During Three Growing Seasons

	1988	1989	1990	Average Uptake Slope
		($\mu g\ g^{-1}$)		($\mu g\ g^{-1}$ (kg $ha^{-1})^{-1}$)
Lettuce				
Control	2.4	n.d.	4.5	0.07
NPK	2.9	n.d.	6.4	0.08
100% compost	3.0	3.6	10.2	0.24
50% compost	2.1	3.9	5.7	0.10
Carrots				
Control	1.3	1.7	2.1	0.03
NPK	1.6	2.5	1.9	0.04
100% compost	2.4	5.8	3.3	0.23
50% compost	2.4	3.2	2.3	0.07
Tomatoes				
Control	n.d.	n.d.	3.2	0.06
NPK	1.8	n.d.	3.9	0.04
100% compost	3.0	2.0	2.8	0.12
50% compost	0.9	2.2	2.9	0.05
Grass				
Control	3.8	4.7	7.6	0.10
NPK	3.3	6.3	3.9	0.08
100% compost	3.9	6.5	8.7	0.30
50% compost	2.9	1.3	6.4	0.09

rates for the compost plots were less than half that of the soil, resulting in higher uptake slopes.

Carrots. Concentrations of Ni in carrots, however, were consistently higher than those of the control and NPK treatment. Combined with lower Ni loading rates, uptake slopes were much greater than from the soil.

Tomatoes and grass. Nickel concentrations in tomatoes and grass were apparently unaffected by treatment. Again, lower loading rates resulted in higher uptake slopes for the 100% compost.

Zinc

The concentrations and uptake rates of Zn in lettuce, carrots, tomatoes, and grass for three growing seasons are presented in Table 9.

Lettuce. Substantially higher concentrations of Zn were found in lettuce even though loading rates were essentially the same. Zinc concentrations in lettuce in the control and NPK treatments were 41 to 149 $\mu g\ g^{-1}$, compared to 123 to 164 $\mu g\ g^{-1}$ in the 50% compost and 250 to 383 $\mu g\ g^{-1}$ in the 100% compost. It appears that Zn became less available with time, as generally there is a decrease in concentration in the second and third years compared to the first year. Uptake slopes vary considerably, with a low of 0.02 for the NPK treatment and a high of 1.19 $\mu g\ g^{-1}$ (kg $ha^{-1})^{-1}$ for the 100% compost.

Carrots. Carrots also showed a significant increase in Zn concentrations in the compost treatments. In addition, uptake slopes were considerably higher in the composts.

Table 9. Zinc Concentrations and Uptake Slopes in Lettuce, Carrots, Tomatoes, and Grass During Three Growing Seasons

	1988	1989	1990	Average Uptake Slope
		($\mu g\ g^{-1}$)		($\mu g\ g^{-1}$ (kg $ha^{-1})^{-1}$)
Lettuce				
Control	149	n.d.	41	0.44
NPK	47	n.d.	46	0.20
100% compost	383	287	250	1.19
50% compost	142	164	123	0.60
Carrots				
Control	23	20	19	0.09
NPK	65	24	16	0.15
100% compost	90	135	85	0.43
50% compost	60	125	30	0.30
Tomatoes				
Control	n.d.	n.d.	22	0.09
NPK	21	n.d.	82	0.18
100% compost	18	23	25	0.09
50% compost	21	22	52	0.13
Grass				
Control	70	60	62	0.28
NPK	46	31	21	0.14
100% compost	101	124	130	0.48
50% compost	68	14	41	0.17

Tomatoes. Zinc concentrations in tomatoes showed little effect of treatment. Also, uptake slopes were lower in tomatoes than any of the other plants.

Grass. Zinc concentrations in grass, as with carrots and lettuce, were considerably higher in the 100% compost treatments. However, in the 50% compost, Zn concentrations were similar to the control and NPK treatment.

SUMMARY AND CONCLUSIONS

Three types of studies were conducted on soil and compost, with the goal of comparing laboratory, field incubation, and plant uptake methods of characterizing metal solubility and availability from compost. Based on the results of these studies, the following conclusions can be made regarding the extractability and retention of metals and the uptake of metals by plants grown in control soil, fertilized soil, a 50% soil/compost mixture, and 100% compost.

Sequential Extraction

The majority of trace metals in composted sewage sludge are not available for plant uptake or leaching through the soil. Distilled water extracts accounted for less than 1% of the total for all metals studied in newly finished compost, and less than 1% of total for Cd, Cr, Cu, and Pb in 2 year-old compost. The only distilled water extracts exceeding 1% were Ni and Zn in the 2 year-old compost.

Exchangeable trace metals varied considerably, however. Amounts of $MgCl_2$ leachable metals were in the order of Cu (0.8 to 0.9%) < Cr (3.5 to 7.7%) < Pb (6.8 to 7.4%) < Ni (12 to 16%) < Zn (15 to 52%) < Cd (32 to 48%). Residual metals (not leached by distilled water, $MgCl_2$, or $Na_4P_2O_7$) is nearly at or well over 50% for all metals, and for Cd, Cu, and Pb greater than 80%.

Trace Metal Retention

Trace metals retention by the compost, or conversely, mass loss of trace metals from compost taken from the field can generally be explained by results from sequential extraction. Those metals that showed little water or $MgCl_2$ leaching (Cr and Cu) were similarly conservative after 9 months in the field. Metal concentrations increased in the compost due to reduction of compost mass by decomposition. Small loss of Cd and Zn and relatively large loss of Ni from the compost occurred after 9 months in the field. Although Cd and Zn concentrations did not change greatly, mass loss of compost suggests corresponding loss of metals. Nickel (documented to be mobile) showed the greatest decrease in concentration (19%), and together with compost decomposition, mass loss of Ni was about 45%.

Plant Uptake

Uptake slopes of plants grown in sludge compost are in the order of Cd > Zn > Ni > Cu > Cr. In general, Cd, Cr and Cu are similarly available for plant uptake in compost (at a significantly lower pH) as compared to the soil, while Ni and Zn are slightly more available. A summary of Cd, Cr, Cu, Ni, and Zn uptake slopes in lettuce, carrots, tomatoes, and grass is presented in Table 10. Plant uptake rates of Cd in the compost treatments were within the range found in the control or fertilized soil for all crops but lettuce. Chromium uptake was higher in lettuce and carrots grown in compost compared to soil, but uptake rates in tomatoes and grass were similar. Copper uptake slopes were unaffected by treatment for all plants. These results were found even though the compost was at an average of 1.4 pH units lower than soil. In contrast, uptake slopes of Ni and Zn by plants grown in compost were higher than in soils for most plants. This suggests that the metals are more available in the compost than in the soil, probably at least partially as a result of the pH difference. Also, compared to the other metals studied, sequential extractions and metal retention portions of this study indicate that these two metals have the highest amount of water soluble plus exchangeable fraction.

Uptake rates of trace metals by the plants in this study are generally in the order of lettuce > grass > carrots > tomatoes. For plants grown in compost treatments, lettuce showed the highest uptake rates for all but Ni (grass was only slightly higher), and tomatoes showed the lowest uptake rates for all but Cu (carrots were only slightly lower).

Table 10. Summary of Cd, Cr, Cu, Ni, and Zn Uptake Slopes in Lettuce, Carrots, Tomatoes, and Grass ($\mu g\ g^{-1}$ $(kg\ ha^{-1})^{-1}$)

	Cd	Cr	Cu	Ni	Zn
Lettuce					
Control	1.05	0.022	0.08	0.07	0.44
NPK	0.61	0.032	0.12	0.08	0.20
100% compost	1.35	0.056	0.11	0.24	1.19
50% compost	1.22	0.027	0.07	0.10	0.60
Carrots					
Control	0.45	0.008	0.04	0.03	0.09
NPK	0.38	0.013	0.05	0.04	0.15
100% compost	0.53	0.033	0.05	0.23	0.43
50% compost	0.49	0.022	0.05	0.07	0.30
Tomatoes					
Control	0.48	0.018	0.07	0.06	0.09
NPK	0.46	0.020	0.11	0.04	0.18
100% compost	0.28	0.023	0.05	0.12	0.09
50% compost	0.23	0.009	0.06	0.05	0.13
Grass					
Control	0.52	0.068	0.07	0.10	0.28
NPK	0.39	0.050	0.07	0.08	0.14
100% compost	0.46	0.046	0.07	0.30	0.48
50% compost	0.29	0.036	0.06	0.09	0.17

REFERENCES

1. Logan, T. J. and A. L. Page (co-chairs). Peer Review of the Standards for the Disposal of Sewage Sludge U.S. SPA Proposed Rule 40 CFR Parts 257 and 503. Cooperative State Research Service Technical Committee W–170. (1989).
2. Barbera, A. "Extraction and Dosage of Heavy Metals from Compost-Amended Soils," in *Compost: Production, Quality, and Use*, de Bertoldi, M., M. P. Ferranti, P. L'Hermite and F. Zucconi, Eds., (London: Elsevier Applied Science. Udine, Italy, 1987) pp. 598–614.
3. Page, A. L., T. J. Logan, and J. A. Ryan, Eds. *Land Application of Sludge* (Chelsea, MI: Lewis Publishers, 1987).
4. Chang, A. C., A. L. Page, J. E. Warneke, and E. Grgurevic. "Sequential Extraction of Soil Heavy Metals Following a Sludge Application," *J. Environ. Qual.* 13(1):33–38 (1984).
5. Korcak, R. F., F. R. Gouin, and D. S. Fanning. "Metal Content of Plants and Soils in a Tree Nursery Treated with Composted Sludge," *J. Environ. Qual.* 8(1):63–68 (1979).
6. Vedy, J. C., T. Dellis, and A. C. M. Bourg. "Biotoxicity of Trace Metals and Composted Sludge/Mineral Substrate Interactions," *Tox. Environ. Chem.* 12(3/4):237–254 (1986).
7. Zasoski, R. J. and R. L. Edmonds. "Water Quality in Relation to Sludge and Wastewater Applications to Forest Land," in *The Forest Alternative for Treatment and Utilization of Municipal and Industrial Wastes*. Cole, D. W., C. L. Henry, and W. L. Nutter Eds., (Seattle: University of Washington Press, 1986) pp. 100–109.

8. Darmody, R. G., J. E. Foss, M. McIntosh, and D. C. Wolf. "Municipal Sewage Sludge Compost-Amended Soils: Some Spatiotemporal Treatment Effects," *J. Environ. Qual.* 12(2):231–236 (1983).
9. Tackett, S. L., E. R. Winters, and M. J. Puz. "Leaching of Heavy Metals from Composted Sewage Sludge as a Function of pH," *Can. J. Soil Sci.* 66:763–765 (1986).
10. Bledsoe, C. S. "Composted Sludge as a Plant Growth Medium," in *Municipal Sludge Application to Pacific Northwest Forest Lands*, Bledsoe, C., Ed. (Seattle: Institute of Forest Resources Publication, 1981) p. 87–92.
11. Falahi-Ardakani, A., J. C. Bouwkamp, F. R. Gouin, and R. L. Chaney. "Growth Responses and Mineral Uptake of Vegetable Transplants Grown in a Composted Sewage Sludge Amended Medium. I. Nutrient Supplying Power of the Medium," *J. Environ. Hort.* 5(3):112–115 (1987).
12. Tunison, K. W., B. C. Bearce, and H. A. Menser. "The Utilization of Sewage Sludge: Bark Screenings Compost for the Culture of Blueberries on Acid Minespoil," in *Land Reclamation and Biomass Production with Municipal Wastewater and Sludge*, Sopper, W. E., et al., Eds., (Penn State University Press, 1982) p. 194–206.
13. Sterrett, S. B., R. L. Chaney, C. W. Reynolds, F. D. Schales, and L. W. Douglass. "Transplant Quality and Metal Concentrations in Vegetable Transplants Grown in Media Containing Sewage Sludge Compost," *HortScience* 17(6):920–922 (1982).
14. Sterrett, S., C. Reynolds, F. Schales, R. Chaney, and L. Douglass. "Transplant Quality, Yield, and Heavy Metal Accumulation of Tomato, Muskmelon, and Cabbage Grown in Media Containing Sewage Sludge," *J. Am. Soc. Hort. Sci.* 108(1):36–41 (1983).
15. Falahi-Ardakani, A., K. A. Corey, and F. R. Gouin. "Influence of pH on Cadmium and Zinc Concentrations of Cucumber Grown in Sewage Sludge," *HortScience* 23(6):1015–1017 (1988).
16. Coleman, M., J. Dunlap, D. Dutton, and C. Bledsoe. "Nursery and Field Evaluation of Compost-Grown Coniferous Seedlings," in Proc. Western Forest Nursery Council Workshop. USDA Forest Service Publication. Olympia, WA (1986).
17. McIntosh, M. S., J. E. Foss, D. C. Wolf, K. R. Brandt, and R. Darmody. "Effect of Composted Municipal Sewage Sludge on Growth and Elemental Composition on White Pine and Hybrid Poplar," *J. Environ. Qual.* 13(1):60–62 (1984).
18. Korcak, R. F. "Effects of Applied Sewage Sludge Compost and Fluidized Bed Material on Apple Seedling Growth," *Commun. Soil Sci. Plant Anal.* 11(6):571–585 (1980).
19. Korcak, R. F. "Growth of Apple Seedlings on Sludge-Amended Soils in the Greenhouse," *Commun. Soil Sci. Plant Anal.* 7(10):1041–1054 (1986a).
20. Korcak, R. F. "Renovation of a Pear Orchard Site with Sludge Compost," *Comm. Soil Sci. Plant Anal.* 17(11):1159–1168 (1986b).
21. Griebel, G. E., W. H. Armiger, J. F. Parr, D. W. Steck, and J. A. Adam. "Use of Composted Sewage Sludge in Revegetation of Surface-Mined Areas," in *Utilization of Municipal Sewage Effluent and Sludge on Forest and Disturbed Land*, Sopper, W. E. and S. N. Kerr, Eds., (University Park: Penn State University Press, 1979) p. 293–305.
22. Johnson, D. W. and D. E. Todd. "Relationships Among Iron, Aluminum, Carbon, and Sulfate in a Variety of Forest Soils," *Soil Sci. Soc. Am. J.* 47:792–800 (1983).

23. Parkinson, J. A. and Allen, S. E. "A Wet Oxidation Procedure Suitable for the Determination of Nitrogen and Mineral Nutrients in Biological Material," *Commun. Soil Sci. Plant Anal.* 6(1):1–11 (1975).

8

Long-Term Behavior of Heavy Metals in Agricultural Soils: A Simple Analytical Model

Karl Harmsen

Institute for Soil Fertility Research, P.O. Box 30003, 9750 RA Haren, The Netherlands

ABSTRACT

A simple model is presented to predict the long-term behavior of heavy metals in agricultural soils. The analytical solutions to the model differ in the assumed relationships between the solubility of heavy metals and their accumulated contents in soils. Four cases are distinguished: (1) precipitation and dissolution reactions determine the solubility of heavy metals in soil, (2) ion exchange or linear adsorption determine the solubility of heavy metals, (3) solution concentrations of heavy metals are related to their accumulated soil contents through Langmuir- or Freundlich-type adsorption equations, and (4) part of the heavy metals in an adsorbed phase is transformed into unavailable (irreversibly adsorbed) form. As an example, the model is applied to assess the long-term behavior of cadmium in agricultural soils in The Netherlands.

INTRODUCTION

In the industrialized countries of the world, heavy metal contents in most agricultural soils have increased significantly over the past century due to the increased use of fertilizers, biocides, and sewage sludges in agriculture and increased atmospheric deposition of heavy metals.[1–12] Present-day observations indicate that the total inputs of heavy metals in soils generally exceed the removal from soils by crop uptake and leaching.[13–18]

Although some of the heavy metals are necessary plant nutrients, high contents of heavy metals in agricultural soils are generally considered a matter of concern to society, as they might adversely affect the quality of agricultural products or leach down to the groundwater.[19–26] As heavy-metal contents are still increasing in most soils, it is of importance to know more about the long-term behavior of heavy metals in agricultural soils, in particular with regard to crop uptake and leaching.[23,27–29]

Crop uptake and leaching depend on conditions of soil, crops, and climate. These factors must be considered when long-term trends in the behavior of heavy metals are to be established.[27,30,31] Also, crops may derive their heavy metals from both atmospheric deposition and from the soil, and it is often difficult to distinguish between these two sources.[32–35] In addition, uptake of heavy metals by crops is influenced by processes occurring in the rhizosphere, which may vary between crops and during the growing season.[36] Therefore, it is difficult to establish long-term trends in crop uptake or leaching of heavy metals from agricultural soils through experimentation or direct measurements.

An alternative approach would be to predict the long-term behavior of heavy metals on the basis of models, making use of existing information on the occurrence and the chemical behavior of heavy metals in agricultural soils.[37] In order to develop such predictive models, one would have to account for the inputs of heavy metals from different sources, and the behavior of heavy metals in soil-crop ecosystems under different climatic conditions. Furthermore, one would have to assess long-term trends in inputs of heavy metals from different sources, and changes in soil, crop, and climatic conditions. In addition, functional relationships between these factors and the processes determining the behavior of heavy metals in soil-crop ecosystems would have to be derived.

Obviously, such comprehensive models would become rather involved. As long-term (historic) datasets on heavy metals in natural ecosystems are scarce,[15,16] these models would be difficult to calibrate or verify. Furthermore, there would still be a number of uncertainties, such as future trends in inputs of heavy metals and changes in land use and environmental conditions.

The objective of this paper is to derive a simple, analytical model that predicts the long-term behavior of heavy metals in agricultural soils, based on knowledge about the chemical behavior of heavy metals in soils. The model uses a limited number of parameters that are relatively easy to obtain from existing information or through direct measurements. By necessity, the applicability of such a model is limited. A simple, analytical model may be suitable, however, to demonstrate the consequences of certain basic assumptions about the behavior of heavy metals in agricultural soils. The basic question addressed herein is how trends in annual rates of crop uptake and leaching develop in time, assuming different relationships between dissolved and accumulated heavy metals in soils, at a constant rate of heavy-metal input.

MODEL DESCRIPTION

The model considers four variables: the input of heavy metals in soil, the soil content, crop uptake (removal), and the amount leached from the top layer. The thickness of the homogeneous top layer depends on the depth of mixing of the soil through cultivation.

The emphasis of this paper is on the behavior of heavy metals with time. Therefore, the variables are presented as their derivatives with time. The annual amount of heavy metals that reaches the soil from different sources is denoted by dq_i/dt, the amount of heavy metals that accumulates annually in the top layer of the soil by dq_s/dt, and the amounts of heavy metals that are removed annually from the soil by crop uptake (crop removal) or leaching from the top layer are denoted by dq_p/dt and dq_l/dt, respectively. All rates are expressed as amount of substance per unit of surface area per unit of time, e.g., kg ha^{-1} y^{-1}. It is assumed that heavy metals that reach the soil either accumulate in the top layer or are removed by crops or through leaching. Hence, for a well-mixed (ploughed) top layer of a soil

$$\frac{dq_i}{dt} = \frac{dq_s}{dt} + \frac{dq_p}{dt} + \frac{dq_l}{dt} \tag{1}$$

It is further assumed that the inputs of heavy metals in agricultural soils were negligibly small until the end of the last century, and that the release of heavy metals in soils by dissolution of soil minerals is negligibly small compared to present-day atmospheric inputs and inputs through fertilizers and other materials. By the end of the last century, the inputs of heavy metals increased as a result of increasing industrial activities, including ore processing; the increased use of fossile fuels; and the use of chemical fertilizers, organic waste materials, and biocides in agriculture. In the model, it is assumed that the inputs of heavy metals in agricultural soils start at $t = 0$ and thereafter remain constant. Hence, for $t \geq 0$

$$\frac{dq_i}{dt} = k_i \tag{2}$$

where k_i is a constant (kg ha^{-1} y^{-1}). Alternatively, it may be assumed that the rate of input of heavy metals increased linearly with time

$$\frac{dq_i}{dt} = k_i t \tag{3}$$

where k_i is a constant (kg ha^{-1} y^{-2}), whose numerical value differs from k_i in Equation 2. In this paper, the emphasis is on Equation 2, as this condition more clearly demonstrates the consequences of the assumed relationships between dissolved and accumulated heavy metals in soils, in the limit for $t \to \infty$.

In order to distinguish between the initial heavy-metal content of soils, $q_s(0)$, and their present-day content, $q_s(t)$, the following quantity is defined

$$\Delta q_s(t) \equiv q_s(t) - q_s(0) \tag{4}$$

that is, $\Delta q_s(t)$ represents the amount of heavy metals that has accumulated in soils between t = 0 and the present time. Equation 1 can now be written in the form

$$\frac{d\Delta q_s}{dt} = k_i - \frac{dq_p}{dt} - \frac{dq_l}{dt} \tag{5}$$

which is the basic equation of the present model. It is further assumed that

$$\frac{dq_i}{dt} > \frac{dq_p}{dt} + \frac{dq_l}{dt} \tag{6}$$

that is, the rate of input is assumed to exceed the rates of removal of heavy metals. This condition is not required to solve Equation 5, but the case of Inequality 6 represents present-day conditions for most heavy metals.

The basic question addressed in this paper is how annual rates of plant uptake and leaching change with time, assuming different relationships between the accumulated heavy-metal contents of agricultural soils, Δq_s, and the solubility of heavy metals. It is assumed that dq_p/dt and dq_l/dt depend on the solubility of heavy metals, as determined by Δq_s, and that they can be expressed as functions of Δq_s. Furthermore, it is assumed that dq_p/dt and dq_l/dt depend in similar ways on the solubility of heavy metals, that is, on Δq_s.

Precipitation and Dissolution (Zero*th*-Order Case)

If precipitation and dissolution reactions determine the solubility of heavy metals in soil, the activities of the free ionic species would be approximately constant and independent of the total amounts of heavy metals in the solid phase of the soil.[38] Ligands in soil solution that form soluble complexes with heavy metals increase the heavy-metal solubility in soils.[39,40] However, if the complexed species do not interact with the solid phase, complexation would not affect the relationship between aqueous heavy metals and the solid phase.

The assumption that the solubility of heavy metals is determined by a solubility product does not imply that adsorption reactions would not occur. In fact, adsorption probably always occurs. In the presence of a solid phase involving heavy metals, however, the solubility (ionic activities) of the heavy metals is ultimately governed by a solubility product and not by an adsorption isotherm.

In the case of precipitation/dissolution, where solubility of heavy metals is not related to their accumulated contents in agricultural soils (i.e., c = constant), rates of crop uptake and leaching would be constant with time

$$\frac{dq_p}{dt} = k_p \tag{7}$$

and

$$\frac{dq_l}{dt} = k_l \tag{8}$$

where k_p and k_l are both constants (kg ha^{-1} y^{-1}). Equations 7 and 8 further assume that soil parameters, other than Δq_s, that may affect the solubility of heavy metals remain constant. This would not apply to seasonal variations, but there should be no long-term changes in pH, redox conditions, organic-carbon contents, etc. Under these conditions, Equation 5 reduces to:

$$\frac{d\Delta q_s}{dt} = k_i - k_p - k_l \tag{9}$$

from which it follows that

$$\Delta q_s = \left(k_i - k_p - k_l\right) t \tag{10}$$

Hence, the rates of change would be constant with time, whereas Δq_s would increase linearly with time if Inequality 6 holds. This situation is approached in the case of leaching of calcium from soils containing lime. If the solubility of calcium is determined by calcium carbonate, and input and soil conditions (e.g., CO_2-levels, pH) remain constant, the rate of leaching of calcium would be constant with time and not related to the accumulated lime content in soil.

Linear Adsorption (First-Order Case)

If ion exchange or linear adsorption determine the solubility of heavy metals, their solution concentrations increase with increasing amounts of heavy metals in the solid phase. Ion exchange and linear adsorption relations result in an approximately linear increase in heavy metal concentrations in soil solution with increasing adsorbed concentrations.[41–43] The presence of ligands that form soluble complexes with heavy metals may result in an increase of dissolved heavy-metal concentrations.[39,44] However, if these complexed heavy-metal species do not interact with the solid phase, they would not affect the relationship between the uncomplexed species (aqueous activities of the free ionic species) and the adsorbed phase. If the amount of ligands available in soil solution is of the same order as the heavy-metal concentrations, the increase in heavy-metal solubility with increasing adsorbed concentrations may be nonlinear.

The linear adsorption of a heavy-metal species from solution may be described by

$$\Delta q_s = kc \tag{11}$$

where k is an adsorption constant, which is dimensionless, if the solution concentration of the heavy-metal species, c, is expressed in the same units as Dq_s, i.e., kg ha^{-1}. In this case, the rates of crop uptake and leaching would be related to the accumulated heavy metal content by

$$\frac{dq_p}{dt} = k_p \Delta q_s \tag{12}$$

and

$$\frac{dq_l}{dt} = k_l \Delta q_s \tag{13}$$

where k_p and k_l are constants (y^{-1}). It may be noted that the numerical values (as well as the dimensions) of k_p and k_l in Equations 12 and 13 are different from those in Equations 7 and 8. The assumption that the rates of crop uptake and leaching increase linearly with Δq_s implies that, if the accumulated soil content increases, the rates of heavy-metal removal increase proportionally.

In this case, Equation 5 reduces to

$$\frac{d\Delta q_s}{dt} = k_i - \left(k_p + k_l\right)\Delta q_s \tag{14}$$

which can be solved for Δq_s:

$$\Delta q_s = \frac{k_i}{k_p + k_l}\left\{l - \exp\left[-\left(k_p + k_l\right)t\right]\right\} \tag{15}$$

assuming that Inequality 6 holds. From Equation 15 it follows that

$$\frac{d\Delta q_s}{dt} = k_i \exp\left[-\left(k_p + k_l\right)t\right] \tag{16}$$

Hence, in the limit for $t \rightarrow \infty$

$$\frac{d\Delta q_s}{dt} \rightarrow 0$$

$$\frac{dq_p}{dt} \rightarrow \frac{k_p k_i}{k_p + k_l}$$

$$\frac{dq_l}{dt} \rightarrow \frac{k_l k_i}{k_p + k_l}$$

and

$$\Delta q_s \rightarrow \frac{k_i}{k_p + k_l}$$

This implies that in the case of linear adsorption, eventually all metals that reach the soil would be removed by crop uptake and leaching, whereas Δq_s would eventually reach a constant value.

Equation 5 may also be solved for a linearly increasing rate of input of heavy metals, according to Equation 3. In that case Equation 5 becomes

$$\frac{d\Delta q_s}{dt} = k_i t - \left(k_p + k_l\right)\Delta q_s \tag{17}$$

If

$$\frac{k_i}{k_p + k_l} > \frac{d\Delta q_s}{dt} \tag{18}$$

Equation 17 can be solved for Δq_s

$$\Delta q_s = \frac{k_i t}{k_p + k_l} - \frac{k_i}{\left(k_p + k_l\right)^2}\left\{l - \exp\left[-\left(k_p + k_l\right)t\right]\right\} \tag{19}$$

From Equation 19 it follows that

$$\frac{d\Delta q_s}{dt} = \frac{k_i}{k_p + k_l}\left\{l - \exp\left[-\left(k_p + k_l\right)t\right]\right\} \tag{20}$$

Hence, for sufficiently large values of t, it follows that

$$\frac{d\Delta q_s}{dt} \rightarrow \frac{k_i}{k_p + k_l}$$

$$\frac{dq_p}{dt} \rightarrow \frac{k_p k_i t}{k_p + k_l} - \frac{k_p k_i}{\left(k_p + k_l\right)^2}$$

$$\frac{dq_l}{dt} \rightarrow \frac{k_l k_i t}{k_p + k_l} - \frac{k_l k_i}{\left(k_p + k_l\right)^2}$$

and

$$\Delta q_s \rightarrow \frac{k_i t}{k_p + k_l} - \frac{k_i}{\left(k_p + k_l\right)^2}$$

In this case the rate of accumulation of heavy metals in soil reaches a constant value, whereas the accumulated content of heavy metals increases linearly with time. The rates of plant uptake and leaching continue to increase with time, proportional to the (linearly increasing) rate of input of heavy metals in soil.

Nonlinear Adsorption (Second-Order Case)

In the case of nonlinear adsorption, solution concentrations of heavy metals increase more than proportionally with increasing adsorbed concentrations. Almost any kind of nonlinear adsorption in soils can be approximated by either a Langmuir or a Freundlich adsorption equation, over a limited range of concentrations or surface coverage.[43,45–47] Although there may be cases where one equation describes the adsorption reaction better than the other, the choice between them is often somewhat arbitrary.

It can be shown that surface-heterogeneity of the adsorbent or species-heterogeneity of dissolved species may result in nonlinear adsorption behavior, which can be approximated by a Freundlich-type adsorption equation.[43,48,49]

With regard to the long-term behavior of heavy metals in soils, the essential point is whether the Gibbs free energy of adsorption can be taken as approximately constant (linear isotherm) or is a decreasing function of surface coverage (nonlinear isotherms). In the latter case, the solubility of heavy metals would increase more than proportionally with Δq_s, which would have consequences for the availability and mobility of heavy metals in soils.

The Freundlich adsorption equation can be written in the form

$$\Delta q_s = k_F c^{1/n} \tag{21}$$

where k_F and n are both constants. The dimension of k_F depends on the exponent (1/n). Values of n (dimensionless) are assumed to be larger than 1. To illustrate the consequences of nonlinear adsorption on the long-term behavior of heavy metals in soils, Equation 5 will be solved for a Freundlich adsorption equation with n = 2, which would be in the range of values to be expected for n in heterogeneous media. In this case, the dimension of k_F is $kg^{1/2}$ $ha^{-1/2}$, if Δq_s and c are expressed in the same units, i.e., kg ha^{-1}. Assuming that the rates of plant uptake and leaching are proportional to the solution concentration of heavy metals as determined by the Freundlich adsorption equation (n = 2), the following expressions can be derived

$$\frac{dp_p}{dt} = k_p \left(\Delta q_s\right)^2 \tag{22}$$

and

$$\frac{dp_l}{dt} = k_l \left(\Delta q_s\right)^2 \tag{23}$$

where the constants k_p and k_l are expressed in kg^{-1} ha y^{-1}. It may be noted that the numerical values of k_p and k_l in Equations 22 and 23 differ from those in the zeroth-order and first-order cases. Inserting Equations 22 and 23 in Equation 5 yields

$$\frac{d\Delta q_s}{dt} = k_i - \left(k_p + k_l\right)\left(\Delta q_s\right)^2 \tag{24}$$

Assuming that Inequality 6 holds, Equation 24 can be solved for Δq_s

$$\Delta q_s = \left(\frac{k_i}{k_p + k_l}\right)^{\frac{1}{2}} \tanh\left(\sqrt{k_i\left(k_p + k_l\right)}\, t\right) \tag{25}$$

where

$$\tanh(x) = \frac{e^{2x} - 1}{e^{2x} + 1} \tag{26}$$

From Equation 25 it follows that

$$\frac{d\Delta q_s}{dt} = k_i \left[\cosh\left(\sqrt{k_i\left(k_p + k_l\right)}\, t \right) \right]^{-2} \tag{27}$$

where

$$\cosh(x) = \frac{1}{2}\left(e^x + e^{-x}\right) \tag{28}$$

Hence, in the limit for $t \rightarrow \infty$

$$\frac{d\Delta q_s}{dt} \rightarrow 0$$

$$\frac{dq_p}{dt} \rightarrow \frac{k_p k_i}{k_p + k_l}$$

$$\frac{dq_l}{dt} \rightarrow \frac{k_l k_i}{k_p + k_l}$$

and

$$\Delta q_s \rightarrow \left(\frac{k_i}{k_p + k_l} \right)^{\frac{1}{2}}$$

Thus, in the case of Freundlich-type adsorption ($n = 2$) of heavy metals in soils, eventually all metals that reach the soil would be removed by crop uptake and leaching, similar to the first-order case. However, in the first-order case, the rates of plant uptake and leaching increased linearly with Δq_s, whereas in the second-order case the increase is proportional to $(\Delta q_s)^2$, that is, the rates of leaching and plant uptake initially tend to be lower than in the first-order case but eventually increase more steeply.

Adsorption and Surface Transformations

So far, a distinction has been made between precipitation/dissolution and adsorption of heavy metals. Adsorption equations refer to an equilibrium state in which the rates of adsorption and desorption are equal; that is, it is assumed that the adsorption process is reversible. In soils, however, reversibly adsorbed heavy metals may be slowly transformed into irreversibly adsorbed forms, through association with reactive surface groups on the adsorbent. These additional bonding mechanisms may include the formation of hydrogen bonds, the coordination with surface groups (ligand exchange), or London-van der Waals interactions between coordinated (organic) ligands and the adsorbent surface. When chemical-bond formation (covalent bonds/electron sharing) is involved, the formation of irreversibly adsorbed forms of heavy metals may be referred to as "surface precipitation."

At the solid-solution interface, concentrations of heavy metals may be much higher than in solution. Also, reactive surface groups (e.g., Si—OH, Al—OH, Fe—OH, Mn—OH) may occur at "concentrations" that exceed those of similar compounds in solution. In addition, electrostatic interactions at the solid-solution interface may result in the (partial) dehydration of hydrated heavy-metal ions or induce the formation of hydrolysis products. Therefore, in the adsorbed phase, reactions may occur that would not occur between similar species in the solution phase.

An essential difference between precipitation from solution and irreversible adsorption is that in the latter case, the reaction is assumed to depend on the heavy-metal concentrations in the adsorbed phase and not on their solution concentrations. If surface transformations occur, adsorption models would apply to the reversibly adsorbed part of the accumulated heavy metals (Δq_{rev}), but not to the irreversibly adsorbed part (Δq_{irr}). Similarly, the rates of crop uptake and leaching would depend on Δq_{rev} rather than on the total accumulated heavy-metal contents.

In order to solve Equation 5 for this case, it will be assumed that

$$\Delta q_s = \Delta q_{rev} + \Delta q_{irr} \tag{29}$$

where the solubility of heavy metals is determined by Δq_{rev} only. Furthermore, it will be assumed that the rate of transformation can be represented by a first-order rate equation, according to

$$\frac{d\Delta q_{irr}}{dt} = k_t \Delta q_{rev} \tag{30}$$

where k_t is a rate constant (y^{-1}). Using Equations 29 and 30, it follows that Equation 5 can be rewritten as

$$\frac{d\Delta q_{rev}}{dt} = k_i - \frac{dq_p}{dt} - \frac{dq_l}{dt} - k_t \Delta q_{rev} \tag{31}$$

As an example, Equation 31 will be solved for the case of linear adsorption, at a constant rate of heavy-metal input (Equation 2)

$$\frac{d\Delta q_{rev}}{dt} = k_i - \left(k_p + k_l + k_t\right)\Delta q_{rev} \tag{32}$$

If

$$k_i > k_p + k_l + k_t \tag{33}$$

Equation 32 can be solved for Δq_{rev}

$$\Delta q_{rev} = \frac{k_i}{k_p + k_l + k_t}\left\{1 - e^{-\left(k_p + k_l + k_t\right)t}\right\} \tag{34}$$

such that Δq_{irr} becomes

$$\Delta q_{irr} = \frac{k_i k_t t}{k_p + k_l + k_t} - \frac{k_i k_t}{\left(k_p + k_l + k_t\right)^2}\left\{1 - e^{-\left(k_p + k_l + k_t\right)t}\right\} \tag{35}$$

Hence, in the limit for $t \rightarrow \infty$

$$\frac{d\Delta q_{rev}}{dt} \rightarrow 0$$

$$\frac{d\Delta q_{irr}}{dt} \rightarrow \frac{k_t k_i}{k_p + k_l + k_t}$$

$$\frac{dq_p}{dt} \rightarrow \frac{k_p k_i}{k_p + k_l + k_t}$$

$$\frac{dq_l}{dt} \rightarrow \frac{k_l k_i}{k_p + k_l + k_t}$$

whereas:

$$\Delta q_{rev} \rightarrow \frac{k_i}{k_p + k_l + k_t}$$

Thus, in the limit for $t \rightarrow \infty$, the rates of plant uptake and leaching would be lower than in the case of linear adsorption only, because of the transformation of reversibly adsorbed forms into irreversibly adsorbed forms. The accumulated amount of reversibly adsorbed heavy metals would eventually reach a constant value.

THE CHEMISTRY OF CADMIUM IN SOILS IN THE NETHERLANDS

The solubility of cadmium in agricultural soils in The Netherlands appears to be largely governed by (nonlinear) adsorption reactions, involving soil organic matter as well as clay minerals and hydrous oxides of Fe, Al, Mn, and Si.[49–53] The binding strength of cadmium in soils has been found to increase strongly with increasing pH.[49,52] This may be due to an increase in charge density on oxide (or organic matter) surfaces, or to the formation of hydrolysis products of aqueous heavy metals, or a combination of both.[49]

Poelstra et al.[53] described the adsorption behavior of cadmium in soils by linear adsorption isotherms, and found that the adsorption constant increased with equilibration time. Chardon[52] described the adsorption behavior of cadmium by a Freundlich adsorption equation (with n = 1.2) for a wide range of agricultural soils in The Netherlands. The constant k_F in the Freundlich equation was found to increase strongly with increasing pH.

Ion Exchange and Specific Adsorption

In an adsorption study with Winsum clay soil, the Cd^{2+}/Ca^{2+} exchange reaction was shown to be reversible, except for Cd^{2+} adsorbed at very low surface coverage (Figure 1). The cation-exchange reaction turned out to be approximately linear, that is, the majority of the adsorption sites showed equal preference for Ca^{2+} and Cd^{2+}. At low degrees of surface coverage by Cd^{2+}, the adsorption behavior deviated from linear cation exchange. This specifically adsorbed Cd^{2+} could not be desorbed by Ca^{2+} in solution, but was fully recovered by extraction with a dilute acid solution (0.01 N HNO_3). It was estimated that the total adsorption capacity of Winsum clay soil was about 200 μeq/g, of which about 15 μeq/g consisted of specific adsorption sites. It was hypothesized that the specific adsorption sites were associated with Si—OH or Al—OH groups on the edges of clay minerals, or with reactive groups associated with hydrous oxides occurring as coatings on clay minerals.[49]

The adsorption of Cd^{2+} on an initially Ca^{2+}-saturated Winsum clay soil, reported by Harmsen,[49] may be described by a combination of a Freundlich adsorption equation

$$q_{Cd}\ (\text{Freundlich}) = 10^{1.43}\left(X_{cd}\right)^{0.275} \tag{36}$$

and a linear ion-exchange equation

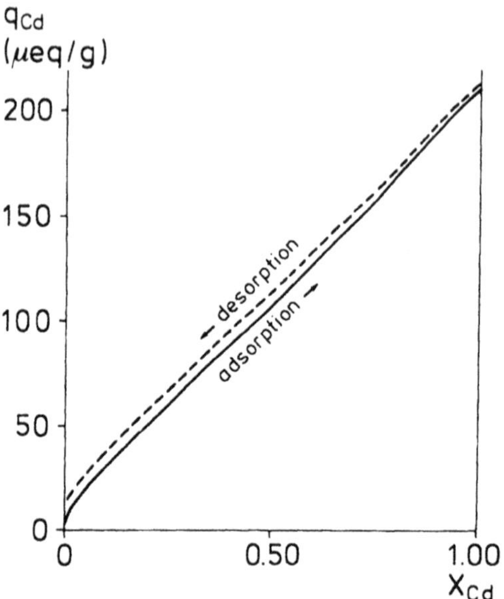

FIGURE 1. Adsorption of Cd^{2+} on an initially Ca^{2+}-saturated Winsum clay soil (solid line), followed by desorption of Cd^{2+} by Ca^{2+} (broken line). The amount of Cd^{2+} adsorbed, q_{Cd} (μeq/g), is plotted against the fractional cadmium concentration in solution (Equation 38).

$$q_{Cd}\ (\text{Ion Exchange}) = 185X_{Cd} \qquad (37)$$

where the fractional cadmium-concentration in solution, X_{Cd}, is given by

$$X_{Cd} = \frac{C_{Cd}}{C_{Cd} + C_{Ca}} \qquad (38)$$

and where the specific adsorption capacity is assumed to be 15 μeq/g, that is, q_{Cd} (Freundlich) increases up to 15 μeq/g and is taken constant thereafter (Figure 2).

Alternatively, the adsorption of cadmium on an initially calcium-saturated Winsum clay soil may be described by a "two-site" model, that is, assuming that the adsorbent consists of a small fraction of "high-selectivity" sites that are highly selective for cadmium, and a large fraction of "low-selectivity" sites that show equal preference for both divalent cationic species, Ca^{2+} and Cd^{2+}. Adsorption on high selectivity sites may be described by

$$q_{Cd}(\text{high}) = \frac{Q_h K_h C_{Cd}}{C_{Ca} + K_h C_{Cd}} \qquad (39)$$

where the adsorption capacity of the high-selectivity sites, Q_h, is taken as 15 μeq/

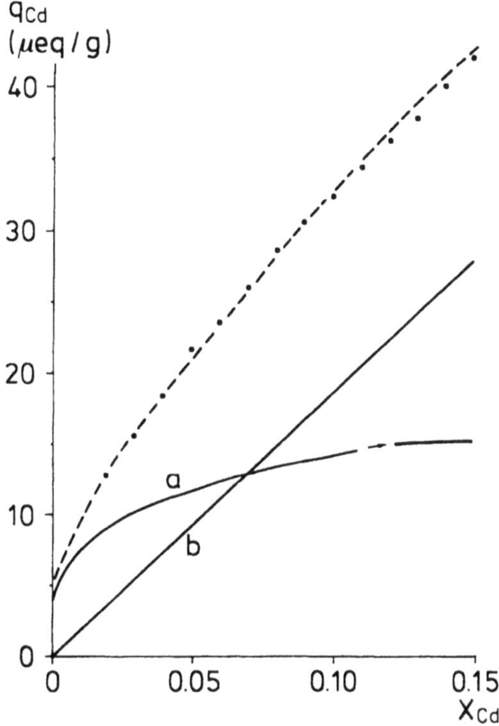

FIGURE 2. Adsorption of Cd^{2+} on an initially Ca^{2+}-saturated Winsum clay soil at low surface coverage by Cd^{2+}. The broken line represents the summation of curve a (Equation 36) and curve b (Equation 37). The specific adsorption capacity (15 μeq/g) is indicated with an arrow. The dots represent experimental measurements.

g and the selectivity coefficient, K_h, as 100. Adsorption on low-selectivity sites may be described by

$$q_{Cd}(\text{low}) = \frac{Q_l K_l C_{Cd}}{C_{Ca} + K_l C_{Cd}} \tag{40}$$

where the adsorption capacity of the low-selectivity sites, Q_l, is taken as 185 μeq/g and the selectivity coefficient, K_l, as 1; that is, Equation 40 reduces to Equation 37. The summation of Equations 39 and 40 gives approximately the same results as the summation of Equations 36 and 37; see Figure 2 (broken line). The conceptual advantage of the two-site model would be that it is continuous over the entire range of cadmium concentrations in solution, whereas the Freundlich adsorption equation would be discontinuous at a value of 15 μeq/g.

From Figures 1 and 2 it follows that specific adsorption of cadmium predominates in the trace region, whereas linear ion exchange occurs at all cadmium concentrations in solution. A specific adsorption capacity of

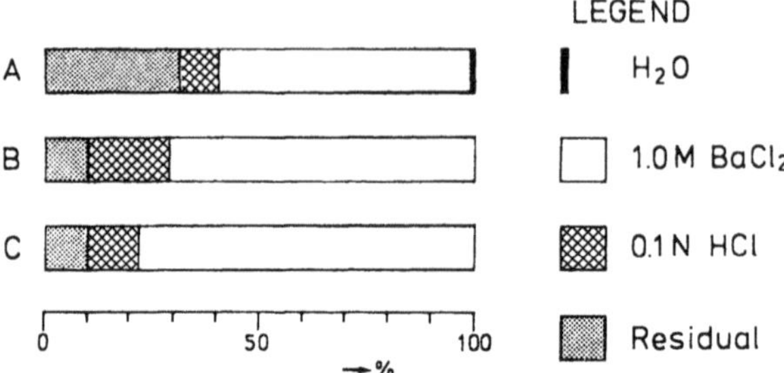

FIGURE 3. Relative amounts (%) of soil cadmium extracted successively with water, a salt solution (1.0 *M* $BaCl_2$) and dilute acid (0.1 *N* HCl), and residual cadmium fractions in the surface layer of soils at (A) Budel, (B) Overpelt, and (C) near the river Geul.

15 μeq/g would by far exceed any conceivable cadmium contents in agricultural soils. Hence, it might be concluded that the binding of cadmium in Winsum clay soil would be predominated by specific adsorption on highly selective adsorption sites. However, it has been shown that specific adsorption is strongly dependent on pH. For example, specific adsorption was observed in a Winsum clay soil at pH = 7.0, whereas no specific adsorption at all occurred in a comparable clay soil at pH = 4.5.[49]

An increased availability of cadmium in near-neutral to slightly acidic soils has been observed in contaminated soils near zinc smelters or in zinc mining areas. For example, in sandy soils near zinc smelters in Overpelt (Belgium) and Budel (The Netherlands), cadmium contents in the surface layer were found to be 24.3 and 9.4 ppm, respectively, whereas in a clay soil along the river Geul (The Netherlands), which had discharged acid mine waters from a former zinc and lead mining area, cadmium contents of 11.5 ppm were found in the surface layer.[49] In all of these soils, salt-extractable cadmium fractions were relatively high, ranging from 60% (Budel) to nearly 80% (Geul), indicating that cadmium in these soils would be rather mobile (Figure 3). Although cadmium contents in these contaminated soils are significantly higher than in most agricultural soils, the data in Figure 3 suggest that in slightly-acidic soils (pH = 5 – 6), elevated cadmium contents may be associated with significant increases in exchangeable forms of cadmium.

For the long-term behavior of cadmium in agricultural soils, this would imply that under neutral to slightly alkaline conditions, the adsorption of cadmium in soil could be described by a Freundlich-type adsorption equation. In this case, the capacity to immobilize cadmium would probably by far exceed any future cadmium inputs. However, under slightly acidic conditions, the adsorption of cadmium in soil might well be near-linear. In this case, the capacity to immobilize cadmium might be considerably lower and increased cadmium contents might result in an increased solubility and mobility of cadmium in soils.

When agricultural soils would be taken out of production, soil pH might decrease as a result of acid deposition. Under such conditions it might be needed to control soil pH through liming, in order to prevent an increase in the solubility and mobility of cadmium (and other heavy metals) accumulated in such soils.

Concentration of Ions vs. Activity of Ions

The notion "concentration" has been used throughout the text, whereas in some cases the correct notion would have been "activity" or "activity of the free ionic species." For example, the precipitation or dissolution of a solid phase, MA(s), involving a heavy metal (M) and an anionic species (A), can be represented by

$$M + A \rightleftharpoons MA(s)$$

where valencies (charge numbers) have been omitted. At equilibrium, the activities of the free ionic species M (a_M) and A (a_A) in solution are related by

$$a_M a_A = K^\circ_{so}$$

where K^o_{so} denotes the solubility product of the solid phase MA(s). Replacing activities by concentrations, it follows that

$$c_M c_A = K_{SO}$$

where

$$K_{SO} = K^\circ_{SO} / \gamma_M \gamma_A$$

and where γ_M and γ_A denote the activity coefficients of the free ionic species M and A, respectively. It may be noted that the activity coefficient of a single ionic species is a theoretical concept rather than a measurable quantity.[54,55] At constant ionic strength of the soil solution, K_{so} would be approximately constant. However, if the ionic strength increases, K_{so} would also increase, i.e., the "concentrations" and, hence, the "solubility" of the species involved would increase, whereas the activities would remain constant.

Similar observations apply to complex formation between aqueous heavy metals and ligands, and to adsorption and ion exchange, which all depend on the activities of the free ionic species in solution, rather than on their "concentrations."

Under natural conditions, the ionic strength of the soil solution varies with moisture conditions, fertilizer application, nutrient uptake, etc. As a result, the activity coefficients of the ionic species will vary and thus their "concentrations" in solution. However, the (seasonal) variations in the activity coefficients of aqueous species are considered of minor importance for the long-term behavior of heavy metals in soils.

The Role of Ligands

Ligands that form soluble complexes with heavy-metal species in solution may have a pronounced effect on the solubility of heavy metals in soils and, therefore, cannot be ignored in the present discussion.

The coordination of a heavy metal in solution (M) and an organic or inorganic ligand (L) can be represented by

$$M+L \rightleftharpoons ML$$

where charge numbers have been omitted. At equilibrium it follows that

$$K^{\circ}_{f} = a_{ML} / a_{M}a_{L}$$

where K^{o}_{f} is the stability constant of complex ML. Replacing activities by concentrations yields

$$K_f = c_{ML}/c_M c_L$$

where

$$K_f = K^{o}_{f\gamma M\gamma L/\gamma ML}$$

If soluble complexes are formed between heavy-metal ions and ligands, the solubility of the heavy metals increases

$$c_t = c_M + c_{ML}$$

where the solubility, c_t, is taken as the sum of the concentrations of all heavy-metal species in solution.

If the activity of a free ionic heavy-metal species in solution is determined by a solubility product, the presence of ligands that form soluble, noninteracting (i.e., no adsorption) complexes would increase the solubility of the heavy metal, but not affect the relation between its ionic activity and the associated solid phase. In this case, the solubility of the heavy metal would ultimately be determined by the concentrations of the anionic species (c_A) and the ligand (c_L) in solution. The concentrations of major anions, such as carbonates or phosphates, would be approximately constant in soil solution and not be affected by the precipitation or dissolution of minor quantities of MA(s). Thus, the presence of ligands would increase the solubility of heavy metals by an amount determined by K_f and c_L.

If the liganded heavy-metal species (ML) were adsorbed in small quantities, the solubility relationship would not be affected. However, if large quantities of ML were adsorbed, the solid phase MA(s) might fully dissolve and disappear from the soil. In that case, the solubility relationship would change and the aqueous activity of M would be determined by an adsorption equation rather than through a solubility product.

In the case of linear adsorption of heavy metals in soil, the presence of ligands that form soluble, noninteracting complexes would increase the solubility but not affect the relationship between the activities of the aqueous and the adsorbed heavy-metal species. If the ligand concentration were much higher than the heavy-metal concentration, the solubility of heavy-metal species ($c_M + c_{ML}$) would increase nearly linearly with the adsorbed concentration of M (q_M). If the initial ligand concentration were of the same order of magnitude as the heavy-metal concentration, the solubility would increase in a nonlinear fashion, that is, the ratio

$$c_{ML}/(c_M + c_{ML})$$

would decrease with increasing q_M.

If the complexed species ML were adsorbed linearly, the total adsorbed concentration ($q_M + q_{ML}$) would increase linearly with the total solution concentration ($c_M + c_{ML}$) and thus with c_M, provided that c_L were much higher than c_M or be buffered through desorption or dissolution. If the initial ligand concentration were of the same order of magnitude as c_M and the complexed species were adsorbed preferentially, this would result in a nonlinear relationship between the total adsorbed concentration ($q_M + q_{ML}$) and the total solution concentration ($c_M + c_{ML}$). This case has been referred to as a "two-species" model, similar to the "two-site" model addressed in a previous section, and it can be shown that the preferential adsorption of complexed heavy-metal species may result in a Freundlich-type adsorption behavior.[49]

It may be concluded that the presence of ligands in soil solution may have a significant effect on the solubility of heavy metals in soil. If the complexed species are not adsorbed, their presence merely increases the solubility of heavy metals in soil but does not affect the relation between the activities of the free ionic heavy-metal species in solution and a solid phase or an adsorbed phase. If the complexed species are adsorbed in soil, this may affect the relationship between the total concentration of heavy-metal species in solution and in an adsorbed phase or a solid phase: precipitates may entirely dissolve and initially linear adsorption may turn into Freundlich-type adsorption behavior.

Plant uptake may also be dependent on the ionic activities of heavy metals in soil solution, rather than on their total concentrations.[55] However, complexed species provide a labile pool of heavy metals in solution, that is, complexed species are decomposed if the activities of the free ionic species decrease due to plant uptake. Therefore, a relation may be expected between the total solubility of all heavy-metal species in solution and plant uptake of heavy metals, provided equilibrium is maintained between the activities of free ionic species and complexed species.

Leaching of heavy metals from the surface layer of agricultural soils would primarily depend on their total solution concentrations. As cationic species would be adsorbed in subsurface layers, the presence of noninteracting species (e.g., $CdSO_4^o$) may significantly increase the mobility of heavy metals in soil upon leaching.

RESULTS AND DISCUSSION

Five parameters would have to be determined experimentally to calibrate the model presented in earlier sections herein. First, the value of $\Delta q_s(t)$, that is, the values of $q_s(0)$ and $q_s(t)$, would have to be measured. The value of $q_s(0)$ can be determined from historic soil samples, taken at the end of the 19th century, or from soil samples taken at sites that have been protected from inputs of heavy metals. The value of $q_s(t)$ would represent the present situation of agricultural soils.

The average annual input of heavy metals in soil, dq_i/dt, would have to be estimated from historic data as well as recent measurements. It will be difficult to estimate dq_i/dt accurately, as long-term datasets on atmospheric deposition or contents of heavy metals in chemical fertilizers are scarce. Approximate values of dq_i/dt can be obtained from data on fertilizer use in agriculture, the use of fossil fuels, and industrial development. In the model, dq_i/dt is assumed to be zero before $t = 0$ and constant thereafter. If more accurate data were available, dq_i/dt could be modified accordingly. In that case, the differential equation would have to be solved in parts, or a computer (simulation) model would have to be developed to solve Equation 5 numerically.

The annual rates of crop uptake, dq_p/dt, and leaching, dq_l/dt, can be determined experimentally for agricultural soils. With regard to crop uptake, one would have to distinguish between heavy metals derived directly from atmospheric deposition and heavy metals taken up from the soil.[35] Leaching of heavy metals from soil is difficult to measure experimentally, because of spatial variability and the low concentrations of these elements in the soil solution of most agricultural soils.

Model Parameter Estimation

The cadmium contents of noncontaminated mineral soils in the Netherlands, $q_s(0)$, have been estimated at 0.08 ppm for sandy soils and 0.25 ppm for clay soils.[56,57] Present cadmium contents of agricultural soils, $q_s(t)$, have been estimated at 0.32 ppm for sandy soils and 0.41 ppm for clay soils.[50] From these data, the average increase in cadmium contents of mineral, agricultural soils was estimated to be 0.20 ppm. Assuming that the top layer of these soils would contain about 2×10^6 kg of soil per ha, it would then follow that $\Delta q_s(t)$ equals approximately 400 g of Cd per ha.

The current annual inputs of cadmium in agricultural soils have been estimated to be 3.5 g ha^{-1} y^{-1} from atmospheric deposition and 3.8 to

5.5 g ha^{-1} y^{-1} with chemical fertilizers or animal manures.[58] From these data, dq_i/dt was estimated to be 8.0 g ha^{-1} y^{-1}. It is further assumed that the annual input of cadmium in agricultural soils is a "step function," which is zero before $t = 0$ and assumes a constant value thereafter (Equation 2).

The amount of cadmium removed annually by crops, dq_p/dt, differs significantly between crops. For a range of arable and horticultural crops the average annual crop uptake would be about 2.5 g ha^{-1} y^{-1} and for grassland about 0.1 g ha^{-1} y^{-1}.[58] For arable crops, such as cereals and potatoes, annual crop uptake would be in the range of 0.5 to 1.5 g ha^{-1} y^{-1}.[51] For the model calculations, an average value of 1.0 g ha^{-1} y^{-1} was used.

The average cadmium concentrations in the groundwater were found to be 0.38 μg l^{-1} at a depth of 10 to 25 m and 0.42 μg l^{-1} below 25 m depth.[59] At a precipitation surplus (precipitation minus evapo-transpiration) of about 3×10^6 l ha^{-1} y^{-1}, this would amount to an annual leaching rate of 1.2 g ha^{-1} y^{-1}. This estimate of dq_l/dt fits in well with other estimates of the amount of cadmium leached annually from the topsoil.[58,60]

Annual inputs of heavy metals in agricultural soils probably gradually increased by the end of the 19th century.[15] From data by Jones et al.[16] it would follow that the cadmium input in soils at Rothamsted Experimental Station (U.K.) increased nearly linearly between 1895 (1.1 g ha^{-1} y^{-1}) and 1950 (6.5 g ha^{-1} y^{-1}), that is, the rate constant in Equation 3 would be about 0.1 g ha^{-1} y^{-2} for that period. After 1950, cadmium inputs increased more steeply, at a rate of about 0.5 g ha^{-1} y^{-1}. In 1980, the annual rate of cadmium input was about 21.7 g ha^{-1} y^{-1}. In principle, Equation 5 could be solved for the zeroth-order and first-order cases, for a stepwise increasing input function (cf. Equation 3). Whether or not a linearly increasing input function is taken in the model does not affect the chemical principles that form the basis of the model presented in this paper. However, the input function does affect the predicted behavior in time of the rates of accumulation, plant uptake, and leaching (cf. earlier section titled "Linear Adsorption [First-Order Case]"), and therefore correct estimates of the input function should be entered in the model, if it is to be used for predictive purposes. In this connection it should be noted that the rate of cadmium input is still expected to increase during the next decade.[61]

The model distinguishes between the initial heavy-metal content in soil, $q_s(0)$, and the accumulated content, $\Delta q_s(t)$. The initial heavy-metal content of soils is thought to be largely made up of heavy metals contained in soil minerals that would dissolve only very slowly in time and thus contribute little to the labile heavy-metal pool in soil. It is assumed that most heavy metals that reach the soil through atmospheric deposition or the use of fertilizers, organic substrates, or biocides enter into the labile heavy-metal pool of the soil; that is, they occur in dissolved form in soil solution or are adsorbed on solid surfaces. The possible transformation of reversibly adsorbed forms of heavy metals into irreversibly adsorbed forms has been dealt with in earlier section titled "Adsorption and Surface Transformation."

The basic assumptions in the model are that a simple relationship exists between the solubility of heavy metals and their accumulated content and that

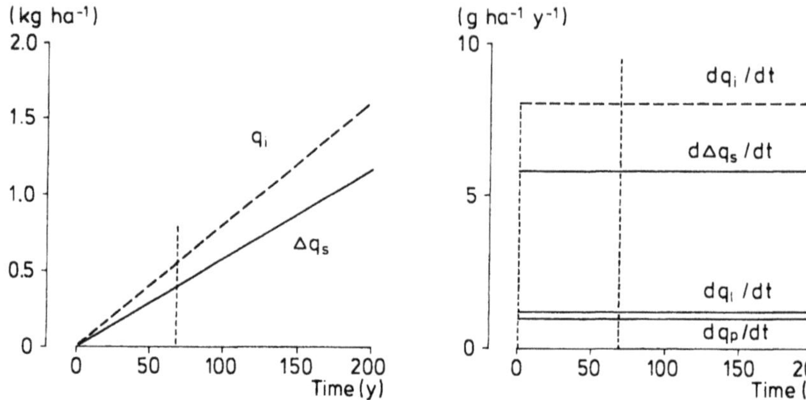

FIGURE 4. Zeroth-order case (Equation 9): total input of cadmium in agricultural soils, $q_i(t)$, accumulated soil content, $\Delta q_s(t)$, (left), annual rates of input, dq_i/dt, accumulation in soil, $\Delta q_s/dt$, leaching, dq_l/dt, and plant uptake of cadmium, dq_p/dt (right), plotted as a function of time. The vertical dotted line at $t = 69$ years represents the present situation.

plant uptake and leaching depend in similar ways on the dissolved concentrations of heavy metals in soils.

Model Results

Although there are few indications that precipitation/dissolution of solid phases containing cadmium would play a role in agricultural soils under aerobic (oxidized) conditions, the zeroth-order case is included for comparison with the first- and second-order cases. In the zeroth-order case (Equation 9), the accumulated soil content of cadmium would increase linearly with time (Figure 4, left). The time when inputs of cadmium in agricultural soils increased to their present level is taken as $t = 0$. In this case, the annual rates of accumulation in soil, leaching, and crop uptake of cadmium would remain constant with time (Figure 4, right).

In the first-order case, represented by Equation 14, the annual rate of accumulation of soil cadmium would decrease with time if the rate of cadmium input remains constant with time (Equation 2). The annual rates of leaching and crop uptake would increase with time (Figure 5). In this case, the model indicates that rates of crop uptake and leaching would double in about 87 years from the present time.

If the annual rate of cadmium input increases linearly with time (Equation 3), the accumulated cadmium content would continue to increase with time (Figure 6, left). The annual rate of cadmium accumulation would eventually reach a constant value, whereas rates of plant uptake and leaching would continue to increase with time (Figure 6, right). In this case, rates of leaching and crop uptake would double in about 61 years from the present time.

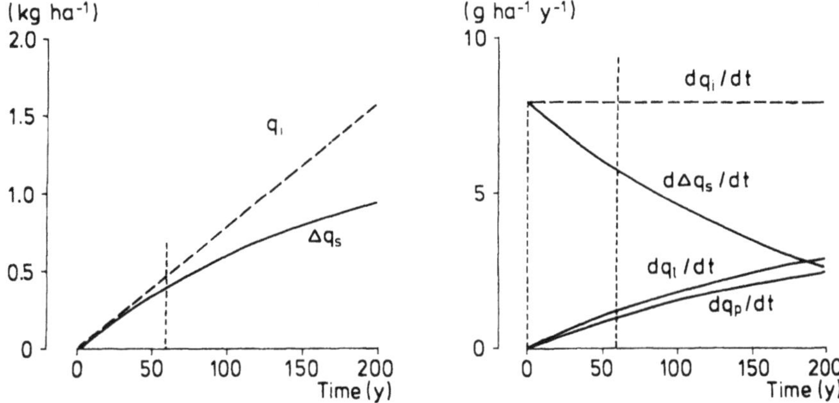

FIGURE 5. First-order case, at a constant rate of cadmium input in soil (Equation 14). For symbols, see Figure 4. The vertical dotted line at $t = 59$ years represents the present situation.

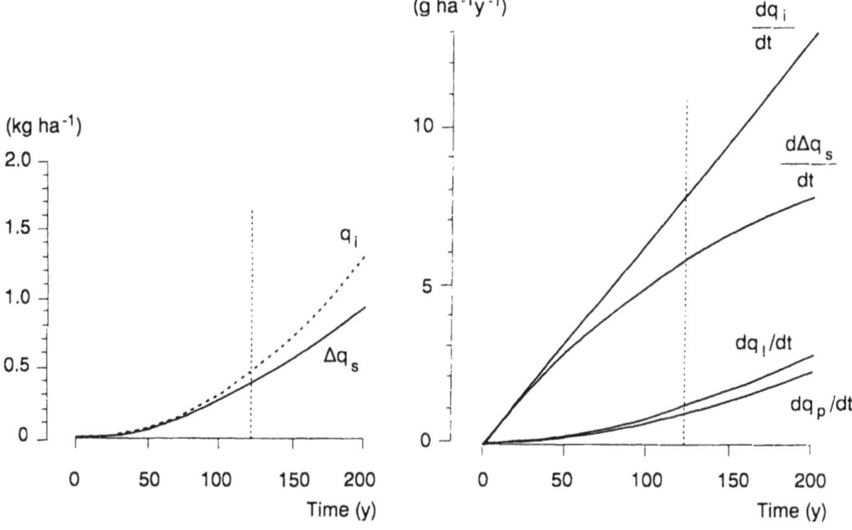

FIGURE 6. First-order case, at a linearly increasing rate of cadmium input in soil (Equation 17). For symbols, see Figure 4. The vertical dotted line at $t = 59$ years represents the present situation.

In the second-order case, represented by Equation 24, the annual rate of soil cadmium accumulation initially decreases slowly and thereafter more steeply (Figure 7). Similarly, leaching and crop uptake initially are virtually zero and thereafter increase rapidly. In this case, it would follow that the annual rates of leaching and crop uptake would double within the next 40 years, assuming that the rate of input would remain constant.

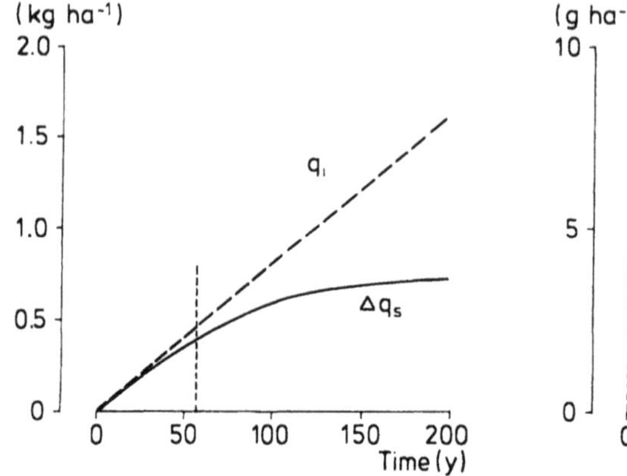

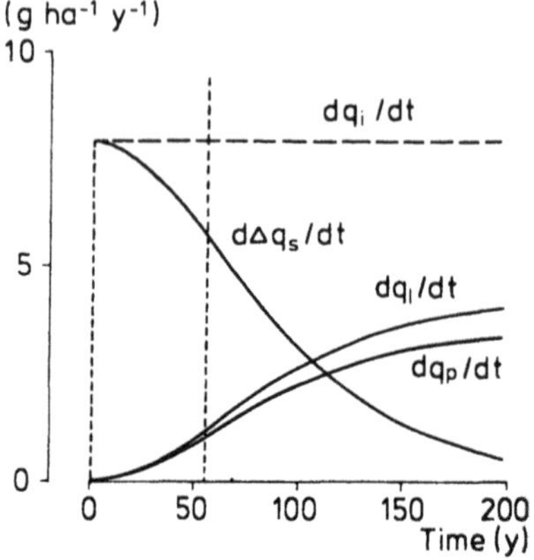

FIGURE 7. Second-order case (Equation 24); see Figure 4. The vertical dotted line at t = 56 years represents the present situation.

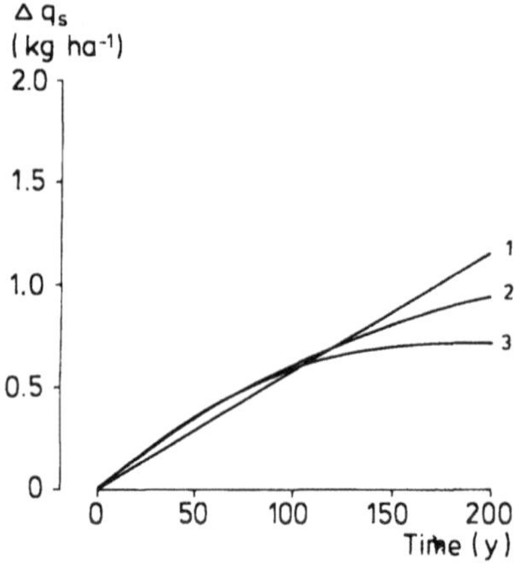

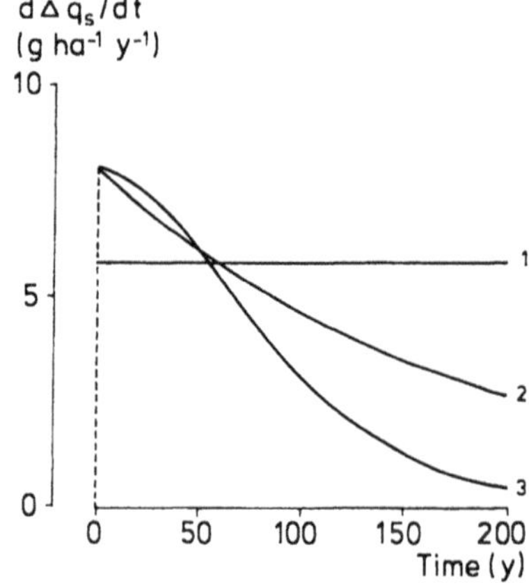

FIGURE 8. Accumulated soil contents of cadmium (left) calculated from (1) Equation (10), (2) Equation (15), and (3) Equation (25). Annual rates of accumulation of cadmium in soil (right) calculated from: (1) Equation (9), (2) Equation (16), and (3) Equation (27).

The long-term behavior of cadmium in soils as predicted by the three solutions to Equation 5 is distinctly different. Initially the total amounts of cadmium accumulated show a similar behavior (Figure 8, left). However, after about 100 years they would start to diverge, in accordance with the annual rates of accumulation (Figure 8, right). Although Curves 2 and 3 (Figure 8, right) will both eventually approach zero, Curve 3 (second-order case) decreases more steeply than Curve 2 (first-order case). Hence, if a Freundlich-type adsorption equation applies to cadmium solubility in soils, a significant decrease in the cadmium accumulation rate may be expected to occur during the next century.

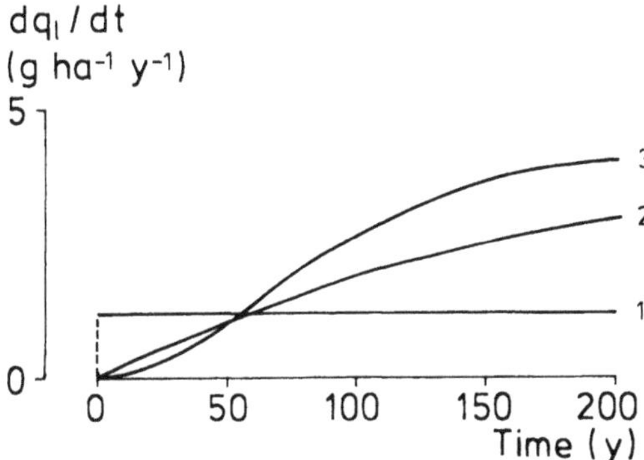

FIGURE 9. Annual rates of leaching of cadmium from the top layer of agricultural soils, calculated from (1) Equation (8), (2) Equations (13) and (15), and (3) Equations (23) and (25).

Similar observations can be made for the rates of plant uptake and leaching. In the first- and second-order cases, the annual leaching rates increase with time (Figure 9). Although Curves 2 and 3 in Figure 9 eventually reach the same asymptotic value, Curve 3 increases more steeply than Curve 2. Crop uptake rates are assumed to behave similarly to leaching rates.

Assuming that the adsorption of cadmium in near-neutral to slightly alkaline soils can be described by a Freundlich adsorption equation, the long-term behavior of the cadmium leaching rate would best be described by Curve 3 in Figure 9. However, at very high soil-cadmium contents, or under slightly acidic conditions, adsorption of cadmium would be nearly linear, that is, close to Curve 2 in Figure 9, assuming in both cases that the rate of cadmium input in soils is approximately constant in time (Equation 2).

Soil-Crop Ecosystems

In soil-crop ecosystems, the relation between the biological availability and the solubility of heavy metals is very complex. For example, the availability of heavy metals to crops depends, among other things, on the distribution and density of the root system, the chemical composition of the soil solution, the distribution of heavy-metal species in the soil, and the structure and transport properties of the soil. Actual uptake of heavy-metal species would further depend on the kinetics of heavy metal release, rates of transport by mass flow and diffusion, and rhizosphere processes, such as the release of chelating compounds by active roots. In turn, these processes are strongly affected by moisture conditions, gas exchange, and soil temperature.

In addition, crop uptake in The Netherlands occurs mainly during spring and summer, when temperatures are relatively high and evapo-transpiration usually exceeds precipitation. Leaching occurs mainly during fall and winter, when temperatures are lower than during the growing season. Therefore, crop uptake and leaching may occur under different crop and climatic conditions.

The chemical processes involved in determining the plant availability and mobility of heavy metals in soils, however, are quite similar. In both cases the precipitation and dissolution of soil minerals, ion exchange, adsorption and desorption processes, the formation of soluble heavy-metal complexes (speciation), the dynamics of organic matter, and soil conditions, such as pH and redox potential, play a role in determining the dynamics of heavy metals in soil. It may be expected that if the solubility of heavy metals in soil increases, the availability to crops and the potential for leaching also increase. In view of the simplified nature of the present model, the assumption that crop uptake and leaching depend on the solubility of heavy metals is therefore considered to be adequate.

Model Predictions

If the solubility of heavy metals in soils as determined by precipitation or dissolution of mineral phases (zeroth-order case), there would hardly be a need to be concerned about the increasing heavy-metal contents of agricultural soils, as the availability and mobility of heavy metals in soils would not increase with time. However, with the possible exception of sulfides, which may form under anaerobic conditions, most mineral phases involving heavy metals seem to be too soluble to persist in agricultural soils. Therefore, the solubility of heavy metals is soils is more likely to be determined by adsorption processes.

The model shows that if a relation exists between the solubility of heavy metals in soils and their accumulated soil contents (first- and second-order cases), the rates of plant uptake and leaching may be expected to increase during the next century, even if the input of heavy metals remains constant. In particular, if the adsorption behavior of heavy metals in soil can be described by a Freundlich adsorption equation, the annual rates of crop uptake and leaching might increase significantly in the near future.

If the model presented in this paper were used for predictive purposes, it would have to be differentiated for heavy-metal species and soil, crop, and climatic conditions. In addition, future trends in heavy-metal inputs and changes in environmental conditions would have to be assessed. If such information were available, it could be included in the present model with the help of computer simulation techniques.

When agricultural soils are (temporarily) taken out of production or when agricultural management practices or environmental conditions change, soil conditions may change in such a way that the availability and mobility of heavy metals increase. Therefore, there is a need for models describing the dynamics of heavy metals in soils as affected by environmental, soil, and crop conditions. On the basis of such models, management strategies for agricultural soils with elevated heavy-metal contents could be designed. Management strategies should aim at minimizing the availability and mobility of heavy metals in soils, thus minimizing their impact on the functioning of soil ecosystems and on the quality of agricultural products as well as the quality of natural waters.

SUMMARY

In the industrialized countries of the world, many agricultural soils contain significant amounts of heavy metals, which present a potential hazard to the quality of food products and the quality of the environment. Therefore, it is important to study the factors and processes that determine the behavior of heavy metals in soils and to develop models that predict their long-term behavior in agricultural soils as a function of soil and crop management and environmental factors. Such models may be used to develop management strategies, aimed at minimizing the negative effects of heavy metals on the quality of agricultural products, the functioning of the soil-crop ecosystem, and the quality of ground and surface waters.

In this paper, a simple model has been presented to predict the long-term behavior of heavy metals in agricultural soils. The analytical solutions to the model differ in the assumed relationships between the solubility of heavy metals and their accumulated contents in soils.

Four cases are distinguished. (1) Precipitation and dissolution reactions determine the solubility of heavy metals in soil. Annual rates of crop uptake and leaching would then remain constant with time, whereas the accumulated content of heavy metals in soil would increase with time, if the inputs of heavy metals exceed leaching and plant uptake from soil. (2) Ion exchange or linear adsorption determine the solubility of heavy metals. Annual rates of crop uptake and leaching would then increase with increasing accumulated contents of heavy metals in soil until they equal the annual inputs of heavy metals from atmospheric deposition, the use of fertilizers, and other sources. (3) Solution concentrations of heavy metals are related to their accumulated soil contents through Langmuir- or Freundlich-type adsorption equations. In this case crop uptake and leaching would initially increase more slowly than in the case of linear ion exchange, but eventually more rapidly, until equilibrium would be reached with annual inputs of heavy metals. (4) Part of the heavy metals in soil is transformed into unavailable (irreversibly adsorbed) forms. In this case plant uptake and leaching will depend on the reversibly adsorbed forms of heavy metals only. Ultimately, the rates of immobilization of heavy metals in soil, plant uptake, and leaching will reach equilibrium with the annual inputs of heavy metals.

As an example, the model is applied to assess the long-term behavior of cadmium in agricultural soils in The Netherlands.

ACKNOWLEDGMENT

Thanks are due to Dr. S. E. A. T. M. Van der Zee (Wageningen Agricultural University) and Dr. W. J. Chardon, Dr. D. L. R. Hesterberg, Dr. J. B. T. M. Roerdink, Dr. W. Salomons, and Dr. W. Van Driel (Institute for Soil Fertility Research) for their critical reading of the manuscript.

REFERENCES

1. Little, P. and M. H. Martin. "A Survey of Zinc, Lead and Cadmium in Soil and Natural Vegetation Around a Smelting Complex," *Environ. Pollut.* 3:241–254 (1972).
2. Lisk, D. J. "Trace Metals in Soils, Plants and Animals," *Adv. Agron.* 24:267–325 (1972).
3. Buchauer, M. J. "Contamination of Soil and Vegetation Near a Zinc Smelter by Zinc, Cadmium, Copper and Lead," *Environ. Sci. Technol.* 7:131–135 (1973).
4. Lagerwerff, J. V., D. L. Brower, and G. T. Biersdorf. "Accumulation of Cadmium, Copper, Lead and Zinc in Soil and Vegetation in the Proximity of a Smelter," in *Proc. Symp. Trace Subst. Environ. Health,* (Columbia: University of Missouri, 1973) pp. 71–78.
5. Andersson, A. "Heavy Metals in Commercial Fertilisers, Manure and Lime," Reports of the Agricultural College of Sweden, Ser. A, No. 283, Uppsala, Sweden (1978).
6. Tjell, J. C. and M. F. Hovmand. "Metal Concentrations in Danish Arable Soils," *Acta Agric. Scand.* 28:81–89 (1978).
7. Van Driel, W. and K. W. Smilde. "Heavy Metal Contents of Dutch Arable Soils," *Landwirtsch. Forsch. Sonderh.* 38:305–313 (1982).
8. Chang, A. C., J. E. Warneke, A. L. Page, and L. J. Lund. "Accumulation of Heavy Metals in Sewage Sludge Treated Soils," *J. Environ. Qual.* 13:87–91 (1984).
9. Pacyna, J. M., A. Semb, and J. E. Hanssen. "Emission and Long-Range Transport of Trace Elements in Europe," *Tellus* 36-B:163–178 (1984).
10. Berrow, M. L. "An Overview of Soil Contamination Problems," in *Proc. Int. Conf. Chemicals in the Environment,* Lester, J. N., R. Perry, and R. M. Sterrit, Eds. (London: Selper, 1986) pp. 543–552.
11. Cawse, P. A. "Trace and Major Elements in the Atmosphere at Rural Locations in Great Britain, 1972–1981," in *Pollutant Transport and Fate in Ecosystems,* Coughtrey, P. J., M. H. Martin, and M. H. Unsworth, Eds., British Ecological Society Special Publication No. 6. (Oxford: Blackwell, 1987) pp. 89–112.
12. Steinnes, E., W. Solberg, H. M. Petersen, and C. D. Wren. "Heavy Metal Pollution by Long-Range Atmospheric Transport in Natural Soils of Southern Norway," *Water Air Soil Pollut.* 45:207–218 (1989).
13. Salmon, L., D. H. F. Atkins, E. M. R. Fisher, C. Healy, and D. V. Law. "Retrospective Trend Analysis of the Content of U.K. Air Particulate Material 1957–1974," *Sci. Total Environ.* 9:161–200 (1978).
14. Nriagu, J. O. "Global Inventory of Natural and Anthrooogenic Emissions of Trace Elements to the Atmosphere," *Nature (London)* 279:409–411 (1979).
15. Jones, K. C., C. J. Symon, and A. E. Johnston. "Retrospective Analysis of an Archived Soil Collection. I. Metals," *Sci. Total Environ.* 61:131–144 (1987).
16. Jones, K. C., C. J. Symon, and A. E. Johnston. "Retrospective Analysis of an Archived Soil Collection. II. Cadmium," *Sci. Total Environ.* 67:75–89 (1987).
17. Van Driel, W. and K. W. Smilde. "Micronutrients and Heavy Metals in Dutch Agriculture," *Fert. Res.* 25:115–126 (1990).
18. Jones, K. C. "Contaminant Trends in Soils and Crops," *Environ. Pollut.* 69:311–325 (1991).

19. Mortvedt, J. J., P. M. Giordano, and W. L. Lindsay, Eds. *Micronutrients in Agriculture* (Madison, WI: Soil Sci. Soc. Am., 1972).
20. Tiller, K. G. "Essential and Toxic Heavy Metals in Soils and Their Ecological Relevance," in *Transactions XIII Congress of the International Society of Soil Science,* Vol. I, (Hamburg, 1986) pp. 29–43.
21. Lagerwerff, J. V. "Lead, Mercury and Cadmium as Environmental Contaminants," in *Micronutrients in Agriculture*, Mortvedt, J. J., P. M. Giordano, and W. L. Lindsay, Eds. (Madison, WI: Soil Sci. Soc. Am., 1972) pp. 593–636.
22. Jaakkola, A., J. Korkman, and T. Juvankoski. "The Effect of Cadmium Contained in Fertilizers on the Cadmium Content of Vegetables," *J. Sci. Agric. Soc. Finl.* 51:158–162 (1979).
23. Mulla, D. J., A. L. Page, and T. J. Ganje. "Cadmium Accumulation and Bio-Availability in Soils from Long-Term Phosphorus Fertilization," *J. Environ. Qual.* 9:408–412 (1980).
24. Mortvedt, J. J., D. A. Mays, and G. Osborn. "Uptake by Wheat of Cadmium and Other Heavy Metal Contaminants in Phosphate Fertilizers," *J. Environ. Qual.* 10:193–197 (1981).
25. Yaron, B., G. Dagan, and J. Goldschmid, Eds. *Pollutants in Porous Media: The Unsaturated Zone Between Soil Surface and Groundwater* (Berlin: Springer-Verlag, 1984).
26. Coughtrey, P. J., M. H. Martin and M. H. Unsworth, Eds. *Pollutant Transport and Fate in Ecosystems*, British Ecological Society Special Publication No. 6. (Oxford: Blackwell, 1987).
27. David, D. J. and C. H. Williams. "Effects of Cultivation on the Availability of Metals Accumulated in Agricultural and Sewage-Treated Soils," *Prog. Water Technol.* 11:257–264 (1979).
28. Smilde, K. W. "Heavy-Metal Accumulation in Crops Grown on Sewage Sludge Amended with Metal Salts," *Plant Soil* 62:3–14 (1981).
29. McGrath, S. P. "Long-Term Studies of Metal Transfers Following Application of Sewage Sludge," in *Pollutant Transport and Fate in Ecosystems*, Coughtrey, P. J., M. H. Martin, and M. H. Unsworth, Eds., British Ecological Society Special Publication No. 6. (Oxford: Blackwell, 1987) pp. 310–317.
30. Cataldo, D. A. and R. E. Wildung. "Soil and Plant Factors Influencing the Accumulation of Heavy Metals by Plants," *Environ. Health Perspect.* 27:149–159 (1978).
31. Adam, A. I. and W. B. Anderson. "Soil Moisture Influence on Micronutrient Cation Availability Under Aerobic Conditions," *Plant Soil* 72:77–83 (1983).
32. Lagerwerff, J. V. "Uptake of Cadmium, Lead, and Zinc by Radish from Soil and Air," *Soil Sci.* 111:129–133 (1971).
33. Hovmand, M. F., J. C. Tjell, and H. Mosbaek. "Plant Uptake of Airborne Cadmium," *Environ. Pollut. Ser. A* 30:27–38 (1983).
34. Harrison, R. M. and M. Chirgawi. "Quantification of Foliar Uptake of Metal Aerosols by Crop Plants," in *Heavy Metals in the Hydrological Cycle*, Astruc, M. and J. N. Lester, Eds. (London: Selper, 1988) pp. 601–604.
35. Dalenberg, J. W. and W. Van Driel. "Contribution of Atmospheric Deposition to Heavy-Metal Concentrations in Field Crops," *Neth. J. Agric. Sci.* 38:369–379 (1990).

36. Merkx, R., J. H. Van Ginkel, J. Sinnaeve, and A. Cremers. "Complexation of Co, Zn and Mn in the Rhizosphere of Maize and Wheat," *Plant Soil* 96:95–107 (1986).
37. Hodgson, J. F. "Chemistry of Micronutrient Elements in Soils," *Adv. Agron.* 15:119–159 (1963).
38. Lindsay, W. L. *Chemical Equilibria in Soils* (New York: John Wiley & Sons, 1979).
39. Hahne, H. C. H. and W. Kroontje. "Significance of pH and Chloride Concentration on Behaviour of Heavy Metal Pollutants: Mercury (II), Cadmium (II), Zinc (II) and Lead (II)," *J. Environ. Qual.* 2:444–450 (1973).
40. Brown, R. M., C. J. Pickford, and W. L. Davison. "Speciation of Metals in Soils," *Int. J. Environ. Anal. Chem.* 18:135–141 (1984).
41. Gaines, G. L., Jr. and H. C. Thomas. "Adsorption Studies on Clay Minerals. II. A Formulation of the Thermodynamics of Exchange Adsorption," *J. Chem. Phys.* 21:714–718 (1953).
42. Bolt, G. H. "Cation-Exchange Equations Used in Soil Science: A Review," *Neth. J. Agric. Sci.* 15:81–103 (1967).
43. Harmsen, K. "Theories of Cation Adsorption by Soil Constituents: Discrete-Site Models," in *Soil Chemistry, B: Physico-Chemical Models*, Bolt, G. H., Ed., (Amsterdam: Elsevier, 1979) pp. 77–139.
44. Davis, J. A. and J. O. Leckie. "Effect of Adsorbed Complexing Ligands on Trace Metal Uptake by Hydrous Oxides," *Environ. Sci. Technol.* 12:1309–1315 (1978).
45. Langmuir, I. "The Adsorption of Gases on Plane Surfaces of Glass, Mica and Platinum," *J. Am. Chem. Soc.* 40:1361–1403 (1918).
46. Freundlich, H. *Kapillarchemie (in German)*, 2nd ed. (Leipzig: Akademische Verlagsgesellschaft, 1922).
47. Ellis, B. G. and B. D. Knezek. "Adsorption Reactions of Micronutrients in Soils," in *Micronutrients in Agriculture*, Mortvedt, J. J., P. M. Giordano, and W. L. Lindsay, Eds. (Madison, WI: Soil Sci. Soc. Am., 1972) pp. 59–78.
48. Halsey, G. D. "The Role of Surface Heterogeneity in Adsorption," *Adv. Catal.* 4:259–269 (1952).
49. Harmsen, K. "Behaviour of Heavy Metals in Soils," *Agric. Res. Reports* 866, Pudoc, Wageningen, The Netherlands (1977).
50. Van Driel, W., B. J. Van Goor, and D. Wiersma. "Cadmium in Cultivated Soils in The Netherlands (in Dutch)," *Bedrijfsontwikkeling* 14:476–480 (1983).
51. Ferdinandus, G. "Calculation of Heavy-Metal Balances for Soils in The Netherlands (in Dutch)," Technische Commissie Bodembescherming, Report TCB A89/01-R, Leidschendam, The Netherlands (1989).
52. Chardon, W. J. "Mobility of Cadmium in Soil (in Dutch)," Ph.D. thesis, Agricultural University, Wageningen, The Netherlands (1984).
53. Poelstra, P., M. J. Frissel, and N. El-Bassam. "Transport and Accumulation of Cd Ions in Soils and Plants," *Z. Pflanzenernaehr. Bodenkd.* 142:848–864 (1979).
54. Lewis, G. N. and M. Randall. *Thermodynamics*, 2nd ed., revised by Pitzer, K. S. and L. Brewer (New York: McGraw-Hill, 1961) p. 310.
55. Sposito, G. "The Future of an Illusion: Ion Activities in Soil Solutions," *Soil Sci. Soc. Am. J.* 48:531–536 (1984).

56. Lexmond, Th. M., Th. Edelman, and W. Van Driel. "Provisional Reference Values and Present Base-Line Contents of a Number of Heavy Metals and Arsenic in the Topsoil of Natural and Agricultural Soils (in Dutch)," Reports and Communications 1986–2, Department of Soil Science and Plant Nutrition, Agricultural University, Wageningen, The Netherlands (1986).
57. Salomons, W. "Provisional Base-Line Contents of Cd, Zn, Ni, Cu and Cr in Sediments in The Netherlands (in Dutch)," Report R–1703, Delft Hydraulics, Delft, The Netherlands (1983).
58. "Cadmium in the Environment (in Dutch)," Tweede Kamer der Staten-Generaal, Vergaderjaar 1983–1984, Volume 18364, Nos. 1–2. The Hague, The Netherlands (1984).
59. Van Duijvenbooden, W., J. Taat, and L. F. L. Gast. "Monitoring Groundwater Quality in The Netherlands (in Dutch)," Final Report, Part 1. State Institute for Public Health and Environmental Hygiene, Leidschendam, The Netherlands (1985).
60. Paul, P. G., J. A. Somers, and D. W. Scholte Ubing. "Inputs of Heavy Metals in Soils in The Netherlands (in Dutch)," *Ingenieur* 93:15–19 (1981).
61. Hutton, M. "A Prospective Atmospheric Emission Inventory for Cadmium — The European Community as a Study Area," *Sci. Total Environ.* 29:29–47 (1983).

9

Atmospheric Deposition of Metals to Agricultural Surfaces

P. M. Haygarth and K. C. Jones

Institute of Environmental and Biological Sciences, Lancaster University, Lancaster, LA1 4YQ, U.K.

ABSTRACT

Atmospheric deposition can be a significant source of heavy metal input to soils and plants in agroecosystems of industrialized countries. This is mostly due to anthropogenic combustion activities, which have substantially enhanced natural emissions of selected heavy metals to the atmosphere. Once emitted, the atmospheric transport of the metal depends upon its chemical properties. Volatile "metalloids" can be transported in a gaseous form or enriched on particles (e.g., Se, Hg, As, and Sb), whereas other metals are transported only in the particle phase (Cd, Pb, and Zn) and may travel long distances before deposition to land. It is argued that atmospheric deposition has resulted in the accumulation of some elements in the surface layers of agricultural soils. There is also evidence that elements adhere to surfaces of crops and, if in the gaseous form, may be absorbed via the foliage into the plant. For Pb, for example, it is shown that about 90% of the total plant uptake is due to deposition from the atmosphere rather than transport from the soil, implying that atmospheric deposition poses a significant source of metal inputs to the foodchain, particularly where background soil levels are relatively low.

INTRODUCTION

Agroecosystems receive inputs of contaminants from the use of fertilizers and agrochemicals, from the application of wastes to soils, and from atmospheric deposition. Clearly these inputs differ in importance for different contaminants and different soils. Long-term changes in soil quality can be induced when inputs are sustained over long periods and when they exceed the rate of contaminant loss from the system. In this context, atmospheric deposition inputs are likely to be of greatest long-term significance and constititute an input to *all* soil-plant systems. In contrast, applications of sewage sludge wastes to land, or the use of specific pesticides will be important inputs to a small proportion of a nation's soil resource. In the U.K., for example, only around 1% of soils receive sewage sludge amendments.

Atmospheric inputs of heavy metals to *soils* are important because they are cumulative, often important over large areas, and difficult to limit and control. Metals are generally immobile in soils. They leach only very slowly, and are inefficiently taken up by roots and translocated to the foliar portion of plants. Consequently they can reside in the surface layers of soils following deposition for hundreds or even thousands of years.

Soil systems differ markedly in their response to these heavy metal inputs. Soils have an ability to buffer against change that is dependent on bulk properties such as texture, organic matter content, and pH. However, there are important implications if the buffering capacity of soils is exceeded; reversing undesirable changes presents difficult management problems. Moreover, the long-term presence of metals in the surface layers of soils can present problems because of continued contamination of the human foodchain, perhaps because of soil ingestion by grazing livestock or the remobilization of metals in wind-blown surface soils and their subsequent transfer onto crop leaf surfaces. In urban areas the persistence of atmospherically derived Pb in surface soils is a cause for continued concern with respect to children's health, given the established importance of the hand-to-mouth transfer of soil Pb. In addition, recent concern has switched to the implications of cumulative atmospherically derived metals on soil fertility through subtle metal effects on microbially mediated nutrient cycling and decomposition processes in surface soils, particularly in woodlands.

Even though the root uptake of atmospherically derived metals from the soils is generally inefficient, this should not be interpreted as inferring that atmospheric inputs of metals to crop plants are insignificant. The contaminant burden of a crop is usually a combination of (1) uptake via the root system, (2) direct foliar uptake and translocation within the plant, and (3) surface-deposited retained contaminant that is associated with fine particulate matter. In many instances, atmospheric inputs to plants (i.e., pathways 2 and 3) are quantitatively far more important than soil inputs.

The U.K. has a long history of metal mining and exploitation, extending back to pre-Roman times. This has provided the incentive for many scientists to study the "legacy" of metal contamination in the U.K., with particular emphasis on the geochemistry and environmental behavior of the elements Pb, Cd, Zn, and Cu in the metalliferous areas of Derbyshire, Wales, and Cornwall. Most of these workers have focussed on the traditional pathways of metal transfer, from rocks — soils — vegetation — livestock, emphasizing element *geo*chemistry. What has perhaps not been fully appreciated, however, is the influence (past and present) of atmospheric transport and deposition in extensively transferring metals from geochemical "hotspots" and from various anthropogenic activities, so that effectively *all* soils in the U.K. are contaminated with metals above their true natural or "historical" background concentration. Indeed, as we show in this chapter, diffuse atmospheric inputs to agroecosystems are still of major importance in influencing crop and soil quality. Furthermore, the history of metal use in the U.K. has played a prominent part in leading to long distance transport and contamination of soils elsewhere in Europe, perhaps most notably in Scandinavia. Clearly, therefore, this subject is of considerable political as well as scientific interest.

The U.K. is an interesting study area for others reasons. There is evidence, for example, that in recent decades the release of metals to the atmosphere has declined. The Clean Air Act was introduced in 1956, ostensibly to tackle the problem of gaseous and particulate emissions contributing to the notorious London Smogs. This signaled a greater awareness of industry as a source of pollution. In addition, recent decades have seen a decline in the traditional reliance on "heavy" industry, coupled with continued improvements in emission control technologies. More recently, the scientific and political debates of the 1970s and 1980s over the contribution and significance of leaded petrol (gasoline) resulted in further improvements in environmental quality. The permitted limit on Pb in petrol dropped from 0.4 to 0.15 g/L at the end of 1985, mirroring the trend that has been seen elsewhere in Europe and in North America. This has provided scientists with the opportunity to track the continued cycling and transfer of Pb in the contemporary environment, to monitor for improvements in environmental quality, and to obtain data on residence times in different environmental compartments. For these reasons the U.K. has been a useful "study site" and "model" for the behavior of heavy metals in the environment. However, many interesting questions still need to be resolved over the environmental chemistry and behavior of heavy metals.

PRINCIPLES AND PROCESSES OF EMISSION TO THE ATMOSPHERE, ATMOSPHERIC TRANSPORT, AND DEPOSITION

This section presents a brief theoretical consideration of the processes of movement and deposition of heavy metals in the atmosphere. To put atmospheric deposition in context and to emphasize the cyclic behavior of metal transport through the atmosphere, *emission* and *transport* are considered prior to the processes of *deposition* onto agricultural surfaces. After considering these largely physical and chemical processes, the propensity for plant uptake will be considered by assessing the *fate* of heavy metals in the agricultural system.

Emissions to the Atmosphere

Emission sources for heavy metals can be either natural or anthropogenic. Nriagu[1] categorized natural sources of atmospheric trace elements as being (1) wind borne soil, (2) sea salt spray, (3) volcanoes, (4) wild forest fires, and (5) biogenic (incorporating continental particulates, continental volatiles, and marine). Natural emissions can occur in gaseous or particulate form. Some elements are methylated (e.g., Se, As, Hg) and can be released from soils, plants, and waters in reduced gaseous forms (e.g., References 2 to 6). Solid phase emissions of metals from natural sources occur as a constituent of fine particles from volcanoes, forest fires, and sea salt sprays or, importantly, as entrained wind-blown soil particles.[7]

Anthropogenic emissions contribute a high proportion of emissions for most heavy metals. Nriagu and Pacyna[8] quantitatively assessed worldwide contamination of air, water, and soils by trace metals from anthropogenic sources. They summarized global emission sources in the following nine categories: (1) coal combustion (including electric utilities and industry and domestic), (2) oil combustion, (3) pyrometallurgical nonferrous metal production (including mining and production of Pb, Cu, Ni, Zn, and Cd), (4) secondary nonferrous metal production, (5) steel and iron manufacturing, (6) refuse incineration (including the burning of municipal waste and sewage sludge), (7) phosphate fertilizers (important for the addition of Cd to soils), (8) cement production, and (9) wood combustion. As with natural sources, anthropogenic emissions of metals will be in both gas and particulate form.[9] Particles from anthropogenic sources are considered to be microscopic amorphous microcrystalline structures ($<50\ \mu m$) (mainly from refuse fly ash) or finer cenospheres ($<20\ \mu m$) (mainly from coal fly ash). Cenospheres are hollow dust particles that encapsulate microspheres and are formed by uneven heating at combustion.[10,11] Metal vapors may be released to the atmosphere from high temperature combustion processes, but probably condense rapidly onto aerosols on leaving the stack. Reviews of heavy metal emissions to the atmosphere are described in Lantzy and Mackenzie,[12] Nriagu,[13] and Salomons.[14]

Transport Through the Atmosphere

Once emitted to the atmosphere metals can be transported in the gaseous or vapor phase, or in a solid form within (or condensed onto) aerosols of varying size. Particulate phase transport is generally the most significant. Transport of metals in the vapor phase depends much more on the chemical properties and emission sources of the metal concerned. For example, for the metalloids As, Hg and Se, transport in the vapor phase tends to be important, whereas less volatile metals such as Pb, Mn, and Zn are transported almost exclusively or completely in the particulate phase.

Our understanding of the atmospheric transport and residence times of the gaseous fractions of Se, As, and Hg is much poorer, possibly because volatiles are more difficult to trap, measure, and speciate. Although there is strong evidence that metalloids are emitted as gaseous forms, their long-term stabilities are unknown in the atmosphere and there has been much speculation that gases are readily oxidized onto fine particles to be transported in the aerosol phase. Prospero et al.[15] point out that particles below 1 μm diameter are formed in a gas to particle conversion process, while particles above this size are formed mechanically (i.e., wind-blown dust, sea salt droplets produced by breaking bubbles, plant fragments, etc.) and argue that this may explain why certain metals tend to be enriched onto finer particles. Stevens et al.[16] categorized aerosol samples from Elkmont, Tennessee, (U.S.) in respect of fine (0 to 2.5 μm) and coarse (2.5 to 15 μm) particles. Al, Si, Ti, V, Fe, and Cu all showed an enrichment on coarser particles, while Zn, As, Se, and Pb were all concentrated on the finer particles.

Fowler[17] characterized two areas of the atmosphere as being important for dry deposition. The *turbulent boundary layer* is the external air mass or *free atmosphere*. Elements may be imported from here into the *laminar boundary* or *viscous sublayer*, the zone immediately surrounding the terrestrial surface where deposition occurs. A complete atmospheric transport cycle for some elements may occur within the viscous sublayer, where gas phase compounds are released from soil, taken up by the overlying vegetation, or vice versa. Obviously this type of transport is extremely "short range"; these pathways will be very important for the cycling and uptake of Se, As, and Hg in agroecosystems. In contrast, traditional approaches to the transport of heavy metals through the atmosphere have concentrated on long range transport (e.g., References 18 to 20). *Long range* metal transport can be defined as occurring over long distances — perhaps hundreds or thousands of kilometers, primarily through the transport of very fine metal-enriched aerosols. Metals that tend to be transported long range undergo a greater dispersion at altitude. Between these two extremes atmospheric transport of metals may occur over "intermediate" distances of perhaps a few meters or several kilometers by "hopping" — the entrainment and redeposition of wind-blown metal-rich surface soil particulates or the release of vapor phase forms out of the viscous sublayer, followed by subsequent entrapment, perhaps following chemical (oxidation) or physical (sorption to aerosols) conversion in the turbulant boundary layer. The short or intermediate range movement of volatile metalloids has recently been established as a pathway meriting further study;[21,22] it is already known to be important in the cycling of persistent semivolatile organochlorines, such as polychlorinated biphenyls (PCBs), DDT, and related compounds. The size of the boundary layer will affect the transport process by affecting how readily elements can escape the viscous sublayer. It in turn is affected by weather conditions, ground surface topography, and seasonal conditions.

Rain splash of soil particles onto vegetation surfaces is arguably an alternative means of atmospheric transport of heavy metals to foliar surfaces and may be of particular concern in soils high in heavy metal content. This is significant because soil particles on crop surfaces could be incorporated into food products

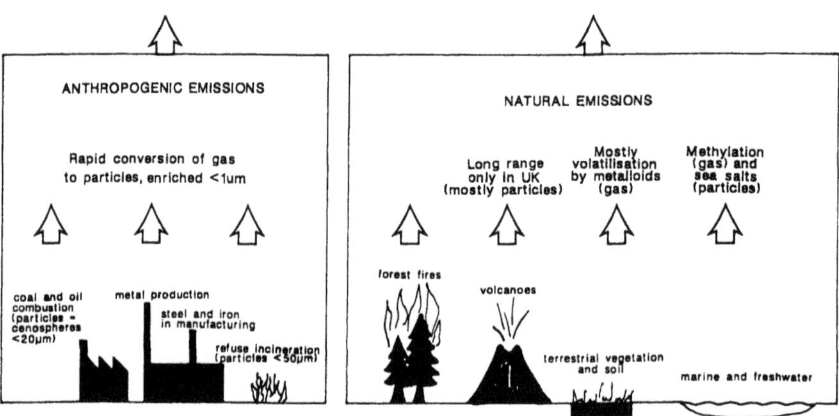

FIGURE 1. Schematic representation of process of heavy metal emission to and transport through the atmosphere.

during processing.[23] Dreicer et al.[24] studied the importance of this mechanism on a number of soil particle sizes and concluded that: (1) most effects were found below 40 cm above the soil surface, (2) particles >105 μm diameter were not detected on plants, and (3) linear relationships were observed for particles <53 μm diameter on vegetation and certain rainfall characteristics. Clearly this mechanism of atmospheric metal transport is extremely short range.

A schematic representation of emission and atmospheric transport of heavy metals is shown in Figure 1.

Processes of Deposition of Heavy Metals onto Agroecosystems

The processes of transfer from air onto the terrestrial surface have been reviewed by Fowler[17,25] and are broadly categorized into wet and dry deposition. Cawse and co-workers at Harwell in the U.K. have made measurements of the long-term annual average concentration of elements in air, dry deposition, and total (dry + wet) deposition at sites in the U.K. (e.g., References 26, 27). This constitutes a most valuable data set on metals in air which has provided much

Table 1. Summary of Data on Heavy Metals in Rural U.K. Air and Deposition

Element	Average Annual Element Concentration[a] ($ng\ m^{-3}$)	Mean Annual Air Particulate EF[b]	Average Annual Total Deposition $mg\ cm^{-2}y^{-1}$	Ratios of Mean Concentrations in Total Deposition for Winter/Summer
Al	122–326	0.31–0.40	24–150	1.3–3.3
Cr	1.6–8.2	2.9–9.3	0.19–0.61	0.6–1.8
Fe	163–449	0.76–0.94	14–57	1.2–1.7
Ni	3.3–60	12–195	0.58–1.1	1.3–1.8
Cu	11–21	81–130	1.7–2.5	0.8–1.5
Zn	39–151	180–240	4.8–10	0.8–1.4
As	1.6–9.8	59–125	0.13–0.40	1.0–2.0
Se	0.8–2.4	760–970	0.028–0.052	0.8–1.9
Cd	<3.3	nd	<0.4	nd
Hg	<0.4	nd	<0.02	nd
Pb	45–159	1250	2.5–3.5	1.3–2.3

Note: Range of values are for 4 UK sites. nd = not determined.

[a] In non-urban air selected as most appropriate for "agricultural" systems. Data for 1972 to 1981.

[b] EF = enrichment factor in non-urban air for 1972 to 1981 data (normalized to Sc).

Adapted from Cawse, P. A., in *Pollutant Transport and Fate in Ecosystems*, Coughtrey, P. J., M. H. Martin, and M. H. Unsworth, Eds. (Oxford: Blackwells, 1987) pp. 89–112.

information on trends (spatial and temporal), as well as the following useful parameters which can be derived from the measurements: (1) percentage elemental composition on a dry weight basis, of the total suspended particulate, (2) the dry deposition velocity (V_g) that relates to particle size, (3) the ratio of dry to total deposition, (4) total and soluble deposition to the ground, (5) washout factors (W), or concentration of element in rain relative to that in air, and (6) enrichment factors (EFs) of elements in air particulate normalized to Sc, i.e., the ratio of an element to Sc in air divided by the ratio of the same elements in average soil. Dry deposition velocities are calculated as

$$\mathrm{Vg\ (cm\ s^{-1})} = \frac{\text{rate of dry deposition } (\mu\text{g cm}^{-1}\text{ s}^{-2})}{\text{concentration in air } (\mu\text{g cm}^{-3})}$$

The following observations can be made from the work of Cawse:

(1) Typically, metal concentrations in air range over several orders of magnitude (see Table 1), for example, 82 to 816 ng m^{-3} air, e.g., Al, Fe, Zn, Pb; 0.8 to 82 ng m^{-3}, e.g., Cr, Ni, Cu, As, Se; <0.8 ng m^{-3}, e.g., Cd, Hg.

(2) EFs also vary over many orders of magnitude, from <1 for some elements (e.g., Al, Fe), up to ~1000 (e.g., Se, Pb) (see Table 1). The mechanism of enrichment has been related to the high volatility of chalcophilic elements such as Zn, Se, Cd, As, and Pb compared to the lithophilic group that includes Al and Fe, with the result that volatile metals condense on the smaller size fraction of air particulate that possesses relatively high surface area. High enrichments

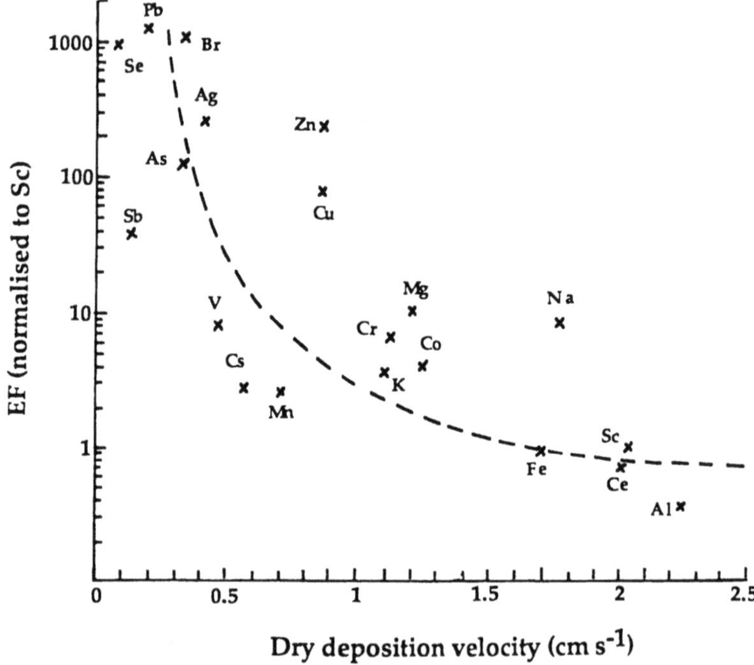

FIGURE 2. Average annual enrichment factors and dry deposition velocities of elements in air particulate at one rural site in the U.K. between 1972 and 1981. Data obtained and adapted from Cawse, P. A., in *Pollutant Transport and Fate in Ecosystems* (Oxford: Blackwells, 1987) pp. 89–112. Note that measurements relate only to *particulate* dry deposition, and do not include the gas phase component for the metalloids.

reflect the influence of industrial activities. The incineration of urban refuse can result in exceptionally high enrichments of metals in particulate emissions.[28]

(3) From the continous measurements of dry deposition, an inverse relationship has been demonstrated between V_g of the elements, which is related to particle size, and the EFs (see Figure 2). Relatively low deposition velocity below 0.5 cm s^{-1} (and hence small particle size) is typical of combustion-derived aerosols of high EF, e.g., As, Se, and Pb, produced by industry or motor vehicles. Higher V_g is typical of elements with low enrichment that originate from soil disturbance, mining, or quarrying, such as Al and Fe.

(4) Many plant nutrients and potentially toxic pollutants of high solubility had enrichments in the range 10 to 700 (normalized to Sc), which is important with respect to mobility and cycling in the terrestrial environment. The solubility of total deposition is generally >80% for Na, Ca, and Se, and in the range 50 to 80% for Cr, As, Ni, Cu, Cd, and Pb. Low solubility is found for Al and Fe.

(5) The majority of elements showed increases in total deposition during the winter period (see Table 1).

(6) The W factor relates the concentration of an element in rain to that in air. It is known that small particles in the range 0.01 to 0.05 μm diameter are scavenged efficiently, which is attributed to their increased hygroscopic component. It is preferable to use soluble deposition data to derive W values, and Cawse[26] reported the following general groupings: low W (<500) Al, Cr, Fe, Se, Pb; intermediate W (500 to 1000) As; high W (1000 to 3000) Ni, Cu, Zn. Low W is associated with a local near to ground source of emission, e.g., Pb from car exhausts or Al from soil dust. Elements with relatively high W values are considered to be present mainly at the rain-forming altitude (~3 km) and are more easily scavenged by rainfall, particularly if associated with small particle size.

An alternative process of deposition is termed *occult* deposition where materials are deposited from clouds and saturated air residing flush with the ground surface, often associated with orographic cloud movements.[29,30] This has been studied extensively on Dunn Fell, an upland and hill grassland environment in the Cumbrian Pennines of Northern England. Deposition of macro elements S and N was found to be highly enhanced by this process, particularly in forests where vegetation acts as a filtering agent, trapping chemicals from the cloud. Although never quantified for metals per se, occult deposition must be considered a significant deposition pathway for metals to upland U.K. agricultural surfaces. Processes of deposition of heavy metals are summarized in Figure 3.

FATE OF ATMOSPHERICALLY DERIVED HEAVY METALS IN AGROECOSYSTEMS

Deposition of heavy metals may result in *uptake* into vegetation — an important route of entry into the food chain. The fate of atmospheric fallout will be either direct plant uptake through the leaf surfaces, uptake via the root system following entry into the soil, or, in the long-term, complete removal from the rooting zone.

The propensity for direct or indirect uptake of wet deposited metals will be governed by metal chemistry, either in the atmosphere or at the plant surface. Presumably, metals with high solubilities will be more easily transported into the plant through the leaf than those of low solubility, but will also tend to be leached away from the leaf surface or the rooting zone by subsequent precipitation[5,31] (see Table 2). pH and redox at plant surfaces will also affect metal mobility and hence propensity for uptake.[32]

Foliar uptake of gaseous species of metalloids will depend on the surface conditions: Fowler[17] showed that V_g varies with height of the canopy, according to whether the leaf stomata are open or closed, and whether the leaf surface is wet or dry.

Interception and retention of aerosols on vegetation has been studied by Chamberlain[31] using radioisotopes as tracers. The proportion of the deposited

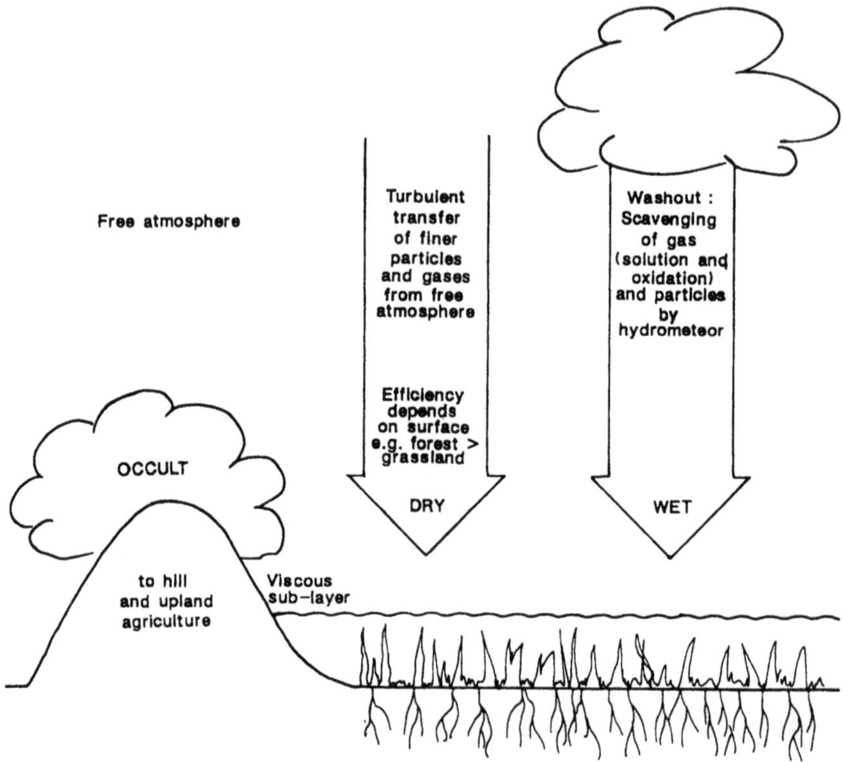

FIGURE 3. Schematic representation of heavy metal deposition to agricultural surfaces.

Table 2. Removal of Dry Deposited Contaminants from Foliar Surfaces by Rainfall

Rainfall (mm)	Contaminant Retained (%)
0.0	100
0.0	84
0.0	71
30.2	32
61.5	13

After Chamberlain, A. C. *Atmos. Environ.* 4:57–78 (1970).

aerosol remaining on the herbage (p) was assumed to be related to an uptake coefficient (μ) (m^2kg^{-1}) by the relation:

$$1-p = \exp - \mu w$$

where w = the herbage density in kg m^{-2} dry matter.

Chamberlain[31] listed five mechanisms that could account for removal from the surface of plants: (1) translocation to the roots, (2) volatilization, (3) leaching

Table 3. Assumptions and Calculated Atmospheric Inputs of Se to a Typical U.K. Pasture Grassland

	Assumptions
Soil mass	200 kg (for 1 $m^2 \times 20$ cm @ 1 g cm^{-3} bulk density)
Soil Se conc.	0.5 mg kg^{-1} (dw) = 100 mg total
Herbage mass	260 g (for 1 m^{-2}, dw aerial biomass)
Herbage Se conc.	0.04 ppm (dw) = 10.4 μg total
Total deposition to herbage	ca. 0.5% of total deposition
Soil volatilisation year $^{-1}$.	0.1% soil Se
Herbage volatilization year $^{-1}$	1% plant Se
Total deposition	0.7 mg m^{-2} year $^{-1}$
Vapor uptake to herbage year $^{-1}$	3% plant Se
Calculated Budget (high input, low loss scenario)	
Herbage	
Inputs	3.5 μg m^{-2}
Losses	0.1 μg m^{-2}
Net gain	3.4 μg m^{-2}
	13 μg kg^{-1}
An increase of	30% year $^{-1}$
Soil	
Inputs	700 μg m^{-2}
Losses	100 μg m^{-2}
Net gain	600 μg m^{-2}
	0.5 μg kg^{-1}
An increase of	0.6% year $^{-1}$

Adapted from Haygarth, P. M., K. C. Jones, and A. F. Harrison. *Sci. Total Environ.* 103:89–111 (1991).

by rain (i.e., removal in solution) (Table 3), (4) removal in particulate form by wind (i.e., saltation), rain, or other disturbances, and (5) dying back or weathering of leaves to their surface layers. Chamberlain extended his ideas to develop the concept of Normalized Specific Activity (NSA) for radioactive elements deposited to vegetation. It is also useful for stable elements and relates the concentration of a contaminant per unit mass to its rate of deposition to the ground:

$$\text{NSA } (\text{m}^2 \text{ days kg}^{-1}) = \frac{\text{Activity (kg}^{-1}\text{ dry weight of crop)}}{\text{Activity deposited (per day m}^{-2}\text{ of ground)}}$$

This is useful for assessments during conditions of continuous deposition from the atmosphere and was developed by Simmonds and Linsley[33] to model interception and retention of deposits from the atmosphere by grain and leafy vegetables.

Pinder et al.[34] developed a filtration model considering the interrelationships between plant biomass, plant surface area, and the interception of particulate deposition by grasses. Predictions compared favorably with data compiled from the literature for a variety of grass species; an exponential relationship between interception and grass biomass was observed (see Figure 4). The accuracy of such models for predicting particulate interception and retention in agricultural systems is reviewed in Pinder et al.[35] Pinder and McLeod[23] used ^{238}Pu tracer in soil to study the fate of rainsplash input mechanisms on surfaces of sunflower, corn, wheat, and soybean and found that leaves were generally more contaminated with ^{238}Pu from soil particles on their surfaces than that derived through the roots.

The fate and processes of uptake of heavy metals are shown in Figure 5.

METHODS FOR ASSESSING ATMOSPHERIC INPUTS

The scope of scientific research methods available for assessing the significance of atmospheric inputs in affecting heavy metal burden of agricultural systems are considered in the following five categories. This section does not aim to review the behavior of specific metals, but emphasizes the *methodology* available to investigate the role and processes of atmospheric metal inputs.

Evidence from the Literature

Mass Balances of Trace Metals

Detailed analysis of existing data can often reveal novel information about the relative importance of exchange pathways. Pacyna[7] suggested three scales for studying anthropogenic metal budgets: global, regional, and local. On a global scale, Nriagu and Pacyna[8] conducted a quantitative assessment of worldwide contamination of air, water, and soils by metals. Nriagu[1] conducted a global inventory of natural sources of atmospheric trace metals. On a regional scale, Ross[36] developed an atmospheric Se budget for the region 30 to 90°N and Pacyna[7] budgeted European anthropogenic emissions of heavy metals to the atmosphere. Haygarth et al.[37,38] calculated a mass balance specifically for atmospheric inputs of Se to "model" grasslands in the U.K. (see later discussion and Table 3). Predictions from this work are currently being tested with field and laboratory work and the initial results imply that theoretical budgeting gives a good indication of the Se balance.[39]

Empirical Relationships Between Soil and Crop Concentrations

By plotting a range of soil (x) and plant concentrations (y), simple linear regressions can be obtained. The value of their intercept, if interpreted with caution, implies a value for atmospheric inputs (=a) (see Figure 6). When the soil concentration = 0, extrapolation of the relationship back to the intercept *does not* pass through plant concentration = 0. In other words, if soil Cd concentration = 0, the plant nevertheless appears to be receiving a residual supply of the metal. If one accepts the limitations and assumptions, i.e., that soil supply remains

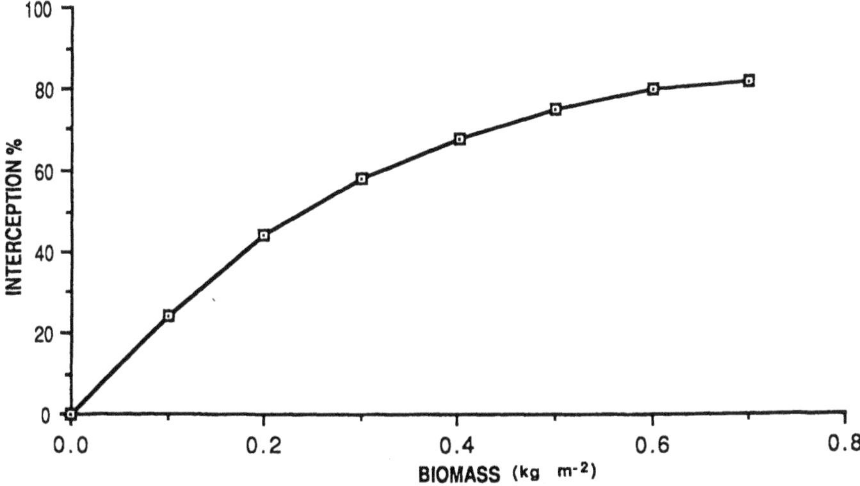

FIGURE 4. Idealized relationship between deposition interception and grass biomass. (After Pinder, J. E., T. G. Ciravolo, and J. W. Bowling. *Health Phys.* 55(1):51–58 (1988).)

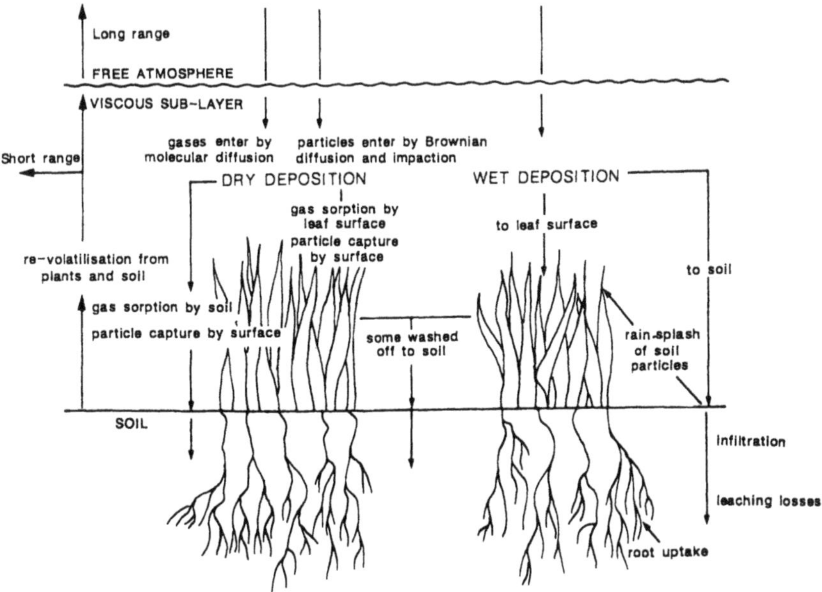

FIGURE 5. Schematic representation of fate and uptake of atmospherically deposited heavy metals by agricultural surfaces.

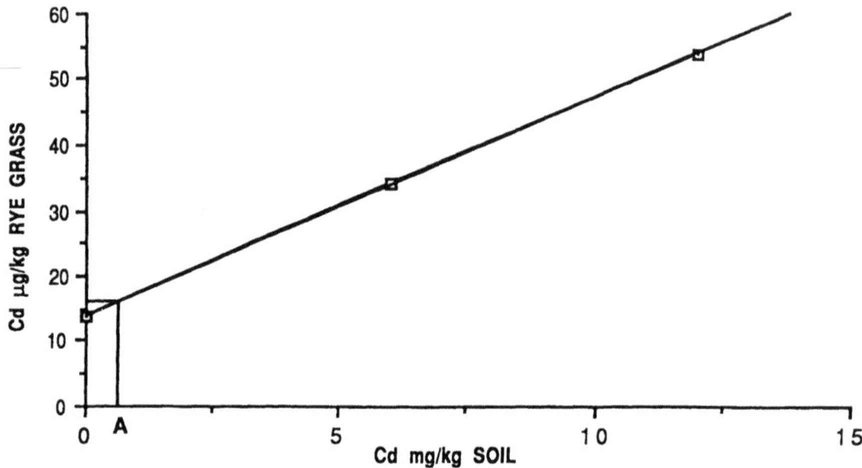

FIGURE 6. Use of regression analysis to derive atmospheric contribution of heavy metals to plants. (Cd data for rye grass, idealized from Davis, *Experimentia* 40(2):117–126 (1984).)

constant and that there are no alternative inputs to the crop, then it can be inferred that this residual supply is from the atmosphere (see References 41 and 42). This frequently appears as a sizable contribution. For example, in Figure 6 the data were obtained from a study where the same soil type had been treated with different amounts of Cd-rich sewage sludge to give exceptionally high Cd concentrations in the soil. The typical concentration of Cd in UK agricultural soils is ~0.5 mg kg^{-1}, shown on Figure 6 as A. By inference, this relates to a typical concentration in ryegrass of ~15 µg kg^{-1} of which, according to Figure 6, a substantial proportion is atmospherically derived.

Evidence for Long-Term Change due to Atmospheric Deposition

Peat Cores

In saturated sites where decay of organic matter is inhibited because of anaerobic conditions, poorly decayed organic plant matter accumulates at a rate that can be quantified.[43] Incremental depth cores can be sampled, dated, and analyzed for metals. The method assumes that: (1) inputs are only from above (i.e., the peat is ombrotrophic), (2) atmospherically derived metals do not leach down through the profile, and (3) no lateral or upward movement of groundwater has occurred. These variables are difficult to control and can lead to difficulties in interpreting data, but accurate long term accumulation rates have been successfully derived with careful interpretation. Usually ^{137}Cs and ^{210}Pb dating techniques are used to derive peat accumulation rates, with ^{14}C measurements used in long-term studies. Peats occur in many sites in the U.K. and studies have shown important temporal trends in deposition[44–47] which can be cross-checked with information from dated sediment cores sampled from rural lake catchments.[48] Extensive use of peat profiles has also been made by Steinnes (e.g., Reference 49) in Scandinavia, and this work is given detailed consideration below.

Soil Profiles

Detailed incremental analysis of soil samples vertically through a profile can also provide insight into the net long-term atmospheric accumulation of metals in agricultural and forest soils. Robberecht et al.[50] successfully applied the method for Se in soils around a nonferrous plant, showing a pronounced enrichment close to the surface (see References 19, 51). The method is subject to the same limitations as peats, but has the advantage of being more representative of an agricultural soil. Unfortunately, however, the method is invalid for many U.K. agricultural sites that have been disturbed by ploughing. Controlled pasture grassland plots at Rothamsted that have been undisturbed for nearly 300 years provide a unique insight into the deposition and movement of heavy metals and trace organic contaminants through structured field soils where the influence of pH and organic matter on contaminant mobility can be investigated.[52]

Archived Samples

Historical collections and carefully stored soils, plants, and air filter samples can be analyzed retrospectively to provide a simple means of assessing long-term time trends in soil and crop composition. This has been extensively carried out for soils and plants on different crop treatments from the untreated plots of the Classical Experiments at Rothamsted Experimental Station in southeast England. Trends of Pb, Cd, and Se have been reported for soil and herbage samples analyzed at regular time increments between the 1850s and present day (see References 41, 42, 53–56). Salmon et al.[57,58] retrospectively analyzed air filter samples collected between 1957 and 1974 for a range of heavy metals at Chilton, Oxfordshire.

Spatial Patterns in Heavy Metal Distribution

National Studies of Metal Distribution Related to Sources

The work of Eiliv Steinnes in Norway provides an excellent focus for this discussion, as he has demonstrated that long-term atmospheric deposition inputs have substantially altered the biogeochemical fluxes of several elements, notably Pb, Cd, As, and Se, in Scandinavian soil-plant systems. The studies dealt with soils that are characterized by low pH and high organic matter contents (>70% of the soils are pH 3.6 to 4.5, with organic matter contents in the range 50 to 90%). These are primarily the podzolic soils and peats that predominate in Scandinavia (e.g., References 49, 59, 60). One of the first observations made by Steinnes and co-workers was of a marked regional difference in the heavy metal distribution of Norwegian soils. The highest heavy metal concentrations were found in largely uninhabited areas with very high annual precipitation. In general, Pb, As, and Cd concentrations in soils are roughly an order of magnitude greater in southern than in northern Norway. The soils also show substantial differences with depth, such that there is a strong surface enrichment of these elements.

Steinnes has argued that these observations reflect regional differences in the atmospheric inputs of heavy metals over centuries. Southern Norway receives greater inputs of heavy metals from the industrial areas of western and eastern Europe than does the north of the country. Centuries ago, inefficient smelting processes would have resulted in substantial atmospheric contamination with metals and sulphurous gases, and the long-range transport of relatively volatile heavy metals in association with fine aerosols.[61] In the U.K., several studies have demonstrated the persistence of metals in surface soils in mineralized areas over many centuries (e.g., Reference 62). It has been estimated that more than 80% of the heavy metal burden of Norwegian air originates from outside the country.[63] Steinnes has also noted marked regional differences in the concentrations of Pb and Cd in vegetation; levels are roughly five times higher in the south than in the north of the country. This may reflect the soil differences and/or the seasonal deposition directly onto the above-ground portion of plants. Studies on Norwegian ombrotrophic peat bogs provide further evidence that long-term inputs are now represented as a cumulative burden in the surface zone. Without dating the peat cores, Steinnes compared the concentrations of metals at different depths in the profile. In the south, for example, Pb and As concentrations are roughly 20 times higher at the surface than at 50-cm depth. Even in the north, there is a slight surface enrichment, indicating that remote locations have not remained unperturbed by these inputs. The picture is different for other elements; there is no difference, for example, between the concentrations of Cu in vegetation from northern and southern Norway. For Se the pattern of deposition is rather more complicated because of the combination of natural inputs of Se emitted from the oceans and the anthropogenic component. Nonetheless, there is still an increase by a factor of 10 from 50-cm depth to the surface in ombrotrophic peats from southeast Norway, although the temporal trends may be complicated by cycling of Se within the peat profile.

The next important question to be addressed concerned the potential transfer of heavy metals from soils and vegetation to the human foodchain. Frøslie et al.[64] compared the metal concentration in lamb liver for grazing animals from different regions of Norway selected to have different cumulative atmospheric deposition loads. They found differences between elements. Pb showed a correlation that was significant at the $P < 0.001$ level; Cd and Se were significant at $P < 0.01$, and As at $P < 0.05$. There was no significant correlation for Mo, Zn, and Cu, probably because internal biological regulation by plants is known to be important for these trace elements. The implications of these regional differences still have to be evaluated with respect to human exposure.

Davies[62] conducted a survey of heavy metals in nearly 700 surface and subsurface soil samples from Wales, an area with a long-term history of metal mining and exploitation. He observed substantial surface enrichment for Cu and Zn. The topsoil (A horizon): subsurface (B horizon) ratios were 2.4:1, 2.1:1,

1.1:1, and 1.1:1 for Cd, Pb, Cu, and Zn, respectively. The median surface concentrations for Cd and Pb were 0.29 and 35 mg kg^{-1}, respectively, and 0.12 and 17 mg kg^{-1} for the subsurface.

The Pb content of agricultural soils is a function of cumulative Pb inputs via atmospheric deposition, local geochemistry, and soil type. Contemporary typical soil Pb concentrations in England and Wales are about 40 mg kg^{-1}. Chamberlain[61] has estimated that approximately 3 mg kg^{-1} may be attributable to inputs from the use of Pb as a fuel additive since 1946, and that as much as 17 mg kg^{-1} may be due to historic emissions of Pb to the U.K. atmosphere since 1700. On this basis, roughly half of the contemporary burden of Pb in U.K. surface soils is of long-term anthropogenic origin. It has been estimated that U.K. agricultural land nationally received about 1540 t Pb $year^{-1}$ in the mid-1980s from atmospheric deposition, which is equivalent to an average input of about 300 g ha^{-1}.[65] Sewage sludge inputs are the only other significant contribution to U.K. agricultural land, accounting for approximately 100 t Pb per annum. Long-term inputs of industrially derived Cd are more difficult to estimate, but contemporary inputs are strongly affected by atmospheric deposition and the presence of Cd as an impurity in phosphate fertilizers. There is good evidence that both these inputs have resulted in the long-term accumulation of Cd in surface soils.[54,55] It seems likely that around 50% of the contemporary Cd burden of UK agricultural soils originated from long-term anthropogenic inputs. It seems reasonable to assume that all soils in industrialized nations are contaminated with certain heavy metals above their pre-industrial levels.

Moss Surveys

Some species of *Sphagnum* moss and lichens (e.g., *Cladonia* spp.) are efficient trappers of atmospheric particulates and thus can be used to provide an index of air concentrations. Andersson and Steinnes[66] used *Hylocomium splendens* to study geographical trends in Hg deposition in Norway. Glooschenko[67] and Burton[68] have reviewed the applicability of these methods. Generally, the ability of mosses to trap metals is a function of stem density (biomass) growth rate, structure of leaves, cation exchange processes, and strength of metal sorption. Harrison and Chirgawi[69] used this method to survey heavy metal deposition around the U.K.

Natural Tracers — The Use of Isotope Ratios

Although this methodology involves technically complex mass spectrometry detection, it has been usefully employed by Dorr and Munnich[70] to assess the migration of atmospherically derived Pb and Cs through European forest soils. Elseewi et al.[71] employed Pb isotope dilution principles to infer atmospheric transport and deposition of petrol derived Pb in the U.S. The methodology is also discussed in Oldfield et al.[43] The applicability of such methods are, however, limited by difficulties in analytical resolution.

Such empirical studies are useful for implying geographical trends in heavy metal deposition, acting as a stimulus for further research. For a more detailed understanding and quantification of the processes involving agricultural surfaces per se, experimental work has to be conducted.

Evidence from Experimental Studies

Experiments, if designed correctly, can yield relevant and detailed quantitative information on the magnitude and behavior of atmospherically derived metals in agricultural systems. The disadvantage is that they often involve artificial situations that do not accurately represent field conditions.

Field Work

The most appropriate and successfully employed method to study the magnitude of heavy metal deposition/inputs onto agricultural crops is the principle of isotope dilution. This involves labeling of soil with an enriched isotope (often radioactive) of metal and growing crop plants in the soil, leaving them exposed to natural processes of atmopheric deposition. Regular harvesting of herbage and analyses for isotope enriched *and* total heavy metal are conducted. The principle is that atmospheric deposition will dilute the quantity of isotope and thus the ratio of enriched metal isotope to total metal (normalized to the soil ratio at day one of the experiment) will be influenced by the contribution atmospheric deposition has made to the crop. The method is currently being employed by Haygarth et al.[39] for Se and has already been sucessfully employed by Tjell et al.[72] for Pb; Hovmand et al.,[73] Harrison and Chirgawi,[69] and Dollard and Davies[74] for Cd, and Mosbaek et al.[75] for Hg. Consideration must be given to the fact that the soil must be extensively disturbed during isotope labeling and that it may take a considerable time for the added tracer to equilibrate with the native soil metal.

Another useful field-based approach to distinguish between the relative importance of air- and soil-derived metals to crop plants involves "soil transplants". Harrison and Chirgawi[76] describe how one homogenized soil was prepared and transferred to a number of sites in the U.K. with different air metal concentrations. They derived concentration factors (CF), defined as:

$$CF = \frac{\text{soil derived metal in plant}}{\text{soil metal concentration}}$$

and air accumulation factors (AAF) where

$$AAF = \frac{\text{air derived metal in plant}}{\text{air metal concentration}}$$

This principle is also currently being employed by Haygarth et al.[77] at seven sites around the U.K., primarily to study the atmospheric contribution to herbage Se contents.

Laboratory Studies

Harrison and Chirgawi[78] complemented their field-based studies by assessing the role of air and soil as contributors of heavy metals to plants by growing vegetables in air cabinets supplied with filtered and ambient air.

Isotopes tracers can also be employed in heavy metal-air exchange experiments. A dual isotope experiment for sealed growth chambers was sucessfully applied by Zeive and Peterson[79] studying foliar uptake of Se. One tracer (^{76}Se) was supplied to the plant from the soil, while another (^{75}Se) was released into the sealed microcosm in gaseous forms generated by microbial cultures. Laboratory experiments using radioisotopes have also been used to measure soil — air volatilisation rates for metalloids (e.g., References 6, 80–82).

SELENIUM DEPOSITION TO GRASSLANDS AS A CASE STUDY IN THE U.K.

It is particularly significant to consider atmospheric inputs of Se to *grassland* agroecosystems because most of Se problems in the U.K. are focused around Se deficiences in the diets of pasture grazing livestock. It appears that the rate of supply of Se from the pasture is generally too low to prevent young lambs, pregnant ewes, and cattle from suffering from Se deficiency syndrome.[83–85] The quantities of Se in U.K. soil are small (generally in the order of <0.5 mg kg^{-1} dw,[38,86] and the transfer rate from soil to pasture plants is very slow (e.g., References 87, 88). Since upward migration of Se from bedrock to soil to plant is sluggish and given that Se is easily volatilized from soil to atmosphere, it seems appropriate to consider the role of the atmosphere in supplying Se to grasslands in the U.K.

Sources of Atmospheric Se for the U.K.

The majority of Se in the U.K. atmosphere comes from natural sources. Natural emissions are predominately generated by volatilization of Se from soils, plants, fresh and marine waters, and sediments brought about by methylation from bacteria and fungi (e.g., *Penicillium* and *Alternaria*) of dimethylselenide ($(CH_3)_2Se$) and related chemical species. Formation depends on a range of environmental factors (e.g., References 2–6). Marine inputs will be of particular significance in the U.K. because of the large coastal area. Volatilization as a means of soil to crop transfer will also be important in rural locations. An inventory of natural Se sources to the atmosphere is presented in Nriagu.[1]

Anthropogenic emissions of Se constitute about 20 to 40% of the U.K. atmospheric burden and are generated primarily from coal combustion and primary metal production. It is commonly speculated that they will be emitted in the vapor phase as SeO_2 (Se IV).[89, 90] Germani and Zoller[9] measured Se emissions from a coal fired plant and found that 59% of total Se emissions were in the gas phase. Anthropogenic sources of Se to the atmosphere are reviewed in Pacyna[7] and Nriagu and Pacyna.[8]

Transport of Atmospheric Se

Although most Se emissions occur in the gas phase, gaseous Se is rapidly scavenged by fine particulate nuclei, generally <1 μm in diameter.[90–92] For natural emissions, which will be predominately dimethylselenide, the process is complete in a few hours.[93] With anthropogenic emissions, the sublimation of SeO_2 and SeO_3 from combustion furnaces will probably be very rapid.[94] Industrial emissions often involve combustion processes that emit large amounts of fine particles, thereby providing a "sink" for gas phase Se. Wadge et al.[11] analyzed Se on particles from 11 coal-fired power plants and a refuse incinerator in the U.K. and found Se to be enriched on cenospheres of diameter <20 μm and amorphous particles <50 μm, respectively.

Recent evidence from research using ^{75}Se tracer in field lysimeters[21,22] suggests that in an unmixed atmosphere simulating the viscous sublayer, a significant proportion of terrestrially generated volatile Se (from grassland plants and soil) undergoes short-range transport and is readily redeposited close to source. This is complimented by the conclusions of Peirson et al.[95] who noted that Se must have a relatively localized atmospheric source because the washout coefficient was low ($W < 500$), suggesting little dispersion at altitude. It is hypothesized that the residual Se volatilized from natural sources which does *not* readily undergo transformation to particle will be brought back to the surface within a short distance of the emission point, either by washout or dry deposition, thereby prodominantly transported *short range* within the viscous sub-layer. This could represent an important transport pathway from soil to plant foliar surfaces, although more research is required to confirm this.

Long-range transport will occur predominantly with air Se particulate derived from anthropogenic combustion sources, emitted into the free atmosphere. As a means of long-range transport from natural sources, Låg and Steinnes[18,19] showed that a considerable proportion of marine-derived Se could be entrained into the free atmosphere (both biological methylation and suspension of sea salt particles), and provide a source of Se that is deposited to terrestrial surfaces a considerable distance inland. Long-range atmospheric inputs will pose a significant import mechanism of Se to grasslands.

The Importance of Atmospheric Deposition as a Source of Se to U.K. Grasslands

Table 3 shows that on the basis of making some realistic assumptions, atmospheric inputs of Se can potentially contribute up to 30% of grass Se uptake and 0.6% of total soil selenium per year. Given that the water-soluble fraction of Se in U.K. soils is known to be as small as 1% of total soil Se,[39] then the relative contribution made by atmospheric inputs must be considerable. These calcula-

Table 4. Changes In Soil Se Concentration Due to Inputs from Long-Range Atmospheric Transport in the Past Century, Rothamsted Experimental Station, Rural Southeast England

Crop Type	Archived Soil Se[a]	Present Soil Se[a]	% Increase
Broadbalk wheat	0.204 (1856)	0.237 (1986)	16
Hoosfield barley	0.243 (1882)	0.285 (1982)	17
Park Grass grassland	0.343 (1876)	0.388 (1984)	13
Barnfield root crops	0.288 (1929)	0.297 (1980)	3

[a] Soil Se analyzed using a nitric-perchloric-hydrochloric acid wet digestion and hydride generation atomic absorption spectrophotometry. Expressed as mg kg^{-1} dw.

Haygarth, P. M., A. I. Cooke, K. C. Jones, A. F. Harrison, and A. E. Johnston. Unpublished data.

tions were published in Haygarth et al.[38] and were made on the basis of a "typical" grassland in the U.K. The theoretical calculations were supported by subsequent analysis of an archived soil collection from the classical experiments at Rothamsted Experimental Station in rural southeast England. Table 4 shows that in three of the four crop treatments, atmospheric deposition increased the soil burden by about 15% in 100 years. This input is interpreted as being due to increased contribution from *long-range* transported Se, particularly reflecting a general increase in industrial Se emissions during the last century.

There is little doubt that atmospheric deposition contributes a significant proportion of the "natural" Se inputs to U.K. agricultural systems. The ability of soil to supply plant-available Se weathered from rocks is known to be sluggish (e.g., Reference 88), especially during reducing conditions associated with anaerobiosis (e.g., Reference 96) or fixation, particularly by organic matter.[87] Thus it may be inferred that with particularly infertile soil types, deposition of Se to crop surfaces or to the crop root zone in soil may be very important and has been shown to be important for grazing animals in Norway.[64]

The significance of Se deposition is emphasised when it is seen in context with soil management practices. Table 5 compares annual atmospheric Se inputs with fertilizer additions and shows that the atmospheric contribution is greatest. In relation to sewage sludge inputs, atmospheric inputs pose a significant input to agricultural systems, particularly if deposition has occurred for long time periods or if sites are of close proximity to emission source.

CONCLUSIONS

(1) Atmospheric deposition has led to a cumulative long-term increase in the heavy metal content of soils in industrialized countries. Nationally and regionally atmospheric deposition may be the most important source of metals to agricultural soils.

(2) Atmospheric inputs are a major source of crop contamination in many areas. This may result from surficial deposits onto the foliage, or uptake and

Table 5. The Importance of Atmospheric vs. Fertilizer Inputs of Se to Agroecosystems in the U.K.

	Calculated Se Input ($\mu g\ m^{-2}$)
300 kg ha^{-1} ammonium nitrate[a]	0.3
800 kg ha^{-1} triple superphosphate[a]	1.06
Annual gain from the atmosphere[b]	600

[a] From calculations by Senesi.[97]
[b] From calculations by Haygarth et al.[38]

translocation of aerially derived metals entering the agroecosystem in particulate form. In the case of metalloids, these pathways may be further enhanced by the direct foliar uptake of gaseous forms.

(3) Further research is required into the deposition of different forms of metals and metalloids to agroecosystems. Studies should attempt to distinguish between wet and dry deposition and should make measurements of the gas phase component for metalloids. Most work to date has considered deposition onto collectors, but workers should work towards addressing the variations in deposition velocities to different (real) surfaces.

(4) Little is known about the fate of atmospherically derived metals and metalloids in soils. Studies should address the cycling and incorporation of this source of metals into the native soil pool and particularly consider the short- and longer-term bioavailability of atmospherically derived metals.

(5) There is growing evidence to suggest that microbially mediated processes may be impeded in soils at heavy metal concentrations somewhat above background. This may be of regional importance where atmospheric deposition has elevated the heavy metal burden of surface soils. These effects may be further enhanced by long-term soil acidification following acid deposition, since metal availability is enhanced under acid conditions.

(6) A number of studies have shown the importance of atmospherically derived metals as a source of crop contamination regionally to agroecosystems, even in rural locations. Future work should consider the metal transfer of atmospherically derived metals to livestock and its contribution to regional differences in human exposure to metals.

ACKNOWLEDGMENTS

Thanks to Jane Rushton and Chris Benefield for graphics work, Andy Cooke for analytical work, and Drs. A. F. Harrison and D. Fowler for comment and advice. P. M. Haygarth is a research student funded by the U.K. Natural Environment Research Council.

REFERENCES

1. Nriagu, J. O. "A Global Assessment of Natural Sources of Atmospheric Trace Metals," *Nature (London)* 338:47–49 (1989).
2. Fleming, R. W. and M. Alexander. "Dimethylselenide and Dimethyltelluride Formation by a Strain of *Penicillium*," *Appl. Microbiol.* 24:424–429 (1972).
3. Chau, Y. K., P. T. S. Wong, B. A. Silverberg, P. L. Luxon, and G. A. Bengert. "Methylation of Se in the Aquatic Environment," *Science* 192:1130–1131 (1976).
4. Thompson-Eagle, E. T., W. T. Frankenberger, and U. Karlson. "Volatilization of Selenium by *Alternaria alternata*," *Appl. Environ. Microbiol.* 55:1406–1413 (1989).
5. Ylaranta, T. "Volatilization and Leaching of Selenium Added to Soils," *Ann. Agric. Fenn.* 21:103–114 (1982).
6. Zieve, R. and P. J. Peterson. "Volatilization of Selenium from Plants and Soils," *Sci. Total Environ.* 32:197–202 (1984).
7. Pacyna, J. M. "Emission Factors of Atmospheric Elements," in *Toxic Metals in the Atmosphere, Volume 17, Advances in Environmental Science and Technology*, Nriagu, J. O. and C. I. Davidson, Eds. (New York: John Wiley & Sons, 1986) pp. 1–32.
8. Nriagu, J. O. and Pacyna, J. M. "Quantitative Assessment of Worldwide Contamination of Air, Water, and Soils by Trace Metals," *Nature (London)* 333:134–139 (1988).
9. Germani, M. S. and W. H. Zoller. "Vapor-Phase Concentrations of Arsenic, Selenium, Bromide, Iodine and Mercury in the Stack of a Coal Fired Power Plant," *Environ. Sci. Technol.* 22:1079–1085 (1988).
10. Fisher, G. L., D. P. Y. Chang, and M. Brummer. "Fly Ash Collected from Electrostatic Precipitators: Microcrystalline Structures and the Mystery of the Spheres," *Science* 192:553–555 (1976).
11. Wadge, A., M. Hutton, and P. J. Peterson. "The Concentrations and Particle Size Relationships of Selected Trace Elements in Fly Ashes from UK Coal Fired Power Plants and a Refuse Incinerator," *Sci. Total Environ.* 54:13–27 (1986).
12. Lantzy, R. J. and F. T. Mackenzie. "Atmospheric Trace Metals: Global Cycles and Assessment of Mans Impact," *Geochem. Cosmochim. Acta* 43:511–525 (1979).
13. Nriagu, J. O. "Global Inventory of Natural and Anthropogenic Emissions of Trace Metals to the Atmosphere," *Nature (London)* 279:409–411 (1979).
14. Salomons, W. "Impact of Atmospheric Inputs to the Hydrospheric Trace Metal Cycle," in *Toxic Metals in The Atmosphere, Volume 17, Advances in Environmental Science and Technology*, Nriagu, J. O. and C. I. Davidson, Eds. (New York: John Wiley & Sons, 1986) pp. 409–467.
15. Prospero, J. M., R. J. Charlson, V. Mohnen, R. Jaenicke, A. C. Delany, J. Moyers, W. Zoller, and K. Rahn. "The Atmospheric Aerosol System: An Overview," *Rev. Geophys. Space Phys.* 21(7):1607–1629 (1983).
16. Stevens, R. K., T. G. Dzubay, R. W. Shaw, W. A. McClenny, C. W. Lewis, and W. E. Wilson. "Characterization of the Aerosol in the Great Smokey Mountains," *Environ. Sci. Technol.* 14:1491–1498 (1980).

17. Fowler, D. "Removal of Sulphur and Nitrogen Compounds from the Atmosphere in Rain and by Dry Deposition," in *Ecological Impact of Acid Precipitation*, Drabløs, D., and A. Tollan, Eds. Proc. International Conferance, Sandefjord, Norway, March 11–14th (1980).
18. Låg, J. and E. Steinnes. "Soil Selenium in Relation to Precipitation," *Ambio* 3:237–238 (1974).
19. Låg, J. and E. Steinnes. "Regional Distribution of Selenium and Arsenic in Humus Layers of Norwegian Forest Soils," *Geoderma* 20:3–14 (1978).
20. Amundssen, C. E., J. E. Hanssen, J. P. Rambaek, A. Semb, and E. Steinnes. "Long Range Transport of Trace Elements to Southern Norway," in *Heavy Metals in the Environment, Vol. l*, Vernet, J. P., Ed. Conf. at Geneva, September1989 (Edinburgh: CEP Consultants Ltd., 1989) pp. 32–35.
21. Haygarth, P. M., K. C. Jones, and A. F. Harrison. "Observation on Short Range Transport of Vapour Phase Se Between Experimental Lysimeters," in *Heavy Metals in the Environment*, Vol. 1. Farmer, J. G., Ed. Conf. at Edinburgh, September 1991 (Edinburgh: CEP Consultants Ltd., 1991), pp. 94–97.
22. Haygarth, P. M., K. C. Jones, and A. F. Harrison. "Short Range Atmospheric Transport of Selenium as a Means of Transfer from Soil to Herbage," in preparation.
23. Pinder, J. E and K. W. McLeod. "Contaminant Transport in Agroecosystems Through Retention of Soil Particles on Plant Surfaces," *J. Environ. Qual.* 17(4):602–607 (1988).
24. Dreicer, M., T. E. Hakouson, G. C. White, and F. W. Whicker. "Rainsplash as a Mechanism for Soil Contamination of Plant Surfaces," *Health Physics* 46(1):177–187 (1984).
25. Fowler, D. "Transfer to Terrestrial Surfaces," *Phil. Trans. R. Soc. London Ser. B.* 305:281–297 (1984).
26. Cawse, P. A. "Trace and Major Elements in the Atmosphere at Rural Locations in Great Britain, 1972–1981" in *Pollutant Transport and Fate in Ecosystems*, Coughtrey, P. J., M. H. Martin, and M. H. Unsworth, Eds. (Oxford: Blackwells, 1987) pp. 89–112.
27. Cawse, P. A. "Deposition of Trace Elements from the Atmosphere in the UK," in *Inorganic Pollution and Agriculture, Ref. Book, 326*, Proc. of a Conference held by A.D.A.S., 1977 (1980).
28. Mitchell, D. J., S. R. Wild, and K. C. Jones. "Arrested Municipal Solid Waste Incinerator Fly Ash as a Source of Heavy Metals to the UK Environment. *Environ. Pollut.* (submitted).
29. Fowler, D., J. N. Cape, I. D. Leith, T. W. Choularton, M. J. Gay, and A. Jones. "The Influence of Altitude on Rainfall Composition at Great Dun Fell," *Atmos. Environ.* 22(7):1355–1362 (1988).
30. Fowler, D., J. N Cape, and M. H. Unsworth. "Deposition of Atmospheric Pollutants on Forests," *Philos. Trans. R. Soc. London Ser. B* 324:247–285 (1989).
31. Chamberlain, A. C. "Interception and Retention of Radioactive Aerosols by Vegetation," *Atmos. Environ.* 4:57–78 (1970).
32. Alloway, B. J. *Heavy Metals in Soils* (London: Blackie, 1990).
33. Simmonds, J. R. and G. S. Linsley. "Parameters for Modelling the Interception and Retention of Deposits from the Atmosphere by Grain and Leafy Vegetables," *Health Phys.* 43(5):679–691 (1982).

34. Pinder, J. E., T. G. Ciravolo, and J. W. Bowling. "The Interrelationship Between Plant Biomass, Plant Surface Area and the Interception of Particulate Deposition by Grasses," *Health Phys.* 55(1):51–58 (1988).
35. Pinder, J. E., K. W. McLeod, and D. C. Adriano. "The Accuracy of Some Simple Models for Predicting Particulate Interception and Retention in Agricultural Systems," *Health Phys.* 56(4):441–450 (1989).
36. Ross, H. B. "An Atmospheric Selenium Budget for the Region 30°N to 90°N," *Tellus* 37B:78–90 (1985).
37. Haygarth, P. M., K. C. Jones, and A. F. Harrison. "A Selenium Budget for Upland Grasslands in the British Isles, with Emphasis on Inputs from and Losses to the Atmosphere," in *Heavy Metals in the Environment, Vol. l*, Vernet, J. P., Ed. Conf. at Geneva, September 1989 (Edinburgh: CEP Consultants Ltd., 1989) pp. 168–171.
38. Haygarth, P. M., K. C. Jones, and A. F. Harrison. "Se Cycling Through Agricultural Grasslands in the UK: Budgeting the Role of the Atmosphere," *Sci. Total Environ.* 103:89–111 (1991).
39. Haygarth, P. M., K. C. Jones, and A. F. Harrison. "The Use of Field Based Experimental Lysimeters and 75-Se to Determine the Se Mass Balance for a Grassland in Rural NorthWest England," in *Heavy Metals in the Environment*, Vol. 2. Farmer, J. G., Ed. Conf. at Edinburgh, September 1991 (Edinburgh: CEP Consultants, Ltd., 1991) pp. 111–114.
40. Davis, R. D. "Cadmium — A Complex Environmental Problem II: Cadmium in Sludges Used as Fertiliser," *Experimentia* 40(2):117–126 (1984).
41. Jones, K. C., C. Symon, P. J. L. Taylor, J. Walsh, and A. E. Johnston. "Evidence for a Decline in Rural Herbage Lead Levels in the U.K.," *Atmos. Environ.* 25A:361–369 (1991).
42. Jones, K. C. and A. E. Johnston. "Significance of Atmospheric Inputs of Lead to Grassland at One Site in the United Kingdom Since 1860," *Environ. Sci. Technol.* 25 (1991).
43. Oldfield, F., P. G. Appleby, P. Crooks, S. Hutchinson, and N. Richardson. "Radioisotope Dating of Heavy Metal Deposition Histories in Sediments and Peats: Problems and Approaches to Their Resolution," in *Heavy Metals in the Environment, Vol. l*, Vernet, J. P., Ed. Conf. at Geneva, September 1989 (Edinburgh: CEP Consultants Ltd., 1989) pp. 457–460.
44. Clymo, R. S., F. Oldfield, P. G. Appleby, G. W. Pearson, P. Ratnesar, and N. Richardson. "The Record of Atmospheric Deposition on a Rainwater-Dependent Peatland" *Philos. Trans. R. Soc.* London (1990).
45. Livett, E., J. A. Lee, and J. H. Tallis. "Lead, Zinc and Copper Analyses of British Blanket Peats." *J. Ecol.* 67:865–891 (1979).
46. Lee, J. A. and J. H. Tallis. "Regional and Historical Aspects of Lead Pollution in Britain," *Nature (London)* 245:216–218 (1973).
47. Sanders, G., J. Hamilton-Taylor and K. C. Jones. Unpublished data.
48. Sanders, G., Jones, K. C., J. Hamilton-Taylor, and H. Dörr. "Concentrations and Deposition Fluxes of Polynuclear Aromatic Hydrocarbons and Trace Metals to the Dated Sediments of a Rural English Lake," *Environ. Toxicol. Chem.* (submitted).

49. Steinnes, E., H. Hovind, and A. Henriksen. "Heavy Metals in Norwegian Surface Waters with Emphasis on Acidification and Atmospheric Deposition," in *Heavy Metals in the Environment, Vol. l*, Vernet, J. P., Ed. Conf. at Geneva, September 1989 (Edinburgh: CEP Consultants Ltd., 1989) pp. 36–39.
50. Robberecht, H., H. Deelstra, D. Vanden Berghe, and R. Van Grieken. "Metal Pollution and Selenium Distributions in Soils and Grass Around a Non-Ferrous Metals Plant," *Sci. Total Environ.* 29:229–241 (1986).
51. Jones, K. C., S. A. Watts, A. F. Harrison, and J. Dighton. "The Distribution of Metals in the Forest Floor of Aged Conifer Stands at a Plantation in Northern England," *Environ. Poll.* 51:31–47 (1988).
52. Jones, K. C. Unpublished data.
53. Jones, K. C., C. J. Symon, and A. E. Johnston. "Retrospective Analysis of an Archived Soil Collection. I. Metals," *Sci. Total Environ.* 61:131–144 (1987).
54. Jones, K. C., C. J. Symon, and A. E. Johnston. "Retrospective Analysis of an Archived Soil Collection. II. Cadmium," *Sci. Total Environ.* 67:75–89 (1987).
55. Jones, K. C. and A. E. Johnston. "Cadmium in Cereal Grain and Herbage from Long-Term Experimental Plots at Rothamsted," *Environ. Pollut.* 57:199–216.
56. Haygarth, P. M., A. I. Cooke, K. C. Jones, A. F. Harrison, and A. E. Johnston. "Temporal Trends in Selenium Deposition to Agricultural Soils at Rothamsted, S.E. England," *Atmos. Environ.* (in preparation).
57. Salmon, L., D. H. F. Atkins, E. M. R. Fisher, C. Healy, and D. V. Law. "Retrospective Trend Analysis of the Content of U. K. Air Particulate Material 1957–1974," *Sci. Total Environ.* 9:161–200 (1978).
58. Salmon, L., D. H. F. Atkins, E. M. R. Fisher, and D. V. Law. "Retrospective Analysis of Air Samples in the UK, 1957–1974," *J. Radioanal. Chem.* 37:867–880 (1977).
59. Steinnes, E. "Heavy Metal Pollution of Natural Surface Soils due to Long-Distance Atmospheric Transport," in *Pollutants in Porous Media: The Unsaturated Zone between Soil Surface and Groundwater*, Yaron, B., G. Dagan, and J. Goldshmid, Eds. (Berlin: Springer-Verlag, 1984) pp. 115–22.
60. Steinnes, E. "Impact of Long-Range Atmospheric Transport of Heavy Metals to the Terrestrial Environment in Norway," in Lead, Mercury, Cadmium and Arsenic in the Environment, Hutchinson, T. C. and K. M. Meema, Eds. (New York: John Wiley & Sons, 1987) p. 107–117.
61. Chamberlain, A. C. "Fallout of Lead and Uptake of Crops," *Atmos. Environ.* 17:693–706 (1983).
62. Davies, B. E. "Baseline Survey of Metals in Welsh Soils," in Proc. 1st Int. Symp. on Geochemistry and Health (Northwood, U.K.: Science Reviews Ltd., 1985) pp. 51–59.
63. Pacyna, J. M., A. Semb, and J. E. Hanssen. "Emission and Long-Range Transport of Trace Elements in Europe," *Tellus* 36B:163–178 (1984).
64. Frøslie, A., G. Norheim, J. P. Rambeak, and E. Steinnes. "Heavy Metals in Lamb Liver: Contribution from Atmospheric Fallout," *Bull. Environ. Contam. Toxicol.* 34:175–182 (1985).
65. Hutton, M. and C. Symon. "The Quantities of Cadmium, Lead, Mercury and Arsenic Entering the UK Environment from Human Activities," *Sci. Total Environ.* 57:129–150 (1986).

66. Andersson, E. M. and E. Steinnes. "Atmospheric Deposition of Mercury in Different Parts of Norway," in *Heavy Metals in the Environment, Vol. l*, Vernet, J. P., Ed. Conf. at Geneva, September 1989 (Edinburgh: CEP Consultants Ltd., 1989) pp. 464–467.
67. Glooschenko, W. A. "Monitoring the Atmospheric Deposition of Metals by Use of Bog Vegetation and Peat Profiles," Nriagu, J. O. and C. I. Davidson, Eds. *Toxic Metals in the Atmosphere* (New York: John Wiley & Sons, Advances in Environmental Science and Technology Series, Vol. 17, 1986) pp. 507–533.
68. Burton, M. A. S. *Biological Monitoring of Environmental Contaminants (Plants): A Technical Assessment* (Monitoring and Assessment Research Centre, University of London, 1986).
69. Harrison R. M. and M. B. Chirgawi. "The Assessment of Air and Soil as Contributors of Some Trace Metals to Vegetable Plants III: Experiments with Field Grown Plants," *Sci. Total Environ.* 83:47–62 (1989).
70. Dorr, H. and Munnich, K. O. "Lead and Cesium Transport Through European Forest Foils," preprint for *Water Air Soil Pollut.* (1990).
71. Elseewi, A. A., R. W. Hurst, and T. E. Davis. "Strontium and Lead Isotopes as Monitors of Fossil Fuel Dispersion," in *Int. Conf. on Metals in Soils Waters, Plants and Animals, 10th Symp. Abst.* No. 308, Orlando, FL (University of Georgia Savannah River Ecology Laboratory, April 30–May 3, 1990).
72. Tjell, J. C., M. F. Hovmand, and H. Mosbaek. "Atmospheric Lead Pollution of Grass Grown in a Background Area in Denmark," *Nature (London)* 280:425–426 (1978).
73. Hovmand, M. F., J. C. Tjell, and H. Mosbaek. "Plant Uptake of Airborne Cadmium," *Environ. Pollut. Ser. A* 30:27–38 (1983).
74. Dollard, G. S. and T. J. Davies. "A Study of the Contribution of Airborne Cadmium to the Cadmium Burdens of Several Vegetable Species," Harwell, U.K. AERE. Report 13256 (June 1989).
75. Mosbaek, H., J. C. Tjell, and T. Sevel. "Plant Uptake of Airborne Mercury in Background Areas," *Chemosphere* 17:1227–1236 (1988).
76. Harrison R. M. and M. B. Chirgawi. "The Assessment of Air and Soil as Contributors of Some Trace Metals to Vegetable Plants II: Translocation of Atmospheric and Laboratory Generated Cadmium Aerosols To and Within Vegetable Plots," *Sci. Total Environ.* 83:35–45 (1989).
77. Haygarth, P. M., Jones, K. C., and Harrison, A. F. Unpublished data.
78. Harrison R. M. and M. B. Chirgawi. "The Assessment of Air and Soil as Contributors of Some Trace Metals to Vegetable Plants I: Use of a Filtered Growth Cabinet," *Sci. Total Environ.* 83:13–34 (1989).
79. Zieve, R. and P. J. Peterson. "Dimethylselenide — An Important Component in the Biogeochemical Cycling of Selenium," in *Trace Substances in Environmental Health—XVIII*, Hemphill, D. D., Ed. (Columbia: University of Missouri, 1984) pp. 262–267.
80. Lewis, B. G., C. M. Johnson, and T. C. Broyer. "Volatile Selenium in Higher Plants; the Production of Dimethylselenide in Cabbage Leaves by Enzymatic Cleavage of Se-Methyl selenomethionine," *Plant Soil* 40:107–119 (1974).
81. Karlson, U. and W. T. Frankenberger. "Accelerated Rates of Selenium Volatilization from California Soils," *Soil Sci. Soc. Am. J.* 53:749–753 (1989).

82. Karlson, U. and W. T. Frankenberger. "Volatilization of Selenium from Agricultural Pond Sediments," *Sci. Total Environ.* 92:41–54 (1990).
83. Cousins, F. B. and I. M. Cairney. "Some Aspects of Selenium Metabolism in Sheep," *Aust. J. Agric. Res.* 12:927–943 (1961).
84. Ullrey, D. E. "The Selenium Deficiency Problem in Animal Agriculture," in *Trace Element Metabolism in Animals–2,* Hoekstra, W. G., J. W. Suttie, H. E. Ganther, and W. Mertz, Eds. (Baltimore: University Park Press, 1975) pp. 275–293.
85. M. A. F. F. "The Selenium Status of Sheep in Britain, map prepared by survey section," A.D.A.S. U.K. (1978).
86. Robberecht, H., D. Vauden Berghe, H. Deel Stra, and R. Van Grieken. "Selenium in the Belgian Soils and its Uptake by Rye Grass," *Sci. Total Environ.* 25:61–69 (1982).
87. Christensen, B. T., F. Bertelsen, and G. Gissel-Nielsen. "Selenite Fixation by Soil Particle-Size Separates," *J. Soil Sci.* 40:641–647 (1989).
88. Neal, R. H. "Selenium," in *Heavy Metals in Soils*, Alloway, B. J., Ed. (Glasgow: Blackie, 1990) pp. 237–260.
89. Pupp, C., R. C. Lao, J. J. Murray, and R. Pottie. "Equilibration Vapour Concentrations of Some Polycyclic Aromatic Hydrocarbons, As_4O_6 and SeO_2 and the Collection Efficiencies of these Air Pollutants," *Atmos. Environ.* 8:915–925 (1974).
90. Cutter, G. A. and T. M. Church. "Selenium in Western Atlantic Precipitation," *Nature (London)* 322:720–722 (1986).
91. Pillay, K. K. S., C. C. Thomas, and J. A. Dondel. "Activation Analysis of Airborne Selenium as a Possible Indicator of Airborne Sulphur Pollutants," *Environ. Sci. Technol.* 5:74–77 (1971).
92. Duce, R. A., B. J. Ray, G. L. Hoffman, and P. R. Walsh. "Trace Metal Concentration as a Function of Particle Size in Marine Aerosols from Bermuda," *Geophys. Res. Lett.* 3:339–342 (1976).
93. Atkinson, R., S. M. Aschmann, D. Hasegawa, E. T. Thompson-Eagle, and W. T. Frankenberger. "Kinetics of the Atmospherically Important Reactions of Dimethylselenide," *Environ. Sci. Technol.* 24:1326–1332 (1990).
94. Ficalora, P. J., J. C. Thompson, and J. L. Margrave. "Mass Spectrometric Studies at High Temperatures — XXVI; The Sublimation of SeO_2 and Se O_3," *J. Inorg. Nucl. Chem.* 31:3771–3774 (1969).
95. Peirson, D. H., P. A. Cawse, L. Salmon and R. S. Cambray. "Trace Elements in the Atmospheric Environment," *Nature (London)* 241:252–256 (1973).
96. Masscheleyn, P. H., R. D. Delaune, and W. H. Patrick. "Transformations of Selenium as Affected by Sediment Oxidation-Reduction Potential and pH," *Environ. Sci. Technol.* 24:91–96 (1990).
97. Senesi, N., M. Polemio, and L. Lorusso. "Content and Distribution of Arsenic, Bismuth, Lithium and Selenium in Mineral and Synthetic Fertilisers and Their Contribution to Soil," *Comm. Soil Sci. Plant Anal.* 10:1109–1126 (1979).

10

Uptake and Accumulation of Metals in Bacteria and Fungi

Nicholas W. Lepp

School of Biological Sciences, Liverpool Polytechnic, Byrom Street, Liverpool L3 3AF, U.K.

ABSTRACT

This review addresses the relationship between trace metals, bacteria, and fungi. Metal uptake by both groups of organisms is fully discussed, with emphasis on reported instances of metal accumulation by pure or mixed species cultures. In the case of bacteria, mechanisms of uptake are considered that include differences observed between gram +ve and gram –ve forms. The relationships of growth phase, metabolic activity, and the production of extracellular chelating agents to the metal uptake and accumulation are discussed, together with the eventual fate of metals in bacterial cells. Differences between fungal and bacterial uptake receive attention, as does the relationship of metal uptake to differences in growth form in selected microfungi. Again, uptake mechanisms, where differing from bacteria, and eventual fate of metals in fungal cells is considered.

Uptake and accumulation of metals in macrofungi is carefully considered; the relationship between metal accumulation in sporophores, substrate, and fungal species is reviewed together with a consideration of the relationship between metal accumulation in sporophores and the growth of these structures. The ecological implications of some notable metal accumulations are discussed.

Table 1. Examples of Elevated Levels of Metal Accumulation in Bacteria and Microfungi

Element	Bacteria	Fungi
Cadmium	*Zooglea ramigera* (40%)[1]	*Rhizopus arrhizus* (3.0%)[2]
Cobalt	*Zooglea* sp. (25%)[1]	—
Copper	*Zooglea ramigera* (40%)[1,3]	*Rhizopus arrhizus* (1.6%)[2]
Chromium	—	*Rhizopus arrhizus* (3.1%)[2]
Lead	*Micrococcus luteus* (49%)[4]	*Rhizopus arrhizus* (10.4%)[2]
Mercury	—	*Rhizopus arrhizus* (5.8%)[2]
Nickel	*Zooglea* sp. (13%)[1]	—
Silver	*Thiobacillus ferrooxidans* (35%)[5]	*Rhizopus arrhizus* (5.4%)[2]
Zinc	—	*Rhizopus arrhizus* (2.0%)[2]

Note: Values in brackets refer to uptake as a percent dry weight. Orginial references should be consulted for experimental details.

The review also addresses the potential of bacteria and fungi for future innovative use in metal recovery from, or decontamination of, a range of effluent materials. Successful uses are documented, together with an evaluation of the problems that still require solutions before such technologies may have wider ranging applications.

INTRODUCTION

Microorganisms exhibit great powers of trace metal accumulation and immbolization. They may be frequently encountered in extreme metal contaminated environments, both natural and man-made, as well as demonstrating the ability to concentrate certain elements from substrates that contain only trace quantities of the element in question (Table 1 gives some examples of elemental bioconcentration). As attention is focused on a range of topics related to metal detoxification, metal removal from effluents, and metal purification, there is growing interest in the potential exploitation of microorganisms in these processes. What started off as scientific curiosity may eventually prove to have widespread industrial and environmental use in future decades. This review highlights our knowledge of metal uptake and accumulation in bacteria, microfungi, and macrofungi, with emphasis on mechanisms of uptake and mechanisms of tolerance to potentially toxic trace metals.

BACTERIA

Gram-positive bacteria possess cell walls with powerful chelating properties.[6–8] Binding of metals to these structures generally obeys the Irving-Williams series for affinity of elements to organic binding sites.[9] Binding sites vary between different species,[6,7] but each species does not appear to possess a range of such sites.[8] Metal-binding to the cell walls may be a biphasic process. Initial interaction between metal ions and reactive groups is followed by inorganic deposition of additional metal.[10] Metals can accumulate in greater than stoichiometric amounts, which cannot be solely accounted for by ion-exchange processes.[4,11] Metal-loaded bacterial cells have been shown to act as nuclei for crystalline metal deposits when mixed into synthetic metalliferous sediments.[12]

The cellular envelopes of Gram-negative bacteria have a very different structure from the cell walls of Gram-positive organisms.[13] These consist of two chemically distinct membrane bilayers (outer and inner) separated by a thin peptidoglycan layer. Studies with purified cell envelopes of *E. coli* K-12 show most metals to be deposited at the polar head group regions of the membranes or along the peptidoglycan layer. Again, a set hierarchy for metal affinity is evident.[13] Changes in cell wall composition that may be attributed to prior organism nutrition have been found to affect metal deposition in *Pseudomonas* and *Bacillus*.[14]

The metabolically independent uptake of metals may account for the most significant proportion of total uptake. Surowitz et al.[15] found up to 90% of total Cd uptake by *Bacillus subtilis* was located in the cell wall, 3 to 4% on the cell membrane, and the remainder in the soluble fraction of the cell.

METABOLICALLY MEDIATED METAL UPTAKE

This process is slower than passive adsorption, demanding the presence of suitable energy sources and ambient conditions. The physiological state of the bacterial cells is also important, as is the nature and composition of the culture medium.[16,17] Several of the elements classed as trace metals are essential to life, and cells may have more or less specific mechanisms for the uptake of elements such as Cu, Co, Fe, Ni, and Zn. Mechanisms for such processes are not known with certainty, but may involve specific carrier systems associated with active ionic fluxes across the cell membranes. If the process is not saturated, there is the potential for significant metal accumulation via this route, greater than for a purely passive association. This may not be true, however, for those organisms producing copious extracellular deposits of polysaccharides with high biosorptive capacities (e.g., *Pseudomonas* sp.).[18]

Specific transport systems for Mn[19,20] and Ni[21–23] have been identified in several bacteria. At low (<40 μmol/l) external Mn levels, the high-affinity systems provides most cell Mn in *E. coli*.[19] At higher levels, Mn enters the cell via the Mg transport system. Cd has been shown to inhibit high affinity Mn transport, and may enter the cell via this system.[20] High affinity Ni transport systems are known to exist in *Alcaligenes entrophus*,[21] *Methanobacterium bryantii*,[22] and *Bradyrhyzobium japonicum*.[23]

Mg-dependent Ni transport occurs in *Enterobacter aerogenes* and *Bacillus megaterium*.[24] In some cases, reduced transport or active efflux of metals has been demonstrated, frequently associated with tolerance phenomena.[25,26]

Longer term studies on metal uptake relate to the ability of growing and

Table 2. Cd Accumulation in Bacteria (μg Cd mg^{-1} protein)

Organism	Cd Uptake	Gram Reaction
Staphylococcus aureus	0.52	+
Streptococcus faecium	0.65	+
Bacillus subtilis	0.79	+
Escherichia coli	2.79	–
Pseudomonas aeruginosa	24.15	–

Data from Morozzi, G., G. Cenci, F. Scardazza, and M. Pitzurra, *Microbios* 48:27–35 (1981).

multiplying cultures of bacteria to accumulate metals. This process can again be modified by a range of factors including the nature of the culture medium, and metal-independent changes in the morphology and physiology of the growing cells. Basic uptake mechanisms identified in short-term studies may be masked by a range of factors including internal detoxification, extracellular precipitation, or the selection for tolerant individuals that may achieve this state by a variety of different mechanisms.

There are several instances where metal tolerant bacterial communities have been isolated from a variety of sources. Charley and Bull[27] isolated a stable community that could tolerate up to 100 mmol l^{-1} Ag and accumulate up to 30% cell dry weight as Ag. Stable communities with similar powers of Cu accumulation have been isolated in metal-contaminated activated sludge.[28] Some *Pseudomonas* species produce a range of reducing compounds, responsible for the deposition of metallic Ag within colonies or on the walls of culture vessels.[29] Gram-negative bacteria are less sensitive to Cd than Gram-positive organisms[30] (Table2).

Differences in growth phase can be important in metal uptake in batch cultured bacteria. Germanium uptake by *Pseudomonas putida* occurs in a biphasic pattern in a catechol-enriched medium; the second uptake phase corresponded to catechol degradation, products of which facilitated Ge transport into cells.[31] Other metals show different growth-dependent uptake patterns; for example, Cu uptake in *E. coli* increased at the end of the exponential growth phase, reached a maximum in the deceleration phase, and declined in the stationary phase of colony growth.[32] In *Citrobacter* sp., Cd uptake occurs as a sharp peak in the mid-exponential phase of colony growth,[33] but in *P. putida*, uptake of this element is maximal in the lag phase.[34]

Within the cell, bacteria may convert metal ions into innocuous forms by precipitation or binding. This is the case for U in *P. aeruginosa*[19] and Hg in *P. oleovorans*.[35] The magnetosomes of magnetotactic bacteria are magnetite crystals (Fe_3O_4) that can account for 3 to 4% of cell dry weight.[36] Bacteria can also synthesize metal-binding proteins that could act as detoxicants. Both *P. putida* and *E. coli* produce low molecular weight (mol wt ≃ 36,000 to 39,000) proteins that bind Cd;[37,38] in the latter case, production is thought to be related to recovery from Cd exposure.

Bacteria are also capable of producing large quantities of extracellular polymers[39] that form either capsules or loose aggregates around individual cells. In many cases, these are of a polysaccharide nature, with anionic properties, and are capable of significant metal cation binding.[3,11] These polymers are strongly implicated in the removal of soluble metal ions by activated sludge biomass; the removal of extracellular polymers from cultures of *Klebsiella aerogenes* considerably reduces its ability to sorb metals.[40] Polysaccharides produced by the common sewage treatment organisms, *Zooglea ramigera* have significant metal-binding properties; 0.3 g Cu and 1.0 g Cd can be sorbed per gram dry weight of polymer.[1,3] Binding is pH dependent; maximum Cu absorption occurs at pH 5.5 and Cd sorption has a pH maximum of 6.5.[3]

Microbial crystallization of metals is a well-known phenomenon. Microbes are implicated in the formation of ferromanganese nodules found on ocean floors, and several bacteria and fungi can promote Mn oxidation, becoming encrusted with manganic oxides.[41] The bacterium *Gallionella* can grow using energy derived from Fe^{2+} oxidation; these organisms form long twisting stalks on which iron oxides are deposited.[42] In *Pediomicrobium*, Mn or Fe oxides are associated with extracellular acidic polysaccharides, accumulating as particles or thread.[42] *Thiobacillus ferrooxidans* adsorbs Au in the cell wall and plasma membrane. Au^{3+} is reduced to Au^0 and visible growth of gold particles has been observed.[43] Sulfate-reducing bacteria, such as *Desulphovibrio*, form sulfide deposits that can immobilize large quantities of metals.[41] *K. aerogenes* may contain up to 2.4% Cd (dry weight), proportional to the sulfide content of the cells.[44] Phosphates may also precipitate metals; Cd,[45] Pb,[4] Hg,[46] and Ag[5] have been reported as microbially associated sulfides.

Some bacteria produce iron-sequestering organic molecules (siderophores)[47] as part of their overall Fe uptake strategy.[48] These can be either phenols/catechols or hydroximates.[47] Ni-requiring bacteria produce Fe-siderophores, but these seem not to be involved in Ni uptake.[49] Gallium has also been shown to exchange for Fe in Fe-siderophores.[50] *P. aeruginosa* produces chelating agents when exposed to U and Th, which possess some molecular similarity to Fe-siderophores.[51]

MICROFUNGI

Fungi can accumulate metals, even from low external concentrations. However, uptake characteristics differ between living and dead cells and, in the case of the former, whether the cells are in active growth.[52,53] As with bacteria, fungi can excrete extracellular substances that can remove metals from solution. Both citric and oxalic acids can fill this role,[54] and the production of H_2S by some yeasts[55] also acts to immobilize many metals. Fungi can release siderophores (qv) that bind Fe and, in certain circumstances, other metals such as Ga and Ni.[52,56]

Biosorption of metal ions to fungal biomass is well known. The process is rapid and is temperature-independent.[57,58] A number of binding groups have been implicated (carboxyl, amine, hydroxyl, phosphate, sulfhdryl);[52,59] this is reflected in the widely different biosorption capacities shown by a range of fungi.[60] Living or dead biomass can act as an efficient biosorptive agent; the magnitude of the phenomenon is directly related to biomass density.[58] Biosorptive capacity can be enhanced by a range of treatments, including powdering[2] (exposes more binding sites) and detergent treatment[61,62] (affects cell permeability).

Many fungal products and byproducts can act as efficient biosorption agents. Such materials include glucans, mannans, melanins, chitins, and chitosans;[63–66] chitin derivatives are very efficient at removing UO_2 ions from solution.[63] Glucans with amino acid or sugar groups have powerful chelating properties,[64] and many fungal phenolic polymers and phenol-containing melanins are equally efficient in this role. Extracellular melanin from pigmented *Aureobasidium pullulans* had a greater binding capacity than intact biomass, and pigmented strains of this organism bound more Cu than nonpigmented counterparts. To date, there has been little attention paid to this aspect of metal/fungal interactions.

Physiologically important metals show metabolically dependent uptake in many fungal species.[17,18] However, due to the high adsorptive capacity of fungal cell walls and low external metal levels, special techniques are frequently required to demonstrate this phenomenon.[67,68] The transport processes are sensitive to conditions that reduce or inhibit the energy metabolism of cells: low substrate concentration, anaerobic conditions, low temperatures, and metabolic inhibitors.[17,57,69] Inhibitors of ATPase actively reduce metal uptake,[17,70] as do membrane depolarizing agents. There are also reports of stimulation of metal uptake following membrane hyperpolarization.[65,71,72] Potassium efflux gradients may or may not be related to metal uptake.[57] Where a relationship exists, a stoichiometry of approximately 1 M^{+}_{in} to 2 K^{+}_{out} is found.[57,73,74] The absence of such a relationship may be metal-induced membrane damage[57,75] or where K efflux is coupled to other energy-utilizing processes.[76,77]

Kinetics of metal uptake into fungal cells frequently resemble the Michaelis-Menten model,[17] but departures from this are quite common.[57,78] This can be due to the presence of separate transport systems with differing affinities for the same element (e.g., biphasic Co uptake[57]). One factor that may have a profound effect on these processes is toxicity. For example, increasing external metal concentration can reduce H^+ efflux,[79] with the attendant consequences for coupled transport systems. An increase in membrane permeability may be induced, leading to uncontrolled K loss from cells. Many low-affinity transport phases may result from this latter phenomenon,[73,79,80] and care should be taken in interpretation of metal uptake isotherms that arise from the application of potentially toxic external metal concentrations.

pH can influence metal uptake.[81] Increasing pH facilitates surface binding of metals,[70] and energy-dependent metal uptake is frequently pH-dependent.[17] Maximal rates are observed between pH 6.0 and 7.0.[82,83]

Mechanisms to account for these phenomena are complex, reflecting a combination of effects on the chemistry of the uptake solution, metal speciation, metal solubility, and changes in the properties of sites of initial ion binding at the cell surface.[17]

Microfungi frequently tolerate elevated metal levels by reducing rates of uptake.[84–86] This can be achieved by exclusion of metals or active regulation of internal levels,[87] or by the production of structures, such as chlamydospores, which show little or now metal uptake.[66,88] These types of mechanism are not always employed. Some fungi tolerate excess internal metal levels,[89,90] possibly as a result of sequestration by organic molecules. There is much evidence that metallothionien synthesis occurs in fungal cells as a response to increased copper uptake.[90,91] Other metals induce different molecules. Cd induces the formation of phytochelatins in *Schizosaccharomyces pombe*.[91] Other Cd-binding proteins are found in *S. cerevisiae*.[92] It is clear that fungi possess a variety of internal mechanisms that can inactivate potentially toxic metals.

Fungal cells also possess the ability to regulate the uptake of essential metals. *Candida utilis* grown on low Zn medium showed greater Zn accumulation than low carbon cells of the same fungus.[93] Mn uptake by *C. utilis* was reduced when grown in a high Mn medium as compared to a normal Mn supply, but Cu uptake was not regulated.[72]

Demand for particular metals is related to the physiological state of the fungal cells; for example, metalloenzyme induction at different stages in the cell cycle. Batch cultures of *C. utilis* showed greater Zn uptake in the lag phase or late exponential phase of growth.[94] Only those elements implicated in enzyme formation or synthesis of other molecules that contain metal ions will be implicated in these processes.

MACROFUNGI

These organisms are predominantly Asco- or Basidiomycetes and produce a distinct fruiting structure, frequently ephemeral, which is often the only indication of the presence of these organisms in an ecosystem.

The mycelium from which the fruiting structures arise has the same basic properties as that found in microfungi, but while there may be distinct differences in morphology between the fungal groups, the function of mycelium is constant across the different fungal orders.

During the last two decades, there has been the progressive realization that a wide range of higher fungi are able to concentrate trace metals in the tissues of their fruiting bodies when growing on nonpolluted substrates. It is now well known, for example, that species of *Agaricus* can bioaccumulate Cd and Hg from soils/composts containing background levels of these elements[95–97] and that *Amanita muscaria* can bioconcentrate Br, Cd, Se, and V from background soils.[95,98–101] Such outstanding examples of natural metal accumulation pose several questions relating the biochemical and physiological significance for the fungus, the potential impact on various trophic levels of the natural ecosystems in which these fungi occur, and the mechanisms by which these metals are biologically concentrated into such ephemeral structures on an annual basis.

BIOLOGY OF MYCELIAL GROWTH AND FRUIT BODY FORMATION

A knowledge of basic aspects of macrofungal physiology is essential for further attempts to understand fungus/metal interactions. This information can be used to identify potential key stages in metal acquisition from the substrate and processes leading to the massive accumulations that have been detected in fruiting bodies of some species.

Macrofungi exist in soils and litter as vegetative hyphae, thought to radiate from an initial inoculum (so-called "fairy-rings" observed in grassland). These can exist as individual hyphae, but some species, notably *Armillarea*, produce aggregations of hyphae known as rhizomorphs,[102] more reminiscent in superficial appearance to a plant root. Nutrient uptake occurs over the whole hyphal surface,[103] but the relative contributions of actively growing hyphae and older mycelium to the overall nutrient intake of fungi is not known.

The hyphae of basidiomycete fungi are septate, with cytoplasmic connections through characteristic pore regions. Cytoplasmic streaming occurs towards the tips of actively growing hyphae, with older hyphae tending to vacuolate. Basidiomycete fruiting bodies arise in the older hyphal regions, initially from an aggregation of hyphae. The fruiting region of the mycelium becomes blocked off from the actively growing margins of the colony, and materials for sporophore growth are mobilized from this demarcated region. Sporophores grow rapidly, given suitable conditions. Knowledge of factors controlling their initiation and growth comes from the study of commercial monocultures of *Agaricus* and *Pleurotus*.[104] For these species, light, temperature, humidity, and CO_2 concentration are important external regulants of sporophore growth. In the field, most temperate fungi fruit in the autumn; this is thought to be due to a favorable combination of air/soil temperatures and humidity, but there is no satisfactory biological explanation for this.[104] Fruiting bodies vary considerably in sized and dry matter content; large *Amanita* and *Boletus* sporophores can weigh up to 500 g fresh weight, whereas the Giant Puff-ball, *Lagermannia*, can reach the weight of 3 kg and has occasionally been mistaken for a sleeping sheep! Much of this biomass is water; typically, *Amanita muscaria* sporophores contain 8% dry matter.[105] The sporophores also lose water to the external environment and create water fluxes within the older hyphae. Fruiting bodies of terrestrial fungi are mostly short-lived, with a life span between 14 and 21 d, depending upon ambient conditions. Wood-decomposing fungi, such as *Polyporus* produce persistent fruit bodies that may remain extant for many years.

There is little information on the distribution of fruit bodies from year to year in the same location. A study of three mycorrhizal fungi associated with planted

Birch (*Betula*) trees in Scotland indicated that one fungal species aggressively expanded (as evidenced by sporophore production) over a 3-year period, the second species remained static, and the third showed a progressive decline. Clearly different species will behave differently with time under the same conditions.[106]

BIOACCUMULATION OF ELEMENTS

The preceding information forms a basic introduction to the biology of fruit body formation in higher fungi. How can this be related to the elevated metal levels encountered in fungal sporophores? Are particular elements biologically concentrated only by fungi? Do fungi exclude others? It is relevant at this point to examine what elements fungi are known to accumulate, from what types of substrate, and to decide on criteria for accumulation. There is very little information on average levels of elements in fungi, due to the lack of widespread systematic investigation, so mean elemental levels found in higher plants have been taken as guidelines, and accumulation has been assumed to occur when metal levels in sporophores are at least fivefold higher than the plant guidelines.

Table 3 lists comparative data for 23 elements, indicating remarkable bioaccumulation for some elements as well as bioexclusion for others. Table 4 lists those fungal genera shown to bioaccumulate two or more elements from nonpolluted soils, together with an indication of their mode of nutrition (mycorrhizal or saprophytic). There is no clear distinction between accumulation in the two groups.

MECHANISMS OF BIOACCUMULATION

Because higher fungi possess a wide ranging vegetative biomass that is not easily detected "in situ" (save for those species that produce rhizomorphs) and because the induction of fruiting bodies in pure culture can be a difficult process (only achieved with relatively few species), controlled studies of uptake and transport of metals to the sporophore in the intact fungus present many problems. Using a model laboratory system,[122] Brunnert and Zadrazil followed uptake of Cd and Hg (applied to an organic substrate) by a range of lignocellulolytic fungi and *Agaricus biosporus*. Some species concentrated up to 75% of applied Cd and 38% of applied Hg in fruiting bodies; *Pleurotus flabellatus* achieved this in 8 d from addition of the metals to its growing substrate. Later studies,[123] using an agar-based system with the fungus *Agrocybe aegerita*, showed that zinc reduced Cd uptake, despite the face that it was poorly absorbed, and translocated by the fungus. No effect was reported of Zn on Hg uptake in the same system.

The speed of elemental accumulation in fruiting bodies poses questions relating to accumulation rates and mobility of elements within the fungal system. Microfungi absorb metals by active and passive processes.[62] It is difficult to

Table 3. Average Plant and Highest Fungal Contents of Selected Trace Metals Derived From Background Soils

	Plants	Highest Fungal Conc.	Ref.
Ag	0.1–0.8	52.0	107
Al	500	427	98
As	0.1–40	427	108
Au	<0.0017	0.78	109
Ba	14	600–700	110
Br	15	128	95
Cd	0.1–2.4	299	98
Co	0.02–1	7.29	111
Cr	0.03–14	2.48	111
Cu	5–20	469	98
Cs	0.2	308	112
Fe	140	4600–6300	113,114,115
Hg	0.005–0.17	80.0	116
I	0.8–2.2	11.0	95
Mn	20–1000	1140–1727	98,117
Mo	0.003–5	2.21	111
Ni	0.02–5	18.9	111
Pb	0.2–20	78.0	96
Rb	20	4700	118
Sb	0.0001–02	0.35	109
Se	0.001–02	47.4	95
Tl	0.03–3	5.5	119
V	0.001–1.5	700	95,98
Zn	1–400	1200	113

Elements accumulated by one or more fungal genera: As, As, Au, Ba, Br, Cd, Co, Cs, Cu, Fe, Hg, I, Rb, Se, V. Accumulation similar to higher plants: Mn, Ni, Pb, Sb, Tl, Zn. Excluded compared to higher plants: Cr, Mo.

Plant element levels from Alloway[120] and Bowen[121] and Brunnert and Zadrazil.[122]

Table 4. Fungal Genera That Have at Least One Species Known to Accumulate More Than One Trace Metal From Unpolluted Soil (Based on Criteria in Table 3)

Genus	Element	Growth Strategy
Agaricus	Ag, Au, Cd, Co, Cu, Hg	S
Amanita	Br, Cd, Cu, Hg, Rb, V	M
Boletus	Ag, Cu, Hg, Rb, Se	M
Calvatia	Ag, Cu	S
Clitocybe	Cu, Hg	S
Coprinus	Cd, Co, Cu	S
Cortinarius	Cd, Cs, Cu, Hg	M
Cytoderma	Cd, Hg	S
Laccaria	Ag, Cd	M
Lactarius	Hg, Rb	M
Lycopordon	Ag, Cu, Hg, Se	S
Macrolepiota	Au, Cu	S
Russula	Cd, Cu, Hg	M
Sarcodon	Hg, I	S
Suillius	Fe, Rb	M

Note: M — Mycorrhizal; S — Saprophytic.

believe that the selective accumulation of elements in higher fungal fruiting bodies can be achieved by passive uptake alone, given the absence of a specialized transport system. However, fruiting bodies lose water to the atmosphere in an unregulated manner; the water fluxes created by this in the mycelial network could serve to accelerate transport. If this were the case, one may expect differences in elemental composition of sporophores with age. At present, data to support this are lacking due to (1) the difficulty in locating developing sporophores in the field and (2) variation of metal content of sporophores within a population. Brunnert and Zadrazil[122] noted increases in sporophore metal content (Cd, Hg) in later crops of lignocellulolytic fungi grown on artificially metal-amended media. At present, there is no clear information on processes of metal uptake and transport in naturally occurring myceliar/sporophore systems to adequately explain the rapid accumulation of elements from background soils.

CHEMICAL FORM OF ELEMENTS

There are few instances of the identification of the chemical form of elements within fungi. The best known example is the molecule amavadine, a vanadium-containing compound found in *Amanita muscaria* and closely related species.[124] There has been much speculation on the possible role of this molecule in *Amanita*; Frausto da Silva[124] has recently suggested that it may function as an electron transfer catalyst or mediator, but conclusive evidence to support such a hypothesis is still required.

The chemical form of Cd and Hg in some fungi has also received investigation. Kojo and Lodenius[125] examined the methyl mercury content of four species of *Agaricus*; in all cases, this fraction accounted for less than 20% of total sporophore Hg. In tissues of *Lagermannia gigantea*, the largest proportion of Hg was bound to high molecular weight molecules, as evidenced by gel filtration. This is in contrast to Cd, where binding to low molecular weight proteins has been observed.

Some authors have assumed that Hg accumulation in fungi could be related to the various sulfur-containing groups (sulfhydryl, disulfide, and methionine) found in proteins.[95,117] Cd-rich proteins do not have high sulfur contents. Kojo and Lodenius[125] found a good correlation between –SH groups and Hg content in mycorrhizal fungi, much less so for Cd.

There is no information on the chemical forms of other elements in sporophores.

ECOLOGICAL SIGNIFICANCE

It is evident that sporophores of certain fungi represent a powerful bioconcentration of potentially toxic elements in otherwise nonpolluted ecosystems. There have been few attempts to investigate the potential ecological significance of this phenomenon, in terms of elemental cycling or food chain contamination. Lepp et al.[105] investigated the plant availability of V and Cd released from decomposing *A. muscaria* sporophores. Cd uptake into foliage from sporophore-amended soil was observed using lettuce (*Lactuca sativa*) as a test crop; sporophore-derived V was taken into roots using the same test system. Calculations based on the soil type (skeletal pararendzina) and Cd content indicated that *A. muscaria* sporophores could circulate up to 10% of the total soil Cd content (0 to 10 cm horizon) over a 14- to 21-d period.

There is little information on transfer of elements from sporophore to herbivore. Fungal fruiting bodies are consumed by mollusks and dipteran larvae. Lodenius[126] investigated Hg levels in larvae feeding on sporophores of saprophytic and mycorrhizal fungi; he found that concentration factors never exceeded 1 in larvae feeding on saprophytes and were frequently much lower for those feeding on mycorrhizal sporophores. It is suggested that efficient excretion prevents bioaccumulation in these organisms, but there is no firm evidence to support this.

WHY DO CERTAIN FUNGI ACCUMULATE METALS?

At present, there is no satisfactory explanation for these phenomena. It has been suggested, for example, that V in *A. muscaria* and related species forms part of a molecule that could act in electron transfer processes. It may be the case that other fungi have metal requirements based on the need for enzymic co-factors, but it is unsubstantiated. Jennings[103] suggests that some fungi may accumulate essential elements such as Fe to enhance their competitive ability in multispecies communities. Such an explanation may also be valid for Cu, but difficult to uphold in the case of nonessential elements such as Cd and Hg.

Fungi with elevated levels of toxic elements in sporophore tissue are not immune from primary consumers; indeed, prized edible species such as *Agaricus* sp. and *Boletus edulis* contain elevated levels of Cd, Hg, and Se.[95–98,116] It could also be suggested that the bioaccumulation of certain elements represents an abbreviated environmental cycle, with rapid reabsorption of elements by hyphae following release from decaying sporophores. Again, rates and patterns of elemental release are not known, although there are several species in which this could be followed experimentally.

It is clear that the sporophores of higher fungi play an active part in the natural biogeochemical cycles of a range of rare and dispersed elements. Closer investigation of the chemical form these elements take in sporophore tissues, together with the form in which they are lost during decomposition, would be of primary importance in determining these cycles. It may be that elements lost from decomposing sporophores are rapidly reabsorbed by fungal hyphae. Competition between these structures and plant roots for inorganic nutrients has not been investigated. In some soils, released metals may be rapidly rebound to organic matter; fungal hyphae may release these during the decomposition of this material and perhaps absorb small metal/organic chelates. Such a phenomenon occurs in Fe uptake following release of siderophores; further investigation is required to investigate the possibility that fungi may have developed powerful selective natural chelates for other elements.[103] The magnitude of accumulation of certain elements and the rapid nature of the process requires study under controlled conditions. It may prove possible to culture fungi on chelator-buffered media, as has been successfully reported for bacteria,[127] thus minimizing the influence of the growth medium on metal uptake patterns, but a greater barrier is the difficulty in making many higher fungi produce sporophores in culture. Studies on primary consumers of fungal sporophores would also be of interest; there is little current information on which to base firm conclusions regarding the possible mobility of sporophore Cd or Hg within particular food webs.

In the few cases where careful consideration of the role of soil factors on metal levels in sporophores has been investigated, it has been concluded that species differences, not soil conditions, are primary regulators of fungal metal content. This is found in sludge-treated sites,[128] naturally mineralized sites,[129] and industrially polluted sites.[130] This would indicate strong selective accumulation of certain elements, possibly by specific mechanisms. Further detailed studies of the relationship between soil metal conditions and sporophore metal content is clearly indicated. The use of a wide ranging species with a demonstrated ability to concentrate one or more elements whose activity in soil is known to be strongly influenced by factors such as pH, CEC, or levels of soluble organic materials would be revealing in this context. The fact that a species such as *A. muscaria*, which has a wide geographical distribution, appears to retain the ability the accumulate V,[99–102,106,125] an element whose activity is highly dependent on soil properties, throughout this wide range indicates that species-specific factors may play an equally important role alongside soil chemical properties in determining the eventual metal burdens of sporophores. Some mycorrhizal fungi have developed tolerances to heavy metals,[131] but many species show marked sensitivity (reflected as a lack of sporophore production) to increasing soil Cu and Zn levels.[132] These complex relationships may prove difficult to elucidate.

APPLICATIONS

There is the potential for the application of bacterial and fungal metal accumulation in a variety of situations. Metal recovery from or metal decontamination of effluents represent two possible applications. However, the use of microorganisms in such situations is very dependent upon their realistic integration into a viable commercial system. Re-use of biomass is a clear priority, and use of dead cells that can be used in multiple sorption-desorption routines may be applicable.[133] This represents a feasible technique for metal recovery, but its efficacy depends on ambient conditions. Acidic solutions can remove metals from biomass[3,19,58,133] at low (<0.1 mol l^{-1}) dilutions, but higher concentrations (>1 mol l^{-1}) can damage biomass and reduce uptake.[84] If effluents are acidic, biomass may not act to remove metals. Organic agents, such as EDTA, can also remove metals from bacteria and fungi,[28] but must be applied in excess to remove strongly complexed metals such as Cu.

Other applications make use of living cells, either free or immobilized or dead, immobilized cells. Mixed microbial cultures have been used for Cu and Ag removal[28] and Cd removal, the latter using a two-stage chemostat system that removed >80% of Cd from an incoming effluent containing 0.8 mg m^{-1} Cd.[134] Metal removal from activated sludge biomass using polysaccharide-producing organisms is more effective in batch than in continuous culture.[1] The economics of frequent replacement of microbial cultures must be accounted for in relation to their routine incorporation into treatment processes.

Free-living microbial biomass is difficult to contain.[135] This limits its use in many potential applications. Immobilization on an inert carrier system opens up new applications and renders the use of batch treatment a more attractive commercial proposition. A variety of materials can be used to entrap cells. *Streptomyces albus,* immobilized in polyacrylamide gel, can remove UO_2, Cu, and Co from solution; these then can be removed with an Na_2CO_3 desorption, and no adverse effects on the organisms were recorded after five successive sorption-desorption cycles.[136,137] Similarly, polyacrylamide-immobilized *Citrobacter* could remove a range of metals from solution supplemented with glycerol-2-phosphate. The bacteria extracellularly liberated HPO_4^{2-} from the latter that precipitated the metals.[138] Unspecified, entrapped microbial biomass is used in systems to recover valuable metals such as Ag and Au, up to 0.8 mmol g^{-1} of the former and 2.0 mmol g^{-1} of the latter have been removed from dilute (10 to 100 mg l^{-1}) metal solutions via fixed bed canisters or fluidized bed systems.[139] Polyacrylamide has low mechanical strength, and laboratory systems based on this material often fail to scale up.[140]

The use of biofilms developed on inert surfaces is one way in which this problem may be overcome. A *Citrobacter* film has been shown to possess metal removing characteristics equivalent to an acrylamide-immobilized system.[140] *Aspergillus oryzae* grown on foam[141] and *Trichoderma reesii* packed in malachite[142] are two examples of fungi grown on inert substrates with good metal-removing potential. Pelletized *A. niger* proved more effective than commercial ion-exchange resin for the removal of U, but disadvantages due to the morphology of the pellets created problems in efficient operation of a fluidized bed system based on this technique.[143]

Microbial biomass has potential application in effluent treatment, metal recovery, and pollution abatement, but the widespread use of this material is dependent upon the development of systems that can operate efficiently and economically. In many cases, the scaling up of a process proves to be impractical. In the future, potential applications of such systems may have to be precisely targeted onto elements of strategic importance, or those that pose major environmental hazards; in all cases, there must be some form of advantage over existing commercial processes.

Fungi and bacteria have evolved unique and complex relationships with metals in the environment, both naturally occurring and polluting. At present, we have scarcely begun to understand these relationships, and our ability to do this, then develop systems based on our knowledge of such phenomena to reclaim, recycle, and decontaminate our wastes and effluents will prove to be a key area of scientific application in future decades.

REFERENCES

1. Norberg, A. and S. Rydin. "Development of a Continuous Process for Metal Recovery by *Zooglea ramigera*," *Biotechnol. Bioeng.* 26:265–268 (1984).
2. Tobin, J. M., D. G. Cooper, and R. J. Neufeld. "Uptake of Metal Ions by *Rhizopus arrhizus* Biomass," *Appl. Environ. Microbiol.* 47:821–824 (1964).
3. Norber, A. B. and H. Persson. "Accumulation of Heavy Metal Ions by *Zooglea ramigera*," *Biotechnol. Bioeng.* 26:239–246 (1984).
4. Tornabene, T. G. and H. W. Edwards. "Microbial Uptake of Lead," *Science* 176:1334 (1972).
5. Pooley, F. D. "Bacteria Accumulate Silver During Leaching of Sulfide Mineral Ores," *Nature (London)* 296:642–643 (1982).
6. Beveridge, T. J., C. W. Fursberg, and R. J. Doyle. "Major Sites of Metal Binding in *Bacillus licheniformis* Walls," *J. Bacteriol.* 150:1438–1448 (1982).
7. Matthews, T. H., R. J. Doyle, and U. N. Streips. "Contribution of Pepidoglycan to the Binding of Metal Ions by the Cell Wall of *Bacillus subtilis*," *Curr. Microbiol.* 3:51–53 (1979).
8. Doyle, R. J., T. H. Matthew, and U. N. Streips. "Chemical Basis for Selectivity of Metal Ions by the *Bacillus subtilis* Cell Wall," *J. Bacteriol.* 143:471–480 (1980).
9. Irving, H. and R. P. J. Williams. "The Stability of Transition Metal Complexes," *J. Chem. Soc.* 3192–3210 (1953).
10. Beveridge, T. J. and R. G. E. Murray. "Sites of Metal Deposition in the Cell Walls of *Bacillus subtilis*," *J. Bacteriol.* 141:876–887 (1980).
11. Beveridge, T. J. and R. G. E. Murray. "Uptake and Retention of Metals by Cell Walls of *Bacillus subtilis*," *J. Bacteriol.* 127:1502–1518 (1976).
12. Beveridge, T. J., J. D. Meloche, W. S. Fyfe, and R. G. Murray. "Diagenesis of Metals Chemically Complexed to Bacteria Laboratory Formation of Metal Phosphates, Sulfides and Organic Condensates in Artificial Sediments," *Appl. Environ. Microbiol.* 45:1094–1108 (1983).
13. Beveridge, T. J. and S. F. Koval. "Binding of Metals to Cell Envelopes of *Escherichia coli* K-12," *Appl. Environ. Microbiol.* 42:325–335 (1981).
14. Baldry, M. G. C. and A. C. R. Dean. "Environmental Change and Copper Uptake by *Bacillus subtilis* ssp. niger and *Pseudomonas fluorescens*," *Biotechnol. Lett.* 3:137–142 (1981).
15. Surowitz, K. G., J. A. Titus, and M. Pfister. "Effects of Cadmium Accumulation on Growth and Respiration of a Cadmium-Sensitive Strain of *Bacillis subtilis* and a Selected Cadmium-Resistant Mutant," *Arch. Microbiol.* 140:107–122 (1984).

16. Borst-Pauwels, G. F. W. H. "Ion Transport in Yeast," *Biochim. Acta* 650:88–127 (1981).
17. Gadd, G. M. "Fungal Responses towards Heavy Metals," in *Microbes in Extreme Environments,* Codd, G. A. and R. A. Herbert, Eds. (London: Academic Press, 1986) pp. 83–110.
18. Strandberg, G. W., S. E. Shumate, and J. R. Parrott. "Microbial Cells as Biosorbents for Heavy Metals. Accumulation of Uranium by *Saccharomyces cerevisiae and Pseudomonas aeruginosa," Appl. Environ. Microbiol.* 41:237–245 (1981).
19. Silver, S. "Transport of Cations and Anions," in *Bacterial Transport,* Rosen, B. P., Ed., (New York: Marcel Dekker, 1978) pp. 221–324.
20. Archibald, F. S. and M. N. Duong. "Manganese Acquistion by *Lactobacillus plantarum," J. Bacteriol.* 158:1–8 (1984).
21. Tabillon, R. and H. Kaltwasser. "Energy-dependent Ni-63 Uptake by *Alcaligenes eutrophus* Strain H-1 and Strain H-16," *Arch. Microbiol.* 113:145–151 (1977).
22. Jarrell, K. F. and G. D. Sprott. "Nickel Transport in *Methanobacterium bryantii," J. Bacteriol.* 151:1195–1203 (1982).
23. Stults, L. W., S. Mallick, and R. J. Maier. "Nickel Uptake in *Bradyrhizobium japonicum," J. Bacteriol.* 169:1398–1402 (1987).
24. Jasper, P. and S. Silver. "Magnesium Transport in Micro-organisms," in *Microorganisms and Minerals,* Weinberg, E. D., Ed. (New York: Marcel Dekker, 1977) pp. 7–47.
25. Laddaga, R. A., R. Bessen, and S. Silver. "Cadmium Resistant Mutant of *Bacillus subtilis* 168 with Reduced Cadmium Transport," *J. Bacteriol.* 162:1106–1110 (1985).
26. Tynecka, Z., Z. Gos, and J. Zajac. "Reduced Cadmium Transport Determined by a Resistance Plasmid in *Staphylococcus aureus," J. Bacteriol.* 147:305–312 (1981).
27. Charley, R. C. and A. T. Bull. "Bioaccumulation of Silver by a Multispecies Community of Bacteria," *Arch. Microbiol.* 123:239–244 (1979).
28. Dunn, G. M. and A. T. Bull. "Bioaccumulation of Copper by a Defined Community of Activated Sludge Bacteria," *Eur. J. Appl. Microbiol. Biotechnol.* 17:30–34 (1983).
29. Belly, R. T. and G. C. Kydd. "Effect of *Pseudomonas* sp. on Ag Reduction by Microbial Cultures," *Dev. Ind. Microbiol.* 23:567–577 (1982).
30. Morozzi, G., G. Cenci, F. Scardazza, and M. Pitzurra. "Cadmium Uptake by Growing Cells of Gram Positive and Gram Negative Bacteria," *Microbios* 48:27–35 (1981).
31. Chmielowski, J. and B. Klapcinska. "Bioaccumulation of Germanium by *Pseudomonas putida* in the Presence of Two Selected Substrates," *Appl. Environ. Microbiol.* 51:1099–1103 (1986).
32. Baldry, M. C. G. and A. C. R. Dean. "Copper Accumulation by Bacteria, Moulds and Yeasts," *Microbios* 29:7–14 (1980).
33. Macaskie, L. E. and A. C. R. Dean. "Cadmium Accumulation by a *Citrobacter* sp.," *J. Gen. Microbiol.* 130:53–62 (1984).
34. Higham, D. P., P. J. Sadler, and D. Scawen. "Cadmium Resistant *Pseudomonas putida* Synthesises Novel Cadmium Proteins," *Science* 225:1043–1046 (1984).

35. Horitsu, H., M. Takagi, and M. Tomoyeda. "Isolation of a Mercuric Chloride Tolerant Bacterium and Uptake of Mercury by Bacterium," *Eur. J. Appl. Microbiol. Biotechnol.* 5:279–290 (1978).
36. Oberhack, M., R. Susmuth, and H. Frank. "Manganese Bacteria from Freshwater," *Z. Natursforsch.* 42:300–306 (1987).
37. Higham, D. P., P. J. Sadler, and M. D. Scawen. "Effect of Cadmium on the Morphology, Membrane Integrity and Permeability of *Pseudomonas putida*," *J. Gen. Microbiol.* 132:1475–1482 (1986).
38. Mitra, R. S. "Protein Synthesis in *Escherichia coli* During Recovery from Exposure to Low Levels of Cd^{2+}," *Appl. Environ. Microbiol.* 47:1012–1016 (1984).
39. Norberg, A. R. and S. O. Enfors. "Production of Extracellular Polysaccharide by *Zooglea ramigera*," *Appl. Environ. Microbiol.* 44:1231–1237 (1981).
40. Brown, M. J. and J. N. Lester. "Role of Bacterial Extracellular Polymers in Metal Uptake in Pure Bacterial Culture and Activated Sludge. I. Effects of Metal Concentration," *Water Res.* 16:1539–1548 (1982).
41. Norris, P. R. and D. P. Kelly. "Toxic Metals Leaching Systems," in *Metallurgical Applications of Bacterial Leaching and Related Microbial Phenomena*, Murr, L. E., A. E. Torma, and J. A. Brierley, Eds. (New York: Academic Press, 1978) pp. 83–102.
42. Ghiorse, W. C. and P. Hirsch. "Ultrastructural Study of Iron and Manganese Associated with Extracellular Polymers of *Pedomicrobium*-like Budding Bacteria," *Arch. Microbiol.* 123:213–226 (1979).
43. Pivovara, T. A., E. D. Korobushkina, S. A. Krasheninnikova, A. E. Rubtsov, and G. I. Karavaiko. "Influence of Gold Ions on *Thiobacillus ferroxidans*," *Microbiologia* 55:774–780 (1978).
44. Aiking, H., K. Kok, H. Van Heerikhuizen, and J. van'Triet. "Adaptation to Cadmium by *Klebsiella aerogenes* Growing in Continuous Culture Proceeds Mainly by the Formation of Cadmium Sulfide," *Appl. Environ. Microbiol.* 44:938–944 (1982).
45. Aiking, H., A. Stijnman, C. van Garderon, H. van Heerikhuizen, and J. van'Triet. "Inorganic Phosphate Accumulation and Cadmium Detoxification in *Klebsiella aerogenes* NCTC-418 Growing in Continuous Culture," *Appl. Environ Microbiol.* 47:374–377 (1984).
46. Aiking, H., H. Govers, and J. van'Triet. "Detoxification of Mercury, Cadmium and Lead in *Klebsiella aerogenes* NCTC-418 Growing in Continuous Culture," *Appl. Environ. Microbiol.* 50:1262–1267 (1985).
47. Neilands, J. B. "Microbial Iron Compounds," *Annu. Rev. Biochem.* 50:715–731 (1981).
48. Raymond, K. N., G. Muller, and B. F. Mayzanke. "Complexation of Iron by Siderophores. A Review of Their Solution and Structural Chemistry and Biological Function," *Top. Curr. Chem.* 123:49–102 (1984).
49. Hausinger, R. P. "Nickel Utilization by Microorganisms," *Microbiol. Rev.* 51:22–42 (1987).
50. Emery, T. "Exchange of Iron by Gallium in Siderophores," *Biochemistry* 25:4629–4633 (1986).

51. Premuzic, E. T., A. J. Francis, M. Lim, and J. Schuber. "Induced Formation of Chelating Agents by *Pseudomonas aeruginosa* Grown in Presence of Thorium and Uranium," *Arch. Environ. Microbial.* 14:759–768 (1985).
52. Gadd, G. M. "Metal Tolerance," in *Microbiology of Extreme Enviroments,* Edwards, C., Ed. (Milton Keynes: Open University Press, 1990) pp. 178–210.
53. Gadd, G. M. "Fungi and Yeast for Metal Accumulation," in *Microbial Mineral Recovery,* Erlich, H. J., J. A. Brierley, and C. L. Brierley, Eds. (New York: Macmillan, 1990) (in press).
54. Murphy, R. T. and J. F. Levy. "Production of Copper Oxalate by Some Copper Tolerant Fungi," *Trans Br. Mycol. Soc.* 81:165–168 (1983).
55. Minney, S. F. and A. V. Quirk. "Growth and Adaptation of *Saccharomyces cerevisiae* at Different Cadmium Concentrations," *Microbios* 42:37–44 (1985).
56. Adjimani, J. P. and T. Emery. "Iron Uptake in *Mycelial sterilia* EP-76," *J. Bacteriol.* 169:3664–3668 (1987).
57. Norris, P. R. and D. P. Kelly. "Accumulation of Cadmium and Cobalt by *Saccharomyces cerevisiae,*" *J. Gen. Microbiol.* 99:317–324 (1977).
58. De Rome, L. and G. M. Gadd. "Copper Adsorption by *Rhizopus arrhizus, Cladosporium resinae* and *Penicillium italicum,*" *Appl. Microbiol. Biotechnol.* 26:84–90.
59. Gadd, G. M. "The Uptake of Heavy Metals by Fungi and Yeasts: The Chemistry and Physiology of the Process and Applications for Biotechnology," in *Immobilization of Ions by Biosorption,* Eccles, H. and S. Hunt, Eds. (Chichester: Ellis Horwood, 1966) pp. 135–147.
60. Gadd, G. M. and L. de Rome. "Biosorption of Copper by Fungal Melanin," *Appl. Microbiol. Biotechnol.* 29:610–617 (1988).
61. Ross, J. S. and C. C. Townley. "The Uptake of Heavy Metals by Filamentous Fungi," in *Immobilization of Ions by Desorption,* Eccles, H. and S. Hunt, Eds. (Chichester: Ellis Horwood, 1986) pp. 49–58.
62. Gadd, G. M., C. White, and L. de Rome. "Heavy Metal and Radionuclide Uptake by Fungi and Yeast," in *Biohydrometallurgy,* Norris, P. R. and D. P. Kelly (Kew: Science and Technology Letters, 1988) pp. 421–435.
63. Sakaguchi, T. and A. Nakajima. "Recovery of Uranium by Chitin and Chitosan Phosphate," in *Chitin and Chitosan,* Mirano, S. and S. Tokura, Eds. (Tottori, Japan: Japanese Society of Chitin and Chitosan, 1982) pp. 177–182.
64. Muzzarelli, R. A. A., F. Bregani, and F. Sigon. "Chelating Abilities of Amino Acid Glucans and Sugar Acid Glucans Derived from Chitosan," in *Immbolization of Ions by Biosorption,*" Eccles, H. and S. Hunt, Eds. (Chichester: Ellis Horwood, 1986) pp. 173–182.
65. Saiz-Jiminez, C. and F. Shafizadeh. "Iron and Copper Binding by Fungal Phenolic Polymers: An Electron Spin Resonance Study," *Curr. Microbiol.* 10:281–286 (1984).
66. Gadd, G. M. and J. L. Mowll. "Copper Uptake by Yeast-Like Cells, Hyphae and Chlamydospores of *Aureobasidium pullulans*" *Exp. Mycol.* 9:230–240 (1985).
67. Thelivenet, A. P. R. and R. J. N. Bindels. "An Investigation into the Feasibility of Using Yeast Protoplasts. To Study the Ion Transport Properties of the Plasma Membrane," *Biochim. Biophys. Acta* 599:587–595 (1980).
68. Gadd, G. M. and C. White. "Copper Uptake by *Penicillium ochrochloron:* Influence of pH on Toxicity and Demonstration of Energy Dependent Copper Influx Using Protoplasts," *J. Gen. Microbiol.* 131:1875–1879 (1985).

69. Paton, W. N. H. and K. Budd. "Zinc Uptake in *Neocosmopora vasinfecta*," *J. Gen. Microbiol.* 72:173–184 (1972).
70. Failla, M. C., C. D. Benedict, and D. Weinberg. "Accumulation and Storage of Zinc by *Candida utilis*," *J. Gen. Microbiol.* 94:23–36 (1976).
71. Parkin, M. J. and J. S. Ross. "The Regulation of Mn^{2+} and Cu^{2+} Uptake by the Cells of the Yeast *Candida utilis* Grown in Continuous Culture," *FEMS Microbiol. Lett.* 37:59–62 (1986).
72. White, C. and G. M. Gadd. "The Uptake and Cellular Distribution of Zinc in *Saccharomyces cerevisiae*," *J. Gen. Microbiol.* 133:727–737 (1987).
73. Fuhrman, G. F. and A. Rothstein. "The Transport of Zn^{2+}, Co^{2+} and Ni^{2+} into Yeast Cells," *Biochim. Biophys. Acta* 463:325–330 (1968).
74. Mowll, J. L. and G. M. Gadd. "Cadmium Uptake of *Aureobasidium pullulans*," *J. Gen. Microbiol.* 130:279–284 (1984).
75. Mowll, J. L. and G. M. Gadd. "Zinc Uptake and Toxicity in the Yeasts *Sporobolomyces roseus* and *Saccharomyces cerevisiae*," *J. Gen. Microbiol.* 129:3421–3425 (1983).
76. Okorov, L. A., L. P. Lichko, and A. Y. Valiakhmetov. "Transmembrane Gradient of K^+ Ions as an Energy Source in the Yeast *Saccharomyces carlsbergensis*," *Biochem. Int.* 6:463–472 (1983).
77. Okorov, L. A., L. P. Lichko, and N. A. Andreeva. "Changes of ATP, Polyphosphate and K^+ Content on *Saccharomyces cerevisiae* During Uptake of Mn^{2+} and Glucose," *Biochim. Int.* 6:481–488 (1983).
78. Borst-Pauwels, G. W. F. H. and A. J. R. Thelivenet. "Apparent Saturation Kinetics of Divalent Cation Uptake in Yeast Caused by a Reduction in the Surface Potential," *Biochem. Biophys. Acta* 771:171–176 (1984).
79. White, C. and G. M. Gadd. "Inhibition of H^+ Efflux and K^+ Uptake and Induction of K^+ Efflux in Yeast by Heavy Metals," *Toxicity Assess.* 2:437–447 (1987).
80. Kuypers, G. A. J. O. and G. M. Roomans. "Mercury-Induced Loss of K^+ from Yeast Cells Investigated by Electron Probe X-Ray Microanalysis," *J. Gen. Microbiol.* 115:13–18 (1979).
81. Gadd, G. M. and A. J. Griffiths. "Influence of pH on Toxicity and Uptake of Copper in *Aureobasidium pullulans*," *Trans. Br. Mycrol. Soc.* 75:91–96 (1980).
82. Theuvenet, A. P. R., G. M. Roomans, and G. W. F. H. Borst-Pauwels. "Intracellular pH and the Kinetics of Rb^+ Uptake by Yeast. Non-Carrier Versus Carrier Mediated Uptake," *Biochim. Biophys. Acta* 469:272–280 (1977).
83. Roomans, G. M., A. P. R. Theuvenet, and G. W. F. H. Borst-Pauwels. "Kinetics of Ca^{2+} and Sr^{2+} Uptake of Yeast. Effects of pH, Cations and Phosphate," *Biochim. Biophys. Acta* 551:187–196 (1979).
84. Joho, M., Y. Sukenddu, E. Egashira, and T. Murumaya. "The Correlation Between Cd^{2+} Sensitivity and Cd^{2+} Uptake in the Strains of *Saccharomyces cerevisiae*," *Plant Cell Physiol.* 24:389–394 (1983).
85. Gadd, G. M., A. Stewart, C. White, and J. L. Mowll. "Copper Uptake by Whole Cells and Protoplasts of a Wild-Type and Copper-Resistant Strain of *Saccharomyces cerevisiae. FEMS Microbiol. Lett.* 24:231–234 (1984).
86. Mohan, P. M., M. P. P. Rudra, and K. S. Sastry. "Nickel Transport in Nickel-Resistant Strains of *Neurospora crassa*," *Curr. Microbiol.* 10:125–128 (1984).
87. Gadd, G. M., J. A. Chudek, R. Foster, and R. H. Reed. "The Osmotic Responses of *Penicillium ochro-chloron*: Changes in Internal Solute Levels in Response to Copper and Salt Stress," *J. Gen. Microbiol.* 130:1969–1975 (1984).

88. Gadd, G. M. and A. J. Griffiths. "Effect of Copper on Morphology of *Aureobasidium pullulans*," *Trans. Br. Mycol. Soc.* 74:387–392 (1980).
89. Bianchi, M. E., M. L. Carbone, and G. Lucchini. "Mn^{2+} and Mg^{2+} Uptake in Mn-Sensitive and Mn-Resistance Yeast Strains," *Plant Sci. Lett.* 22:345–352 (1981).
90. Butt, T. R. and D. J. Ecker. "Yeast Metallothionein and Applications in Biotechnology," *Microbiol. Rev.* 51:351–364 (1987).
91. Beltramini, M. and K. Lerch. "Spectroscopic Studies on *Neurospora* Copper Metallothionen," *Biochemistry* 22:2043–2048 (1983).
92. Joho, M., C. Yamananka, and T. Murayama. "Cd^{2+} Accommodation by *Saccharomyces cerevisiae*," *Microbios* 45:169–179 (1986).
93. Lawford, H. G., J. R. Pik, G. R. Lawford, T. Williams, and A. Kligerman. "Hyperaccumulation of Zinc by Zinc-Depleted *Candida utilis* Grown in Chemostat Culture," *Can. J. Microbiol.* 26:71–76 (1980).
94. Failla, M. L. and E. D. Weinberg. "Cyclic Accumulation of Zinc by *Candida utilis* During Growth in Batch Culture," *J. Gen. Microbiol.* 99:85–97 (1977).
95. Byrne, A. R., V. Ravnik, and L. Kosta. "Trace Element Concentrations in Higher Fungi," *Sci. Total Environ.* 6:65–78 (1976).
96. Kuusi, T., K. Laaksovirta, H. Liukkonen-Lilja, M. Lodenius, and S. Piepponen. "Lead, Cadmium and Mercury Contents of Fungi in the Helsinki Area and in Unpolluted Control Areas," *Z. Lebensm. Uters. Forsch.* 173:261–267.
97. Lodenius, M., T. Kuusi, K. Laaksovirta, H. Liukkonen-Lilja, and S. Peipponen. "Lead, Cadmium and Mercury Content of Fungi in Mikkeli, SE Finland," *Ann. Bot. Fenn.* 18:183–186 (1981).
98. Tyler, G. "Metals in Sporophores of Basidiomycetes," *Trans. Br. Mycol. Soc.* 74:41–49 (1980).
99. Watkinson, J. H. "A Selenium Accumulating Plant of the Humid Regions: *A. muscaria*," *Nature (London)* 202:1239–1240 (1964).
100. Ter Meulen, E. V. "Sur la Repartition du Molybdene dans la Nature," *Rev. Trav. Chim. Pays-Bas* 50:491–505 (1931).
101. Bertrand, D. "Le Vanadium chez les Champignons et Plus Spécialement chez les Amanites," *Bull. Soc. Chim. Biol.* 25:194–197 (1943).
102. Thompson, W. "Distribution, Development and Functioning of Mycelium Cord Systems of Decomposer Basidiomycetes of the Deciduous Woodland Floor," in *The Ecology and Physiology of the Fungal Mycelium*, Jennings, D. H. and A. D. M. Rayner, Eds. (Cambridge: CUP, 1984) pp. 185–214.
103. Jennings, D. H. "The Presidential Address: The Medium is the Message," *Trans. Br. Mycol. Soc.* 89:1–11 (1987).
104. Manachere, G. "Conditions Essential for Controlled Fruiting of Macromycetes — a Review," *Trans Br. Mycol. Soc.* 75:255–270 (1980).
105. Lepp, N. W., S. C. S. Harrison, and B. G. Morrell. "A Role for *Amanita muscaria* L. in the Circulation of Cadmium and Vanadium in a Nonpolluted Woodland," *Environ. Geochem. Health* 9:61–64 (1987).
106. Ford, E. D., P. A. Mason, and J. Pelham. "Spatial Patterns of Sporophore Distribution Around a Young Birch Tree in Three Successive Years," *Trans Br. Mycol. Soc.* 75:287–296 (1980).
107. Byrne, A. R., M. Deremelj, and T. Vakselj. "Silver Accumulation by Fungi," *Chemosphere* 10:815–821 (1979).
108. Byrne, A. R. and M. Tusek-Znidaric. "Arsenic Accumulation in the Mushroom *Laccaria amethystina*," *Chemosphere* 12:1113–1117.

109. Allen, R. O. and E. Steinnes. "Concentration of Some Potentially Toxic Metals and Other Trace Elements in Wild Mushrooms from Norway," *Chemosphere* 4:371–378 (1978).
110. Kalin, M. and P. M. Stokes. "Macrofungi on Uranium Mill Tailings — Association and Metal Contents," *Sci. Total Environ.* 19:83–94 (1981).
111. Mutsch, P., O. Horak, and H. Kinzel. "Trace Elements in Higher Fungi," *Z. Pflanzenphysiol.* 94:1–10 (1979).
112. Seeger, R. and P. Schweinshaut. "Occurrence of Caesium in Higher Fungi," *Sci. Total Environ.* 19:253–276 (1981).
113. Hinneri, S. "Mineral Elements of Macrofungi in Oak-Rich Forests on Lenholm Island, Inner Archipeligo of SW Finland," *Ann. Bot. Fenn.* 12:135–140 (1975).
114. Ohtonen, T. R. "Mineral Concentration in Some Edible Fungi and Their Relation to Fruit Body Size and Mineral Status of the Substrate," *Ann. Bot. Fenn.* 19:203–209 (1982).
115. Autio, S. "Metsämaan Happamusis Ja Sientien Metallipitor-suudet," Ph.D. thesis, Dept. of Environmental Conservation, University of Helsinki, Finland (1988).
116. Stijve, T. and R. Besson. "Mercury, Cadmium, Lead and Selenium Content of Mushroom Species Belonging to the Genus *Agaricus*," *Chemosphere* 5:151–158 (1976).
117. Vogts, K. S. and R. L. Edmonds. "Patterns of Nutrient Concentration in Basidiocarps in Western Washington," *Can. J. Bot.* 58:694–698 (1980).
118. Tyler, G. "Accumulation and Exclusion of Metals in *Collybia peronata* and *Amanita rubescens*," *Trans Br. Mycol. Soc.* (1982).
119. Seeger, R. and M. Gross. "Thallium in Higher Fungi," *Z. Lebensm. Unters. Forsch.* 173:9–15 (1981).
120. Alloway, B. J., Ed. *Heavy Metals in Soils* (Glasgow: Blackie, 1990).
121. Bowen, H. J. M. *Trace Elements in Biochemistry* (London: Academic Press, 1966).
122. Brunnert, H. and F. Zadrazil. "The Translocation of Mercury and Cadmium into the Fruiting Bodies of Six Higher Fungi. A Comparative Study on Species Specificity in Five Lignocellulolytic Fungi and the Cultivated Mushroom *Agaricus bisporus*," *Eur. J. Appl. Microbiol. Biotechnol.* 17:358–364 (1983).
123. Brunnert, H. and F. Zadrazil. "The Influence of Zinc on the Translocation of Cadmium and Mercury in the Fungus *Agrocybe aegerita* (a model system)," *Agnew. Bot.* 59:469–477 (1985).
124. Frausto da Silva, J. J. R. "Vanadium in Biology — The Case of the *Amanita* Toadstools," *Chem. Speciation Bioavailability* 1:139–150 (1989).
125. Kojo, M.-R. and M. Lodenius. "Cadmium and Mercury in Macrofungi — Mechanisms of Transport and Accumulation," *Angew. Bot.* 63:279–292 (1989).
126. Lodenius, M. "Mercury Content of Dipterous Larvae Feeding on Macrofungi," *Ann. Entomol. Fenn.* 47:63–64 (1981).
127. Angle, J. S. and R. L. Chaney. "Cadmium Resistance Screening in Nitrilotriacetate-Buffered Minimal Media," *Appl. Environ. Microbiol.* 55:2101–2104 (1989).
128. Zabowski, D., R. J. Zasoki, W. Litke, and J. Ammirati. "Metal Content of Fungal Sporocarps from Urban, Rural and Sludge-Treated Sites," *J. Environ. Qual.* 19:372–377 (1990).
129. Bargagli, R. and F. Baldi. "Mercury and Methyl Mercury in Higher Fungi and Their Relation with Substrate in a Cinnabar Mining Area," *Chemosphere* 13:1059–1071 (1984).

130. Gast, C. H., E. Janson, J. Bierling, and L. Hannstra. "Heavy Metals in Mushrooms and Their Relationship with Soil Characteristics," *Chemosphere* 17:789–799 (1988).
131. Colpaert, J. V. and J. A. Van Assche. "Heavy Metal Tolerance in Some Ectomycorrhizal Fungi," *Functional Ecol.* 1:415–421 (1987).
132. Ruhling, Å., E. Bååth, A. Nordgren, and B. Söderström. "Fungi in Metal Contaminated Soil Near the Gusum Brass Mill, Sweden," *Ambio* 13:34–36 (1984).
133. Tsezos, M. "Removal of Uranium from Biological Adsorbents. Desorption Equilibrium," *Biotechnol. Bioeng.* 26:973–981 (1984).
134. Houba, C. and J. Remacle. "Removal of Cadmium by Micro-organisms in a Two Stage Chemostat," *Appl. Environ. Microbiol.* 47:1158–1160 (1984).
135. Tsezos, M. "Adsorption by Microbial Biomass as a Process for Removal of Ions from Process or Waste Solutions," in *Immobilization of Ions by Bio-sorption,* Eccles, H. and S. Hunt, Eds. (Chichester: Ellis Horwood, 1986) pp. 201–218.
136. Nakajima, A., T. Horikoshi, and T. Sakaguchi. "Studies on the Accumulation of Heavy Metal Elements in Biological Systems. 21. Recovery of Uranium by Immobilized Microorgansims. *Eur. J. Appl. Microbiol. Biotechnol.* 16:88–91 (1982).
137. Nakajima, A. and T. Sakaguchi. "Selective Accumulation of Heavy Metals by Microorgansims," *Appl. Microbiol. Biotechnol.* 24:59–64 (1986).
138. Macaskie, L. E., J. M. Wates, and A. C. R. Dean. "Cadmium Accumulation by a *Citrobacter* sp. Immobilised on Gel and Solid Supports. Applicability to the Treatment of Liquid Wastes Containing Heavy Metal Cations," *Biotechnol. Bioeng.* 30:66–73 (1987).
139. Hutchings, S. R., M. S. Davidson, J. A. Brierley, and C. L. Brierley. "Micro-organisms in Reclamation of Metals," *Annu. Rev. Microbiol.* 40:311–336 (1986).
140. Macaskie, L. E. and A. C. R. Dean. "Use of Immobilized Biofilm of *Citrobacter* sp. for the Removal of Uranium and Lead from Aqueous Flows," *Enzyme Microb. Technol.* 9:2–4 (1987).
141. Kiff, R. J. and D. R. Little. "Biosorption of Heavy Metals by Immobilized Fungal Biomass," in *Immobilization of Ions by Biosorption,* Eccles, H. and S. Hunt, Eds. (Chechester: Ellis Horwood, 1986) pp. 71–80.
142. Townsley, C. C., I. S. Ross, and A. S. Atkins. "Copper Removal from a Simulated Leach Effluent Using the Filamentous Fungi *Trichoderma viride,*" Eccles, H. and S. Hunt, Eds. (Chichester: Ellis Horwood, 1986) pp. 159–170.
143. Yakubu, N. A. and A. W. L. Dudeney. "Biosorption of Uranium with *Aspergillus niger,*" Eccles, H. and S. Hunt, Eds. (Chichester: Ellis Horwood, 1986) pp. 183–200.

11

Metals in Invertebrate Animals of a Forest Ecosystem*

Mechthild Roth

University of Ulm, Ecology and Morphology of Animals (Biology III), Albert-Einstein-Allee 11, D-7900 Ulm/Donau, Germany

ABSTRACT

In the most important invertebrate animals (Annelida, Arthropoda) of a virtually unpolluted spruce stand in southern Germany, the concentrations of 18 elements were determined by atomic-spectrometrical methods: Al, Ba, Ca, Cd, Co, Cr, Cu, Fe, K, La, Mg, Mn, Na, Ni, P, Pb, V, Zn. In all species the highest concentrations were found for macronutrients; K, Na, Ca, Mg, and P represent between 2 and 11% of invertebrate dry matter. The lowest concentrations were manifested for essential or potentially toxic trace metals such as Co, V, and La.

The elemental composition of forest invertebrates was influenced by species- or group-specific morphological structures, physiological processes, and ecological life strategies. High concentrations of Ca, Mg, Mn, and partly of P were found in animals with mineral-containing exoskeleton (e.g., Diplopoda, Oribatida, Insecta). Above average concentrations of Cu were typical for invertebrates, which use hemocyanin as the respiratory pigment (e.g., spiders). High con-

---

* Dedicated to Prof. Dr. W. Funke on the occasion of his 60th birthday.

centrations of K and Na were found in web-building spiders, which kill their prey with toxins.

The elemental composition of invertebrate animals was often influenced by the concentration of elements in their food. This was especially true for the levels of K (and Mg) in phytophagous insects, feeding on physiologically active plant tissues and for the concentrations of essential and potentially toxic metals in edaphic animals feeding on "metal-enriched" fungal mycelia (Oribatida, Collembola, Diptera).

The determination of accumulation factors in food chains of invertebrates, beginning with litter, needles, bark, and roots of spruce yielded the following. Apart from a few exceptions, all primary consumers (phytophagous and detritophagous invertebrates) accumulated P and the essential metals Na, K, Mg, Zn, and Cu. In addition to these macro- and micronutrients, only few primary consumers enriched further elements. This was especially true for Cd in litter decomposers (Lumbricidae, Enchytraeidae, Oribatida, Diptera). Most zoophagous invertebrates (secondary consumers) concentrated neither essential nor potentially toxic elements. An enrichment of elements was only manifested for Zn and Cu in all spiders, for Cd, Ca, V, and La in few spiders and zoophagous beetles.

INTRODUCTION

Invertebrate animals contribute a substantial part to the zoocoenoses of forest ecosystems. Annelids and arthropods, which constitute some of the most important invertebrate groups in terrestrial ecosystems, are present in most forest soils in numbers exceeding 100,000 individuals/m^2.[1–4] These enormous population densities, associated with high species numbers and biomass, demonstrate the potential impact of invertebrate animals on the element budget of forest ecosystems.

Apart from a few exceptions, systematic analysis of element concentrations and element fluxes in terrestrial ecosystems, including a wider spectrum of metals, has been restricted in the past to the inventory of elements in soil and vegetation. The community of invertebrate animals, integrated in the element budget of coniferous and deciduous forests as heterotrophic consumers, has almost totally been excluded. This is in spite of the fact that they contain a pool of macronutrients and essential or potentially toxic trace metals that is transmitted through food webs and finally, via decomposition, gains access to the nutrient pool of autotrophic plants.

In many cases this exclusion is due to the lack of knowledge about the structure and dynamics of the invertebrate animal community in terrestrial ecosystems. Investigations on the taxonomic, spatial, and trophic structure of these communities are thus a prerequisite to understand the contribution of invertebrates to the budget and cycling of elements in forest ecosystems.

The present study reports on element-analytical investigations that have been carried out on the invertebrate fauna of a spruce stand in southern Germany. In former long-term studies, this stand has been well characterized with respect to species spectrum, dominance structure, and trophic relationships of the invertebrate zoocoenosis.[4,5] The purpose of this investigation was to determine element concentrations in the most important invertebrate species of the forest, in relation to their systematic and trophic level. The spectrum of elements included a range of metals, such as macronutrients and essential or potentially toxic trace metals.

In the literature few data are available on the transport of heavy metals through invertebrate food webs.[6–8] Therefore, food chains of invertebrates, beginning with needle litter, spruce needles, bark, and roots of the trees, have been analyzed for possible accumulation of elements. Thereby, data were screened for organisms that might serve as indicators for heavy metal pollution.

SITE DESCRIPTION

Study Area

Investigations on metals in invertebrate animals have been conducted on the dominant fauna of arthropods and annelids in an 80-year-old monoculture of Norway spruce (*Picea abies* L. Karsten). The study site (U1) is situated about 613 m above sea level at the southeast edge of the "Swabian Alb" (southern Germany), approximately 3 km northwest of Ulm (lat. 48°25′N, long. 9°57′E).

Overstocking of trees has led to sparsely developed shrub and herb layers. Only at the edge of the forest or at scattered sites does the vegetation cover consist of, e.g., elder trees (*Sambucus racemosa*) and wood sorrel (*Oxalis acetosella*).

The region of the "Swabian Alb" is under the influence of a cool-temperate subatlantic climate. Mean temperature differences of nearly 20°C between summer and winter demonstrate the continental influence. The mean annual air temperature in U1 is 7.7°C, the mean annual precipitation is 745.4 mm/year. The area is characterized by sedimentary rocks. Soils developed from this bedrock are classified as typical gley-like soils, derived from a well distinct para-brown soil type. The humus form is of the mor-type with well distinguished layers of litter, fermentation, and humification (O_L, O_F, O_H). The pH_{KCl} of the humus horizon ranges from 3.0 to 4.3.

Except for Pb, the U1 site is relatively uncontaminated with metal pollutants. The area is influenced predominantly by winds from the sparsely settled and industrialized peripheral region of the "Swabian Alb." The industrial zone of Ulm is situated outside of the prevailing wind direction. Therefore, metal emissions of industrial plants — in addition to the deposition of elements from atmospheric long-range transport — are not to be expected. This assumption is confirmed — except for Pb — by low concentrations of potentially toxic elements in the humus layer of U1, a zone that is characterized by a high potential for the fixation of heavy metals. The concentrations of Zn, Cd, and Cu in the humus layer of U1 are far below the metal levels of forest ecosystems that are subject to substantial aerial deposition of metals from smelters or from atmospheric long-range transport (Table 1).[9,10] In the humus layer of U1 Zn, Cd, and

Table 1. Concentrations (ppm) of Zn, Pb, Cu, and Cd in the Humus Layers (O_L, O_F, O_H) of U1 and Various Forest Ecosystems with Different Degrees of Metal Contamination

	U1			Walker Branch[a]		Haw Wood[b]			Midger Wood[c]		Bärhalde[d]
	O_L	O_F	O_H	O_L	O_F/O_H	O_L	O_F	O_H	O_L	O_F/O_H	$O_L/O_F/O_H$
Zn	90	59	55	56	59	644	1450	2348	67	182	140
Pb	15	65	116	31	37	293	1027	1808	31	85	134
Cu	6	16	18	—	—	44	67	88	4	14	36
Cd	0.2	0.5	0.2	0.3	0.6	6	22	52	0.8	1.4	3

[a] Relatively uncontaminated mixed deciduous forest in eastern Tennessee, U.S.[11]
[b] Oak-hazel wood in southeastern England with substantial aerial deposition of metals from a smelter.[9]
[c] Uncontaminated oak-hazel wood in southeastern England.[9]
[d] Spruce stand in the southern "Black Forest," affected by atmospheric long-range transport emissions of heavy metals.[10]

Cu reach levels of relatively uncontaminated forest ecosystems in the U.S. and in England.[9,11] Relatively high amounts of Pb are found in the litter layers of the U1 site. The Pb-burden originates from the emissions of a highway which is 2 km away and in the prevailing wind direction.

The Invertebrate Fauna of the Study Area

The Taxonomic Structure of the Invertebrate Community

Contrary to the low diversity of the vegetation, U1 is characterized by a rich community of invertebrate animals. The most important invertebrate metazoa, living in terrestrial ecosystems, are Nematoda, Gastropoda (Pulmonata), Annelida, Crustacea (Isopoda), and other Arthropoda.

In soil-acidic coniferous forests, with insufficient saturation of bases, shell-building Gastropoda, Isopoda, and also Diplopoda and Lumbricidae are present in low population densities. The species and individual spectrum of invertebrate animals is dominated by Nematoda, Enchytraeidae, numerous Insecta, and Arachnida. Investigations with quantitative sampling methods have shown that the dominance structure of these important taxonomic groups of annelids and arthropods is determined by a few species. These species occur every year in high population densities and are characteristic for the ecosystem "spruce stand" (Figure 1).[4]

The Trophic Structure of the Invertebrate Community

Despite a scarcity of direct knowledge of trophic relationships between invertebrate animals of forest ecosystems, most species of arthropods and annelids of U1 may be associated with well-defined feeding types. Indirect indications of the trophic level results from the structure of the mouth parts, from feeding experiments in the laboratory, and from the analysis of gut contents.

Among edaphic organisms, species involved in decomposition processes of litter are of special importance for element cycling in forests. In U1, decomposition is accomplished mainly by the activities of the most abundant species of Collembola (*Entomobrya nivalis*, *Lepidocyrtus lignorum*, *Tomocerus minor*, *Orchesella flavescens*), Enchytraeidae (*Cognettia sphagnetorum*), oribatid mites (e.g., *Hermannia gibba*, *Platynothrus peltifer*, *Phthiracarus piger*), of acid-tolerant Lumbricidae (*Lumbricus rubellus*) and Diplopoda (*Julus scandinavius*), and, above all, of the larvae of Tipulidae (e.g., *Tipula scripta*), Sciaridae (*Corynoptera trispina*, *Ctenosciara hyalipennis*, *Epidapus atomarius*, *Lycoriella fucorum*), and Cecidomyiidae (Figure 2).

Many of the above specified litter decomposers, especially Collembola, cryptostigmatic mites, Enchytraeidae, and Cecidomyiidae, are temporarily feeding on microphytes, such as fungal hyphae, fungal sporocarps, and algae, that are partly associated with mineralization processes in the humus horizons. Thus, very often saprophagy is difficult to separate from mycetophagous or phytophagous feeding behavior.

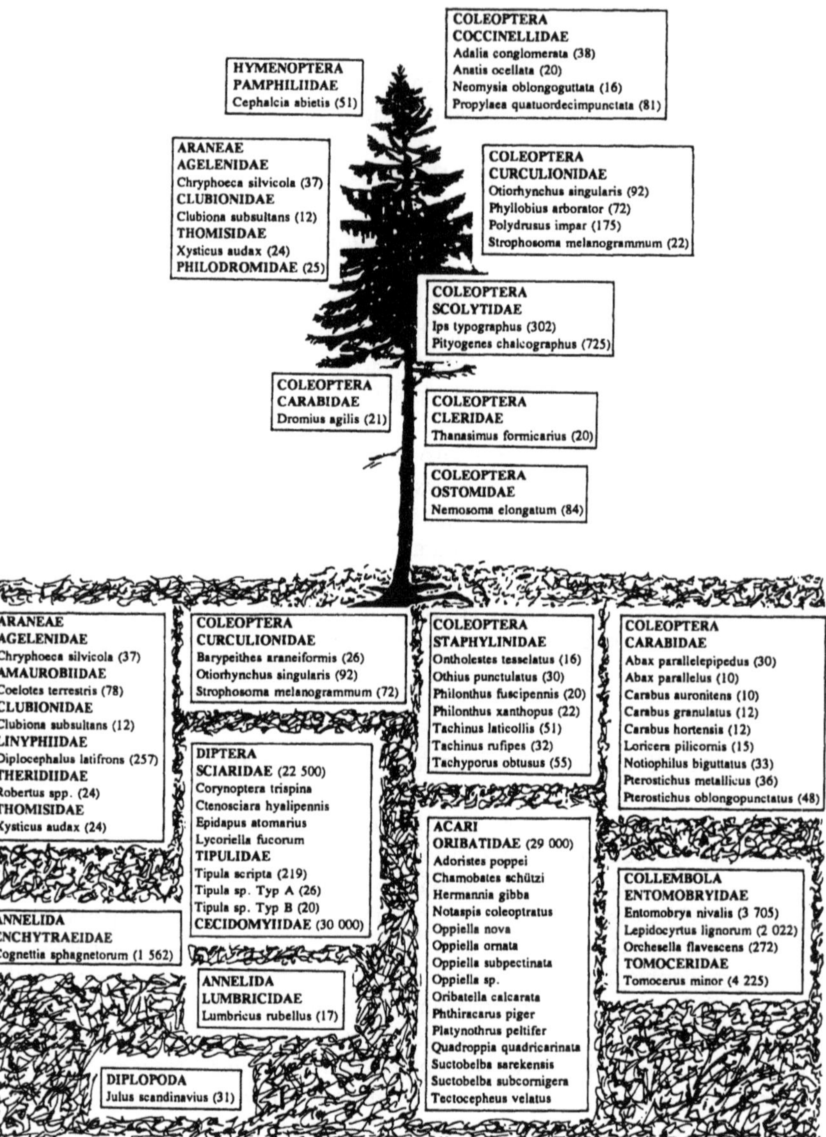

FIGURE 1. The taxonomic and spatial structure of the fauna of dominant invertebrate-species (Annelida, Arthropoda) in U1 — numbers of chemically analyzed individuals are mentioned in parenthesis.

In spruce stands with sparsely developed herb and shrub layers, the phytophagy is almost exclusively represented by consumers of spruce needles (e.g., *Cephalcia abietis*, Hymenoptera, Pamphiliidae), bark (e.g., *Ips typographus, Pityogenes chalcographus*, Coleoptera, Scolytidae) and roots (larvae of weevils, e.g., *Barypeithes araneiformis, Otiorhynchus singularis, Polydrusus impar, Phyllobius arborator, Strophosoma melanogrammum*).

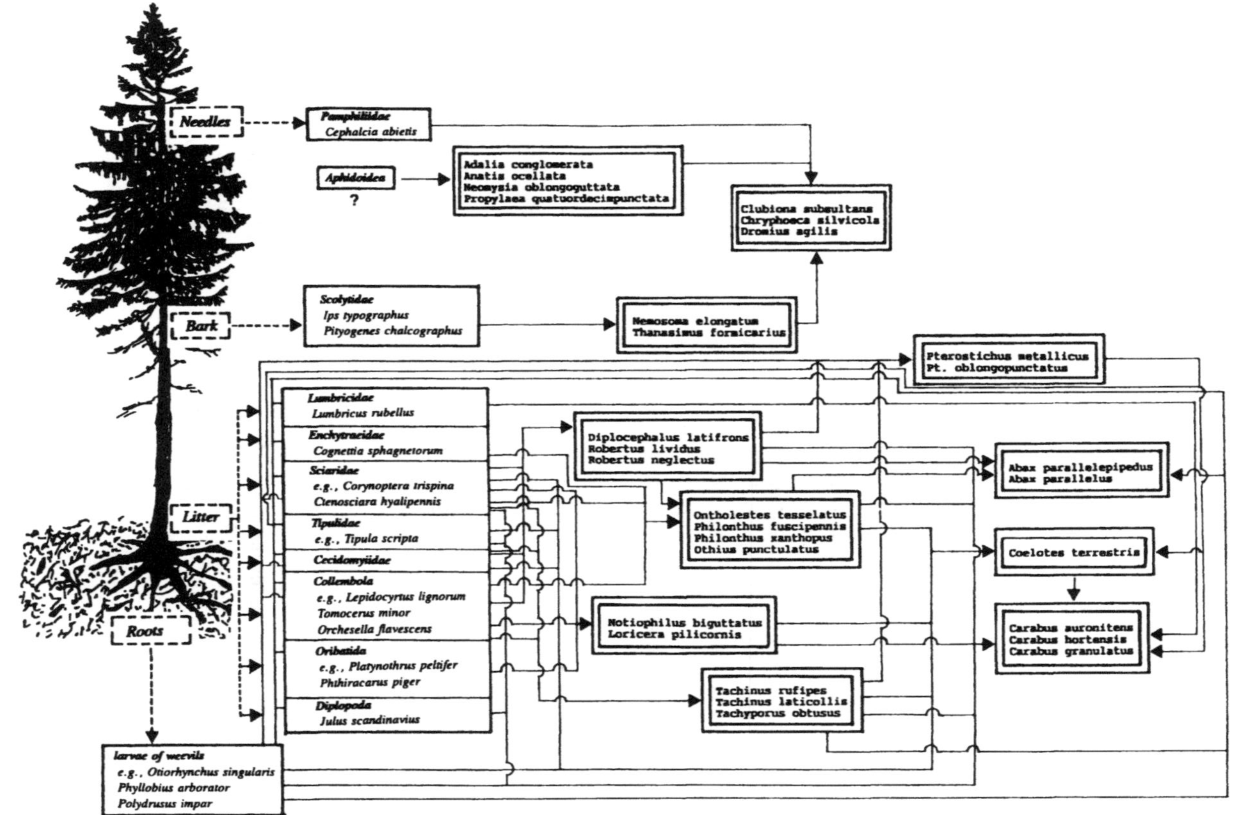

FIGURE 2. The trophic structure of the invertebrate fauna in U1 — food web of the spruce ecosystem with dominant species of *phytophagous*, *detritophagous*, and zoophagous invertebrates.

Because of their high population densities, spiders and predatory beetles (Carabidae, Staphylinidae, and Coccinellidae) are the most important zoophagous invertebrates of the U1 site. In contrast to some phytophagous insects, most predators are polyphagous, as illustrated in the complex invertebrate food web structure (Figure 2).

Diplocephalus latifrons (Araneae, Linyphiidae), the most abundant spider of the soil surface at the U1 site, as well as the species representing the genus *Robertus*, *R. lividus*, *R. neglectus*, (Araneae, Theridiidae) are feeding predominantly on surface-dwelling Collembola and smaller Diptera (Sciaridae, Cecidomyiidae) that got caught in their webs. The substantially larger spider *Coelotes terrestris* (Amaurobiidae) is a top-carnivore. Their prey consists of Diptera, Curculionidae, middle-sized Carabidae (*Notiophilus biguttatus*, *Loricera pilicornis*, *Pterostichus metallicus*, *Pt. oblongopunctatus*) and Staphylinidae (*Ontholestes tesselatus*, *Philonthus fuscipennis*, *Philonthus xanthopus*, *Tachinus laticollis*, *T. rufipes*, *Tachyporus obtusus*).[12] *Clubiona subsultans* (Araneae, Clubionidae), and *Chryphoeca silvicola* (Araneae, Agelenidae) are feeding on arthropods of the crown and trunk region of the trees.

With the exception of *Dromius agilis*, all abundant Carabidae are searching for their prey on the soil surface. *Notiophilus biguttatus* and *Loricera pilicornis* specialize in capturing Collembola.[13,14] The prey spectrum of the genus *Pterostichus* consists of smaller Staphylinidae (*Tachinus* spp., *Tachyporus* sp.), Araneae (*D. latifrons*, *Robertus* spp.), Diptera (Sciaridae, Cecidomyiidae), Collembola, and Curculionidae.[15] A similar variety of prey-objects, including additionally larger-sized Staphylinidae (*Ontholestes tesselatus*, *Philonthus* spp.), Carabidae (*N. biguttatus*, *L. pilicornis*, *Pterostichus* spp.), Diptera (Tipulidae), and Annelida is characteristic for *Abax* spp. and *Carabus* spp.[16,17]

In laboratory tests *Philonthus fuscipennis* prefers less sclerotized insects.[18] In the spruce stand Collembola, Diptera, Enchytraeidae, and smaller Araneae belong to the prey spectrum of *Philonthus* spp. and *Ontholestes tesselatus*. According to gut-analysis and serological precipitation-tests, *Othius punctulatus* (Coleoptera, Staphylinidae) is feeding on Enchytraeidae, Oribatida, Collembola, and Diptera.[17] According to Lipkow, the smaller-sized Staphylinidae (*Tachinus* spp. and *Tachyporus obtusus*) prefer Diptera and Collembola (Figure 2).[19]

Close feeding relationships exist between phloeophagous insects and zoophagous beetles of the trunk region. Immature and adult animals of *Thanasimus formicarius* (Coleoptera, Cleridae) and *Nemosoma elongatum* (Coleoptera, Cleridae) are living carnivorously below the bark of scolytid-inhabitated trees. They feed on larvae and adults of *I. typographus* and *P. chalcographus*.

MATERIAL AND METHODS

Material

Because of the lack of knowledge of the taxonomic and biological structure of the nematod-fauna in U1, element-analytical investigations have been restricted to the significant groups of Arthropoda and Annelida in the spruce stand.

Great importance has been linked to Lumbricidae, Enchytraeidae, Araneae, Oribatida, Diplopoda, Collembola, Coleoptera, Diptera, and Hymenoptera.

Among these groups, species, dominant in individual numbers or biomass, have been analyzed for their metal contents (see Figure 1). Altogether, about 100,000 individuals of invertebrate species have been analyzed for essential or potentially toxic metals.

Methods

Sampling of Invertebrate Animals

For the representative sampling of arthropods and annelids, different methods have been employed, which had proven to be efficient in former studies.[20]

The arthropod fauna of tree-crown and trunk region was captured by tree-silhouettes, equipped with polyethylene sampling boxes at different heights. Soil photoeclectors and pitfall traps, following the pattern of Barber, served to sample the surface-dwelling meso- and macrofauna, especially Insecta, Araneae, and Diplopoda.[21] Scolytidae and their predators were captured with pheromone traps, using Pheroprax and Chalcoprax (Merck, Darmstadt) for *I. typographus* and *P. chalcographus*, respectively. Edaphic Annelida, Araneae, and Oribatida were extracted from litter or soil samples, using the methods of O'Connor, MacFadyen, Kempson, and Dunger.[22–25]

During the capture period (April to November 1986 and 1987), the sampling containers were emptied daily in order to obtain living individuals. To avoid the falsification of element concentrations owing to gut contents, Arthropoda and Annelida were stored alive in the laboratory for 2 to 3 d.

Sampling of Plant Material

Needle litter and spruce roots were collected in monthly intervals during the growing season of 1986 and 1987. In order to obtain representative material, the sample sites were determined within a randomized sample screen.[20] Needle litter was taken from the litter and fermentation layer of the mor-humus-horizons. Fine roots (diameter: 1 to 2 mm) and very fine roots (diameter: <1 mm) were excavated from the humification layer, the zone with most of the spruce roots and their feeders.

For the determination of element concentrations in spruce needles, five trees were cut down in October 1987. As shown in several studies, the concentrations of elements in spruce needles change with the position of the branches in the canopy.[26] To arrive at representative samples of spruce needles, branches from the first seven whorls were collected. Needles were sampled and separated according to age (1, 2, and 3-years-old). Needles of the same age were pooled for every tree.

Pieces of bark, the feeding substrate of scolytids, were cut from the five trees, just below the tree crown. This was done with a ceramic knife (Boker, Solingen) in order to prevent metal contamination during sample preparation.

Sample-Preparation Technique

Animals were killed by freezing. Dust particles and other surface contaminations, visible under 20× power, were removed from the waxy cuticle with an artist's brush. Dirt, sometimes adhering some carbid beetles, was removed in an ultrasonic bath with distilled water. In the same manner spruce roots were freed from soil particles.

After lyophilization and homogenization the samples were ashed in teflon autoclaves (Bergmann, Tübingen) using either concentrated (65% v/v) nitric acid (Suprapur, Merck), or a combination of nitric acid and hydrofluoric acid (HNO_3: HF = 9:1).[27]

Because of the small size of many invertebrate species, individuals of one or more species were pooled for analysis. This was especially necessary for *C. sphagnetorum*, most Araneae, all Collembola, Sciaridae, Cecidomyiidae, Hymenoptera, and several Coleoptera. The number of individuals per sample was kept as low as possible to obtain a number of replicates sufficient for the statistical analysis of the data. The determination of element concentrations in single individuals was conducted with *Coelotes terrestris* (Araneae), *Abax parallelepipedus*, *A. parallelus*, *Anatis ocellata*, *Carabus auronitens*, *C. hortensis*, *C. granulatus*, *Pterostichus metallicus*, *Pt. oblongopunctatus*, *Philonthus fuscipennis*, *Ph. xanthopus*, *Ontholestes tesselatus*, *Otiorhynchus singularis* (Coleoptera), *Julus scandinavius* (Diplopoda), *Lumbricus rubellus* (Lumbricidae), and *Tipula* spp. (Diptera).

Analytical Methods

Detection of most elements was made by sequential atomic emission spectroscopy with inductively coupled plasma as the excitation source, using a Perkin-Elmer ICP 5500 B.[28] Trace levels of elements were measured by graphite furnace atomic absorption spectrometry (Perkin Elmer: AAS 1100, HGA 400, AS 40). The conditions were similar to those recommended for the stabilized temperature platform furnace.[29] Detailed information about instrumentation and operating parameters are given by Roth-Holzapfel.[20]

Verification of Analytical Methods

A simple way to distinguish methodical errors of the analytical technique from the inherent variability of biological samples is the analysis of Standard Reference Materials (SRM) under comparable conditions. To verify the analytical procedure used for the determination of element concentrations in invertebrates and in spruce tissues, three SRMs of the National Institute of Standards and Technology (Washington D.C.) were used: bovine liver (SRM 1577a), orchard leaves (SRM 1571), and pine needles (SRM 1575). The element concentrations found in our laboratory in the SRMs were in good agreement with the certified values (recovery rates: 90 to 110%). Round-robin testing of invertebrate sample solutions, partly with different methods (e.g., instrumental neutron activation analysis, flame atomic absorption spectrometry) confirmed the results of element determination in the animals.

RESULTS AND DISCUSSION

Spectrum of Elements

In the most abundant and significant species of invertebrate animals of U1, the concentrations of 18 elements (Al, Ba, Ca, Cd, Co, Cr, Cu, Fe, K, La, Mg, Mn, Na, Ni, P, Pb, V, Zn) were determined. These elements were selected for their relevance for the metabolism of invertebrates and their function as potentially toxic substances from anthropogenic sources.

The range of elements analyzed included alkaline and alkaline earths with macronutrient functions (Na, K, Mg, Ca), and metals such as Pb and Cd with well-established toxic effects on invertebrates.[30] Of special interest have been the transition metals (Co, Cr, Cu, Fe, Mn, Ni, V, Zn), since they influence the metabolism as essential or toxic trace metals (survey of literature in Hopkin).[30]

Since the beginning of industrialization, most of these transition metals, including Pb, have been released in Central Europe into the atmosphere in increasing amounts.[31] Depending on particle size, meteorological and orographical factors, even remote ecosystems are affected by the emissions of these metal pollutants.[32] Because of their high filtering capacity for airborne pollutants, coniferous forest ecosystems function as a sink for metal aerosols.[33] Harmful influences on various organisms have been reported by different authors and it is difficult to find totally uncontaminated forest ecosystems in industrialized countries.[34–39]

In this context the increased mobilization of heavy metals and especially of Al from acidic forest soils with increased acid precipitation has to be considered.[40]

As example for an anion (PO^{-4}), which is involved in biochemical reactions, and which thus constitutes a possible partner of metals in biological compounds, the concentrations of P have been additionally determined in invertebrate animals and their food.

Concentrations of Elements in Invertebrate Animals of U1

The analytical investigations on invertebrate animals of U1 confirm the variety of elements, which are typically detectable in biological samples.[41–43] Because of the low degree of environmental contamination by potentially toxic metals at the U1 site, the elemental content of arthropods and annelids reflect relative lack of industrial deposition. The elements detected in forest annelids and arthropods range between 10^{-1} to 10^{5} ppm (= mg/kg dry matter, see Figure 3).

The highest concentrations of elements were found in all invertebrate groups of the spruce stand for macronutrients. Together with P, the alkaline and alkaline-earth metals represent between 2 and 11% of invertebrate dry matter. Hereby, in addition to P, the highest proportion is contributed by Ca in Diplopoda

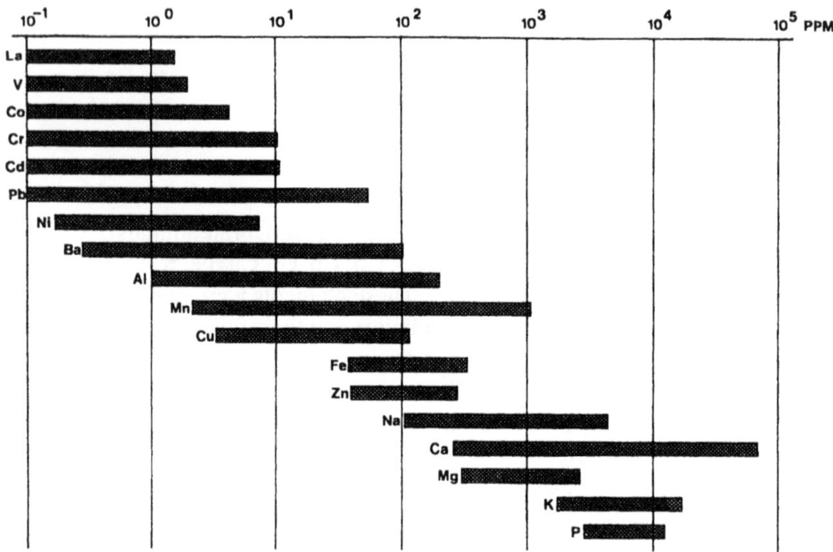

FIGURE 3. Ranges of element concentrations in the invertebrate fauna (Annelida, Arthropoda) of U1.

(>80,000 ppm), Oribatida (>18,000 ppm), and Lumbricidae, and by K in Enchytraeidae (>5,000 ppm), Araneae (>9,000 ppm), and most insects (>15,000 ppm).

All macronutrients are essential parts of animal cells and participate in fundamental metabolic reactions (e.g., formation and transmission of nerve cell potentials, contractions of striated muscles, energy transfer at the cell level, stabilization of ribosome structure, regulation of cell permeability, and viscosity of protoplasm). Thus, the concentrations of Na, Ca, Mg, K, and P exceed, constantly, in Annelida, Arachnida, Diplopoda, and Insecta, levels of 100, 400, 500, 3,000, and 4,500 ppm, respectively (Figure 3).

The highest concentrations of Ca are found in *J. scandinavius* (86,327 ppm), Oribatida (18,847 ppm), *L. rubellus* (13,838 ppm), and in *I. typographus* (23,759 ppm). Apart from *L. rubellus*, the above-mentioned species, including in addition some Diptera, Curculionidae, and Araneae, possess increased levels of Mg (*I. typographus*: 4,544 ppm; *J. scandinavius*: 3,205 ppm; Oribatida, Sciaridae, Cecidomyiidae, *Robertus* spp., *O. singularis*, *P. impar*: >2,000 ppm), and apart from cryptostigmatic mites, including Diptera, Collembola, Araneae, and Hymenoptera, increased levels of P (*J. scandinavius*: 19,982 ppm, *I. typographus*: 16,589 ppm; Sciaridae, Cecidomyiidae, *C. abietis*, *E. nivalis*, *L. lignorum*, *T. minor*, *Robertus* spp., *D. latifrons*, *C. terrestris*, *Ch. silvicola*, *L. rubellus*: ≥10,000 ppm).

In most invertebrates, the concentrations of K vary between 5,000 and 8,000 ppm. Exceptions are *J. scandinavius*, which shows lowest K values (3,126 ppm), as well as species of Araneae and Insecta with remarkably higher levels of potassium. Among these species are Hymenoptera (*Cephalcia abietis*: 28,605 ppm), Collembola (>13,000 ppm), Diptera (Cecidomyiidae: 14,863 ppm, Sciaridae: 9,529 ppm), Scolytidae (>8,788 ppm), and the spiders *C. terrestris*, *D. latifrons*, *Robertus* spp. (>10,000 ppm), *X. audax*, and the Philodromidae (>9,000 ppm).

Contrary to the macronutrients, essential or potentially toxic trace metals make up only 0.05 to 0.2% of invertebrate dry matter. Except for Mn, the levels of the trace metals are far below 1,000 ppm in all investigated species (Tables 2 and 3).

As already noted for Ca and P, the highest concentrations of Mn are found in *J. scandinavius* (1,116 ppm) and in *I. typographus* (626 ppm). Oribatid mites (589 ppm) and Cecidomyiidae (515 ppm) also possess increased levels of Mn (Tables 2 and 3).

The micronutrients Fe and Zn are found in forest annelids and arthropods in nearly identical ranges of concentration (Figure 3). The levels vary for Fe between 61 ppm (*Ph. xanthopus*) and 584 ppm (oribatid mites) and for Zn between 63 ppm (*L. pilicornis*) and 523 ppm (*P. chalcographus*). Additionally, increased concentrations of Zn are found in Araneae (e.g., *C. terrestris*: 514 ppm, *Robertus* spp.: 451 ppm), Lumbricidae (382 ppm) and some Coleoptera (*I. typographus*, *Th. formicarius*, *N. elongatum*: >320 ppm). Fe levels above the average found for Articulata exist also in Enchytraeidae (423 ppm), Collembola (*O. flavescens*: 318 ppm), Diplopoda (*J. scandinavius*: 241 ppm), Araneae (*D. latifrons*: 226 ppm), and in Lumbricidae (*L. rubellus*: 212 ppm).

The concentrations of Cu vary in most insects between 20 and 40 ppm. Levels remarkably above the average are found in *J. scandinavius* (146 ppm), in the springtail *T. minor* (100 ppm), and in many spiders (*C. terrestris*, *Ch. silvicola*, *Robertus* spp.: >120 ppm). On the other hand, Annelida contain considerably lower amounts of Cu (5.3 to 12.1 ppm).

Relatively wide ranges of concentrations of Al and Ba are incorporated in forest invertebrates. The concentrations of Al vary between 1 ppm (*Tipula* sp. Typ A) and 331.5 ppm (*C. sphagnetorum*), the concentrations of Ba between 0.5 ppm (*Ch. silvicola*, *A. conglomerata*) and 101.5 ppm (*I. typographus*). The mean level of Al is 58 ppm, the mean level of Ba 11 ppm. Among others, high concentrations of Al and Ba are found in Curculionidae and Oribatida (Table 2 and 3).

Even lower concentrations of Cr and Ni are found in Annelida and Arthropoda. In most cases, the levels are around 1 ppm. Increased amounts of Cr are found in *C. sphagnetorum* (10.2 ppm), in Oribatida (4.7 ppm), and in all Collembola (>2.9 ppm), except *O. flavescens*.

Lead contents of forest invertebrates range from 0.1 ppm (*Ph. arborator*) to 76 ppm (*C. sphagnetorum*). In contrast to the Annelida (*L. rubellus* has elevated levels of Pb, too), most arthropods contain amounts considerably below 5 ppm. Most insect levels are even below 1 ppm.

In the majority of invertebrates the concentrations of Cd lie below 1 ppm. Essentially higher levels are only shown by Annelida (*L. rubellus*: 8.6 ppm, *C. sphagnetorum*: 3.3 ppm), Araneae (*C. terrestris*: 12 ppm, all other spiders >2.2

Table 2. Concentrations of Essential or Potentially Toxic Elements in Species of Annelida, Araneae, Oribatida,[a] Diplopoda, Collembola, Hymenoptera, and Diptera of the Spruce Stand (U1) — Mean Values (ppm) ± Confidence Interval (ppm), $p \leq 0.05$

	Al	Ba	Cd	Co	Cr	Cu	Fe	La	Mn	Ni	Pb	V	Zn
ANNELIDA													
L. rubellus	6.0± 2.9	2.3± 1.4	8.6±2.8	1.3±0.4	1.2±0.5	5.3± 1.2	212± 70	0.6±0.2	147± 32	3.4± 1.1	57.6±21.4	1.0± 0.6	382 ± 87
C. sphagnetorum	331.5± 357	15.9±15.9	3.3±0.1	0.6±0.2	10.2± 12	12.1± 2.7	423±262	0.9±1.1	356±411	1.3± 0.6	76.0±63.6	3.7± 0.8	233± 12
ARANEAE													
Ch. silvicola	9.7± 3.3	0.5± 0.1	2.5±1.1	0.1±0.1	0.2±0.1	127.5±13.4	159± 35	0.1±0.01	95± 8	0.3± 0.2	1.3± 0.8	1.5± 2	329± 70
C. terrestris	7.2± 2.0	1.5± 0.7	12.0±3.3	0.1±0.01	0.6±0.5	131.3±13.3	163± 15	0.1±0.01	212± 35	0.5± 0.4	0.3± 0.3	0.2± 0.1	514± 22
C. subsultans	10.2± 1.2	1.3± 0.1	2.8±0.5	0.2±0.1	0.4±0.2	85.0± 5.1	152± 11	0.1±0.01	73± 14	2.3± 1.1	1.8± 0.2	0.4± 0.1	351± 9
D. latifrons	23.3± 5.3	3.0± 1.0	3.1±0.3	0.3±0.1	1.1±0.6	87.6±28.8	226± 27	0.3±0.1	103± 29	1.7± 1.1	4.7± 1.4	0.6± 0.3	328± 28
Philodromidae	16.7± 3.3	2.0± 0.2	2.4±0.5	0.1±0.01	0.4±0.2	65.0± 6.0	180± 27	0.4±0.2	96± 8	2.6± 1.3	1.2± 0.3	0.5± 0.3	266± 86
Robertus spp.	45.0	11.0	2.2	0.1	0.5	124.0	192	0.1	248	0.9	4.4	0.1	451
X. audax	22.3± 7.1	3.3± 3.6	2.5±0.5	0.1±0.1	0.2±0.1	81.3±21.0	108± 24	0.3±0.4	122± 19	0.7± 0.1	0.9± 0.2	0.4± 0.1	264± 51
ORIBATIDA	76.7± 2.9	38.4± 2.7	1.2±0.1	0.5±0.05	4.7±0.9	33.7± 2.2	584± 6	0.5±0.2	589± 13	1.9± 0.8	38.6± 1.8	0.2± 0.1	316± 47
DIPLOPODA													
J. scandinavius	177.7±93.8	21.8± 4.1	0.9±0.4	0.3±0.2	1.3±0.5	145.9±23.7	241± 52	0.5±0.3	1116±268	1.3± 0.6	5.5± 1.2	1.1± 0.5	212± 52
COLLEMBOLA													
E. nivalis	27.1± 9.8	3.2± 0.6	0.7±0.2	1.6±0.8	2.9±0.8	36.1± 2.8	84± 5	0.1±0.01	78± 6	2.0± 1.1	9.1± 2.3	0.5± 0.2	193± 22
L. lignorum	52.3±13.7	12.9± 7.7	1.0±0.5	1.0±0.5	3.0±2.8	49.7± 6.3	136± 9	0.1±0.1	216± 61	0.5± 0.1	5.2± 3.6	0.1± 0.1	161± 15
O. flavescens	41.0	6.4	0.1	0.6	0.6	57.0	318	<0.1	76	2.7	1.3	<0.1	102
I. minor	74.4± 8.9	4.4± 1.1	0.4±0.2	0.5±0.1	3.4±0.5	99.7± 6.4	130± 25	0.1±0.1	117± 8	1.8± 0.9	7.5± 2.6	0.1± 0.1	133± 7
HYMENOPTERA													
C. abietis	20.5± 7.4	4.5± 0.8	0.1±0.01	0.1±0.01	0.5±0.1	17.0± 1.7	77± 10	0.2±0.2	33± 5	0.8± 0.4	0.8± 0.2	0.3± 0.1	116± 3
DIPTERA													
Sciaridae	22.9± 3.4	15.4± 2.1	2.3±0.7	0.3±0.1	0.9±0.1	32.4± 1.1	145± 12	0.2±0.05	26± 5	0.3±0.01	3.1± 0.2	0.2± 0.1	165± 10
Cecidomyiidae	8.6± 2.7	4.8± 0.9	5.5±0.5	0.8±1.0	1.0±0.3	35.8± 7.7	92± 5	0.2±0.1	515± 61	1.2± 0.7	4.7± 2.1	0.2± 0.2	233± 33
T. scripta	8.7± 2.9	9.6± 2.6	0.4±0.1	0.2±0.1	1.1±0.3	22.1± 1.6	111± 8	0.5±0.2	19± 4	1.9± 0.8	0.5± 0.2	0.5± 0.2	115± 9
Tipula sp. Typ A	1.0± 0.1	1.5± 0.3	0.5±0.3	0.1±0.01	1.0±0.9	23.1± 3.5	110± 13	0.1±0.02	16± 3	0.3± 0.2	0.5± 0.4	0.1± 0	115± 21
Tipula sp. Typ B	14.9± 1.5	1.6± 1.6	0.8±0.4	0.2±0.1	1.4±1.0	26.0± 2.6	98± 1	0.3±0.02	13± 2	3.9± 5.2	0.4± 0.1	1.3± 0.1	149± 21

[a] Element concentration of Oribatida are related to the species mentioned in Figure 1.

Table 3. Concentrations of Essential or Potentially Toxic Elements in Species of Coleoptera of the Spruce Stand (U1) — Mean Values (ppm) ± Confidence Interval (ppm), $p \leq 0.05$

	Al	Ba	Cd	Co	Cr	Cu	Fe	La	Mn	Ni	Pb	V	Zn
CARABIDAE													
A. parallelepipedus	62.8±14.4	0.9±0.2	0.9±0.3	0.1±0.05	0.4±0.1	14.0± 1.0	77±10	0.2±0.05	31±10	0.8±0.3	1.9±0.7	0.2±0.1	115± 6
A. parallelus	88.3±48.1	1.3±0.5	0.3±0.1	0.1±0.05	0.3±0.1	15.3± 3.8	88±31	0.2±0.1	59±37	0.5±0.2	0.9±0.2	0.2±0.2	147±20
C. auronitens	36.7±34.6	0.8±0.3	0.5±0.3	0.2±0.1	0.2±0.1	26.3±11.1	68±28	0.1±0.05	56±26	0.6±0.3	0.4±0.2	0.1±0.05	121±12
C. hortensis	42.8±23.9	4.3±1.1	2.3±0.7	0.1±0.1	0.4±0.2	27.7± 4.3	101±25	0.2±0.1	90±29	1.1±0.3	1.1±0.5	0.1±0.05	153±23
C. granulatus	44.3±17.3	0.8±0.2	1.4±0.4	0.2±0.1	0.4±0.2	30.8± 8.5	88±14	0.3±0.2	53±18	1.0±0.3	0.4±0.3	0.2±0.1	126± 8
D. agilis	18.0	2.0	0.6	0.2	0.2	20.0	107	0.2	4	3.1	0.4	0.2	144
N. biguttatus	75.0	2.7	0.1	0.1	0.4	21.0	89	0.1	67	0.7	0.8	1.3	99
Pt. metallicus	55.7±22.7	1.2±0.4	0.4±0.1	0.2±0.1	0.3±0.2	18.3± 2.8	70±11	0.3±0.2	40± 8	0.7±0.3	0.7±0.3	0.1±0.05	158±12
Pt. oblongopunctatus	91.1±27.1	2.5±0.8	0.6±0.2	0.1±0.05	0.4±0.4	24.0± 2.8	103±16	0.2±0.1	54±10	0.4±0.2	1.4±0.4	0.3±0.2	132±12
L. pilicornis	100.5±65.8	2.1±0.1	0.1±0.05	0.1±0.05	0.2±0.1	16.0± 4.2	114±32	0.1±0.05	25± 5	0.8±0.1	1.9±2.3	0.8±0.1	63± 1
CLERIDAE													
Th. formicarius	23.0	16.0	2.3	0.1	1.6	63.0	78	0.4	157	1.2	0.3	0.1	322
COCCINELLIDAE													
A. ocellata	16.4± 7.2	1.0±0.5	0.5±0.1	0.5±0.5	0.3±0.1	18.0± 1.8	94±15	0.3±0.1	92±23	0.5±0.4	1.2±0.5	0.6±0.1	156±22
P. quatuordecimpunc.	19.0± 3.4	1.5±0.5	0.3±0.1	0.2±0.1	0.3±0.1	18.2± 0.9	83± 6	0.1±0.05	56± 4	0.9±0.9	0.9±0.2	0.3±0.2	172± 6
N. oblongoguttata	18.5± 4.9	1.1±0.8	0.3±0.1	0.2±0.02	0.1±0.01	19.3± 1.1	91± 3	0.1±0.01	73±17	1.1±1.3	1.3±0.1	0.4±0.2	150±16
A. conglomerata	12.5± 4.9	0.5±0.05	0.6±0.1	0.2±0.05	0.1±0.05	21.5± 3.5	99±28	0.1±0.05	72± 6	1.2±0.9	1.9±0.5	0.7±0.2	186±36
CURCULIONIDAE													
P. imper	56.6± 9.8	17.6±4.4	0.8±0.1	0.3±0.1	0.3±0.1	43.5± 2.3	123± 6	0.3±0.2	188±10	1.4±0.2	0.9±0.3	0.2±0.1	167± 8
Ph. arborator	38.5± 5.0	1.6±0.1	0.3±0.05	0.8±0.05	0.6±0.6	25.5± 0.4	111± 7	0.6±0.1	243± 8	1.8±0.6	0.1±0.01	0.1±0.01	221±10
S. melanogrammum	160.0	67.0	0.4	0.4	2.6	39.5	166	0.5	291	2.6	6.0	1.0	209
O. singularis	271.8±51.4	36.3±4.8	0.6±0.1	0.2±0.05	1.0±0.2	19.4± 1.5	192±26	0.5±0.2	128±10	1.8±0.5	2.9±0.4	0.6±0.1	103± 4
B. aranciformis	127.0	67.0	0.3	0.3	1.3	33.3	137	2.7	383	0.7	8.0	0.1	127
OSTMIDAE													
N. elongatum	126.0	4.0	1.6	0.2	0.8	94.0	190	0.5	34	4.4	2.0	0.2	393
SCOLYTIDAE													
P. chalcographus	152.7± 5.3	33.1±1.3	0.7±0.1	0.5±0.05	0.9±0.1	57.7± 1.2	197± 8	0.3±0.1	111± 2	0.6±0.1	4.2±0.4	0.5±0.1	523±10
I. typographus	80.3±30.6	101.5±9.0	0.8±0.2	0.6± 0.1	0.9±0.1	43.1± 2.2	130±38	0.5±0.2	626±80	1.0±0.3	2.2±0.8	0.5±0.2	377±17
STAPHYLINIDAE													
T. rufipes	51.0	2.3	0.5	0.1	0.4	21.0	184	0.7	51	0.9	0.5	0.4	181
T. laticollis	13.0	1.5	0.6	0.1	0.7	29.0	106	0.3	39	0.1	0.3	0.2	190
Ph. fuscipennis	13.5±10.6	0.9±0.1	0.2±0.1	0.1±0.04	0.7±0.2	23.5±13.4	64± 6	0.8±0.7	31± 1	1.2±1.5	0.3±0.1	0.4±0.1	227±29
Ph. xanthopus	3.0± 1.0	3.3±0.1	3.3±0.1	0.1±0.01	1.9±2.6	18.5±10.6	61± 1	0.4±0.05	25± 2	1.1±0.7	1.0±0.5	1.4±1.8	197±42
O. tesselatus	9.9± 1.9	2.1±1.0	0.4±0.2	0.1± 0.1	0.3±0.2	19.6± 2.0	82±10	0.2±0.1	30± 7	0.1±0.05	1.9±2.1	0.1±0.04	179±16
T. obtusus	13.0	2.0	0.5	0.1	1.1	32.0	85	0.1	34	1.3	8.5	0.1	111
O. punctulatus	43.0	10.0	0.5	0.1	2.7	16.0	81	0.5	38		1.1	0.1	238

ppm), Diptera (Sciaridae: 2.3 ppm, Cecidomyiidae 5.5 ppm), some zoophagous Carabidae (*C. hortensis*: 2.3 ppm), and by the predators of Scolytidae (*Th. formicarius*: 2.3 ppm, *N. elongatum*: 1.6 ppm).

Among the elements investigated, Co, V and La show the lowest concentrations in all representatives of invertebrates. More than 50% of the species contain levels ≤0.2 ppm, more than 80% levels ≤0.5 ppm. Higher amounts (>1 ppm) are only found for V in Annelida (*C. sphagnetorum*), Araneae (*Ch. silvicola*), Coleoptera (*N. biguttatus, Ph. xanthopus*), Diptera (*Tipula* spp.), for Co in Collembola (*E. nivalis*) and Lumbricidae as well as for La in Curculionidae (*B. araneiformis*).

Factors Influencing the Elemental Composition of Invertebrate Animals

For the most part, the elemental composition of forest invertebrates is influenced by species-specific morphological structures, or has to be evaluated as species- or group-specific adaptations to physiological processes or ecological life strategies.

An example for morphological structures is the participation of metals in the formation of exoskeletons of arthropods. Especially high concentrations of macronutrients are found in animals with mineral-containing exoskeletons.[44] Thus, at the U1 site, *J. scandinavius* (Diplopoda) shows, e.g., the highest levels of Ca, Mg, and P. Cuticular "encrustations" with alkaline-earth salts are probably responsible for the high concentrations of Ca and Mg in *I. typographus*, an endophloeic bark-beetle, and in oribatids of the spruce stand. Ca-containing "encrustations", as they usually serve crustaceans to strengthen their cuticula, have already been found by several authors in various developmental stages of different insects and mites.[45–47] In contrast to Diplopoda and Scolytidae, the low P content of cryptostigmatic mites indicates a fixation of macronutrients, predominantly as carbonates.

The high levels of alkaline earth metals in *J. scandinavius*, *I. typographus*, and in oribatid mites are related to increased concentrations of Mn, indicating the participation of manganese in the formation of exoskeleton structures. Hopkin similarily refers to similarities in the chemical behavior of Mg, Ca, and Mn: Mn and the alkaline earth metals are attached to class A metals, preferring chemical bonds with oxygen and being part of intracellular type-A-granules, which are found in different terrestrial invertebrates.[30] The assumed analogies between Mn, Ca, and Mg are also verified by the comparison of ionic radii and neurophysiological experiments.[48]

Considerable amounts of Ca and P are also found in edaphic invertebrates, which tolerate a broader range of pH values and which use Ca, predominantly as carbonate or phosphate, to buffer fluctuations of acid concentrations. A member of this group at the U1 site is *L. rubellus*. This earthworm possesses complex calciferous glands, consisting of $CaCO_3$-secretory lobes and nonsecretory pouches, which contain distinct calcite concretions.[49] Together with the chloragog cells, another storage organ for Ca, P, and further metals in Lumbricidae, the calciferous glands are integrated in the pH balance system.[50]

The elemental composition of invertebrates is often influenced by the concentrations of elements in their food. Increased levels of K (and Mg) are, e.g., typical for phytophagous insects, feeding on physiologically active plant tissues.[51] Thus, *C. abietis*, a monophagous hymenoptera feeding on spruce needles, shows at the U1 site the highest concentrations of K. According to Florkin and Jeuniaux, K and Mg play an important part as osmoregulatory active substances in the hemolymph of higher developed insect orders.[52] This fact may represent an evolutionary adaptation to the availability of cations in spermatophytic plants.[53]

Another example of nutritional influences on the elemental composition of animals is found in mycetophagous invertebrates. Fungi represent the greatest part of microbial biomass in terrestrial soils and contain substantial amounts of inorganic nutrients.[54,55] Fungal rhizomorphs and fungal sporocarps accumulate enormous amounts of Ca, K, Na, P, and essential trace elements from the humus layers.[54,56] Thus, Todd and co-workers reported up to 9 mg Ca/g dm in soil-borne fungal hyphae of different forest ecosystems.[57] Many edaphic organisms are feeding, at least partly, selectively on these fungi. Symbiotic organisms, able to release nutrients from fungal mycelia, have already been found in the gut of many soil-living organisms.[55]

The increased macro- and micronutrient concentrations of cryptostigmatic mites, of Collembola, Sciaridae, and Cecidomyiidae are surely due to their feeding of fungal mycelia and confirm the mycetosaprophagy of these animals. However, saprophagous animals such as Sciaridae may eat fungal material perhaps more or less by chance as a natural part of the litter substrate and thus incorporate, involuntarily high levels of P and metal ions.

Within the Collembola, *O. flavescens*, a species which feeds to 90% on macrophytic wastes, contains the lowest amounts of Mg, K, P, Na, Cd, Cr, La, V, Pb, and Zn; see Table 2 and 4). *O. flavescens* is living predominantly in the litter layer (O_L) of the humus horizons.[58] The O_L-layer consists of rather undecomposed needles and possessess low microbial activity. Essentially higher concentrations of macronutrients of Cr, Zn, Pb, and Cd are found in *L. lignorum* and *T. minor* at the U1 site. Both species live mostly in the fermentation and humification layers of the humus horizon. In U1 these zones are strongly intersected with fungal mycelia.

The increased concentrations of Mg, K, P, Cd, Co, Cr, Pb, V, and Zn in the springtail *E. nivalis* may be due to fungal feeding in the litter horizons or to the grazing on layers of algae on tree trunks. This springtail is changing regularly between litter habitat and trunk region.[59]

Above average concentrations of Na, K, and P are also found in spiders of the spruce stand.

The importance of Na for the ionic balance of spider hemolymph was mentioned by Sutcliffe.[60] Just as in other organisms, also in spiders, K and Na participate in osmoregulation.[61] Additionally, the increased levels of Na and K

Table 4. Concentrations of Macronutrients in Species of Collembola of the Spruce Stand (U1) — Mean Values (ppm) ± Confidence Interval (ppm), $p \leq 0.05$

	Mg	Ca	K	P	Na
Entomobrya nivalis	1,118 ± 21	2,227 ± 287	16,846 ± 387	12,251 ± 339	5,484 ± 900
Lepidocyrtus lignorum	1,042 ± 3	3,854 ± 331	14,692 ± 1106	11,973 ± 107	6,768 ± 634
Orchesella flavescens	580	2936	13307	9783	6732
Tomocerus minor	1,148 ± 28	2,419 ± 122	16,116 ± 755	11,652 ± 223	4,916 ± 338

in spiders are most likely related to specific morphological, physiological, or ecological adaptations of this zoophagous arachnid group to their ecological niches. In comparison to the other investigated species, spiders are characterized by silk-producing organs and species-specific venoms. Schildknecht et al. found the potassium salts of nitric acid and dihydrogenphosphoric acid to be essential parts of silk fibers of Araneidae.[62] As hydrophilic substances, they stabilize the adhesive properities of fibers and prevent them from microbial destruction.[63] In addition, K and Na are involved in pumps that participate, at least at Araneidae, in the regulation of silk water content. The same could be valid for web-building spider species of U1 (*Ch. silvicola*, *C. terrestris*, *D. latifrons*, *Robertus* spp.) and therefore explain the high levels of K (and Na) in the spider fauna of U1.

All spiders of U1 kill their prey with neuro-, cyto-, or hemolytic toxins, injected with the chelicerae. According to Mommsen, the venom of *Tegenaria* sp. (Agelenidae) contains enormous amounts of K, exceeding even the levels of K in the hemolymph by a factor of 50.[64]

As Na is the only element occurring in secondary consumers (e.g., Araneae, Carabidae, Staphylinidae) in generally higher concentrations than in primary consumers, a relation between Na concentration and carnivorous feeding behavior of invertebrates is evident.[51] To what extent different physiological demands for Na in phytophagous, saprophagous, and zoophagous invertebrates are really responsible for the differences in the Na concentration of this trophic groups needs to be examined in experimental studies. As Na (and K) are essential for the formation and transmission of nervous excitation, differences in relation to the structure and function of the nervous system, e.g, in the structure and organization-level of receptors and neurons are possible.

Because of their high toxicity, associated with an increased release from anthropogenic sources, Pb and Cd play an important part among the investigated metals.

The concentrations of Pb in forest invertebrates reflect the spatial distribution of Pb in the different components of forest vegetation (Figure 4). Accordingly, high amounts of Pb are found in edaphic animals, especially in organisms living in the fermentation and humification layers of the humus horizons. These zones are known for a high fixation potential of heavy metals in their organic fraction, which is combined with long residence times.[9] Among edaphic animals with high Pb-levels are Annelida (*L. rubellus*, *C. sphagnetorum*), Diplopoda (*J.*

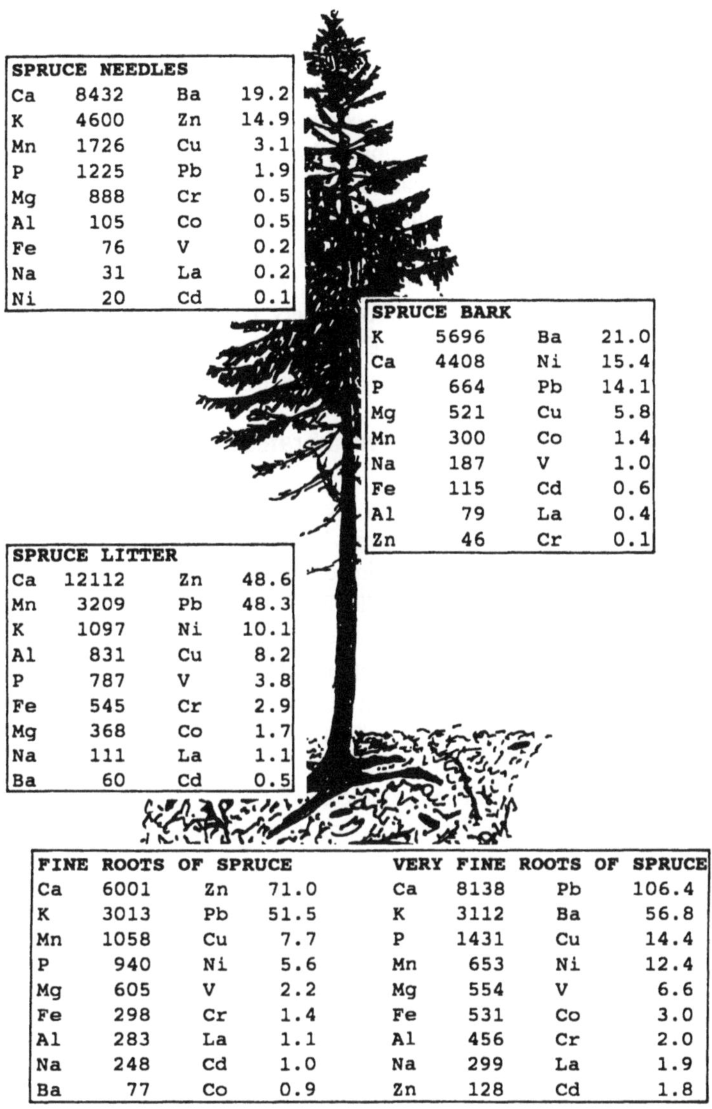

SPRUCE NEEDLES			
Ca	8432	Ba	19.2
K	4600	Zn	14.9
Mn	1726	Cu	3.1
P	1225	Pb	1.9
Mg	888	Cr	0.5
Al	105	Co	0.5
Fe	76	V	0.2
Na	31	La	0.2
Ni	20	Cd	0.1

SPRUCE BARK			
K	5696	Ba	21.0
Ca	4408	Ni	15.4
P	664	Pb	14.1
Mg	521	Cu	5.8
Mn	300	Co	1.4
Na	187	V	1.0
Fe	115	Cd	0.6
Al	79	La	0.4
Zn	46	Cr	0.1

SPRUCE LITTER			
Ca	12112	Zn	48.6
Mn	3209	Pb	48.3
K	1097	Ni	10.1
Al	831	Cu	8.2
P	787	V	3.8
Fe	545	Cr	2.9
Mg	368	Co	1.7
Na	111	La	1.1
Ba	60	Cd	0.5

FINE ROOTS OF SPRUCE				VERY FINE ROOTS OF SPRUCE			
Ca	6001	Zn	71.0	Ca	8138	Pb	106.4
K	3013	Pb	51.5	K	3112	Ba	56.8
Mn	1058	Cu	7.7	P	1431	Cu	14.4
P	940	Ni	5.6	Mn	653	Ni	12.4
Mg	605	V	2.2	Mg	554	V	6.6
Fe	298	Cr	1.4	Fe	531	Co	3.0
Al	283	La	1.1	Al	456	Cr	2.0
Na	248	Cd	1.0	Na	299	La	1.9
Ba	77	Co	0.9	Zn	128	Cd	1.8

FIGURE 4. Concentrations (ppm) of macronutrients, essential and potentially toxic metals in litter and vegetation components of spruce in U1.

scandinavius), Collembola, and Diptera, inhabiting the fermentation and humification layers of the humus horizons (*L. lignorum, T. minor*, Sciaridae, Cecidomyiidae), as well as Curculionidae with rhizophagous larvae, living as adults in the litter layers. Zoophagous spiders, feeding predominantely on Pb-enriched Diptera and Collembola (*D. latifrons, Robertus* spp.) contain also increased Pb-concentrations.

Within these invertebrates, the highest concentrations of Pb are found in Annelida, which are characterized by the chloragog cells (a lumbricid storage organ for ingested Pb) and the mucopolysaccharid layers of the cuticle (a storage organ for surface-adsorbed Pb).[50,65] In contrast to some insects, which remove toxic metals by molting, the extraordinarily high contents of Pb in annelids may be due to the lack of this detoxification mechanism.

Higher levels of Pb are also found in invertebrates populating tree trunks and increasingly exposed to long-term emissions of Pb. This is especially true for endophloeic Scolytidae (*I. typographus, P. chalcographus*), some of their predators (*N. elongatum*), as well as for spiders, hunting for prey on the trunk region of the trees (*Ch. silvicola, C. subsultans*, Philodromidae) and for Collembola, grazing on epiphytic algae (*E. nivalis*).

Accumulation of Elements in Invertebrates

Regionally, the deposition of heavy metals contributes substantially to the anthropogenic burden of the environment. Therefore, intense efforts have been made to screen for organisms that indicate the heavy metal contaminations of their habitat by an accumulation of pollutants.

Investigations on the accumulation of elements in arthropods and annelids have been carried out on important trophic relationships of the spruce stand. Among them are hymenopters, feeding on spruce needles, as well as weevils with rhizophagous larvae. The food chain beginning with spruce bark, with phloeophagous scolytids as primary consumers and carnivores of the trunk region as components of higher trophic levels, has also been investigated. Because of their function in element cycling, special attention was paid to litter decomposers and their predators.

Reliable parameters for the understanding of the behavior of elements on their passage through food chains are the factors of accumulation (A_f). These are calculated by comparing the element concentrations in animals and in their food. $A_f>1.0$ are indicative of element enrichment in invertebrates. $A_f<1.0$ suggest the discrimination of elements from one trophic level to the next.

The results confirm significant differences between primary and secondary consumers of the spruce stand with regard to element accumulation.

Accumulation of Elements in Primary Consumers

Primary consumers at U1 include litter decomposers and invertebrates, which feed on different components of spruce. Therefore, the basis for the calculation of accumulation factors of needle-feeding hymenopters, phloeophagous scolytids, curculionids with rhizophagous larvae, as well as of detritophagous annelids and arthropods are the element concentrations in spruce litter, in needles, bark, and roots of the trees (Figure 4).

Apart from a few exceptions, all primary consumer groups show identical patterns of accumulation. Phytophagous and detritophagous invertebrates accu-

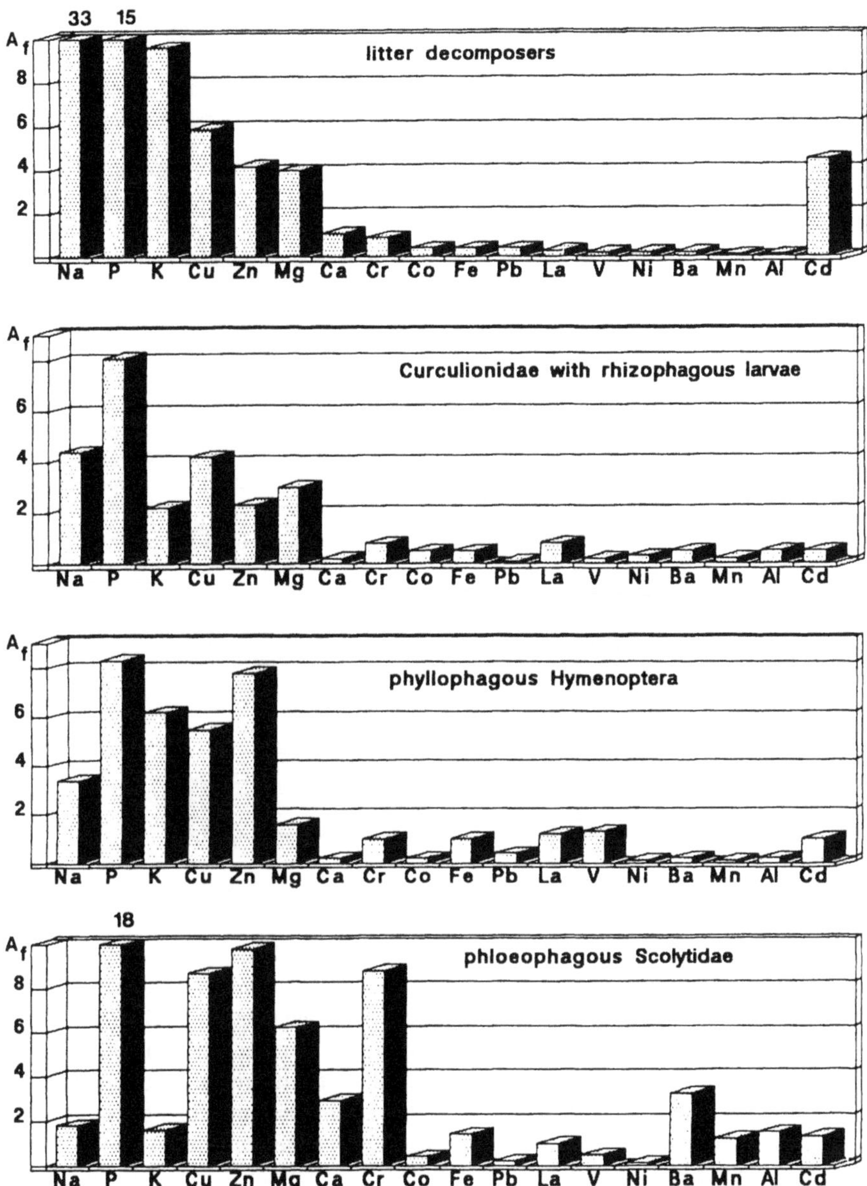

FIGURE 5. Accumulation factors (A_f) of elements in different primary consumer groups.

mulate the macronutrients Na, K, Mg, and P as well as the essential trace metals Zn and Cu (Figure 5). That is also true for trophic groups with high taxonomic diversity, such as litter decomposers (including Lumbricidae, Enchytraeidae, oribatid mites, Diplopoda, Collembola, and Diptera). Except for Cu in Lumbricidae, all detritophagous animals accumulate the above-specified macro-

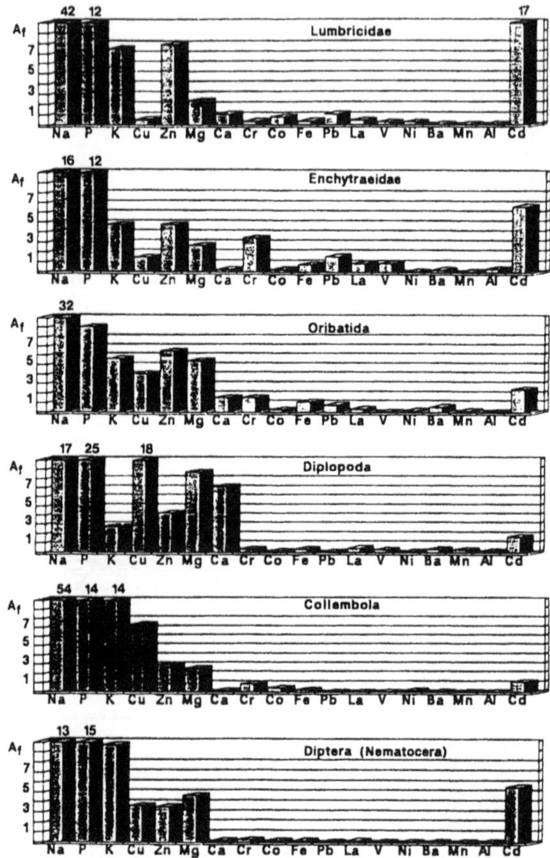

FIGURE 6. Accumulation factors (A_f) of elements in detritophagous groups.

and micronutrients (Figure 6). As investigations on Collembola (as well as on Tipulidae and Curculionidae with rhizophagous larvae) have shown, this pattern of accumulation is detectable down to the species level (Figure 7). Differences between the various primary consumer groups and, within the same trophic level, between taxonomic groups or species exist only in the degree of accumulation. Thus, on the average of the trophic level, the accumulation factors of Na vary between 1.9 (phloeophagous Scolytidae) and 33.0 (litter decomposers; Figure 5). The accumulation factors of P are in the range between 8.1 (Curculionidae with rhizophagous larvae) and 18.0 (phloeophagous Scolytidae). The highest accumulation factors for K ($A_f = 9$) are found in litter decomposers, the highest factors for Mg ($A_f = 6.2$), Zn ($A_f = 9.8$), and Cu ($A_f = 8.7$) in bark feeders.

The accumulation of these macro- and micronutrients down to the species level of primary consumer groups confirms a physiological demand for Na, K, Mg, P, Zn and Cu in invertebrate metabolism that is in excess to the supply via plant nutrition.

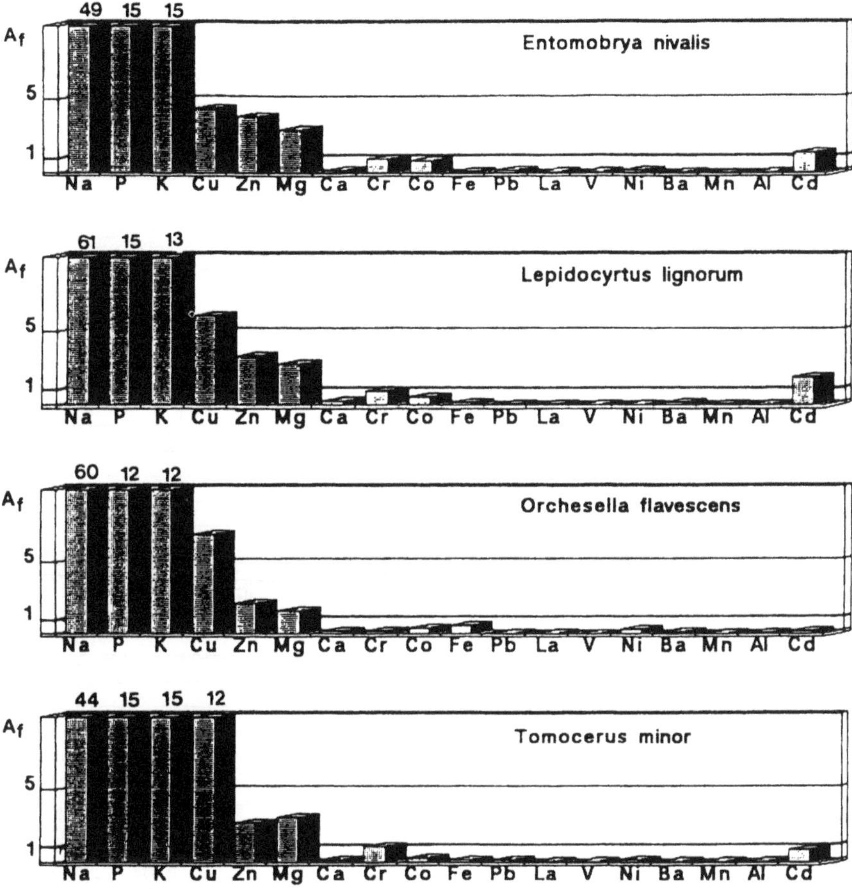

FIGURE 7. Accumulation factors (A_f) of elements in different species of Collembola.

In addition to these macro- and micronutrients, some primary consumers accumulate further elements, such as e.g., Ca, Ba and Cr (bark-inhabiting Scolytidae). These types of accumulation are species-specific and restricted only to a few representatives within the trophic level. Of special interest in this connexion is the accumulation of Cd in litter-decomposers (Figure 6). Lumbricidae, Enchytraeidae, oribatid mites and Diptera may function as accumulative indicators for Cd contaminations in forest ecosystems.

Accumulation of Elements in Secondary Consumers

Spiders and, among the beetles, Carabidae and Staphylinidae belong in U1 to the most important predators of the soil surface. Their prey consists mainly of saprophagous invertebrates and of curculionids with rhizophagous larvae. In most secondary consumers of U1 the accumulation factors vary about 1.0

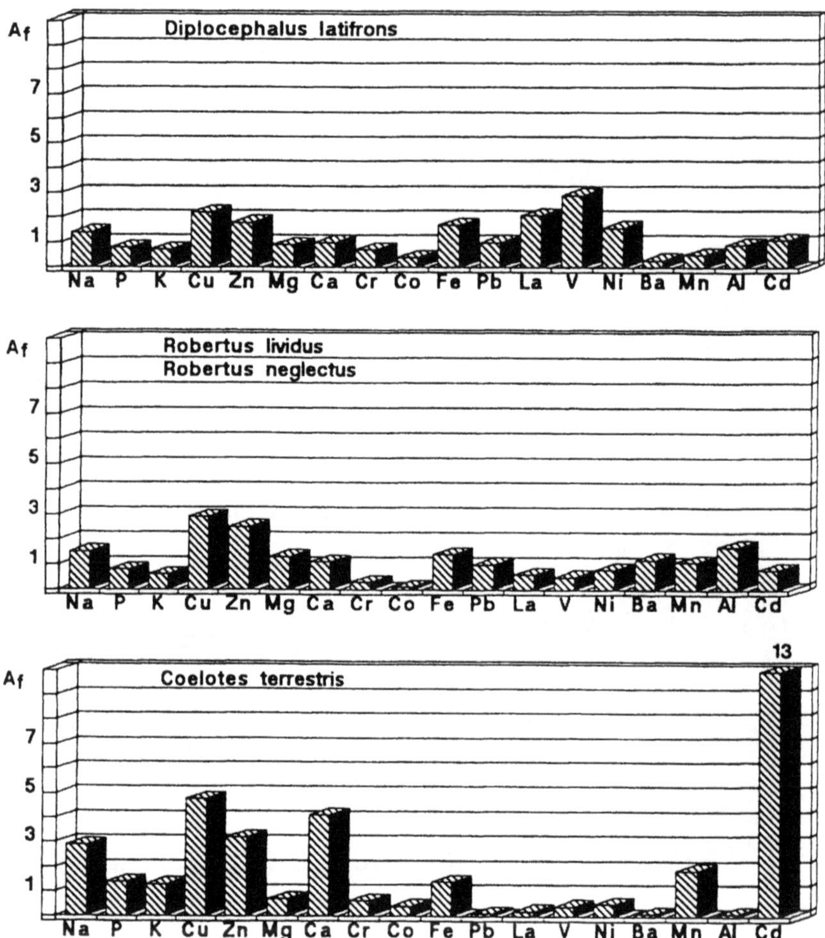

FIGURE 8. Accumulation factors (A_f) of elements in dominant species of Araneae.

(Figure 8, 9, and 10). Therefore, the transfer of elements from primary consumers to secondary consumers is realized mostly without an enrichment of elements. This is true, with few exceptions, for macronutrients and essential and potentially toxic trace metals. The supply of elements in secondary consumers via zoophagous feeding behaviour seems to be sufficient for most carnivorous species.

As mentioned above for secondary consumers, the accumulation factors of most elements vary in the spider species about 1.0 (Figure 8). An enrichment of elements is only manifested for Zn and Cu. This reflects, in the case of Cu, the increased physiological demand of spiders for Cu, functioning as part of the respiratory pigment hemocyanin. In addition to these elements, La and V are

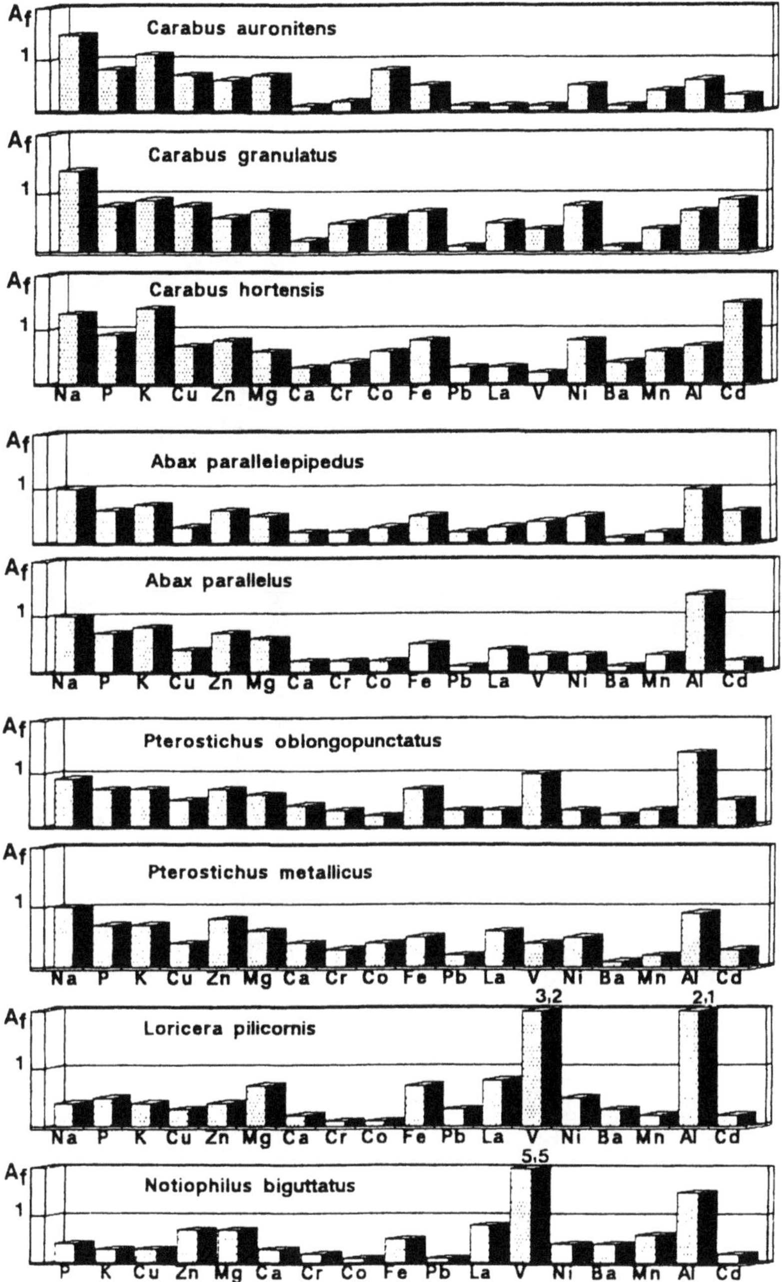

FIGURE 9. Accumulation factors (A_f) of elements in dominant species of Carabidae.

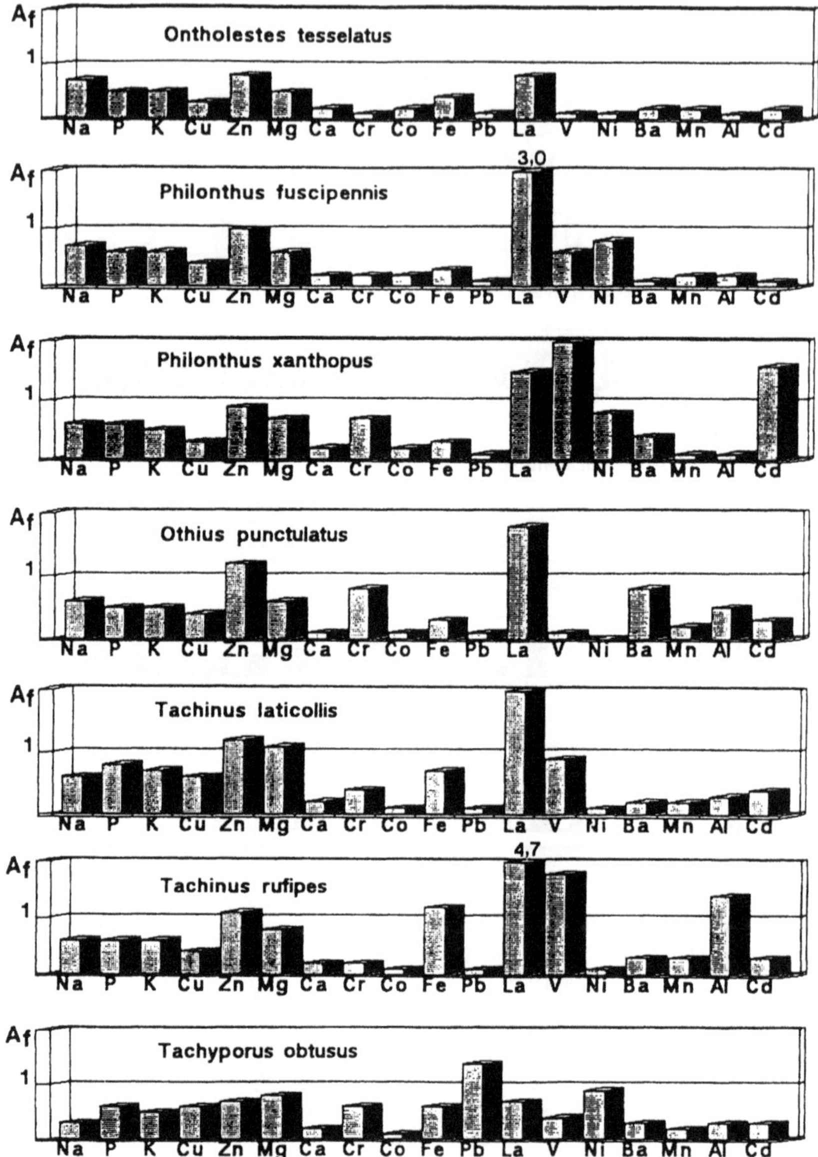

FIGURE 10. Accumulation factors (A_f) of elements in dominant species of Staphylinidae.

accumulated by *D. latifrons*, a species feeding predominatly on Collembola and Diptera. *C. terrestris* accumulates Na, Ca, and especially Cd to a considerable extent (Figure 8). Because of the enormous enrichment of Cd, *C. terrestris* may serve as an indicator for Cd-pollutions of the environment.

A similar behavior with regard to the accumulation of elements has been manifested for the zoophagous beettles of the soil surface. In most species of Carabidae and Staphylinidae, the accumulation factors vary about 1.0. Within carabid beetles a clear enrichment of elements was only manifested for V in the collembolophagous species *L. pilicornis* and *N. biguttatus* and for Al in *L. pilicornis* (Figure 9). Within the rove-beetles La and perhaps V are the only elements that show tendencies for accumulation (Figure 10).

In contrast to other investigations, no accumulation of lead was manifested within the invertebrate food web.[7] Neither in primary consumers nor in secondary consumers of the spruce stand was an accumulation of lead established. This is despite of the fact, that the U1 site is characterized by elevated levels of Pb especially in the humus layers and in spruce bark.

REFERENCES

1. Crossley, D. A., Jr. and K. K. Bohnsack. "Long-Term Ecological Study in the Oak Ridge Area," *Ecology* 41:628–638 (1960).
2. Dindal, D. L. *Soil Biology Guide* (New York: John Wiley & Sons, 1990) p. 1349.
3. Ellenberg, H., R. Mayer, and J. Schauermann. *Ökosystemforschung. Ergebnisse Des Solling-Projektes.* (Stuttgart: Ulmer, 1986) p. 507.
4. Funke, W. "Tiergesellschaften im Ökosystem 'Fichtenforst' (Protozoa, Metazoa — Invertebrata) — Indikatoren von Veränderungen in Waldökosystemen," KfK-PEF Report 9 (1986) pp. 1–150.
5. Roth, M. "Die Coleopteren im Ökosystem 'Fichtenforst'. I. Ökologische Untersuchungen," *Zool. Beitr. N.F.* 29(2):227–294 (1985).
6. Carter, A. "Cadmium, Copper, and Zinc in Soil Animals and Their Food in a Red Clover System," *Can. J. Zool.* 61:2751–2757 (1983).
7. Price, P. W., B. J. Rathcke, and D. A. Gentry. "Lead in Terrestrial Arthropods: Evidence for Biological Concentration," *Environ. Entomol.* 3(3):370–372 (1974).
8. Williamson, P. and P. R. Evans. "Lead: Levels in Roadside Invertebrates and Small Mammals," *Bull. Environ. Contam. Toxicol.* 8:280–288 (1972).
9. Martin, M. H., E. M. Duncan, and P. J. Coughtrey. "The Distribution of Heavy Metals in a Contaminated Woodland Ecosystem," *Environ. Pollut.* 3B:147–157 (1982).
10. Zöttl, H. W. "Heavy Metal Levels and Cycling in Forest Ecosystems," *Experientia* 41:1104–1113 (1985).
11. Van Hook, R. I., W. F. Harris, and G. S. Henderson. "Cadmium, Lead and Zinc Distribution and Cycling in a Mixed Deciduous Woodland," *Ambio* 6:281–286 (1977).
12. Albert, R. "Untersuchungen zur Struktur und Dynamik von Spinnengesellschaften verschiedener Vegetationstypen im Hoch-Solling," *Hochschulsammlung Biologie* 16 (1982).
13. Bauer, Th. "Predation by a Carabid Beetle Specialized for Catching Collembola," *Pedobiologia* 24:169–179 (1982).
14. Ernesting, G. and J. W. Jensen. "Interspecific and Intraspecific Selection by the Predator Notiophilus biguttatus (F.) Concerning Two Collembolan Prey Species," *Oecologia (Berlin)* 33:173–183 (1978).

15. Koehler, H. "Nahrungsspektrum und Nahrungskonnex von Pterostichus oblongopunctatus (F.) und Pterostichus metallicus (F.) (Coleoptera, Carabidae)," *Proceedings of the Sixth Annual Conference of the German Ecological Society* (The Hague: Junk, 1976) pp. 103–111.
16. Thiele, H. U. "Carabid Beetles in Their Environments," in *Zoophysiology and Ecology*, Farner, D. S., Ed. (Berlin: Springer-Verlag 1977) pp. 1–169.
17. Dennison, D. F. and I. D. Hodkinson. "Structure of the Predatory Beetle Community in a Woodland Soil Ecosystem. I. Prey Selection," *Pedobiologia* 25:109–115 (1983).
18. Eghtedar, E. "Zur Biologie und Ökologie der Staphyliniden Philonthus fuscipennis Mannh. und Oxytelus rugosus Gran., " *Pedobiologia* 10:169–170 (1970).
19. Lipkow, E. "Biologisch-ökologische Untersuchungen an Tachyporus-Arten und Tachinus rufipes," *Pedobiologia* 6:140–170 (1966).
20. Roth-Holzapfel, M. "Elementanalytische Untersuchungen an Anneliden und Arthropoden eines Fichtenbestandes," KfK-PEF Report (86), pp. 1–136 (1991).
21. Barber, H. S. "Traps of Cave Inhabiting Insects," *J. Elisha Mitchell Sci. Soc.* 46:259–266 (1931).
22. O'Connor, F. B. "Extraction of Enchytraeid Worms from Coniferous Forest Soil," *Nature (London)* 175:815–816 (1955).
23. MacFadyen, A. "Methods of Investigation of Productivity of Invertebrates in Terrestrial Ecosystems," in *Secondary Productivity of Invertebrates in Terrestrial Ecosystems*, Petrusewicz, C., Ed. (Polish Academy of Sciences, 1962) pp. 17–49.
24. Kempson, D., M. Lloyd, and R. Ghelardi. "A New Extractor for Woodland Litter," *Pedobiologia* 3:1–21 (1963).
25. Dunger, W. and H. J. Fiedler. *Methoden der Bodenbiologie* (Stuttgart: Fischer, 1989) p. 432.
26. Hüttl, R. "'Neuartige' Waldschäden und Nährelementversorgung von Fichtenbeständen (Picea abies Karst.) in Südwestdeutschland," *Freib. Bodenkundl. Abhandl.* 16:1–192 (1985).
27. Kotz, L., G. Kaiser, P. Tschöpel, and G. Tölg. "Aufschluß biologischer Matrices für die Bestimmung sehr niedriger Spurenelementgehalte bei begrenzter Einwaage mit Salpetersäure unter Druck in einem Teflongefäß," *Z. Anal. Chem.* 260:207–209 (1972).
28. Schrader, W., Z. Grobenski, and H. Schulze. "Einführung in die AES mit dem induktiv gekoppelten Plasma (ICP)," *Angew. Atom-Spektroskopie* 28:1–38 (1981).
29. Völlkopf, U. and Z. Grobenski. "A Concept of Interference-Free Graphit Furnace AAS," *Angew. Atom-Spektroskopie* 30:1–15 (1983).
30. Hopkin, S. P. *Ecophysiology of Metals in Terrestrial Invertebrates* (London: Elsevier Applied Science, 1989) p. 366.
31. Nürnberg, H. W., P. Valenta, V. D. Nguyen, M. Gödde, and E. Urano de Carvalho. "Studies on the Deposition of Acid and Ecotoxic Heavy Metals with Precipitates from the Atmosphere," *Fresenius Z. Anal. Chem.* 317:314–323 (1984).
32. Burkitt, A., P. Lester, and G. Nickless. "Distribution of Heavy Metals in the Vicinity of an Industrial Complex," *Nature (London)* 238:327–328 (1972).
33. Ulrich, B. "Deposition von Säure und Schwermetallen aus Luftverunreinigungen und ihre Auswirkungen in Waldökosystemen," in *Metalle in der Umwelt*, Merian, E., Ed. (Weinheim: VCH, 1986) pp. 163–170.

34. Bengtsson, G. and S. Rundgren. "Ground Living Invertebrates in Metal Polluted Forest Soils," *Ambio* 13:29–33 (1984).
35. Coughtrey, P. J., C. H. Jones, M. H. Martin, and S. W. Shales. "Litter Accumulation in Woodlands Contaminated by Pb, Zn, Cd and Cu," *Oecologia (Berlin)* 39:51–60 (1979).
36. Freedman, B. and T. C. Hutchinson. "Effects of Smelter Pollutants on Forest Leaf Litter Decomposition Near a Nickel-Copper Smelter at Sudbury, Ontario," *Can. J. Bot.* 58:1722–1736 (1980).
37. Jackson, D. R. and A. P. Watson. "Disruption of Nutrient Pools and Transport of Heavy Metals in a Forested Watershed Near a Lead Smelter," *J. Environ. Qual.* 6(4):331–338 (1977).
38. O'Neill, R. V., B. S. Ausmus, D. R. Jackson, R. I. Van Hook, P. Van Voris, C. Washburne, and A. P. Watson. "Monitoring Terrestrial Ecosystems by Analysis of Nutrient Export," *Water Air Soil Pollut.* 8:271–277 (1977).
39. Tyler, G. "Effects of Heavy Metal Pollution on Decomposition Rates and Mineralisation Rates in Forest Soils," in *Proceedings of the International Conference on Heavy Metals in the Environment, Toronto, Canada* (Edinburgh: CEP Consultants, 1975) pp. 217–226.
40. Ulrich, B., R. Mayer, and P. K. Khanna. "Deposition von Luftverunreinigungen und ihre Auswirkungen in Waldökosystemen im Solling," in *Schriften aus der Forstlichen Fakultät der Universität Göttingen und der Niedersächsischen Forstlichen Versuchsanstalt* 58 (Frankfurt: Sauerländer's Verlag, 1979) p. 291.
41. Bowen, H. J. M. *Environmental Chemistry of the Elements* (London: Academic Press, 1979) p. 333.
42. Lieth, H. H. F. and B. Markert. *Aufstellung und Auswertung ökosystemarer Element-Konzentrationskataster. Eine Einführung* (Berlin: Springer-Verlag, 1988) p. 193.
43. Adriano, D. C. *Trace Elements in Terrestrial Environment* (Berlin: Springer-Verlag, 1986) p. 531.
44. Reichle, D. E., M. H. Shanks, and D. A. Crossley, Jr. "Calcium, Potassium, and Sodium Content of Forest Floor Arthropods," *Ann. Entomol. Soc. Am.* 62 (1):57–62 (1969).
45. Crossley, D. A., Jr. "Oribatid Mites and Nutrient Cycling," in *Biology of Oribatid Mites*, Dindal, D. L., Ed. (Syracuse: SUNY College of Environmental Science and Forestry, 1977) pp. 71–85.
46. Clark, E. W. "A Review of Literature on Calcium and Magnesium in Insects," *Ann. Entomol. Soc. Am.* 51:142–154 (1958).
47. Gilby, A. R. and J. W. Mc Kellar. "The Calcified Puparium of a Fly," *Insect Physiol.* 22:1465–1468 (1976).
48. Hagiwara, S. and S. Miyazaki. "Ca and Na Spikes in Egg Cell Membrane," in *Cellular Neurobiology*, Hall, Z., R. Kelly, and C. F. Fox, Eds. (New York: Alan R. Liss, 1977) pp. 147–158.
49. Morgan, A. J. and C. Winters. "The Elemental Composition of the Chloragosomes of Two Earthworm Species (Lumbricus terrestris and Allolobophora longa) Determined by Electron Probe X-Ray Microanalysis of Freeze Dried Cryosections," *Histochemistry* 73:589–598 (1982).
50. Morgan, A. J. "The Localization of Heavy Metals in the Tissues of Terrestrial Invertebrates by Electron Microprobe X-Ray Analysis," *Scanning Electron Microsc.* 4:1847–1865 (1984).

51. Schowalter, T. D. and D. A. Crossley, Jr. "Forest Canopy Arthropods as Sodium, Potassium, Magnesium and Calcium Pools in Forests," *For. Ecol. Manage.* 7:143–148 (1983).
52. Florkin, M. and C. Jeuniaux. "Hemolymph Composition," in *The Physiology of Insecta*, Vol. 5, Rockstein, M., Ed. (New York: Academic Press, 1974) pp. 255–307.
53. Mullins, D. E. "Chemistry and Physiology of the Hemolymph," in *Comprehensive Insect Physiology, Biochemistry and Pharmacology*, Vol. 3, Kerkut, G. A. and L. I. Gilbert, Eds. (Oxford: Pergamon Press, 1985) pp. 355–401.
54. Cromack, K., Jr., R. L. Todd, and C. D. Monk. "Patterns of Basidiomycetes Nutrition Accumulation in a Conifer and Deciduous Forest Litter," *Soil Biol. Biochem.* 7:265–268 (1975).
55. Cromack, K., Jr., P. Sollins, R. L. Todd, D. A. Crossley, Jr., W. M. Fender, R. Fogel, and W. Todd. "Soil Microorganisms — Arthropod Interactions: Fungi as Major Calcium and Sodium Sources," in *the Role of Arthropods in Forest Ecosystems*, Mattson, W. J., Ed. (New York: Springer-Verlag, 1977) pp. 78–84.
56. Stark, N. "Nutrient Cycling Pathways and Litter Fungi," *Bioscience* 22:355–360 (1972).
57. Todd, R. L., K. Cromack, Jr., and J. C. Stormer, Jr. "Chemical Exploration of the Microhabitat by Electron Probe Microanalysis of Decomposer Organisms," *Nature (London)* 243:544–546 (1973).
58. Anderson, B. J. M. and J. N. Healey. "Seasonal and Interspecific Variation in Major Components of Gut Contents of Some Woodland Collembola," *J. Anim. Ecol.* 41:359–368 (1972).
59. Allmen, H. and J. Zettel. "Populationsbiologische Untersuchungen zur Art Entomobrya nivalis (Collembola)," *Rev. Suisse Zool.* 89:919–926 (1982).
60. Sutcliffe, D. W. "The Chemical Composition of Haemolymph in Insects and Some Other Arthropods, in Relation to Their Phylogeny," *Comp. Biochem. Physiol.* 9:121–135 (1963).
61. Pulz, R. "Thermal and Water Regulations," in *Ecophysiology of Spiders*, Nentwig, W., Ed. (Berlin: Springer-Verlag, 1987) pp. 26–55.
62. Schildknecht, H., P. Kunzelmann, D. Krauss, and C. Kuhn. "Über die Chemie der Spinnwebe I., " *Naturwissenschaften* 59:98–99 (1972).
63. Tillinghast, E. K. "The Chemical Fractionation of the Orb Web of Agriope Spiders," *Insect Biochem.* 14:115–120 (1984).
64. Mommsen, T. P. "Zusammensetzung und Funktion von Extraintestinaler Verdauungsflüssigkeit und Gift einer Spinne (Tegenaria atrica CL. Koch, Agelenidae)," Dissertation, University of Freiburg, Germany (1977).
65. Fleming, J. R. and K. S. Richards. "Localization of Absorbed Heavy Metals on the Earthworm Body Surface and Their Retrieval by Chelation," *Pedobiologia* 23:415–418 (1982).

12

Metal Tolerance In Plants: Signal Transduction and Acclimation Mechanisms

Jonathan R. Cumming[1] and A. Brian Tomsett[2]

[1] Department of Botany, University of Vermont, Burlington, VT 05405
and
[2] Department of Genetics and Microbiology, University of Liverpool, Liverpool, U.K. L69 3BX

ABSTRACT

The responses of plants to metal exposure have received considerable attention in toxicological and evolutionary contexts. The biochemical and genetic events involved in acclimation responses of plants to metals are less understood. Our intent is to review the responses of plants to metal exposure, propose pathways by which metal "signals" in the environment are transduced to biochemical responses, and to present examples of metal-tolerance mechanisms based on such pathways. Many metals impact cell physiological function, altering cytoplasmic pH, trans-plasma membrane electrical potential, electrical currents, and ion fluxes. Such alterations are known to affect basic cell activity, including transport functions, cytoplasmic enzyme activity, and gene expression, and may activate/induce metal tolerance mechanisms. Exclusion of metals from the cytoplasm may involve changes in the relative proportions of membrane components, altered transport kinetics of ion uptake systems, or the operation of specific ion transport mechanisms in the plasma membrane and tonoplast that maintain cytoplasmic metal levels below toxic concentrations.

The mobility or toxicity of free metal species in the rhizosphere may be altered by cellular exudation of various organic compounds, limiting the challenge to the plant cell. Within the cell, metals may directly alter the activity of cytoplasmic enzymes or may induce *de novo* production of new gene products that are involved in metal tolerance. Metal-chelating organic ligands, including organic acids, phytochelatins, and metallothioneins, may reduce the chemical activity of toxic ionic species in the cytoplasm, thereby reducing the impact of toxic metals on critical metabolic functions. A discussion of signals and responses is designed to draw attention to the molecular basis of metal tolerance metabolism in plants.

INTRODUCTION

The growth and survival of plants in nature depends on their ability to respond to prevailing environmental conditions. Physiological processes and biochemical pathways leading to the maintenance or recovery of cellular homeostasis in cells under stress form the basis of stress-tolerant genotypes. Pathways that allow acclimation to metal stress are not as well known as the physiological and biochemical aberrations that are mediated by metal exposure. This stems from the simplicity of measuring plant responses to a stressor and the difficulty in determining which responses reflect active, tolerance mechanisms. In these pages, we will investigate changes in plant metabolism activated/induced by metal exposure and propose pathways by which plants respond to the presence of metals in the environment and transduce such "signals" into orchestrated biochemical responses conferring metal tolerance. Signals may result from perturbations of cell function mediated by metal exposure or may result specifically from interactions between metal ions and specific metabolic pathways within the cell. Figure 1 highlights a dichotomous pathway for stress-tolerant and stress-sensitive genotypes exposed to any given stress. Stress-tolerant genotypes, through various physiological and biochemical processes responding to the metal signal, alter the impact of the metal on cell function and become resistant to the presence of the metal in the environment. We believe an approach targeting signal transduction will help focus researchers of metal tolerance in plants by drawing attention to the complexity of physiological and biochemical events accompanying metal exposure that are similarly involved in normal plant growth, development, and gene expression.

GENERAL ASPECTS OF PLANT RESPONSE TO STRESS

In Levitt's[1,2] treatise on plant response to stress, two strategies were outlined that form the basis of much of our concepts of plant adaptation and sensitivity to stress. Plants may possess mechanisms leading to the *avoidance* of thermodynamic equilibria with the applied stress or may exhibit *tolerance* of thermodynamic conditions dictated by the prevailing stress.

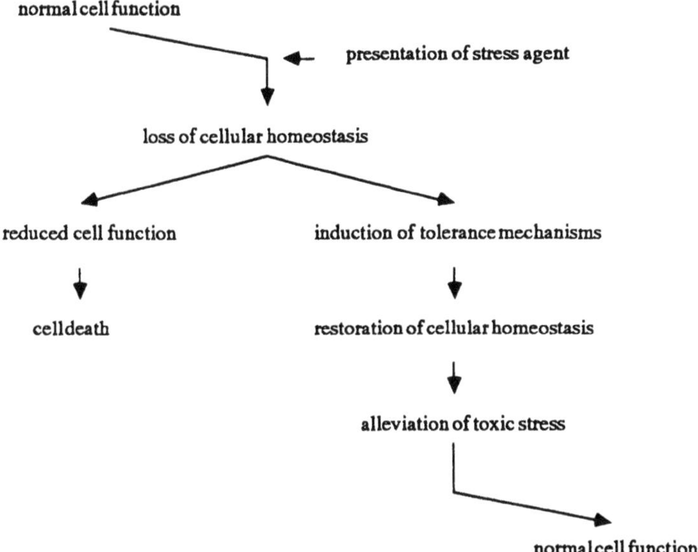

FIGURE 1. Schematic of response of plant function to perturbation and the events leading to the re-acquisition of homeostasis.

It is difficult to imagine a situation wherein a plant, or more specifically the cytoplasm of a plant cell, comes into thermodynamic equilibrium with a metal in the environment. Differences in pH between the soil solution and the cytoplasm and interactions between metals and biological ligands within the cytoplasm would seem to maintain a constant disequilibrium between various ionic species and chelated forms of a metal in the environment and within the cell. More important are the equilibria (1) between free metal ions at the cell surface and bound to membrane constituents, (2) between metals in the environment and within the cytoplasm, (3) between free metal ions in the cytoplasm and metal bound with biological ligands within the cell, and (4) the ability of the plant cell to alter these equilibria as to maintain normal metabolic function. Thus, mechanisms of metal tolerance, such as those outlined in Table 1 and to be discussed at length later in this chapter, function to *avoid* thermodynamic equilibria between the cell and toxic metals in the environment.

Metal Tolerance: Acquired or Constitutive?

Also of interest is the concept of *constitutive* vs. *induced* metal tolerance. Again, our historical precedent has been based on Levitt's teachings and we tend to believe that there are "classes" of genotypes, those that are sensitive, sometimes intermediate, or tolerant to a given stress. These classifications are often based on a "tolerance index," e.g., root growth potential of plants under metal exposure, and miss potential shifts in metabolism noted in Figure 1. As an example of acquired metal tolerance, Cumming et al.[17] found that Al^{3+}-tolerant

Table 1. Strategies and Mechanisms for Avoidance of Thermodynamic Equilibria with Toxic Metals in the Environment

Strategy	Mechanisms
Exclusion	Membrane selectivity
	Altered ion transporter kinetics
	Altered membrane composition
	Compartmentation
	Efflux external environment
	Sequestration in vacuole
Speciation	Extracellular chelation
	Intracellular chelation
	Organic acids and similar ligands
	Phytochelatins
	Metallothioneins

and Al^{3+}-sensitive cultivars of *Phaseolus vulgaris* were *equally* sensitive to Al^{3+} in the *short-term* (Figure 2). However, plants of the tolerant cultivar overcame this initial sensitivity and resumed normal root growth (and, coincidently, shoot growth) under exposure to Al^{3+}. Such patterns are not easily detected, or may not exist, for other metals, but this example reflects the concept that Al^{3+} tolerance in *Phaseolus vulgaris* is a condition apparently acquired after exposure to Al^{3+}. Also suggestive of metal tolerance as an acquired/induced trait are reports where exposure to low, nonlethal concentrations of a metal increase plant tolerance to subsequent exposure to normally lethal concentrations of that metal.[18,19] The tolerant genotypes, through various mechanisms that restore normal cell function, are able to come to a new, functional thermodynamic state with regard to the metal in the environment. Such acquired tolerance has at its basis one or more mechanisms that detoxify metals, either in the environment or in the cytoplasm, that can be evoked by the appropriate physiological or biochemical signal.

Unraveling the metabolic signals involved in acclimation responses requires that we understand the physiochemical responses of plant cells to stress prior to the investigation of tolerance metabolism. There is an increasing literature that suggests that metabolic responses to stresses as seemingly divergent as heat shock, oxidative stress, wounding, and some metal stresses are similar.[20–24] This indicates that there may be specific biochemical or biophysical signals created by these stresses that signal the cell to initiate tolerance metabolism. Unless the physical, chemical, and metabolic changes that occur in a cell under stress are understood, the means by which the cell avoids/accommodates such changes are unlikely to be recognized as specific mechanisms of metal tolerance.

SENSING METALS IN THE ENVIRONMENT

The similarities evident among many stress responses may have at their centers several related alterations in cellular homeostasis. Different metals, differing greatly in their reactions with the cell surface and with biological ligands in the cytoplasm,[25] may deflect homeostasis in similar directions and

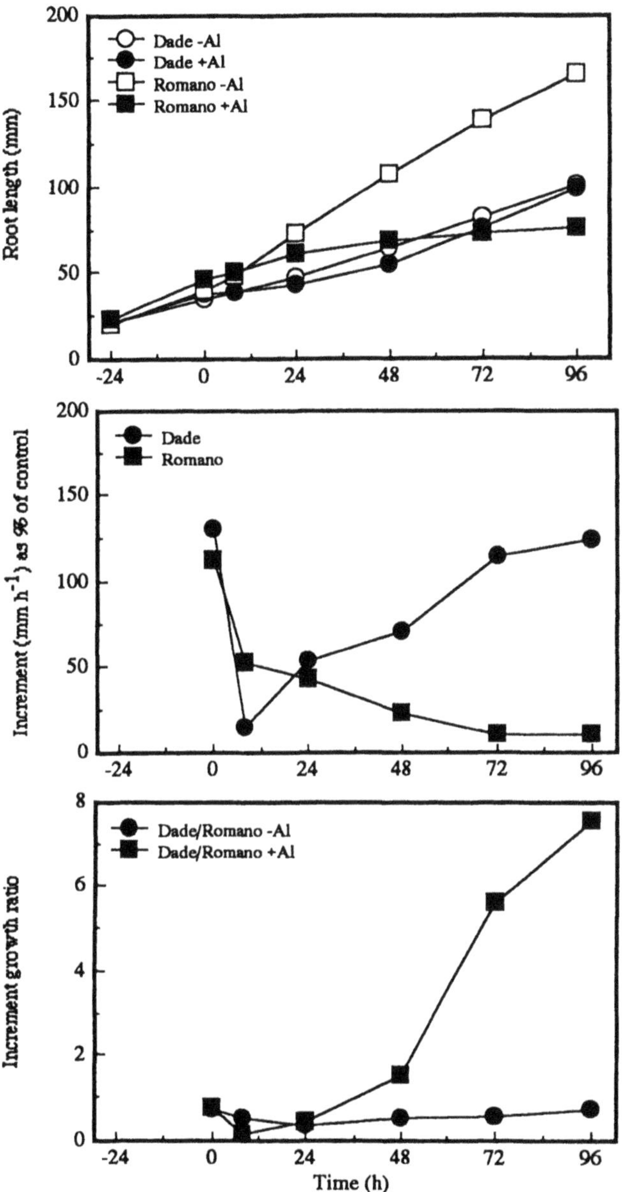

FIGURE 2. Root growth responses of *Phaseolus vulgaris* cultivars exposed to 10 μM Al^{3+} in solution culture (Al^{3+}-tolerant "Dade" and Al^{3+}-sensitive "Romano" cultivars). Total root length over time (top), increment growth over time as a percent of control for each cultivar (middle), and ratio of Dade to Romano increment as affected by Al^{3+} (bottom). Measurement of total length is not refined enough to distinguish the initial response and subsequent acclimation to Al in Dade seedlings; this is highlighted in growth as a percent of control. Increment ratio (bottom) indicates that Dade and Romano growth is similarly affected by Al^{3+} during the first 24 h of exposure (see Cumming et al.[17] for experimental details).

activate/induce similar cell responses. Taken in this light, plant response to metals can potentially be modulated by any one of several common signals that normally control cellular biochemical reactions. Such signals may be *indirect* and *nonspecific*, in which case metal exposure activates/induces a general cellular stress that leads to an accompanying cellular response. For example, Cd^{2+} exposure led to the expression of subsets of heat shock poly(A^+) RNAs in seedlings of *Glycine max*[21] and exposure to Cd^{2+}, Cu^{2+}, and Zn^{2+} conferred heat shock tolerance to seedlings of *Zea mays*,[23] indicating that these metals activate/ induce a suite of general stress responses in plant systems. There may also be *direct* yet *nonspecific* signals, where a metal that is normally involved in cell metabolism exerts posttranslational control over cytoplasmic enzymes and increases the production of end-products that may bind and detoxify the metal. Such was the case for the activation of phytochelatin synthase by Cd^{2+} in *Rauvolfia serpentina*[26] and by Cu^{2+} in *Schizosaccharomyces pombe*.[27]

Not all signals are the result of interactions of metals with membrane or cytoplasmic constituents leading to a loss of homeostasis. In several microorganisms, there exist *direct* and *specific* signals for the activation/induction of metal tolerance metabolism, where the presence of a specific metal ion triggers metal-responsive regulatory operons and subsequently leads to the production of novel mechanisms that reduce the impact of the metal on cell function.[28]

Two-Component Signal Transduction

Plant cells may potentially activate/induce metal tolerance metabolism in response to one or a group of perturbations of homeostasis.

Cytoplasmic Proton Concentration

Cell compartmentation allows for, among other things, the separation of labile biological ligands, enzymes, and reactions, many of which exhibit distinct pH optima for conformation and activity.[29] In addition to being a substrate and a by-product of metabolism, protons have the potential to communicate information concerning the energy status of the cell.[30] The buffering capacity of the cytoplasm is relatively low in comparison to the intensity of change of proton concentrations mediated by physiological processes (such as ion transport).[29] Thus, perturbation of cytoplasmic pH by metal ion exposure may lead to changes in several key processes, which may not only orchestrate metal toxicity but activate metal tolerance metabolism as well.

Cytoplasmic pH is maintained within strict bounds by the activity of the plasma membrane H^+-ATPase,[31–34] potentially H^+/Na^+ and H^+/K^+ antiports in some systems,[29,35] H^+/K^+ symports,[36] and organic acid production/catabolism within the cytoplasm[30] (see Kurdkjian and Guern[29] for review). The activity of many of these systems is dependent on cytoplasmic pH. Under normal conditions, cation (e.g., K^+) uptake stimulates H^+ efflux, increases cytoplasmic pH, and enhances carboxylation reactions and organic acid accumulation. In con-

Table 2. Effect of Short-Term Metal Exposure on H^+ Efflux (as a Percentage of Treatment Without Metals)

Metal	Concentration	H^+ Efflux	Citation
Al^{3+}	5 μ*M*	178	55[a]
	5 μ*M*	97	55[b]
Mn^{2+}	2 m*M*	407	53[c]
Zn^{2+}	1 m*M*	1325	5[d]
	2 m*M*	343	53
Ni^{2+}	0.1 m*M*	315	14[e]
	1 m*M*	578	14
	1 m*M*	1496	5
	2 m*M*	511	53
Co^{2+}	1 m*M*	614	5
	2 m*M*	426	53
Cd^{2+}	5 μ*M*	124	10[f]
	100 μ*M*	50	10

[a] Al^{3+}-tolerant ("Atlas") cultivar of *Triticum aestivum*.
[b] Al^{3+}-sensitive ("Scout") cultivar of *Triticum aestivum*.
[c] Efflux from roots of *Zea mays* in the presence of fusicoccin.
[d] Efflux from roots of *Zea mays* compared to roots exposed to 1 m*M* K^+.
[e] Efflux from roots of *Zea mays*.
[f] Efflux from roots of *Zea mays*.

trast, under metal exposure, cation (e.g., Zn^{2+}, Ni^{2+}, Co^{2+}) exposure/uptake reduces cytoplasmic pH and stimulates H^+ efflux, perhaps via changes in the activity of pumps or conductances of channels that may be linked to cytoplasmic pH or the transmembrane electrical potential (Table 2). In this case, decarboxylation reactions prevail and organic acids are catabolized.[5,29,37]

Are there metal tolerance mechanisms keyed to such changes in cytoplasmic pH? This question remains unanswered, although there are examples of metabolic changes correlated with changes in cytoplasmic pH and similar responses in plant systems exposed to metals. In many microorganisms, cytoplasmic acidification leads to a cessation or delay of the events of mitosis[38–40] or a deflection in the course of differentiation.[38,41] There are also metabolic events that may be controlled by internal proton concentrations. In *Saccharomyces cerevisiae*, for example, cytoplasmic acidification stimulated cAMP production that subsequently activated cAMP-dependent protein kinases that phosphorylate target enzymes.[42–44] There is thus a signal cascade initiated by cytoplasmic proton concentration leading to a specific cell response modulated through enzyme phosphorylation.

Little is known of the relationships between cellular and metabolic events and cytoplasmic pH in higher plants.[29] Treatment of *Allium cepa* roots with vanadate, which inhibits the plasma membrane proton ATPase, led to a reduction in root elongation and a parallel reduction in mitotic index.[45] Phosphorylation of a membrane-bound 33 kDa polypeptide was suppressed by cytoplasmic acidification in cell lines of *Acer pseudoplatanus, Nicotiana tabacum*, and *Daucus carota*.[46,47]

Given this conceptual framework, one can postulate that alterations in cell division, metabolism, and growth, common responses of plant systems exposed to metals, may result from cytoplasmic acidification associated with metal exposure/uptake. This cytoplasmic acidification, however, may facilitate the activation/induction of metal tolerance metabolism and subsequent recovery of homeostasis. Heat shock protein synthesis and intracellular acidification were correlated in *Saccharomyces cerevisiae*,[48] heat shock increased metal tolerance in *Triticum aestivum*,[24] metal exposure induced subsets of heat shock poly(A^+) RNAs in *Glycine max*,[21] and metal exposure induced the production of various proteins in *Zea mays*, which subsequently conferred tolerance to heat shock.[23] Thus, intracellular signal transduction may involve cytoplasmic H^+ concentration and may lead to acquired stress tolerance.[20]

Transmembrane Electrical Potential

Coupled to the maintenance of cytoplasmic pH through the operation of plasma membrane-bound H^+-ATPases, the cell electrical potential (PD) represents a controlling factor over many cell phenomena.[31–34] Depolarization of the PD may be a mechanism whereby external stimuli are transduced into metabolic responses through the operation of voltage-gated ion channels,[49] especially those involving Ca^{2+} fluxes (see below).[50,51] For example, cAMP dependent enzyme activity in *Neurospora crassa, Saccharomyces cerevisiae*, and *Mucor racemosus* was dependent on PD,[52] although conflicting data exist for *Saccharomyces cerevisiae*.[42–44]

The transport of cationic metals across the plasma membrane will, by nature of their positive charge, depolarize cell PD.[10,53] In addition, metals may depolarize the PD by altering membrane fluidity, thereby altering H^+-ATPase activity, or may potentially interact directly with membrane transport proteins or channels, altering ion fluxes carrying charge.[54] Depolarization of the cell PD by metal exposure may lead to an increase in the activity of the plasma membrane H^+-ATPase, if this enzyme system is unaffected by the metal, and the transmembrane electrical potential may recover or hyperpolarize.[49] Either the initial depolarization or the subsequent hyperpolarization may act as a transducing agent, translating metal exposure to an altered metabolic state, modulating the influx of Ca^{2+} via voltage-gated channels, efflux of K^+ via voltage-gated channels, and, potentially, the efflux of organic acids (Figure 3).[49]

Roots of an Al^{3+}-tolerant cultivar of *Triticum aestivum* exposed to Al^{3+} exhibited enhanced H^+ efflux and altered K^+ fluxes under Al^{3+} exposure,[55] which may reflect controlled responses to alterations of PD or plasma membrane structure or function mediated by Al^{3+}. Similar patterns have been noted for Al^{3+} in *Hordeum distichion*[15] and *Pisum sativum*.[13] Short-term exposure of *Zea mays* root segments to Ni^{2+}, Zn^{2+}, Co^{2+}, or Mn^{2+} led to reductions in cell PD and elevated proton extrusion (Table 2).[5,14,53] On the other hand, Kennedy and Gonsalves[10] reported variable PD and H^+ efflux patterns in roots of *Zea mays* exposed to Zn^{2+}, Cd^{2+}, Cu^{2+}, Pb^{2+}, and Hg^{2+}, dependent on the duration of

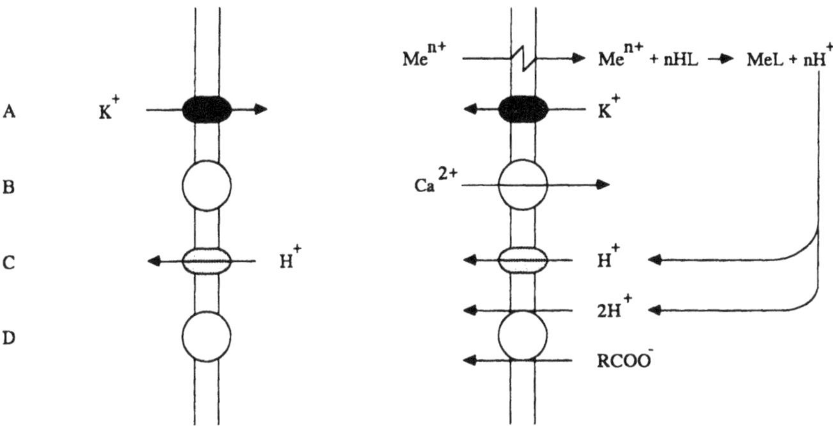

FIGURE 3. Activity of plasma membrane ion pumps and channels under control (left) and metal ion (right) exposure. Transport of metals across the plasma membrane depolarizes the cell electrical potential and acidifies the cytoplasm. These physiological changes initiate several responses, including K^+ efflux via voltage gated influx/efflux channels (A), Ca^{2+} influx via voltage gated channels (B), elevated proton extrusion via H^+-ATPase (C), and organic acid/proton efflux via a cotransport system (D).

exposure and the presence of other ions in the bathing solution. Such variable reports for these different metals probably reflect their differential uptake into the cell and impacts on membrane components.

Electrical Currents

In addition to the overall electrical status of the cell as reflected by the PD, physiological electrical status/signals may be differentially separated across the surface of a cell or a complex tissue. Electric currents may play a major role in differentiation and growth of cells and tissues.[56] Developmental events are often preceded by changes in the electrical field properties of a cell or tissue in a wide range of organisms.[57–60]

Electrical fields develop from asymmetric distribution of ion pumps and channels in the plasma membrane of a cell or from differential ion utilization by developing cells in a complex tissue. These asymmetries are often induced by external stimuli and lead to differential distribution of metabolites within the cell that can thus direct growth and development.[61] The ionic constitution of these electrical currents apparently differs with species, and may involve H^+, K^+, Cl^-, HCO_3^-, and Ca^{2+}. In many cases, the current carried by Ca^{2+} is small, but Ca^{2+} may control the fluxes of other ions carrying current.[57,60]

Metal exposure may affect ion currents, hence signals and responses, by impacting voltage gated channels or by altering the fluxes of ions that carry current.[62] In an Al^{3+}-sensitive cultivar of *Triticum aestivum*, Al^{3+} inhibited inward currents at the root tips in a time-course similar to that of the inhibition of root elongation.[63] Inward, whole-cell Cl^- currents were inhibited by Zn^{2+} in *Asclepias tuberosa*.[64] In developing pollen of *Lilium longiflorum*, La^{3+} exposure inhibited tube elongation.[57] Lanthanum blocks Ca^{2+} channels, and other metals may act similarly.[65] For example, Cd^{2+} adsorption into the cell walls *of Pinus resinosa* pollen displaced cell wall-bound Ca^{2+} and subsequently inhibited germination.[66] Exposure of developing pollen to A23187, a Ca^{2+} ionophore, also inhibited growth,[57] suggesting not only that Ca^{2+} fluxes/concentrations must be maintained within strict bounds for normal development, but that modulation of ion currents by Ca^{2+} may represent an important lesion (and possible signal) in cells exposed to metals in the environment.

Calcium-Related Signals

Cytoplasmic Ca^{2+} levels and Ca^{2+}-related processes may represent a signal for the expression of metal toxicity and, potentially, metal tolerance metabolism in plant cells. Calcium, Ca^{2+}-binding proteins (including calmodulin) and related plasma membrane phosphoinositides are believed to be involved in signal transduction and subsequent modulation of cellular metabolism, and thus may represent a signal transduction system in plants under metal exposure. The importance of Ca^{2+} in signal transduction processes has received considerable attention in the past decade, and the reader is referred to the recent compendium by Leonard and Hepler[67] and reviews by Hepler and Wayne,[50] Marme and Dieter,[68] and Einspahr and Thompson.[69]

Intracellular Ca^{2+} concentration is maintained within strict bounds through efflux or sequestration of Ca^{2+} in intracellular compartments.[70] This is accomplished by a number of membrane-bound Ca^{2+}-ATPases and voltage-gated channels that maintain cytoplasmic Ca^{2+} concentration at about 100 n*M* under normal conditions.[71] The activity of these transport systems, hence cytoplasmic Ca^{2+} concentration, may be modulated by external stimuli, cell PD, membrane phosphoinositide turnover, and/or hormones.[50,68,69]

Calcium directly, or indirectly through calmodulin, modulates the activity of a number of key enzymes that may direct cell metabolism, such as PEP carboxylase, NAD kinase, membrane-bound Ca^{2+}-ATPase, and protein kinases,[68] and is implicated in a wide variety of cell processes.[50,71] Given this pivotal role in metabolic regulation, Siegel and Haug[72] and Cheung[73] have postulated that metals may alter cell metabolism (and activate/induce metal tolerance metabolism?) by impacting normal Ca^{2+} fluxes and cytoplasmic concentrations.

Calcium is intimately involved in cell division and elongation processes in plant systems,[61,74–76] processes that are often perturbed by metal exposure. During mitosis, internal concentrations of free and bound Ca^{2+} cycle, suggesting that Ca^{2+} may be integrally involved in directing the mitotic process. Many tip growing and differentiation processes are also affected by Ca^{2+} availability and Ca^{2+} gradients across the cell.[50] Manipulations that reduce the availability of external Ca^{2+} for uptake or that block Ca^{2+} uptake channels (e.g., EGTA and

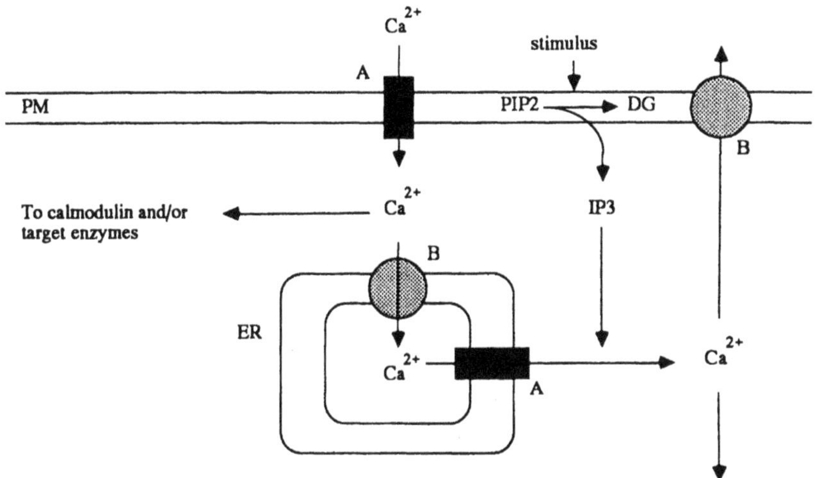

FIGURE 4. Calcium cycling among external, cytosolic, and endoplasmic reticular pools. Under normal conditions, receptors in the plasma membrane (PM) induce inositide turnover within the PM (PIP2, phosphatidylinositol 4,5-bisphosphate; IP3, inositol triphosphate; DG, diacylglycerol) leading to the release of IP3 which subsequently stimulates Ca^{2+} release from internal pools. Voltage gated channels in the endoplasmic reticulum (ER) and PM open, leading to elevated cytoplasmic Ca^{2+} concentrations. Calcium in the cytoplasm can be sequestered in the ER or extruded from the cell by membrane-bound ATPases or can influence cytoplasmic enzyme activity or following binding to calmodulin (see Einspahr and Thompson[69] for further details).

La^{3+}) arrest mitosis in the metaphase-anaphase transition[75,77] and inhibit Golgi activity and secretory processes.[3,78] In contrast, treatments that enhance internal Ca^{2+} concentration (e.g., Ca^{2+} ionophores, Ca^{2+} channel agonists, and elevated external Ca^{2+} concentrations) speed the metaphase-anaphase transition and stimulate cell differentiation.[77]

The biochemical Ca^{2+} pathways involved in signal transduction under normal physiological conditions are depicted in Figure 4. External stimuli induce inositide turnover within the plasma membrane, the end-products of which stimulate Ca^{2+} release from internal pools. The resultant increase in cytoplasmic Ca^{2+} then impacts cell metabolism via interactions with protein kinases and calmodulin.[69] Under metal exposure, transient depolarization of the transmembrane PD by metals may open voltage-gated channels in the plasma membrane, allowing Ca^{2+} influx. This may represent a signal for the activation/induction of metal tolerance metabolism: Ca^{2+} is involved in modulating several metal responsive systems in plants including callose deposition and organic acid metabolism (see below). It is not known whether the activity of additional metal tolerance mechanisms can be linked to cellular Ca^{2+} fluxes.

Interactions Among Potential Signals

From the above discussion it is evident that our understanding of the signals involved in inducing metal tolerance systems is in its infancy. Bennet and Breen[62] recently presented the concept of an "aluminum signal" involved in the transduction of Al^{3+} exposure in higher plants. In this case, the presence of Al^{3+} is perceived by root cap cells and translated into growth or regulatory responses involving the transduction of signals between cell types in the root tip. Such a signal would be a diffusible element, and these authors suggested the involvement of hormones in directing the response of roots to Al^{3+}. Although specific for one metal, one can envisage this or related processes controlling root/cell response to a number of metals. Interactions between metals, Ca^{2+}, cell PD, and hormones may act to control metabolic/growth responses of cells/tissues under metal stress conditions. Little is known of such interactions, and their investigation will help elucidate the signals controlling plant response to metals in the environment.

Direct Signal Transduction

For some metals, there are direct interactions between the metals and enzymatic or transcriptional systems that lead to enhanced activity or *de novo* production of metal tolerance mechanisms.

Phytochelatins (PCs) are metal binding polypeptides produced by phytochelatin synthase, a cytosolic enzyme system that is controlled by the concentration (activity) of certain metals in the cytoplasm.[79] Phytochelatin synthesis is stimulated by a wide variety of metals, and the equilibrium reactions between free metals in the cytoplasm, PCs, and other metal-requiring apoenzymes are critical in trace metal homeostasis.[79,80] Under normal conditions, PCs accumulate when metals are freely available, act as a metal storage system, and are catabolized as metals are assimilated into other enzyme systems.[80] The PC system is *nonspecific* in response to elevated metals, and will respond to metals having high affinity for sulfhydryl ligands.[26,27,81]

In contrast to PC synthesis, metal exposure in some cases induces *de novo* transcription and translation of metal tolerance machinery. Metallothionein (MT) synthesis in animals is inducible by a variety of heavy metals (including Cd^{2+}, Zn^{2+}, Cu^{2+}) at concentrations just below the level causing cell toxicity.[82] This induction results from a stimulation of MT gene transcription by metals: *cis*-acting DNA sequences, termed metal regulatory elements (MREs), upstream of the MT genes are essential to metal inducibility of gene transcription. Presumably, *trans*-acting DNA-binding proteins, which "sense" the heavy metals, bind at the MREs to stimulate transcription; as yet, such factors have not been characterized, except in *Saccharomyces cerevisiae*. It is unclear whether a single protein mediates induction or whether multiple regulatory proteins are involved. As will be discussed later in this chapter, the characterization of plant MT genes is not as advanced.

There are also cases where *direct* and *specific* metal signals induce specific detoxification systems. In these cases, metal-responsive regulatory proteins alter gene expression in the presence of specific metals. The metal ion-protein interaction leads to a physical change in the receptor protein that propagates the signal by promoting gene transcription by RNA polymerase. Enhanced mRNA production and subsequent translation leads to the production of the metal detoxification system. For example, the Hg^{2+} detoxification system in *Escherichia coli*, mercuric ion reductase, relies on activation of the promotor protein by Hg^{2+}, which alters the affinity of this protein for DNA and subsequently allows downstream portions encoding for the reductase system to be transcribed by RNA polymerase.[28]

These examples of direct signal transduction and induction of metal tolerance metabolism are well characterized for several metals in microbiological systems.[28] It is yet to be determined whether similar systems are operable in higher plants.

MECHANISMS OF METAL TOLERANCE

The transduction of a metal signal into metabolic responses is the crucial step determining the survival of a cell/tissue/organism. Here, the genetic constitution of the individual will determine what responses will occur and, ultimately, whether the organism is to survive and reproduce. Evolutionary aspects of metal tolerance have received considerable attention[83–88] and it is on a long-term time scale that natural selection of genotypes capable of activating one or more metal tolerance mechanisms will lead to the formation of metal-tolerant ecotypes, populations, and metalliferous species. What appears to be critical is the breadth of genetic base from which metal stress-tolerant genotypes can be selected.[84,85] For the purposes of this review, we will assume that natural selection under such conditions is favoring individuals with genetic codes for metal stress-tolerance machinery outlined below. However, many of our examples are selected from short-term studies, as these have tended to be more mechanistic in nature, and comparison of mechanisms operating in field-collected and in laboratory-treated plant material will be needed to verify this assumption.

Exclusion and External Metal Detoxification Mechanisms

Taylor[89] reviewed examples of metal exclusion and their potential importance as tolerance mechanisms in plants (see also Baker[90]). However, metal exclusion is not a general feature of metal-tolerant organisms, as there are many examples of organisms that accumulate high concentrations of metals in root and foliar tissue.[84,86,87] It should be stressed, however, that tissue concentration may not be a good indicator of biochemical exclusion per se, as metals may be bound to external binding sites or may be complexed or compartmentalized within the cell (see below). Some metals, many of which are essential micronutrients, may traverse the plasma membrane via nonspecific cationic carriers, necessitating the operation of internal homeostatic mechanisms. Others may enter the cytoplasm as anions, cations, or neutral or variously charged hydroxy species via protein channels that cannot discriminate between similarly charged species[28] or may cause nonbilayer conformations in plasma membrane lipids, facilitating their transport into the cytoplasm.[91–94]

The Plasma Membrane. The structure and function of the plasma membrane may be modified directly by metals or by the cell in response to metal exposure. The former response may be considered a toxic lesion of metal exposure and has been recently reviewed.[54,95,96] Membrane conductance and ion uptake are affected by a variety of biological and biophysical factors including the selectivity of ion transporters, surface charge, fatty acid composition, and sterol content of the plasma membrane.[92,96–102] Inherent differences in, or capacity of a tolerant genotype to alter, membrane composition (or local surface environment) would lead to differential response to metal exposure.

There are reports of differential metal accumulation associated with metal tolerance that apparently reflect differences in plasma membrane properties. Copper resistance in metal-tolerant strains of *Saccharomyces cerevisiae*[103] and *Chlorella vulgaris*[104] was associated with reduced Cu^{2+} influx in comparison to wild-type strains, a pattern also exhibited for Cd^{2+} in the genera Peltigera[105] and Euglena.[106] In *Saccharomyces cerevisiae* and *Peltigera membranacea*, Cu^{2+} tolerance was associated with altered Cu^{2+} transport characteristics, suggesting that tolerance resulted from differential affinity of divalent cation carriers for nutrient cations over Cu^{2+}.[103,105]

In higher plant systems, there are also reports of altered membrane processes conferring metal tolerance. In *Silene cucubalis*, Cu^{2+}-tolerance was associated with reduced Cu^{2+} uptake into roots.[12] Arsenate (AsO_4^{3-}) uptake into roots of *Holcus lanatus* occurred via the phosphate (presumably PO_4^{3-}) transport system, and an AsO_4^{3-}-tolerant genotype exhibited a much lower affinity for AsO_4^{3-}, leading to lower metal uptake.[107] It is evident that additional studies on metal ion uptake systems are necessary to elucidate the potential of membrane-bound carrier specificity/kinetics in modulating toxic metal accumulation.

Little information is available on inherent differences or changes in membrane structure in response to metals. Differential plasma membrane binding of Al^{3+} was noted in *Triticum aestivum*, with membranes from Al^{3+}-tolerant genotypes exhibiting a lower affinity for Al^{3+} in comparison to Al^{3+}-sensitive genotypes.[108] Xu and Patterson[109] noted differences in the ratio of sitosterol:stigmasterol in *Glycine max* suspension cells exposed to Cd^{2+}, but attributed this to a toxic effect on sterol conversion rather than controlled cell response to stress. Steer[110] and Thompson[96] have reviewed membrane turnover

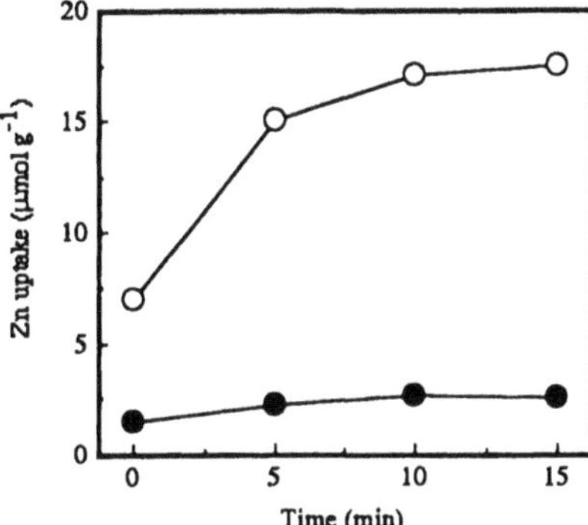

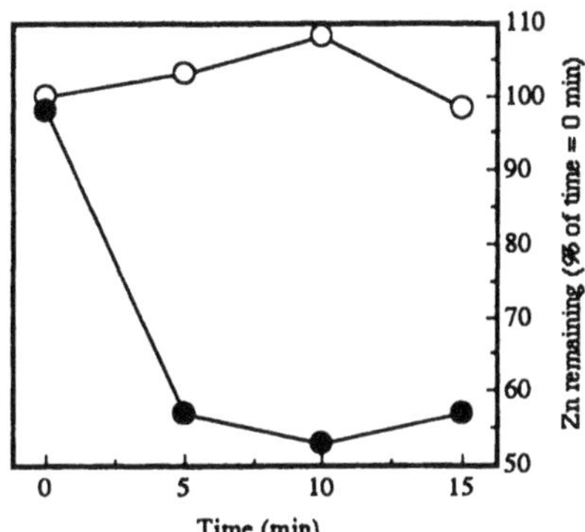

FIGURE 5. Net effect of the activity of Zn^{2+}-ATPase on cellular Zn^{2+} levels in *Alcaligenes eutrophus.* (Left) Accumulation of $^{65}Zn^{2+}$ by Zn^{2+}-sensitive (open symbols) and Zn^{2+}-tolerant (closed symbols) strains. (Right) Retention of $^{65}Zn^{2+}$ by Zn^{2+}-sensitive and -tolerant strains after transfer to nonlabeled solution (after Neis and Silver,[111] with permission from the American Society for Microbiology).

and the mechanisms of membrane response to environmental stress. Controlled changes in membrane composition would seem to represent an important response to adverse environmental conditions. Indeed, there is information on differences (changes?) in glycolipid:phospholipid ratio[98] and proportion of membrane associated sterols[97,98] in roots under salt stress. As sterols affect phospholipid packing density and membrane conductance, similar differences/changes may influence the transport/conductance of metals across the plasma membrane. Documentation of altered, yet stable, membrane composition after metal exposure and subsequent removal of the stress would elucidate whether such changes are indeed toxic lesions or controlled cellular responses to metal exposure.

Metal Ion-Specific Efflux Pumps. The operation of metal ion-specific efflux pumps would effectively reduce the net uptake, intracellular exposure, and potential interactions between toxic metals and vital cellular components. Metal ion efflux systems have been well documented in prokaryotic systems.[28] Tolerance to Zn^{2+} in *Alcaligenes eutrophus*, for example, required prior exposure to Zn^{2+}, which led to transcription, translation, and subsequent function of a plasma membrane-bound Zn^{2+}-ATPase that transported Zn^{2+} out of the cell.[111] Strains containing the genetic information for this system exhibited reduced intracellular Zn^{2+} that resulted from the activity of the efflux pump (Figure 5). Similar systems conferred resistance to Cd^{2+}, Zn^{2+}, Co^{2+}, Ni^{2+}, and CrO_4^{2-} in *Alcaligenes eutrophus*, and strains of *Pseudomonas*, and *Streptococcus*.[111–113] The operation of these systems was dependent on metabolic energy, and treatments that reduced the availability of ATP enhanced the intracellular accumulation of the metals.[28]

Although the existence of such systems in higher plants has not been elucidated, one may postulate the activity of metal ion efflux pumps in cases where metabolic poisons similarly enhance cellular accumulation of metals. The exclusion of Al^{3+} from roots of *Triticum aestivum*,[114] *Brassica oleracea*, *Lactuca sativa*, *Pennisetum clandestinium*,[115] *Raphanus sativus*, *Fagopyrum esculentum*, *Cucumis sativus*, and *Oryza sativa*[116] was highly dependent on maintenance of normal metabolic function, and stimulation of Al^{3+} uptake occurred when roots were treated with metabolic inhibitors. Hence, these plant systems control *net* metal accumulation via some metabolically-dependent mechanism, which may involve metal efflux.

Sequestration in Subcellular Compartments. Sequestration of metals in subcellular compartments where no sensitive metabolic activities occur would negate the interaction of metals with metabolically important components in the cytoplasm. The vacuole is potentially the largest sink for accumulated metals and this compartment contains high concentrations of organic ligands that may chelate metals in soluble or insoluble forms.[117]

Zinc-tolerant clones of *Deschampsia caespitosa* exhibited more effective Zn^{2+} compartmentation in the vacuole in comparison to nontolerant clones, especially under elevated Zn^{2+} exposure.[118] Thurman and Rankin[119] and Godbold et al.[120] found citrate to be the primary organic acid in roots of this species. Although the accumulation of this organic acid was stimulated by Zn^{2+} exposure in *Deschampsia caespitosa*, endogenous organic acid concentrations were sufficient to complex absorbed Cd^{2+} and Zn^{2+} in *Nicotiana tabacum* suspension cells where, in contrast, metal ion exposure did not stimulate organic acid production.[121] Mathys[122] noted elevated malate concentrations in aerial tissues of Zn^{2+}-tolerant ecotypes of *Silene cucubalis* and *Rumex acetosa* in comparison to nontolerant ecotypes (although differences in *Agrostis tenuis* were small), and formulated a model (Figure 6, bottom) whereby malate acts as a shuttle to remove excess Zn^{2+} from the cytoplasm to the vacuole, where it is complexed by a variety of terminal ligands, including citrate, oxalate, mustard oils, and anthocyanins. While attractive, there are problems with this model, notably that elevated production of organic acids should confer tolerance to some other metals, which is not the case for *Deschampsia caespitosa*.[119]

In addition to the malate shuttle hypothesis, Vogeli-Lange and Wagner[123] proposed a similar system for metal-binding polypeptides in *Nicotiana rustica* (Figure 6, top). In this case, Cd^{2+} stimulates the production of binding proteins (see below) that complex the metal and transfer it to the vacuole, where the complex dissociates and the metal forms terminal complexes with organic acids in this compartment. In addition, Van Steveninck et al.[124] noted the vacuolar deposition of Zn^{2+} as globular bodies with K^+, Mg^{2+}, and PO_4^{3-} (phytate) in roots of *Deschampsia caespitosa*. Metal-organic ligand sequestration in the vacuole has also been noted for Ni^{2+}.[125,126]

Some metals have been localized in subcellular spheroids within the cytoplasm and within the endoplasmic reticulum. Together, one could postulate that these endomembrane systems are acting as a sequestration and/or extrusion system. Ksiazek et al.[127,128] found that Pb^{2+} accumulated primarily in the cell

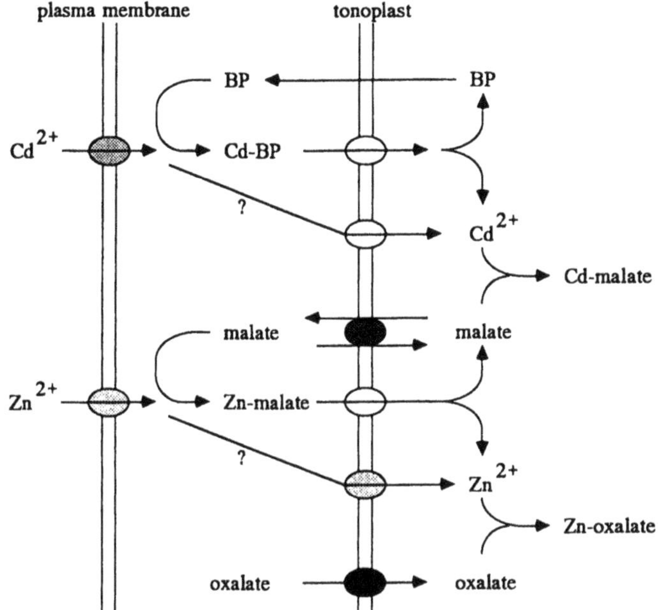

FIGURE 6. Model for metal tolerance via sequestration of metals within the vacuole. Metals entering the cytoplasm may initiate the production of metal-binding polypeptides (BP) (top) or may complex with endogenous organic acids (bottom) that shuttle the metals to the vacuole. The transported complexes dissociate once in the more acidic matrix of the vacuole, and terminal complexes form other endogenous ligands (after Vogeli-Lange and Wagner[123] and Mathys[122]).

walls and between the cell wall and plasma membrane in suspension cells and leaf tissue of *Populus* spp., although intracellular Pb^{2+} was located in pinocytotic vesicles. Similar patterns have been noted in *Raphanus sativus*[129] and *Euglena*.[130] Although these authors suggested that these vesicles remove Pb^{2+} to the vacuole, little Pb^{2+} was located in the vacuoles of these species, suggesting that the vesicles may act to discharge the toxic metal to the external environment.

Organic Ligand Exudation. Organisms not only respond to prevailing environmental conditions but may also modify their immediate surroundings as a matter of course or in response to some extracellular signal. The production and extrusion of extracellular biomolecules may effectively alter the availability of metals by altering the chemical speciation of metals in the environment.[131–133]

The mechanisms of soluble organic ligand exudation are not well understood, although there is evidence that efflux is coupled to the cell PD and stimulated by cell depolarization.[134–136] The chemical composition, in particular the dissociable moieties, of exudates will greatly affect what metals are chelated external to the cell. Metal chelation often reduces metal uptake[137–140] and subsequent toxicity,[116,139,141] reflecting loss of metal reactivity with critical cell constituents.[142,143]

Qualitative or quantitative differences in exudate production may confer differential metal tolerance to specific genotypes, although there is little information on such distinctions. Butler et al.[104] noted that Cu^{2+}-tolerant strains of *Chlorella vulgaris* produced more extracellular organic material with higher stability constants for Cu^{2+} in comparison to exudates produced by nontolerant strains. As examples of the importance of assessing normal root-associated microflora in determining the metal tolerance or sensitivity of a species, ectomycorrhizal symbionts have been found to confer Zn^{2+} and Al^{3+} tolerance in *Betula*[144] and *Pinus*[145] by production of fungal-specific extracellular metal-binding materials (see also Baker and Walker[87]).

In addition to soluble exudates, mucilage at the root tip represents a strong sink for metals, and was indeed crucial in protecting the metal-sensitive root meristem from Al^{3+}.[146,147] Removal of root cap mucilage facilitated entry of Al^{3+} into root tissues and increased the sensitivity of plants to Al^{3+}.[146,147] However, in *Vigna unguiculata*, Al^{3+}-sensitive genotypes produced mucilage with a *higher* Al^{3+} binding capacity than Al^{3+}-tolerant genotypes,[146] a pattern that raises questions as to the importance of mucilage in conferring *differential* metal tolerance among genotypes.

Cells may additionally synthesize callose in response to stress.[148] Callose production is mediated by the plasma membrane-bound enzyme β-1,3-glucan synthase, the activity of which appears to be modulated by cell electrical potential, plasma membrane architecture, and cell Ca^{2+} levels.[149–153] These modulating properties make this system ideal as a potential component of metal tolerance metabolism. Elevated callose production has been noted in roots of *Glycine max* in response to Al^{3+}, Co^{2+}, Ni^{2+}, and Zn^{2+} [149] and in suspension cells of *Glycine max, Phaseolus vulgaris, Nicotiana tabacum,* and *Petroslinum hortense* exposed to chitosan (a polycation).[150] Treatment of cells with EGTA, which effectively removes Ca^{2+} from external cell pools, did not stimulate callose production. Thus, activation of β-1,3-glucan synthase appears to be in response to increased intracellular Ca^{2+} that results from depolarization of cell PD under metal exposure.[153]

Internal Metal Detoxification Mechanisms

Metal ions do traverse the plasma membrane and must subsequently be detoxified by one or more mechanisms in the cytoplasm if normal cell function is to continue. Many biological ligands have the capacity to bind metals, but the potential reactions and stability constants for metal-ligand complexes under biological conditions are not well known. Some patterns of ligand production (accumulation) may represent the end-products of altered metabolism whereas others may represent concerted cell responses to metals in the cytoplasm.

Organic Ligand Accumulation. The very characteristics of metals that make them toxic, i.e., high charge density and stability constants for biological ligands, may allow biological systems to chelate and detoxify metal ions in the cytoplasm. It is difficult, however, to separate toxic effects of metals on enzyme systems from changes in metabolite pools directed by the cell in response to metal stress.

Carboxylate, sulfhydryl, and phosphate moieties of biological ligands, and the affinities and ligand exchange rates of different metals for these ligands, will determine the relative distribution of free and bound metals in the cytoplasm.[25,154] Indeed, once inside the plant, most metals will be found as other than free ionic species. For example, using biological ligand concentrations measured in xylem exudates of *Yucca flaccida*, computer-generated estimates of metal speciation indicated that Al and Fe were found primarily as hydroxides, Cu as a glutamine complex, Mn as an asparagine complex, and Zn as a complex with alanine.[155]

There are few examples where tissue levels of organic ligands correlate with genetic tolerance to metals in the environment. In Ni^{2+} accumulating species, Ni^{2+} was bound and sequestered in the vacuole as a citrate complex.[125,126] However, in cultivars of *Triticum aestivum*, Mn^{2+} and Al^{3+} tolerances were not associated with tissue concentrations of organic acids,[4,156] and there was no evidence for altered organic acid synthesis in *Nicotiana tabacum* exposed to Cd^{2+} or Zn^{2+} [121] or in differentially Zn^{2+}-tolerant ecotypes of *Deschampsia caespitosa* exposed to Zn^{2+}, Co^{2+}, Hg^{2+} or Ni^{2+}.[119]

Metal-Binding Proteins and Peptides. Recently, much of the attention of those interested in metal tolerance has centered on the chelation of metal ions by metal-binding proteins and peptides. This has almost certainly been due to the comparable information from animal and fungal studies. For example, when mammalian cell cultures are subjected to stepwise selection to increasing concentrations of Cd^{2+}, the resultant Cd^{2+}-resistant cells have a higher capacity for metallothionein (MT) synthesis[82,157] due to amplification of the MT genes.[158] Such studies prompted a search for these proteins in plants.

Early studies of "metallothionein-like" proteins in plants, however, led to documentation of relatively simple metal-binding peptides more closely resembling glutathione than gene-encoded polypeptides. These peptides have been the subject of several recent reviews[79,159,160] and are most often referred to as "phytochelatins" (PCs) or by the structural name poly-(γ-glutamyl cysteinyl)-glycines ($[\gamma EC]_nG$).

Phytochelatins. The structure of PCs ($[\gamma EC]_nG$) is similar to glutathione (γECG), and their biosynthesis is enzymatic (rather than via transcription and translation), as evidenced by the γ-peptide bond between the carboxyl group of the side chain of glutamate and the amino group of cysteine rather than the more usual α-peptide bond found in proteins. The molecules range in size from 5 amino acids ($[\gamma EC]_2G$) to 23 amino acids ($[\gamma EC]_{11}G$),[161] although des-Gly peptides lacking the C-terminal glycine have been observed.[162] Phytochelatin synthesis is induced by a wide range of metals[79,161] and yet the binding of metals to phytochelatins has only been reported for Cd^{2+}, Cu^{2+}, Pb^{2+}, and Zn^{2+}. The nature of the metal-PC complexes *in vivo* has not been defined; this is compli-

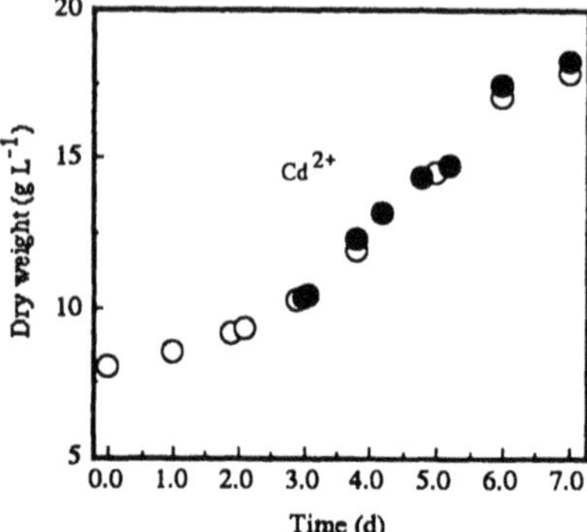

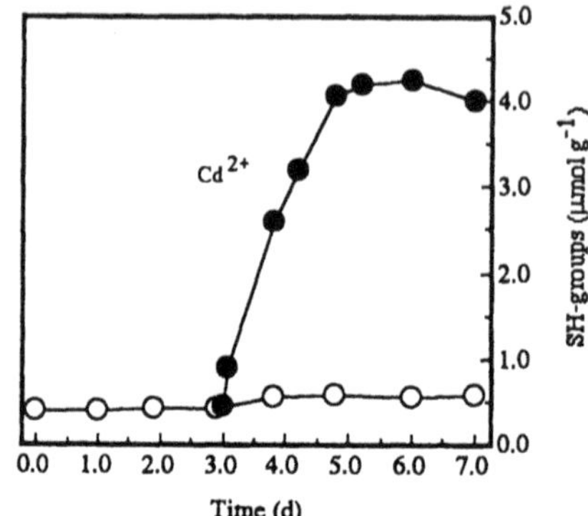

FIGURE 7. The induction of Cd^{2+} tolerance in *Rauvolfia serpentina* cell suspension cultures. (Left) Rapid PC synthesis (Cd^{2+}, solid symbols; controls, open symbols) leads to intracellular binding of Cd^{2+} with no changes in growth patterns after Cd^{2+} exposure. (Right) Time course for PC production after the addition of Cd^{2+} as measured by SH-group concentrations of crude cell extracts (after Grill et al.,[26] with permission from *Naturwissenschaften*).

cated by their size heterogeneity. However, it is clear that complexes differ by the presence/absence of sulfide or sulfite ions.[79] Such acid-labile sulfur in the complex may enhance metal sequestration.

Grill and co-workers[161] have examined over 200 species of plants, and PCs are present in all of them as well as in certain fungi.[163] The production of PCs, and the subsequent binding of excess intracellular metals, in response to metal exposure is very rapid and apparently allows the cell system to continue normal physiological function with little or no perturbation of normal growth (Figure 7).

Metallothioneins. The evidence for the presence of gene-encoded MTs in higher plants is not yet definitive; however, two lines of investigation indicate that plant MTs exist. First, although the discovery of PCs among impure fractions of "metallothionein-like" proteins led researchers to concentrate on PC purification, Reese and Wagner[164] reported that a Cu-complex from tobacco suspension cultures did not appear to be PC-like in comparison to Cd-PC isolated from the same tissue. Tomsett et al.[165] observed two Cu^{2+}-induced Cu^{2+}-binding components from *Mimulus guttatus*, which are not impure PC complexes,[166,167] that have amino acid compositions more similar to fungal MTs. Although these components have not been purified, these results indicate that gene-encoded MT-like proteins exist in plants, at least in response to Cu^{2+}.

Second, reports of cDNA clones whose predicted protein products have domains similar to animal/fungal MTs support the existence of MTs in higher plants (Figure 8).[168–171] These (approximately) eight kDa proteins are unusual in that they each possess an N-terminal and a C-terminal MT-like domain separated by an amino acid sequence totally unlike any MT. The MT-like domains can be defined by their cysteines, in the characteristic cys-x-cys arrangement (where x

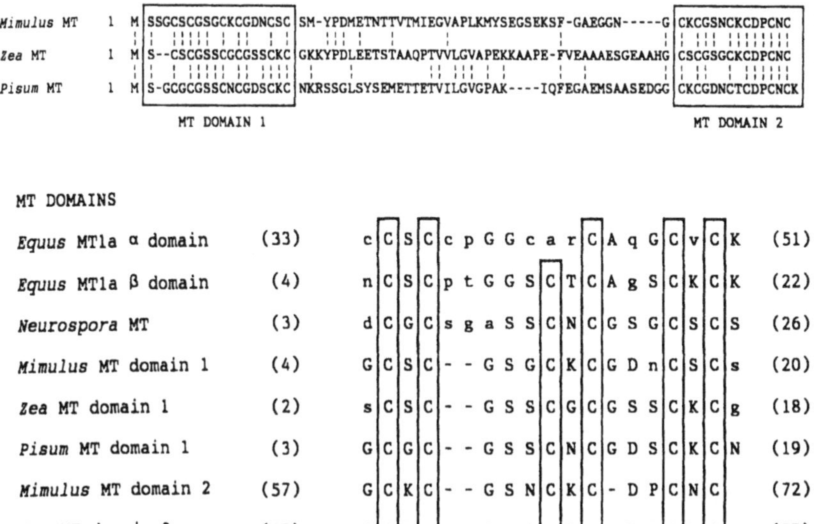

FIGURE 8. (a) Amino acid sequence alignment of the putative *Mimulus guttatus*,[168] *Pisum*,[169] and *Zea*[170] metallothioneins. In (b), domains 1 and 2 of *Mimulus guttatus, Pisum sativa,* and *Zea mays* MTs are compared with the α and β domains of Equine Mt and with the single domain of *Neurospora crassa* MT. The inclusion of (–) in the sequence indicates a gap introduced for optimal alignment and (|) denotes amino acid homology.

= an amino acid other than cysteine) used to define other MTs,[172] and they can be aligned (although not perfectly) with animal and fungal MTs (Figure 8). There can be little doubt that with their high cysteine content, these proteins will bind significant amounts of metals, although this property has yet to be demonstrated.

What is clear from the above discussion is that the involvement of MTs in metal homeostasis/detoxification/tolerance should receive further investigation. Studies of the yeast *Candida glabrata*[162] have revealed that it synthesizes PCs in response to Cd^{2+}, but, when challenged with Cu^{2+}, it produces two MT-like proteins. This difference may also be true for plants: the MT-like proteins are produced *specifically* in response to Cu^{2+} exposure, whereas PCs are *nonspecifically* synthesized in response to this and other metals in the cytoplasm.[165]

PCs, MTs, and Metal Tolerance. The function of both PCs and MTs and their potential involvement in metal tolerance in plants is not yet understood. It is clear that PCs are a major cellular component for metal ion homeostasis and detoxification. Their synthesis is rapid, being detected only minutes after Cd^{2+} addition, because production is mediated by a constitutive enzyme, phytochelatin synthase (see below).[173] Recent work[123] has also shown that PCs are located in vacuolar compartments and hence may represent a shuttle system for the transfer of metals from the cytoplasm to the vacuole, where the metals would be sequestered from critical metabolic systems in the cytoplasm.

It is unclear whether elevated PC synthesis represents a mechanism of metal tolerance. Steffens[79] reviewed the available evidence and concluded that PC overproduction is an unlikely mechanism for metal tolerance, owing to the energy required for sulfate reduction to support PC synthesis. However, it does seem likely that it is one (perhaps an essential) component of a complex mechanism required by metal-tolerant plant cells. The evidence for this results from experiments using an inhibitor of PC synthesis, buthionine sulphoximine (BSO). The combination of BSO plus Cd^{2+} in the growth medium, at concentrations that do not greatly reduce plant growth when added individually, is lethal.[79,167] This indicates a key role for PCs in metal detoxification, if not as a tolerance mechanism in naturally adapted plants.

Multiple Mechanisms and Co-Tolerances

Multiple metal-stress tolerance mechanisms may have evolved over time in response to a given metal in the environment.[84,86] For example, under metal ion exposure, reduced toxicant uptake by transporters that are more specific for nutrient ions (over toxic metal ions) will protect the cell as well as the operation of metal ion efflux pumps. Thus, the evolution of these divergent mechanisms would allow individuals of the same species to function equally well under metal ion exposure. In addition, the biochemical/biological function of such mechanisms may further impart metal co-tolerances, when the action of a detoxification mechanism may not be specific for a given metal but will act similarly on groups of metals with related properties.

Silver and Misra[28] recently reviewed metal tolerance mechanisms in microorganisms, where the genetics of these systems and the specific activity of the metal-tolerance machinery are well described. There are a number of systems for Cd^{2+} tolerance in *Staphylococcus aureus*. Cell lines carrying the *cadA* gene exhibit tolerance via Cd^{2+} efflux from the cell (mediated by a Cd^{2+}-efflux ATPase[28] or a H^+/Cd^{2+} antiporter[174]). The *cadB* gene confers tolerance via production of an, as of yet, uncharacterized metal-binding agent.[175] Interestingly, the *cadC* system also confers Cd^{2+}-tolerance by Cd^{2+} efflux; the *cadC* efflux pump is specific for Cd^{2+} whereas *cadA* confers tolerance to Cd^{2+} as well as Zn^{2+}.[28] Similarly, in *Alcaligenes eutrophus*, a number of plasmids carry genetic codes for efflux systems for Co^{2+}, Zn^{2+}, Cd^{2+}, and/or Ni^{2+},[111] and multiple metal tolerances and co-tolerances for Ni^{2+}, Co^{2+}, and Zn^{2+} have been reported for *Neurospora crassa*.[176,177]

As metal tolerance and detoxification systems are not as well described in higher plant systems, much of our knowledge is inferred from growth responses, gross tissue metal concentrations, and incidental accumulation of metabolites. However, there are patterns that suggest that different metal tolerance mechanisms for the same metal may operate in higher plants and that these may impart metal co-tolerances. Woolhouse[86] has outlined four potential Cu^{2+}-tolerance strategies that support this view. These include (1) altered ion uptake systems, (2) cell wall binding, (3) compartmentation of soluble complexes, and (4) compartmentation of insoluble complexes.

In *Silene cucubalis*, Cu^{2+} exposure stimulated the production of Cu^{2+}-binding compounds (PCs) that may detoxify Cu^{2+} within the cell.[178] Similarly, Rauser and Curvetto[179] noted the production of metallothioneins in roots of a Cu^{2+}-tolerant clone of *Agrostis gigantea* upon exposure to Cu^{2+}. In contrast, some Cu^{2+}-tolerant clones of *Silene cucubalis* exhibit reduced Cu^{2+} uptake and accumulation,[12] suggesting the operation of alternative tolerance mechanisms such as altered ion uptake or efflux systems. Cadmium tolerance in *Euglena gracilis* was the result of altered ion carrier affinity for divalent cations that decreased Cd^{2+} uptake into cells.[106] These altered carrier kinetics also conferred tolerance to Co^{2+} and Zn^{2+}.[180]

The production of metal-chelating ligands high in thiol groups might also render co-tolerance to metals such as Cd^{2+} and Zn^{2+}. Phytochelatins were synthesized in response to (and bound) Cd^{2+} in suspension cells of *Lycopersicon esculentum*,[181,182] *Rauvolfia serpentina*,[81,126] and in roots of *Silene cucubalis*;[91] Grill et al.[80] found phytochelatin synthesis in *Rauvolfia serpentina* correlated with media Zn^{2+} and Cu^{2+} concentrations. Furthermore, Verkleij and Prast[178] and Verkleij and Bast-Cramer[183] found co-tolerance to Cd^{2+} and Cu^{2+} and Cd^{2+} and Zn^{2+}, respectively, in *Silene vulgaris*, suggesting that certain metal-tolerance mechanisms can confirm tolerance to more than one metal in higher plants (see Thurman and Rankin;[119] Cox and Hutchinson;[184,185] Kishinami and Widholm[186] for additional examples).

METAL ION REGULATION OF GENE/ENZYME ACTIVITY: EXAMPLES

A number of heavy metals, including Zn^{2+} and Cu^{2+}, are required as micronutrients in biological systems, but these same metals are toxic at elevated concentrations. Cellular responses to metals must, therefore, distinguish between patterns of gene expression for processing essential metal components of normal metabolism from the protective functions required when metals are present at toxic levels. A comparison of these circumstances reveals that the activity of some processes is induced by the presence of metals, either by activation of constitutive enzyme systems or by *de novo* protein synthesis, whereas the activity of other processes is repressed. Many of these changes are nonspecific effects of metal exposure, such as the production of subsets of heat shock proteins,[21,22] and are presumably the result of indirect triggering of other stress/homeostatic processes. Although our understanding of the nature of such metal ion regulation is not well defined, several examples in which the specific regulation of cell activity occurs by the presence of metals have been elucidated.

Phytochelatins are synthesized by the recently discovered enzyme γ-glutamylcysteine dipeptidyl transpeptidase, also termed phytochelatin synthase.[173] The enzyme has been purified from *Silene cucubalus* and used to examine metal ion regulation of PC synthesis. Metals such as Cd^{2+} rapidly stimulate PC synthesis, in some cases within 5 min of metal exposure,[181,187] and the enzyme system responsible for such a rapid response must be constitutive in nature. Loeffler et al.[173] have shown that metal "induction" is thus a post-translational control mechanism, whereby metals activate pre-existing protein.

Phytochelatin synthase mediates the reaction:

$$\text{Glu-Cys-Gly} + \text{Glu-Cys-Gly} \rightarrow (\gamma\text{Glu-Cys})_2\text{Gly} + \text{Gly}$$

i.e.,

$$\text{glutathione} + \text{glutathione} \rightarrow PC_2 + \text{glycine.}$$

After a lag period, higher order peptides (PC_3, PC_4, etc.) can be formed from the addition of dipeptides to the newly synthesized PC_2 molecules. This enzyme has an absolute requirement for heavy metals: incubation of phytochelatin synthase with glutathione led to PC synthesis only upon addition of Cd^{2+}. Furthermore, the reaction ceases when all the metals are complexed with PCs, but can be restarted by the addition of more metal. Synthesis can also be terminated at any stage by the addition of excess metal-free PC or EDTA.[173] Thus the metal regulation is a simple negative feedback loop: metal stimulates PC synthase and the product of the reaction (PC) removes the metal thus terminating enzyme activity. Since the substrate, glutathione, is always available within plant cells, this tolerance response is rapid and is a *direct*, although *nonspecific*, response to the inducing metal.

In contrast, metallothionein synthesis in *Saccharomyces cerevisiae* is *specific* to Cu^{2+} and Ag^{2+}.[82,188,189] Like other fungi, *Saccharomyces* produces a Cu^{2+}-thionein that, although capable of binding Cd^{2+} and Zn^{2+} *in vitro*, is synthesized only in response to Cu^{2+} and Ag^{2+} *in vivo*. The regulation of MT synthesis is transcriptional: at low physiological Cu^{2+} concentrations, MT synthesis is just sufficient to bind free ions, but under conditions of Cu^{2+} stress, MT synthesis is elevated to high levels to protect against Cu^{2+} toxicity. The transcriptional regulation involves a *trans*-acting activator protein, called *ace1*, that binds directly to distinct copper regulatory sequences upstream of the MT (*cup1*) gene in the presence of Cu^{2+} ions. This binding does not occur in the absence of Cu^{2+}. Copper activates the *ace1* protein by inducing a change in its conformation that stimulates its binding to DNA, leading to enhanced production of downstream gene products (MTs). Thus, *ace1* is a unique DNA-binding protein, having distinct features from DNA-binding protein previously classified from a wide spectrum of eukaryotes. Common features for DNA-binding domains are the helix-turn-helix motif, the leucine zipper, and zinc fingers.[190] Zinc fingers are loops within the protein in which zinc is chelated via two cysteine and two histidine residues or via four cysteine residues. Such structures have been conserved through a wide range of species and in many different regulatory circuits. The *ace1* protein does not conform to these models and appears to have adopted a structural Cu^{2+}-binding motif from metallothionein, the gene it regulates. Presumably, this evolutionary adaptation lies at the heart of the metal specificity of this metal tolerance system.

EPILOGUE

The environment rarely presents optimal conditions to an organism. The transduction of stress signals and activation/induction of tolerance metabolism involved in the restoration of homeostasis often results in tolerance to more than the original stressor. This suggests that there are common chemical or biophysical pathways involved in stress signal transduction and that there are a suite of metabolic responses that subsequently increase the organisms' hardiness. Future study of such pathways should focus on the roles of intracellular pH and Ca^{2+}, and associated *trans*-membrane gradients, as modulators of tolerance metabolism as well as the potential existence of metal-ion specific tolerance mechanisms in higher plants.

Given the diverse physical and chemical characteristics of various metals, their different affinities for biological ligands and potential sites of interaction in the rhizosphere, at the plasma membrane or in the cytoplasm, a series of different physiological and biochemical pathways are most likely responsible for differential responses of plants to metal exposure. Many of these may be fortuitous end-products of perturbed biochemical function, and should be recognized as such. However, the elucidation of ion-specific detoxification mechanisms in some biological systems amply indicates that genetic information encoding for specific metal detoxification mechanisms exists and calls for future investigation in other plant systems.

The search for specific modes of metal toxicity and mechanisms of metal tolerance has inadvertently missed an integrated approach to the study of metal tolerance metabolism in plants. Mechanistic studies involved in elucidating the short-term responses of plants to metals in the environment may bear little semblance to mechanisms in populations under intense selective pressures. As such, metal tolerance in metal-tolerant ecotypes, populations, or species may involve many, some, or none of the mechanisms outlined to date. With our increased understanding of plant signal transduction and gene expression, we will better appreciate the impacts of metals on plant systems and the responses of plants to cellular perturbations resulting from metal exposure.

REFERENCES

1. Levitt, J. *Responses of Plants to Environmental Stresses I. Chilling, Freezing and High Temperature Stresses* (New York: Academic Press, 1980) p. 497.
2. Levitt, J. *Responses of Plants to Environmental Stresses II. Water, Radiation, Salt, and Other Stresses* (New York: Academic Press, 1980) p. 606.
3. Bennet, R. J., C. M. Breen, and M. V. Fey. "The Effects of Aluminum on Root Cap Function and Root Development in *Zea mays* L.," *Environ. Exp. Bot.* 27:91–104 (1987).
4. Burke, D. G., K. Watkins, and B. J. Scott. "Manganese Toxicity Effects on Visible Symptoms, Yield, Manganese Levels, and Organic Acid Levels in Tolerant and Sensitive Wheat Cultivars," *Crop Sci.* 30:275–280 (1990).
5. Cocucci, S. M. and S. Morgutti. "Stimulation of Proton Extrusion by K^+ and Divalent Cations (Ni^{2+}, Co^{2+}, Zn^{2+}) in Maize Root Segments," *Physiol. Plant.* 68:497–501 (1986).
6. Fuhrer, J. "Ethylene Biosynthesis and Cadmium Toxicity in Leaf Tissue of Beans (*Phaseolus Vulgaris* L.)," *Plant Physiol.* 70:162–167 (1982).
7. Godbold, D. L., W. J. Horst, H. Marschner, J. C. Collins, and D. A. Thurman. "Root Growth and Zn Uptake by Two Ecotypes of *Deschampsia Caespitosa* as Affected by High Zn Concentrations," *Z. Pflanzenphysiol.* 112:315–324 (1983).
8. Jensen, P. and S. Adalsteinsson. "Effects of Copper on Active and Passive Rb Influx in Roots of Winter Wheat," *Physiol. Plant.* 75:195–200 (1989).
9. Keck, R. W. "Cadmium Alteration of Root Physiology and Potassium Ion Fluxes," *Plant Physiol.* 62:94–96 (1978).
10. Kennedy, C. D. and F. A. N. Gonsalves. "The Action of Divalent Zinc, Cadmium, Mercury, Copper and Lead on the Trans Root Potential and H^+ Efflux of Excised Roots," *J. Exper. Bot.* 38:800–817 (1987).
11. Lindberg, S. and G. Wingstrand. "Mechanism for Cd^{2+} Inhibition of (K^+ Mg^{2+})ATPase Activity and K^+($^{86}Rb^+$) Uptake in Roots of Sugar Beet (*Beta vulgaris*)," *Physiol. Plant.* 63:181–186 (1985).
12. Lolkema, P. C. and R. Vooijs. "Copper Tolerance in *Silene cucubalus*. Subcellular Distribution of Copper and its Effects on Chloroplasts and Plastocyanin Synthesis," *Planta* 167:30–36 (1986).
13. Matsumoto, H. and T. Yamaya. "Inhibition of Potassium Uptake and Regulation of Membrane Associated Mg^{2+}-ATPase Activity of Pea Roots by Aluminum," *Soil Sci. Plant Nutr.* 32:179–188 (1986).
14. Morgutti, S., G. A. Sacchi, and S. M. Cocucci. "Effects of Ni^{2+} on Proton Extrusion, Dark CO_2 Fixation and Malate Synthesis in Maize Roots," *Physiol. Plant.* 60:70–74 (1984).
15. Veltrup, W. "The *In Vivo* and *In Vitro* Effects of Ca^{2+} and Al^{3+} Upon ATPases from Barley Roots," *J. Plant Nutr.* 6:349–361 (1983).
16. Witton, B. A. and F. H. A. Shehata. "Influence of Cobalt, Nickel, Copper, and Cadmium on the Blue-Green Alga *Anacystis nidulans*," *Environ. Pollut.* (A) 27:275–281 (1982).
17. Cumming, A. B., J. R. Cumming, and G. J. Taylor. "Patterns of Root Respiration Associated with the Induction of Aluminum Tolerance in Bean (*Phaseolus vulgaris* L.)," *J. Exp. Bot.* (in press).
18. Aniol, A. "Induction of Aluminum Tolerance in Wheat Seedlings by Low Doses of Aluminum in Nutrient Solution," *Plant Physiol.* 75:551–555 (1984).

19. Brown, H. and M. H. Martin. "Pretreatment Effects of Cadmium on the Root Growth of *Holcus lanatus* L., " *New Phytol.* 89:621–629 (1981).
20. Weitzel, G., U. Pilatus, and L. Rensing. "Similar Dose Response of Heat Shock Protein Synthesis and Intracellular pH Change in Yeast," *Exp. Cell Res.* 159:252–256 (1985).
21. Czarnecka, E., L. Edelman, F. Schoffl, and J. L. Key. "Comparative Analysis of Physical Stress Responses in Soybean Seedlings Using Cloned Heat Shock cDNAs," *Plant Mol. Biol.* 3:45–58 (1984).
22. Lin, C.-Y., J. K. Roberts, and J. L. Key. "Acquisition of Thermotolerance in Soybean Seedlings," *Plant Physiol.* 74:152–160 (1984).
23. Bonham-Smith, P. C., M. Kapoor, and J. D. Bewley. "Establishment of Thermotolerance in Maize by Exposure to Stresses Other Than a Heat Shock Does Not Require Heat Shock Protein Synthesis," *Plant Physiol.* 85:575–580 (1987).
24. Orzech, K. A. and J. J. Burke. "Heat Shock and the Protection Against Metal Toxicity in Wheat Leaves," *Plant Cell Environ.* 11:711–714 (1988).
25. Nieboer, E. and D. H. S. Richardson. "The Replacement of the Nondescript Term 'Heavy Metals' by a Biologically and Chemically Significant Classification of Metal Ion," *Environ. Pollut.* (B) 1:3–26 (1980).
26. Grill, E., M. H. Zenk, and E. L. Winnacker. "Induction of Heavy Metal-Sequestering Phytochelatins by Cadmium in Cell Cultures of *Rauvolfia serpentina*," *Naturwissenschaften* 72:432–433 (1985).
27. Reese, R. N., R. K. Mehra, E. B. Tarbet, and D. R. Winge. "Studies on the -Glutamyl Cu-Binding Peptide from *Schizosaccharomyces pombe*," *J. Biol. Chem.* 263:4168–4192 (1988).
28. Silver, S. and T. K. Misra. "Plasmid-Mediated Heavy Metal Resistances," *Annu. Rev. Microbiol.* 42:717–743 (1988).
29. Kurkjian, A. and J. Guern. "Intracellular pH: Measurement of and Importance in Cell Activity," *Annu. Rev. Plant Physiol. Plant Mol. Biol.* 40:271–303 (1989).
30. Felle, H. "Short-Term pH Regulation in Plants," *Physiol. Plant.* 74:583–591 (1988).
31. Spanswick, R. M. "Electrogenic Ion Pumps," *Annu. Rev. Plant Physiol.* 32:267–289 (1981).
32. Spanswick, R. M. "The Role of H^+-ATPases in Plant Nutrient Transport," in *Frontiers of Membrane Research in Agriculture*, St. John, J. B., E. Berlin, and P. C. Jackson, Eds. (Totowa, NJ: Rowman and Allenheld, 1985) pp. 243–256.
33. Sze, H. "H^+-Translocating ATPases of the Plasma Membrane and Tonoplast of Plant Cells," *Physiol. Plant.* 61:683–691 (1984).
34. Poole, R. J. "Plasma Membrane and Tonoplast," in *Solute Transport in Plant Cells and Tissues*, Baker, D. A. and J. L. Hull, Eds. (New York: John Wiley & Sons, 1988) pp. 83–105.
35. Zilberstein, D., V. Agmon, S. Schuldiner, and E. Padan. "The Sodium/Proton Antiporter is Part of the pH Homeostasis Mechanism in *Escherichia coli*," *J. Biol. Chem.* 257:3687–3691 (1982).
36. Blatt, M. R. and C. L. Slayman. "Role of 'Active' Potassium Transport in the Regulation of Cytoplasmic pH by Nonanimal Cells," *Proc. Natl. Acad. Sci. U.S.A.* 84:2737–2741 (1987).
37. Guern, J., Y. Mathieu, and A. Kurkdjian. "Phosphoenolpyruvate Carboxylase Activity and the Regulation of Intracellular pH in Plant Cells," *Physiol. Veg.* 21:855–866 (1983).

38. Morisawa, M. and R. A. Steinhardt. "Changes in Intracellular pH of *Physarum* Plasmodium During the Cell Cycle and in Response to Starvation," *Exp. Cell Res.* 140:341–351 (1982).
39. Gillies, R. J., K. Ugurbil, J. A. den Hollander, and R. G. Shulman. "^{31}P NMR Studies of Intracellular pH and Phosphate Metabolism During Cell Division Cycle of *Saccharomyces cerevisiae*," *Proc. Natl. Acad. Sci. U.S.A.* 78:2125–2129 (1981).
40. Kay, R. R., D. G. Gadian, and S. R. Williams. "Intracellular pH in *Dictyostelium*: a ^{31}P Nuclear Magnetic Resonance Study of its Regulation and Possible Role in Controlling Cell Differentiation," *J. Cell Sci.* 83:165–179 (1986).
41. Gross, J. D., J. Bradbury, R. R. Kay, and M. J. Peacey. "Intracellular pH and the Control of Cell Differentiation in *Dictyostelium Discoidium*," *Nature (London)* 303:244–245 (1983).
42. Caspani, G., P. Tortora, G. M. Hanozet, and A. Guerritore. "Glucose Stimulated cAMP Increase may be Mediated by Intracellular Acidification in *Saccharomyces cervisiae*," *FEBS Lett.* 186:75–79 (1985).
43. Thevelieu, J. M., M. Beullens, F. Honshoven, G. Hoebeck, K. Detremerie, J. A. den Hollander, and A. W. H. Jans. "Regulation of the cAMP Level in the Yeast *Saccharomyces cervisiae*: Intracellular pH and the Effect of Membrane Depolarization Compounds," *J. Gen. Microbiol.* 133:2191–2196 (1987).
44. Valle, E., L. Bergillus, S. Gaseon, F. Parra, and S. Remos. "Trehalase Activation in Yeasts: Mediated by an Initial Acidification," *Eur. J. Biochem.* 154:247–251 (1986).
45. Hildago, A., P. Nauas, and G. Garcia-Herdago. "Growth Inhibition by Vanadate in Onion Roots," *Environ. Exp. Bot.* 28:131–136 (1988).
46. Basso, B., P. Rosa, and L. Tognoli. "Microsome Protein Phosphorylation in *Acer pseudoplatanus* Cell as Influenced by Intracellular Alkalinization," *Plant Cell Environ.* 12:191–196 (1989).
47. Tognoli, L. and B. Basso. "The Fusicoccin Stimulated Phyosphorylation of a 33 kDa Polypeptide in Cells of *Acer pseudoplatanus* as Influenced by Extracellular and Intracellular pH," *Plant Cell Environ.* 10:233–239 (1987).
48. Weitzel, G., U. Pilatus, and L. Rensing. "The Cytoplamic pH, ATP and Total Protein Synthesis Rate During Heat Shock Protein Inducing Treatments in Yeast," *Exp. Cell Res.* 170:64–79 (1987).
49. Hedrich, R. and J. I. Schroeder. "The Physiology of Ion Channels and Electrogenic Pumps in Higher Plants," *Annu. Rev. Plant Physiol.* 40:539:569 (1989).
50. Hepler, P. K. and R. O. Wayne. "Calcium and Plant Development," *Annu. Rev. Plant Physiol.* 36:397–439 (1985).
51. Rincon, M. and J. B. Hanson. "Controls on Calcium Ion Fluxes in Injured or Shocked Corn Root Cells: Importance of Proton Pumping and Cell Membrane Potential," *Plant Physiol.* 67:576–583 (1986).
52. Trevillyan, J. M. and M. L. Pall. "Control of Cyclic Adenosine 3′,5′-Monophosphate Levels by Depolarizing Agents in Fungi," *J. Bacteriol.* 138:397–403 (1979).
53. Marre, M. T., G. Romani, M. Cocucci, M. M. Moloney, and E. Marre. "Divalent Cation Influx, Depolarization of the Transmembrane Electrical Potential and Proton Extrusion in Maize Root Segments," in *Plasmalemma and Tonoplast: Their Functions in the Plant Cell*, Marme, D., E. Marre, and R. Hertel, Eds. (Amsterdam: Elsevier Biomedical Press, 1982) pp. 3–13.

54. Cumming, J. R. and G. J. Taylor. "Mechanisms of Metal Tolerance in Plants: Adaptations for Exclusion of Metal Ions From the Cytoplasm," in *Stress Responses in Plants: Adaptation and Acclimation Mechanisms*, Alscher, R. G. and J. R. Cumming, Eds. (New York: Wiley-Liss, 1990) pp. 329–356.
55. Miyasaka, S. C., L. V. Kochian, J. E. Shaff, and C. D. Foy. "Mechanisms of Al^{3+} Tolerance in Wheat. An Investigation of Genotypic Differences in Rhizosphere pH, K^+ and H^+ Transport, and Root-Cell Membrane Potentials," *Plant Physiol.* 91:1188–1196 (1989).
56. Jaffe, L. F. "Control of Development by Ionic Currents," in *Biological Structures and Couple Flow*, Oplatha, A. and J. M. Balaban, Eds. (New York: Academic Press, 1983) pp. 445–456.
57. Weisenseel, M. H. and L. E. Jaffe. "The Major Growth Current Through Lily Pollen Tubes Enters as K^+ and Leaves as H^+," *Planta* 133:1–7 (1977).
58. Weisenseel, M. H., A. Dorn, and L. F. Jaffe. "Natural H^+ Currents Traverse Growing Roots and Root Hairs of Barley (*Hordeum vulgare* L.)," *Plant Physiol.* 64:512–518 (1979).
59. Miller, A. L. and N. A. R. Gow. "Correlation Between Root-Generated Ionic Currents, pH, Fusicoccin, Indolacetic Acid, and Growth of the Primary Root of *Zea mays*," *Plant Physiol.* 89:1198–1206 (1989).
60. Miller, A. L., J. A. Raven, J. I. Sprent, and M. A. Weisenseel. "Endogenous Ion Currents Traverse Growing Roots and Root Hairs of *Trifolium repens*," *Plant Cell Environ.* 9:79–83 (1986).
61. Weisenseel, M. H. and R. M. Kircherer. "Ionic Currents as Control Mechanism in Cytomorphogenesis," in *Cytomorphogenesis in Plants*, Kiermayer, O., Ed. (New York: Springer-Verlag, 1981) pp. 379–399.
62. Bennet, R. J. and C. M. Breen. "The Aluminum Signal: New Dimensions to Mechanisms of Al Tolerance," 2nd Int. Symp. on Plant-Soil Interactions at Low pH, Beckley, WV (1990).
63. Kochian, L. V. and J. E. Shaff. "Investigating the Relationship Between Aluminum Toxicity, Root Growth and Root-Generated Ion Currents," *Plant Physiol.* submitted.
64. Schauf, C. L. and K. J. Wilson. "Properties of Single K^+ and Cl^- Channels in *Asclepias tuberosa* Protoplasts. *Plant Physiol.* 85:41–418 (1987).
65. Pooviah, B. W. and A. S. N. Reddy. "Calcium Messenger System in Plants," *CRC Crit. Rev. Plant Physiol.* 6:47–103 (1987).
66. Strickland, R. C., W. R. Chaney, and R. J. Lamoreaux. "Cadmium Uptake by *Pinus resinosa* Ait. Pollen and the Effect on Cation Release and Membrane Permeability," *Plant Physiol.* 64:366–370 (1979).
67. Leonard, R. T. and P. K. Hepler, Eds. *Calcium in Plant Growth and Development: Proceedings of the 13th Annual Riverside Symposium in Plant Physiology, Current Topics in Plant Physiology Vol. 4.* (Rockville, MD: American Society of Plant Physiologists, 1990) p. 205.
68. Marme, D. and P. Dieter. "Role of Ca^{2+} and Calmodulin in Plants," in *Calcium and Cell Function*, Vol. 4, Cheung, W. Y., Ed. (New York: Academic Press, 1983) pp. 263–311.
69. Einspahr, K. J. and G. A. Thompson, Jr. "Transmembrane Signalling Via Phosphatidylinositol 4,5-Bisphophate Hydrolysis in Plants," *Plant Physiol.* 93:361–366 (1990).

70. Briskin, D. P., L. H. Gildensoph, and S. Basu. "Characterization of the Ca^{2+}-Transporting ATPase of the Plant Plasma Membrane Using Isolated Membrane Vesicles," in *Calcium in Plant Growth and Development: Proceedings of the 13th Annual Riverside Symposium in Plant Physiology*, Leonard, R. T. and P. K. Hepler, Eds. (Rockville, MD: American Society of Plant Physiologists, 1990) pp. 46–54.
71. Kauss, H. "Some Aspects of Calcium-Dependent Regulation in Plant Metabolism," *Annu. Rev. Plant Physiol.* 38:47–72 (1987).
72. Siegel, N. and A. Haug. "Calmodulin-Dependent Formation of Membrane Potential in Barley Root Plasma Membrane Vesicles: A Biochemical Model of Aluminum Toxicity in Plants," *Plant Physiol.* 59:285–291 (1983).
73. Cheung, W. Y. "Calmodulin: Its Potential Role in Cell Prolification and Heavy Metal Toxicity," *Fed. Proc.* 43:2995–2999 (1984).
74. Wolniak, S. M., P. K. Hepler, and W. T. Jackson. "Ionic Changes in the Mitosis Apparatus at the Metaphase/Anaphase Transition," *J. Cell Biol.* 96:598–605 (1983).
75. Hepler, P. K. "Calcium Restriction Prolongs Metaphase in Dividing *Tradescantia* Stamen Hair Cells," *J. Cell Biol.* 100:1363–1368 (1985).
76. Ruth, J. B., J. L. Kotenko, and J. H. Miller. "Role of Asymmetric Cell Division in Pteridophyte Differentiation. II. Effect of Ca^{2+} on Asymmetric Cell Division, Rhizoid Elongation and Antheridium Differentiation in *Jittaria gemmae*," *Am. J. Bot.* 75:1755–1764 (1988).
77. Hepler, P. K., D. Zhang, and D. A. Callaham. "Calcium and the Regulation of Mitosis," in *Calcium in Plant Growth and Development: Proceedings of the 13th Annual Riverside Symposium in Plant Physiology*, Leonard, R. T. and P. K. Hepler, Eds. (Rockville, MD: American Society of Plant Physiologists, 1990) pp. 93–110.
78. Bennet, R. J. and V. Bandu. "The Ultrastructure Response of Root Cap Cells to the Ca^{2+} Chelating Agent EGTA," *Elect. Microsc. Soc. S. Afr. Proc.* 19:51–52 (1989).
79. Steffens, J. C. "The Heavy Metal-Binding Peptides of Plants," *Annu. Rev. Plant Physiol. Plant Mol. Biol.* 41:55–575 (1990).
80. Grill, E., J. Thurmann, E. L. Winnacker, and M. H. Zenk. "Induction of Heavy-Metal Binding Phyochelatins by Inoculation of Cell Cultures in Standard Media," *Plant Cell Rep.* 7:375–378 (1988).
81. Grill, E., E.-L. Winnacker, and M. H. Zenk. "Phytochelatins: The Principal Heavy-Metal Complexing Peptides of Higher Plants," *Science* 230:674–676 (1985).
82. Hamer, D. H. "Metallothionein," *Annu. Rev. Biochem.* 55:913–955 (1986).
83. Antonovics, J., A. D. Bradshaw, and R. G. Turner. "Heavy Metal Tolerance in Plants," *Adv. Ecol. Res.* 7:1–85 (1971).
84. Baker, A. J. M. "Metal Tolerance," *New Phytol.* 106(Suppl):93–111 (1987).
85. Bradshaw, A. D. "Adaptation of Plants to Soils Containing Toxic Metals — A Test for Conceit," in *Origins and Development of Adaptation*, Ciba Foundation Symposium 102 (London: Pitman, 1984) pp. 4–14.
86. Woolhouse, H. W. "Toxicity and Tolerance in the Responses of Plants to Metals" in *Physiological Plant Ecolgy III: Responses to the Chemical and Biological Environment*, Lange, O. L., P. S. Nobel, C. B. Osmond, and H. Ziegler, Eds. (New York: Springer-Verlag, 1983) pp. 246–300.

87. Baker, A. J. M. and P. L. Walker. "Ecophysiology of Metal Uptake by Tolerant Plants," in *Heavy Metal Tolerance in Plants: Evolutionary Aspects*, Shaw, A. J., Ed. (Boca Raton, FL: CRC Press, 1990) pp. 155–214.
88. Baker, A. J. M. and J. Proctor. "The Influence of Cadmium, Copper, Lead and Zinc on the Distribution and Evolution of Metallophytes in the British Isles," *Plant Syst. Evol.* 173:91–108 (1990).
88a. Shaw, A. J., Ed. *Evolutionary Aspects of Heavy Metal Tolerance in Plants* (Boca Raton, FL: CRC Press, 1990).
89. Taylor, G. J. "Exclusion of Metals from the Symplasm: A Possible Mechanism of Metal Tolerance in Higher Plants," *J. Plant Nutr.* 10:1213–1222 (1987).
90. Baker, A. J. M. "Accumulators and Excluders — Strategies in the Response of Plants to Heavy Metals," *J. Plant Nutr.* 3:643 (1981).
91. Haug, A. "Molecular Aspects of Aluminum Toxicity," *CRC Crit. Rev. Plant Sci.* 1:345–373 (1984).
92. Thibaud, J.-B., C. Romieu, R. Gibrat, J.-P. Grouzis, and C. Grignon. "Local Ionic Environment of Plant Membranes: Effects on Membrane Functions," *Z. Pflanzenphysiol.* 114:207–213 (1984).
93. Green, D. E., M. Fry, and G. Blondin. "Phospholipids as the Molecular Instruments of Ion and Solute Transport in Biological Membranes," *Proc. Natl. Acad. Sci. U.S.A.* 77:257–261 (1980).
94. Cullis, P. R. and B. De Kruijff. "Lipid Polymorphism and the Functional Roles of Lipids in Biological Membranes," *Biochim. Biophys. Acta* 559:399–420 (1979).
95. Haug, A. and C. R. Caldwell. "Aluminum Toxicity in Plants: The Role of the Root Plasma Membrane and Calmodulin," in *Frontiers of Membrane Research in Agriculture*, St. John, J. B., E. Berlin, and P. C. Jackson, Eds. (Totowa, NJ: Rowman and Allanheld, 1985) pp. 359–381.
96. Thompson, G. A., Jr. "Mechanisms of Membrane Response to Environmental Stress," in *Frontiers of Membrane Research in Agriculture*, St. John, J. B., E. Berlin, and P. C. Jackson, Eds. (Totowa, NJ: Rowman and Allanheld, 1985) pp. 347–357.
97. Douglas, T. J. "NaCl Effects on 4-Desmethylsterol Composition of Plasma-Membrane-Enriched Preparations from Citrus Roots," *Plant Cell Environ.* 8:687–692 (1985).
98. Brown, D. J. and F. M. DuPont. "Lipid Composition of Plasma Membranes and Endomembranes Prepared from Roots of Barley (*Hordeum vulgare* L.). Effects of Salt," *Plant Physiol.* 90:955–961 (1989).
99. Brotherus, J. R., P. C. Jost, O. H. Griffith, J. F. W. Keana, and L. E. Hokin. "Charge Selectivity at the Lipid-Protein Interface of Membranous Na^+,K^+-ATPase," *Proc. Natl. Acad. Sci. U.S.A.* 77:272–276 (1980).
100. Caldwell, C. R. and A. Haug. "Divalent Cation Inhibition of Barley Root Plasma Membrane-Bound Ca^{2+}-ATPase Activity and its Reversal by Monovalent Cations," *Physiol. Plant.* 54:112–118 (1982).
101. Douglas, T. J. and R. R. Walker. "4-Desmethylsterol Composition of Citrus Rootstocks of Different Salt Exclusion Capacity," *Physiol. Plant.* 58:69–74 (1983).
102. Douglas, T. J. and S. R. Sykes. "Phospholipid, Galactolipid, and Free Sterol Composition of Fibrous Roots from Citrus Genotypes Differing in Chloride Exclusion Ability," *Plant Cell Environ.* 8:693–699 (1985).

103. Gadd, G. M., A. Stewart, C. White, and J. L. Mowll. "Copper Uptake by Whole Cells and Protoplasts of a Wild-Type and Copper Resistant Strain of *Saccharaomyces cerevisiae*," *FEMS Microbiol. Lett.* 24:231–234 (1984).
104. Butler, M., A. E. J. Haskew, and M. M. Young. "Copper Tolerance in the Green Alga, *Chlorella vulgaris*," *Plant Cell Environ.* 3:119–126 (1980).
105. Beckett, R. P. and D. H. Brown. "The Relationship Between Cadmium Uptake and Heavy Metal Tolerance in the Lichen Genus *Peltigera*," *New Phytol.* 97:301–311 (1984).
106. Bariaud, A., M. Bury, and J. C. Mestre. "Mechanism of Cd^{2+} Resistance in *Euglena gracilis*," *Physiol. Plant.* 63:382–386 (1985).
107. Meharg, A. A. and M. R. MacNair. "An Altered Phosphate Uptake System in Arsenate-Tolerant *Holcus lanatus* L., " *New Phytol.* 116:29–35 (1990).
108. Caldwell, C. R. "Analysis of Aluminum and Divalent Cation Binding to Wheat Root Plasma Membrane Proteins Using Terbium Phosphorescence," *Plant Physiol.* 91:233–241 (1989).
109. Xu, S. and G. Patterson. "The Biochemical Effects of Cadmium on Sterol Biosynthesis by Soybean Suspension Culture," *Curr. Top. Plant Biochem. Physiol.* 4:245–248 (1985).
110. Steer, M. W. "Plasma Membrane Turnover in Plant Cells," *J. Exp. Bot.* 39:987–996 (1988).
111. Nies, D. H. and S. Silver. "Plasmid-Determined Inducible Efflux is Responsible for Resistance to Cadmium, Zinc, and Cobalt in *Alcaligenes eutrophus*," *J. Bacteriol.* 171:896–900 (1989).
112. Efstathiou, J. D. and L. L. McKay. "Inorganic Salts Resistance Associated with a Lactose-Fermenting Plasmid in *Streptococcus lactis*," *J. Bacteriol.* 130:257–265 (1977).
113. Bopp, L. H., A. M. Chakrabarty, and H. L. Ehrlich. "Chromate Resistance Plasmids in *Pseudomonas fluorescens*," *J. Bacteriol.* 155:1105–1109 (1983).
114. Zhang, G. and G. J. Taylor. "Kinetics of Aluminum Uptake by Excised Roots of Aluminum-Tolerant and Aluminum-Sensitive Cultivars of *Triticum aestivum* L.," *Plant Physiol.* 91:1094–1099 (1989).
115. Huett, D. O. and R. C. Menary. "Aluminum Uptake by Excised Roots of Cabbage, Lettuce, and Kikuyu Grass," *Aust. J. Plant Physiol.* 6:643–653 (1979).
116. Wagatsuma, T. "Effect of Non-Metabolic Conditions on the Uptake of Aluminum by Plant Roots," *Soil Sci. Plant Nutr.* 29:323–333 (1983).
117. Marschner, H. *Mineral Nutrition in Higher Plants* (New York: Academic Press, 1986).
118. Brookes, A., J. C. Collins, and D. A. Thurman. "The Mechanism of Zinc Tolerance in Grasses," *J. Plant Nutr.* 3:695–705 (1981).
119. Thurman, D. A. and J. L. Rankin. "The Role of Organic Acids in Zinc Tolerance in *Deschampsia caespitosa*," *New Phytol.* 91:629–635 (1982).
120. Godbold, D. L., W. J. Horst, J. C. Collins, D. A. Thurman, and H. Marschner. "Accumulation of Zinc and Organic Acids in Roots of Zn Tolerant and Non-Tolerant Ecotypes of *Deschampsia caespitosa*," *J. Plant Physiol.* 116:59–69 (1984).
121. Krotz, R. M., B. P. Evangelou, and G. J. Wagner. "Relationships Between Cadmium, Zinc, Cd-Peptide, and Organic Acid in Tobacco Suspension Cells," *Plant Physiol.* 91:780–787 (1989).

122. Mathys, W. "The Role of Malate, Oxalate, and Mustard Oil Glucosides in the Evolution of Zinc-Resistance in Herbage Plants," *Physiol. Plant.* 40:130–136 (1977).
123. Vogeli-Lange, R. and G. J. Wagner. "Subcellular Localization of Cadmium and Cadmium-Binding Peptides in Tobacco Leaves: Implication of a Transport Function for Cd-Binding Peptides," *Plant Physiol.* 92:1086–1093 (1990).
124. Van Steveninck, R. F. M., M. E. Van Steveninck, D. R. Fernando, W. J. Horst, and H. Marschner. "Deposition of Zinc Phytate in Globular Bodies of *Deschampsia caespitosa* Ecotypes; a Detoxification Mechanism?" *J. Plant Physiol.* 131:247–257 (1987).
125. Lee, J., R. D. Reeves, R. R. Brooks, and T. Jaffre. "Isolation and Identification of a Citrato-Complex of Nickel from Nickel-Accumulating Plants," *Phytochemistry* 16:1503–1505 (1977).
126. Brooks, R. R., S. Shaw, and A. Asensi Marfil. "The Chemical Form and Physical Function of Nickel in Some Iberian *Aylssum specum,*" *Physiol. Plant.* 51:167–170 (1981).
127. Ksiazek, M., A. Wozny, and F. Mlodzianowski. "Effect of $Pb(NO_3)_2$ on Poplar Tissue Culture and the Ultrastructural Localization of Lead in Culture Cells," *For. Ecol. Manage.* 8:95–105 (1984).
128. Ksiazek, M., A. Wozny, and R. Siwecki. "The Sensitivity of Poplar Leaves to Lead Nitrate and the Intracellular Localization of Lead," *Eur. J. For. Pathol.* 14:113–122 (1984).
129. Lane, S. D. and E. S. Martin. "An Ultrastructural Examination of Lead Localization in Germinating Seeds of *Raphanus sativus,*" *Z. Pflanzenphysiol.* 107:33–40 (1982).
130. Silverburg, B. A. "Ultrastructural Localization of Lead in *Strigeoclonium tunue* (Chlorphyceae, Ulotrichales) as Demonstrated by Cytochemical and X-Ray Microanalysis," *Phycologia* 14:265–274 (1975).
131a. Clarkson, D. T. "Factors Affecting Mineral Nutrient Acquisition by Plants," *Annu. Rev. Plant Physiol.* 36:77–115 (1985).
131. Rovira, A. D. "Plant Root Exudates," *Bot. Rev.* 35:35–57 (1969).
132. Merckx, R., J. H. van Ginkel, J. Sinnaeve, and A. Cremers. "Plant-Induced Changes in the Rhizosphere of Maize and Wheat. I. Production and Turnover of Root-Derived Material in the Rhizosphere of Maize and Wheat," *Plant Soil* 96:85–93 (1986).
133. Merckx, R., J. H. van Ginkel, J. Sinnaeve, and A. Cremers. "Plant-Induced Changes in the Rhizosphere of Maize and Wheat. II. Complexation of Cobalt, Zinc, and Manganese in the Rhizosphere of Maize and Wheat," *Plant Soil* 96:95–107 (1986).
134. Kaczorowski, G. J. and H. R. Kaback. "Mechanism of Lactose Translocation in Membrane Vesicles from *Escherichia coli* I. Effect of pH on Efflux, Exchange, and Counterflow," *Biochemistry* 18:3691–3697 (1979).
135. Otto, R., A. S. M. Sonnenberg, H. Veldkamp, and W. N. Konings. "Generation of an Electrochemical Proton Gradient in *Streptococcus cremoris* by Lactate Efflux," *Proc. Natl. Acad. Sci. U.S.A.* 77:5502–5506 (1980).
136. Konings, W. N. "Generation of Metabolic Energy by End-Product Efflux," *Trends Biol. Sci.* 10:317–319 (1985).
137. Bartlett, R. J. and D. C. Riego. "Effect of Chelation on the Toxicity of Aluminum," *Plant Soil* 37:419–423 (1972).

138. Coombes, A. J., D. A. Phipps, and N. W. Lepp. "Uptake Patterns of Free and Complexed Copper Ions in Excised Roots of Barley (*Hordeum vulgare* L.C.V. Zephyr)," *Z. Pflanzenphysiol. Bd.* 82:435–439 (1977).
139. Taylor, G. J. and C. D. Foy. "Differential Uptake and Toxicity of Ionic and Chelated Copper in *Triticum aestivum*," *Can. J. Bot.* 63:1271–1275 (1985).
140. Clarke, S. E., J. Stuart, and J. Sanders-Loehr. "Induction of Siderophore Activity in *Anabaena* spp. and Its Moderation of Copper Toxicity," *Appl. Environ. Microbiol.* 53:917–922 (1987).
141. Hue, N. V., G. R. Craddock, and F. Adams. "Effect of Organic Acids on Aluminum Toxicity in Subsoils," *Soil Sci. Soc. Am. J.* 50:28–34 (1986).
142. Shi, B. and A. Haug. "Uptake of Aluminum by Lipid Vesicles," *Toxicol. Environ. Chem.* 17:337–349 (1988).
143. Suhayda, C. G. and A. Haug. "Organic Acids Reduce Aluminum Toxicity in Maize Root Membranes," *Physiol. Plant.* 68:189–195 (1986).
144. Denny, H. J. and D. A. Wilkins. "Zinc Tolerance in *Betula* spp. IV. The Mechanism of Ectomycorrhizal Amelioration of Zinc Toxicity," *New Phytol.* 106:545–553 (1987).
145. Cumming, J. R. and L. H. Weinstein. "Aluminum-Mycorrhizal Interactions in the Physiology of Pitch Pine Seedlings," *Plant Soil* 125:7–18 (1990).
146. Horst, W. J., A. Wagner, and H. Marschner. "Mucilage Protects Root Meristems from Aluminium Injury," *Z. Pflanzenphysiol.* 105:435–444 (1982).
147. Bennet, R. J., C. M. Breen, and M. V. Fey. "Aluminium Uptake Sites in the Primary Root of *Zea mays* L., " *S. Afr. J. Plant Soil* 2:1–7 (1985).
148. Fincher, G. B. and B. A. Stone. "Metabolism of Noncellulosic Polysaccharides," in *Encyclopedia of Plant Physiology 13B, Plant Carbohydrates ii. Extracellular Carbohydrates*, Tanner, W. and F. A. Loewus, Eds. (New York: Springer-Verlag, 1981) pp. 68–132.
149. Wissenmeier, A. H., F. Klotz, and W. J. Horst. "Aluminum Induced Callose Synthesis in Roots of Soybean (*Glycine max* L.)," *J. Plant Physiol.* 129:487–492 (1987).
150. Kohle, H., W. Jeblick, F. Poten, W. Blaschek, and H. Kauss. "Chitosan-Elicited Callose Synthesis in Soybean Cells as a Ca^{2+} Dependent Process," *Plant Physiol.* 77:544–551 (1985).
151. Bacic, A. and D. P. Delmer. "Stimulation of Membrane Associated Polysaccharide Synthetases by a Membrane Potential in Developing Cotton Fibers," *Planta* 152:346–351 (1981).
152. Jacob, S. R. and D. H. Northcote. "*In Vitro* Glucan Synthesis by Membranes of Celery Petioles: The Role of the Membrane in Determining the Type of Linkage Formed," *J. Cell Sci. Suppl.* 2:1–11 (1985).
153. Kauss, H. and W. Jeblick. "Synergistic Activation of 1,3-β-Glucan Synthase by Ca^{2+} and Polyamines," *Plant Sci.* 43:103–107 (1986).
154. Martin, R. B. "Bioinorganic Chemistry of Metal Ion Toxicity," in *Metal Ions in Biological Systems, Vol. 21, Concepts on Metal Ion Toxicity*, Siegle, H. Ed. (New York: Marcel Dekker, 1986) pp. 21–65.
155. Mullins, G. L., L. E. Sommers, and T. L. Housley. "Metal Speciation in Xylem and Phloem Exudates," *Plant Soil* 96:377–391 (1986).
156. Foy, C. D., E. H. Lee, C. A. Coradetti, and G. J. Taylor. "Organic Acids Related to Differential Aluminium Tolerance in Wheat (*Triticum aestivum*) Cultivars," in *Plant Nutrition — Physiology and Applications*, van Beusichem, M. L., Ed. (The Hauge, Netherlands: Kluwer Academic Publishers, 1990) pp. 381–389.

157. Tomsett, A. B. and D. A. Thurman. "Molecular Biology of Metal Tolerances of Plants," *Plant Cell Environ.* 11:383–394 (1988).
158. Beach, L. and R. D. Palmiter. "Amplification of the Metallothionein I Gene in Cadmium-Resistant Mouse Cells," *Proc. Natl. Acad. Sci. U.S.A.* 78:2110–2114 (1981).
159. Rauser, W. E. "Phytochelatins," *Annu. Rev. Biochem.* 59:61–86 (1990).
160. Robinson, N. J. "Metal-Binding Polypeptides in Plants," in *Heavy Metal Tolerance in Plants: Evolutionary Aspects*, Shaw, A. J., Ed. (Boca Raton, FL: CRC Press, 1990) pp. 195–214.
161. Grill, E., E.-L. Winnacker, and M. H. Zenk. "Phytochelatins, a Class of Heavy Metal-Binding Peptides From Plants, Are Functionally Analogous to Metallothioneins," *Proc. Natl. Acad. Sci. U.S.A.* 84:439–443 (1987).
162. Mehra, R. K., E. B. Tarbet, W. R. Gray, and D. R. Winge. "Metal-Specific Synthesis of Two Metallothioneins and -Glutamyl Peptides in *Candida glabrata*," *Proc. Natl. Acad. Sci. U.S.A.* 85:8815–8819 (1988).
163. Gekeler, W., E. Grill, E.-L., Winnacker, and M. H. Zenk. "Survey of the Plant Kingdom for the Ability to Bind Heavy Metals Through Phytochelatins," *Z. Naturforsch.* 44c:361–369 (1989).
164. Reese, R. N. and G. J Wagner. "Properties of Tobacco (*Nicotiana tabacum*) Cadmium-Binding Peptide(s)," *Biochem. J.* 241:641–647 (1987).
165. Tomsett, A. B., D. E. Salt, J. R. DeMiranda, and D. A. Thurman. "Metallothioneins and Metal Tolerance," *Aspects Appl. Biol.* 22:365–372 (1989).
166. Jones, S. Unpublished results (1990).
167. Sewell, A. Unpublished results (1990).
168. de Miranda, J. R., M. G. Thomas, D. A. Thurman, and A. B. Tomsett. "Metallothionein Genes from the Flowering Plant *Mimulus guttatus*," *FEBS Lett.* 260:277–280 (1990).
169. Evans, I. M., L. N. Gatehouse, J. A. Gatehouse, N. J. Robinson, and R. R. D. Croy. "A Gene from Pea (*Pisum sativum*) with Homology to Metallothionein Genes," *FEBS Lett.* 262:29–32 (1990).
170. de Framond, A. Unpublished results (1990).
171. Chongpraditnum, P. and M. Chino. Unpublished results (1990).
172. Kagi, J. H. R. and A. Schaffer. "Biochemistry of Metallothionein," *Biochemistry* 27:8509–8515 (1988).
173. Loeffler, S., A. Hochberger, E. Grill, E.-L. Winnacker, and M. H. Zenk. "Termination of the Phytochelatin Synthase Reaction Through Sequestration of Heavy Metals by the Reaction Product," *FEBS Lett.* 258:42–46 (1989).
174. Tynecka, Z., Z. Gos, and J. Zajac. "Energy-Dependent Efflux of Cadmium Coded by a Plasmid Resistance Determinant in *Staphylococcus aureus*," *J. Bacteriol.* 147:313–319 (1981).
175. Perry, R. D. and S. Silver. "Cadmium and Manganese Transport in *Staphylococcus aureus* Membrane Vesicles," *J. Bacteriol.* 150:973–976 (1982).
176. Mohan, P. M. and K. Sivarama Sastry. "Interrelationship in Trace-Element Metabolism in Metal Toxicity in Nickel-Resistant Strains of *Neurospora crassa*," *Biochem. J.* 212:205–212 (1983).
177. Venkateswerlu, G. and K. Sivarama Sastry. "Interrelationships in Trace-Element Metabolism in Metal Toxocities in a Cobalt-Resistant Strain of *Neurospora crassa*," *Biochem. J.* 132:673–680 (1973).
178. Verkleij, J. A. C. and J. E. Prast. "Cadmium Tolerance and Co-Tolerance in *Silene vulgaris* (Moench.) Garcke [=S. cucubalus (L.) Wib.]," *New Phytol.* 111:637–645 (1989).

179. Rauser, W. E. and N. R. Curvetto. "Metallothionein Occurs in Roots of *Agrostis* Tolerant to Excess Copper," *Nature (London)* 287:563–564 (1980).
180. Bariaud, A. and J. C. Mestre. "Heavy Metal Tolerance in a Cadmium Resistant Population of *Euglena gracilis*," *Bull. Environ. Contam. Toxicol.* 32:597–601 (1984).
181. Scheller, H. V., B. Huang, E. Hatch, and P. B. Goldbrough. "Phytochelatin Synthesis and Glutathione Levels in Response to Heavy Metals in Tomato Cells," *Plant Physiol.* 85:1031–1035 (1987).
182. Steffans, J. C. and W. Williams. "Heavy Metal Tolerance in Tomato Cells," *Plant Biol.* 4:109–118 (1987).
183. Verkleij, J. A. C. and W. B. Bast-Cramer. "Co-Tolerance and Multiple Heavy Metal Tolerance in *Silene cucubalus* from Different Heavy-Metal Sites" in *Proceedings International Conference on Heavy Metals in the Environment, Athens*, Lekkas, T. D., Ed. (Edinburgh: CEP Consultants, 1985) pp. 174–176.
184. Cox, R. M. and T. C. Hutchinson. "Multiple Metal Tolerances in the Grass *Deschampsia caespitosa* (L.) Beauv. From the Sudbury Smelting Area," *New Phytol.* 84:631–647 (1980).
185. Cox, R. M. and T. C. Hutchinson. "Metal-Cotolerance in the Grass *Deschampsia caespitosa*," *Nature (London)* 279:231–237 (1979).
186. Kishinami, I. and J. M. Widholm. "Selection of Copper and Zinc Resistant *Nicotina plumbaginifolia* Cell Suspension Cultures," *Plant Cell Physiol.* 27:1263–1268 (1986).
187. Robinson, N. J., R. L. Ratcliff, P. J. Anderson, E. Delhaize, J. M. Berger, and P. J. Jackson. "Biosynthesis of Poly(γ-Glutamyl-Cysteinyl) Glycines in Cadmium Resistant *Datura innoxia* Cells," *Plant Sci.* 56:197–204 (1988).
188. Butt, T. R. and D. J. Ecker. "Yeast Metallothionein and Applications to Biotechnology," *Microbiol. Rev.* 51:351–364 (1987).
189. Furst, F., S. Hu, R. Hackett, and D. Hamer. "Copper Activates Metallothionein Gene Transcription by Altering the Conformation of a Specific DNA-Binding Protein," *Cell* 55:705–717 (1988).
190. Struhl, R. "Helix-Turn-Helix, Zinc-Finger, and Leucine Zipper Motifs for Eukaryotic Transcriptional Regulatory Proteins," *TIBS* 14:137–140 (1989).

13

Dissipation of Soil Selenium by Microbial Volatilization

W. T. Frankenberger, Jr. and U. Karlson

Department of Soil and Environmental Sciences, University of California, Riverside, CA 92521

ABSTRACT

Selenium (Se) contamination from agricultural drainage water has been blamed for wildlife deaths and grotesque deformities of hatchlings at Kesterson Reservoir (Merced County, CA). Some approaches in reducing the toxic Se concentrations have been proposed, but most of these tend to be costly and/or ineffectual (e.g., physical removal and disposal). One approach, which is based on biomethylation of Se, is being considered as a remedial alternative. This study involves the development of an in situ biomethylation process based upon landfarming principles to detoxify seleniferous sediments. Optimum management practices were identified to accelerate Se volatilization, a naturally occurring component of Se cycling in the biosphere. This bioremediation technique is highly dependent on specific carbon amendments, temperature, moisture, aeration, and activators (cofactors). The biotechnology prototype developed could be applicable for cleanup of polluted sediments on irrigated farms throughout the western U.S. where selenium is a growing problem.

INTRODUCTION

Microbial production of alkylselenides is recognized as an important process affecting the mobility and toxicity of selenium (Se) in the soil environment. This microbial transformation links the global cycle of Se through production of dimethylselenide (DMSe) and lesser quantities of dimethyldiselenide (DMDSe). Selenium is emitted into the atmosphere in substantial quantities through high temperature (fossil fuel burning, industrial activities [smelting], volcanism, and burning vegetation) and low temperature processes (aerosol generation at the sea surface and biomethylation by soils and plants). In the northern hemisphere, it is estimated that up to 9000 metric tons of Se per year are emitted from these sources. The soil microbial component of this estimate is approximately 1300 metric tons per year.[1,2] Gaseous loss of Se from soil was first observed without establishing the mechanism involved.[3] Formation of DMSe by bread mold was observed prior to the discovery of Se volatilization from soil.[4,6] Since then, microbial production of DMSe has been reported from microbial isolates enriched from a variety of natural environments, including sewage,[7,8] soil, [9–14] lake sediment,[15] decaying plant material,[16] and pond water.[17,18] Fungi predominate among the Se-methylating microorganisms in soil,[5,6,9,12,13] although some bacterial isolates have also been identified, particularly in aquatic ecosystems.[18]

Evolution of methylated Se has been monitored during incubation of soil samples,[11–14,19–25] lake sediments,[15] and sewage.[24] The major volatile Se species reported in these studies was DMSe, with smaller quantities of DMDSe observed in some of the experiments.[22,24] The majority of incubation studies employed aerobic conditions. Experiments under waterlogged conditions typically produced smaller quantities of volatile Se.[15,22–24] To this date, many of the studies concerning the microbial production of gaseous Se were conducted exclusively in the laboratory. The seleniferous substrates employed to promote volatile alkyl-Se included inorganic and organic Se sources. Selenite [Se(IV)] and selenate [Se(VI)] were added to fungal isolates,[4-8] soil samples,[20–23] sewage,[24] water,[18] or sediment,[15] while elemental Se (Se^o) was applied to soil[22,24] and sewage samples.[24] Organic Se compounds (Se-cystine, Se-methionine, selenourea) serve as substrates when added to soil[14,19,22] and sediment samples,[15] and to microbial isolates.[19] The addition of organic C sources tend to enhance the production of volatile Se from soils.[22,23,25] Bioremediation of Se-contaminated soils or sediments through microbial Se methylation was not considered as a possibility until recently.

This review focuses on the application of microbial volatilization as a bioreclamation technique to detoxify Kesterson Reservoir, near Fresno, California. In addition, we report on studies conducted in the laboratory using Se-amended soils and native seleniferous sediments.

LABORATORY STUDIES

The volatility of alkylselenides depends on their vapor pressures along with their water solubility, soil adsorptivity, stability, and existing wind speed and turbulence. Studies were conducted to determine the vapor pressure of DMSe and DMDSe and the solubility of DMSe.[28] The vapor pressure was determined by an isoteniscope method, while solubility was calculated from the headspace vapor density. The vapor pressure for DMSe and DMDSe at 25°C was calculated at 30.5 and 0.38 kPa, respectively. The vapor pressure of DMSe is approximately five times that of ethanol and one half that of ether. Raising the soil temperature from 10 to 25°C approximately doubles the vapor pressure of DMSe, while raising the temperature from 25 to 40°C doubles it again, i.e., producing a vapor pressure four times as high as that at 10°C. Obviously, the Se field emission rates are highly dependent upon soil and air temperatures. The heat of evaporation of DMSe and DMDSe is 31.9 and 74.9 kJ mol^{-1}, respectively. The solubility of DMSe was calculated at 13.8 mg mL^{-1} water at 25°C. Being highly soluble in water, DMSe may be retained by water films that could decrease its rate of dissipation. This retention could be overcome by certain management schemes such as irrigation with wetting and drying cycles.

^{75}Se Trace Studies

We recently developed a new method to monitor volatile Se as a microbial metabolite when soils were exposed to ^{75}Se.[11] Labeled inorganic Se was applied to soil, and gaseous products were monitored in a continuous flow system by trapping with activated carbon traps and counting the recovered ^{75}Se. The method is well suited for feasibility studies with a limit of detection being 8 ng Se and a relative standard deviation ranging from 1.3 to 4.1%. Recovery of the added label was 97.5%.

Karlson and Frankenberger[12] found that the rate of Se volatilization was enhanced considerably upon incorporation of pectin to a seleniferous soil. As much as 9% of the added Se was recovered in the volatile form after 15 d of incubation upon treatment with pectin. The addition of trace elements such as Mo, Hg, Cr, and Pb (25 mmol kg^{-1} soil) inhibited volatilization of Se while As, B, and Mn had little effect. The presence of Co, Ni, and Zn in soil at moderate to high levels (5 to 25 mmol kg^{-1} soil) dramatically stimulated volatilization of Se. High nitrogen applications without carbon inhibited volatilization. Inoculation with isolates of Se-methylating fungi including *Acremonium falciforme, Penicillium citrinum,* and *Ulocladium tuberculatum* facilitated Se volatilization in autoclaved soils and enhanced it in nonsterile soils.

The fraction of Se volatilized per unit of time was dependent upon the Se concentration, soil properties, Se species, and carbon amendment.[13] Without the addition of C, volatilization rates were up to an order of magnitude higher with Se(IV) as the Se source, when compared with Se(VI). Carbon addition in the form of pectin accelerated Se evolution 2- to 130-fold, which was more pronounced with Se(IV). With pectin amendments, Se(VI) was volatilized almost as rapidly as Se(IV). With three pectin amendments over a period of 118 d, total Se volatilization ranged from 11.3 to 51.4% of the added Se. A minimum Se threshold for alkylselenide production was not found applying Se additions as low as 10 μg kg^{-1} soil.

Gas Chromatographic Studies

Further work was conducted to determine environmental factors affecting microbial volatilization of Se.[14] A seleniferous sediment collected from Pond 4 at Kesterson Reservoir was assayed for DMSe production by incubating for 120 h at 22°C with various treatments. DMSe was determined by gas chromatographic analysis and identified by gas chromatography-mass spectrometry. The optimum buffer pH for volatilization was found to be pH 8, which is within the same range as the Kesterson sediments. The optimum water content appears to be at field capacity (–33 kPa). Warm temperatures enhanced DMSe production ($temp_{opt} = 35°C$). The methylating activity was approximately 17.8-fold greater at 35°C than at 5°C. The average Q_{10} was calculated at 2.60. Organo-Se as well as inorganic Se compounds are readily used as substrates for Se volatilization. Selenomethionine enhanced the volatilization reaction much more (6.2-fold) than any other organo-Se substrate (e.g., selenoethionine, selenoguanosine, selenopurine, and selenourea) (Figure 1). Specific carbohydrates and amino acids were found to stimulate DMSe production. Glucose as well as fructose, cellobiose, chitin, galacturonic acid, and methionine promoted volatile Se production. Among all the compounds tested, proteins (e.g., casein, albumin, and gluten) enhanced biomethylation of Se more than any other treatment (Figure 2). Albumin (0.1 g C kg^{-1} soil) applied to the seleniferous sediment promoted gaseous Se production 19.3-fold over the unamended moist control.

Animal Toxicity

Dimethylselenide is 500 to 700 times less toxic to rats than aqueous Se(IV) and Se(VI) ions.[26,27] A study on the toxicity of inhaled DMSe was investigated with 85 adult rats exposed to four concentrations (0, 1607, 4499, and 8034 ppm) for 1 h.[28] Not a single animal was killed by the gaseous vapors of DMSe. After exposure, the animals were observed for a 1-week period for clinical abnormalities. All the animals appeared normal during the 7-d observation period. The exposed and control rats were sacrificed and their major tissues and organs were examined. There was a slight increase in the lung weight after one day of exposure. There was also a slight injury to the spleen with increased spleen protein and RNA at the highest concentration (8034 ppm), but normal recovery was noted later. This effect appeared to be more of an irritation than an injury. Histological sections of the lung, liver, spleen, thymus, lymph nodes, pancreas, and adrenal gland were examined and appeared normal. Protein, DNA, and RNA content of the liver and lungs were measured to quantify the minor inflammation response. Changes in the liver and lungs appeared to be minor, and after 7 d the organs had completely recovered. The Se content in the lungs and serum were slightly elevated at day one post-exposure, but it appears that the half-life of DMSe is very short and the compound is eliminated mainly via the lung. The data indicate that inhaled DMSe vapors are not toxic to the rat.

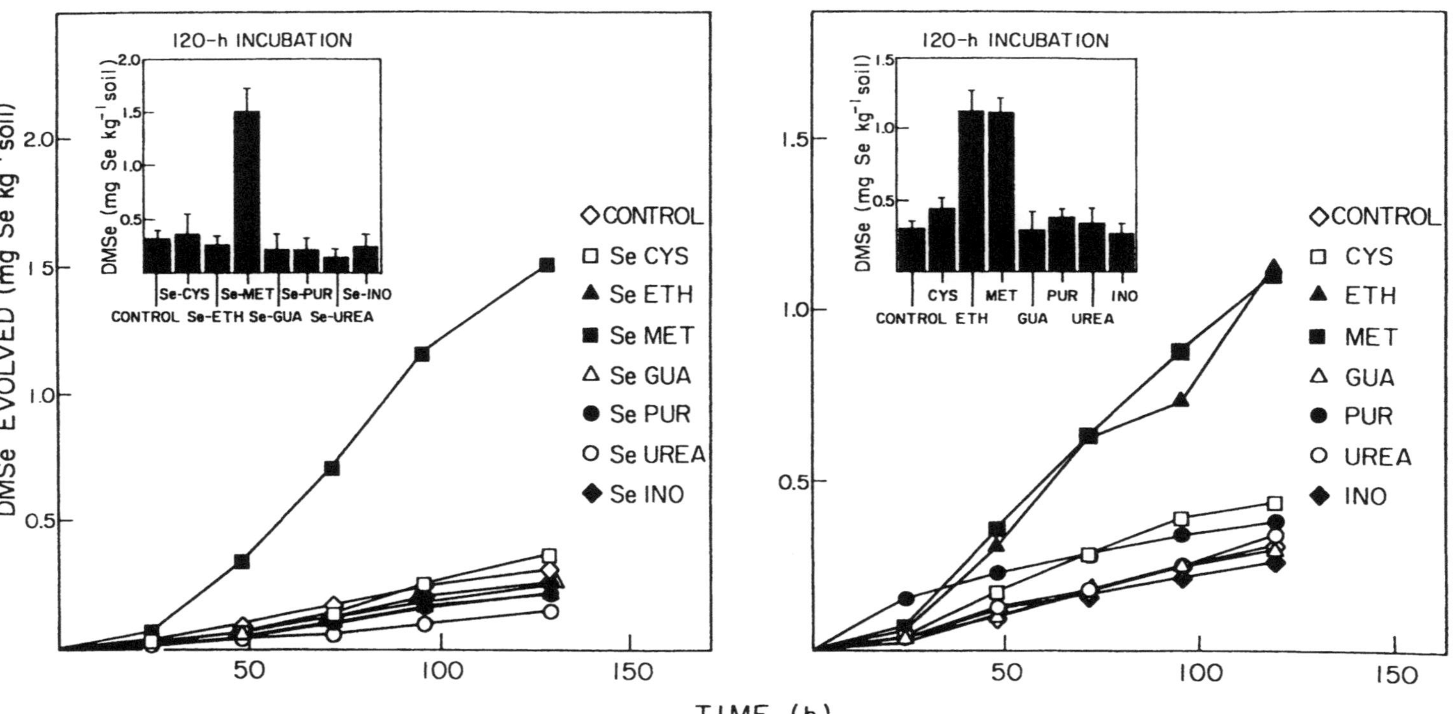

FIGURE 1. Influence of organo-selenium compounds on dimethylselenide (DMSe) production from soil. Se CYS = seleno-DL-cystine; Se ETH = seleno-DL-ethionine; Se MET = seleno-DL-methionine; Se GUA = 6-selenoguanosine; Se PUR = 6-selenopurine; Se UREA = selenourea; Se INO = selenoinosine. (From Frankenberger, W. T., Jr. and U. Karlson. *Soil Sci. Soc. Am. J.* 53: 1435–1442 (1989). With permission.)

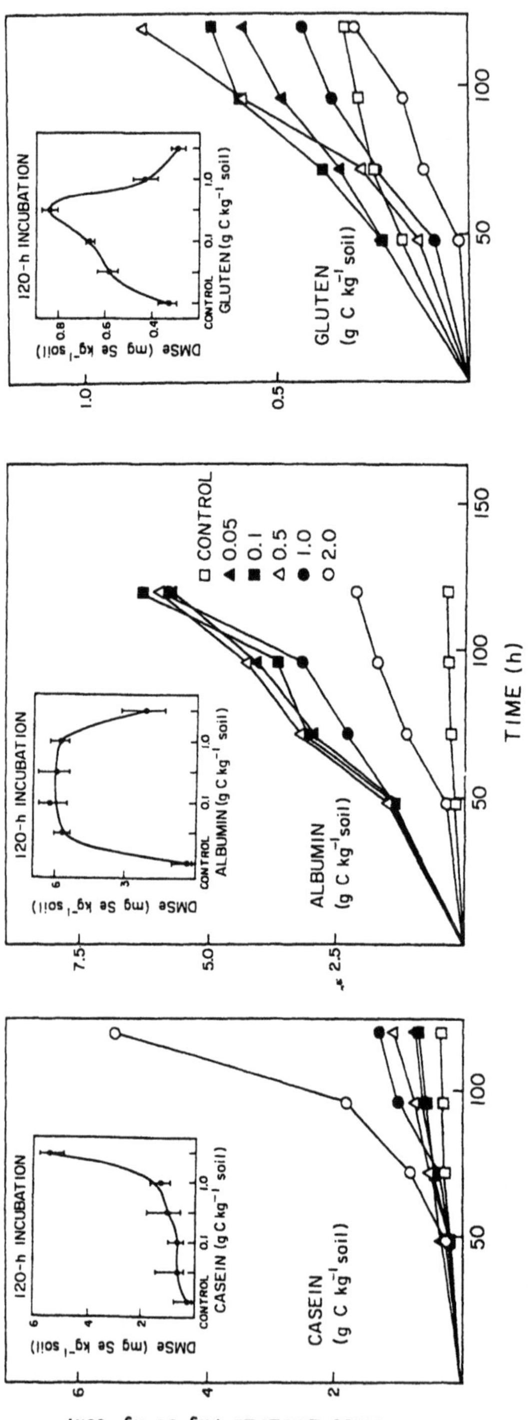

FIGURE 2. Influence of proteins (casein, albumin, and gluten) on dimethylselenide (DMSe) production from sediment. (From Frankenberger, W. T., Jr. and U. Karlson. *Soil Sci. Soc. Am. J.* 53: 1435–1442 (1989). With permission.)

MICROCOSM STUDIES

Microbial volatilization of Se was evaluated as a means to detoxify Se-contaminated dewatered sediments obtained from the field. Sediment samples containing 60.7 (Kesterson Reservoir) and 9.0 mg Se kg^{-1} (Peck evaporation ponds) were incubated for 273 d in closed systems located in the glasshouse, simulating environmental temperatures.[29] Volatile Se was trapped from a continuous air-exchange flow system using activated carbon. Various economical and readily-available organic and inorganic amendments were tested for their capacity to enhance microbial volatilization, including *Citrus* (orange) peel, *Vitis* (grape) pomace, feedlot manure, barley straw, chitin, pectin, $ZnSO_4$, $(NH_4)_2SO_4$, and an inoculum of *Acremonium falciforme* (an active Se-methylating fungus). With the Kesterson sediment, the highest Se removal (44.0%) resulted with the combined application of citrus peel plus $ZnSO_4$ followed by citrus peel alone (39.6%) and citrus peel combined with $ZnSO_4$, $(NH_4)_2SO_4$, and *A. falciforme* (30.1%). Cattle manure (19.5%), pectin (16.4%), chitin (9.8%), and straw plus NH_4NO_3 (8.8%) had less pronounced effects. Grape pomace (3.0%) inhibited volatilization. Upon application of water alone, 14.0% of the native Se was volatilized in 273 d. With the Peck sediments, the highest amount of Se removed was observed with chitin (28.6%), cattle manure (28.5%), and citrus peel alone (27.3%).

The differences in the effectiveness of each treatment between the two seleniferous sediments may be a result of the residual N content of the sediments. The Kesterson sediment was high in organic C and N, and added N inhibited volatilization of Se. The Peck sediments were low in organic C and N, and N-rich materials tended to accelerate Se volatilization. Inoculation with *A. falciforme* did not enhance Se evolution from either sediment, suggesting that there was sufficient microflora capable of producing gaseous Se.

FIELD STUDIES

Field experiments were carried out at Pond 4 of Kesterson Reservoir starting July 28, 1987, to assess the potential of microbial volatilization as a bioremediation technique to remove Se from a highly-contaminated seleniferous sediment. Figure 3 illustrates schematically the principal factors (carbon amendments, aeration, moisture, and temperature) involved in this process. An area measuring 35 × 35 m was disked until the upper 10 to 15 cm of soil was loosened.

Subplots consisting of 1.2 × 1.2 m squares with 2.5 m borders were staked out in a randomized block design. Selenium concentrations ranged from 10 to 209 mg kg^{-1} (median = 39 mg kg^{-1}). There were 16 subplots each with various treatments. Treatments consisted of the application of water alone or with cattail straw, cattle manure, orange peel, casein, and gluten. Some plots were treated with fertilizers such as $(NH_4)_2SO_4$ and/or $ZnSO_4$. The application rates of

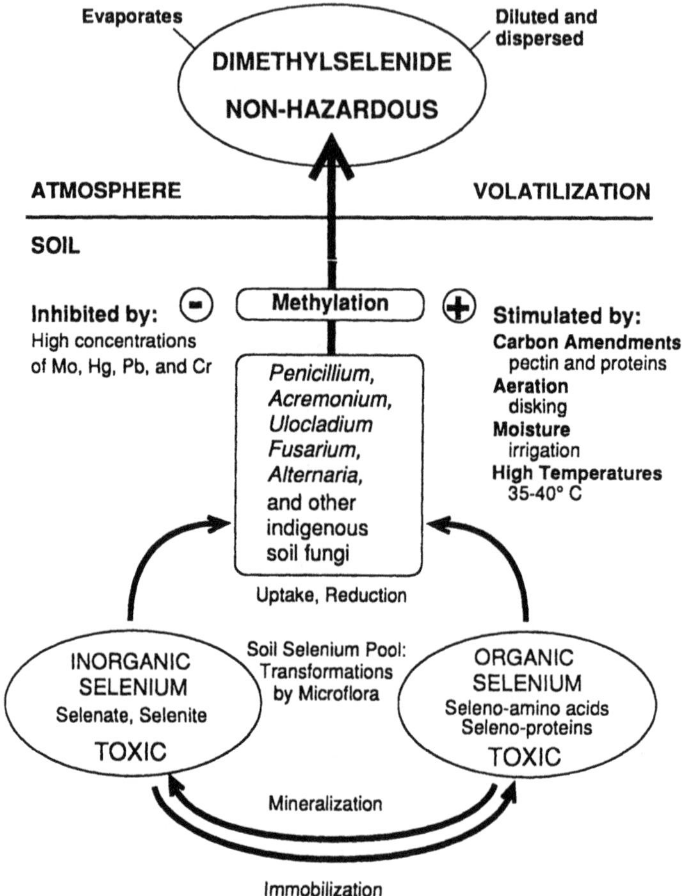

FIGURE 3. Factors affecting volatilization of selenium from soil or sediment.

inorganic and organic amendments are specified in Table 1. Each of the subplots was tilled once a week and was sprinkler irrigated up to once per day to keep the upper few centimeters of the sediment aerated and moist, because this is the most active zone for methylation of Se.

Emanation of Volatile Selenium

This bioremediation operation relies upon a net release of gaseous DMSe from sediment pores to the atmosphere. Two important factors affecting volatilization of Se are temperature and wind velocity. A thermal differential permits gas exchange of DMSe between the atmosphere and sediment air at the immediate soil surface. The optimum temperature for DMSe production occurs at 35 to 40°C.[14] The turbulence of high winds causes renewal of the sediment atmosphere, particularly of barren sediments. Wind speeds of 15 mph can

Table 1. Organic and Inorganic Materials Used as Amendments in the Field Study

Material	Source	Application Rate
Citrus (orange) peel	Sunkist Growers, Tipton, CA	67 Mg ha^{-1}
Nitrogen + orange peel	Ammonium sulfate	5 g N kg^{-1} orange peel
Zinc + orange peel	Zinc sulfate	1 g Zn kg^{-1} orange peel
Cattle manure	Harris Feed Co., Coalinga, CA	200 Mg ha^{-1}
Barley straw (chopped)	Local	22 Mg ha^{-1}
Typha latisolia (cattail straw) + N	Local	11 Mg ha^{-1} plus NH_4SO_4 at 1 g N kg^{-1} cattail straw

penetrate several centimeters into sediments and mulches.[30] Even without mass flow, fluctuations in air pressure at the soil surface result in considerable mixing, enhancing transport beyond that due to diffusion.

Seasonal variability of gaseous Se emission was evident with high volatilization rates recorded in the late spring and summer months (Figure 4). The greatest emission of gaseous Se with all treatments was recorded in July through September 1988, which correlates with the high soil temperatures and high Se concentration. Less volatile Se was released in the fall and winter months of October 1988 through February 1989. It was noted that the gaseous emission flux increased again in the spring months of 1989, corresponding to the high soil temperatures. However, the magnitude of this emission was not the same as in the summer months of 1988, possibly due to the lower Se concentration available for methylation. The temporal variability in gaseous Se evolution was most likely attributed to variation in daily temperature, moisture, availability of organic carbon, and the Se content.

The average emission rates for each of the treatments applied to the plots at Kesterson Reservoir are listed in Table 2. Irrigation and tillage alone (subplots 56, 57, and 62) resulted in an average volatile Se emission of 16.1 μg Se m^{-2} h^{-1} during this 2-year field study. This rate is considerably higher (6.2-fold) than the average background level (2.6 μg Se m^{-2} h^{-1}). Moisture is obviously a limiting factor for volatilization of Se from these sediments. Amendments with cattail (*Typha latisolia*) plus N and cattle manure were less effective than the moist-only treatment. Nitrogen fertilizer had been added with the cattail straw to enhance decomposition of this native vegetation. Both the cattail straw plus N and the cattle manure treatments included an exogenous N source that could have been responsible for the lower emission rates of volatile Se. *Citrus* (orange) peel was chosen as an amendment because of its high pectin content (30 to 35%). Karlson and Frankenberger[12] reported that pectin is an active component in promoting microbial volatilization of Se. Application of citrus peel to subplots 54, 58, and 61 dramatically enhanced Se methylation with an average emission

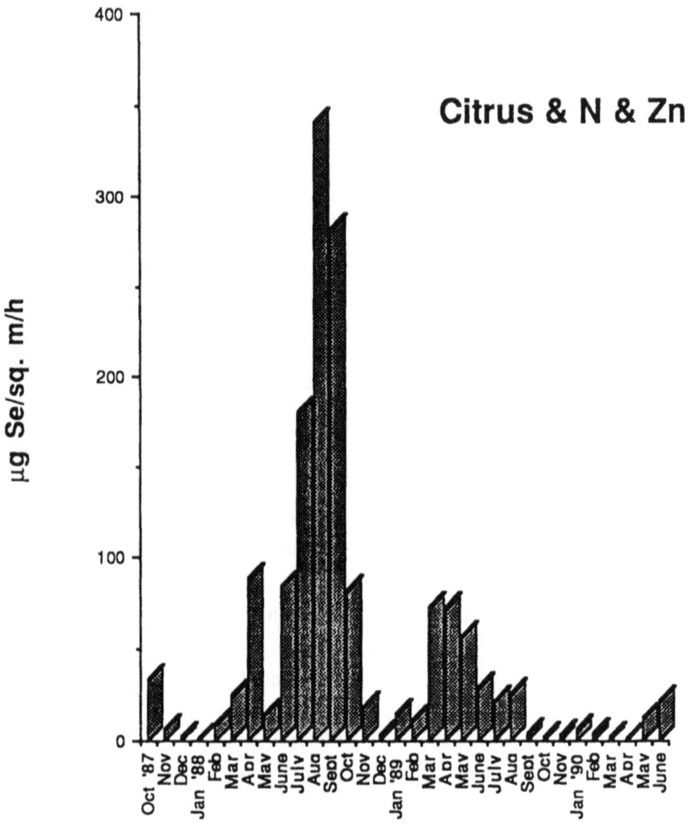

FIGURE 4. Seasonal flux of the average volatile Se emission from Pond 4 sediments upon treatment with citrus peel + N + Zn.

Table 2. Evolution of DMSe from Pond 4 Sediments, Kesterson Reservoir

Treatments	Subplots	Average Emission	Highest Emission
		(μg Se m^{-2} h^{-1})	
Moist	56, 57, 62	16.1	140
Cattail straw + N	66	10.7	39
Cattle manure	51, 53, 60	11.2	116
Citrus peel	54, 58, 61	32.5	297
Citrus peel + N + Zn	63, 64	110.1	808
Casein	65	49.8	749
Gluten	52, 55, 59	44.7	391
Background		2.6	8

flux of 32.5 μg Se m^{-2} h^{-1} during this 2-year field study. This is approximately a twofold enhancement over the moist-only treatment. One subplot (63) was treated with citrus, N plus Zn to promote Se volatilization (Figure 4). The combined application of these amendments enhanced volatilization more than any other treatment. The highest emission rate recorded from this subplot was 808 μg Se m^{-2} h^{-1}. The average emission reading over the entire field study was 110.1 μg Se m^{-2} h^{-1} that is 42-fold greater than the background control and 6.8-fold greater than the moist-only treatment. The application of casein, a milk protein, also promoted methylation of Se with an average emission rate of 49.8 μg Se m^{-2} h^{-1}. The highest emission flux recorded with this treatment was 749 μg Se m^{-2} h^{-1}. Gluten, a wheat protein intermixed with starchy endosperm, is composed of approximately 80% protein and 7% lipids. The application of gluten to subplots 52, 55, and 59 resulted in a flux of volatile Se emission immediately after application on June 25, 1988. The peak of activity lasted approximately 4 weeks. The emission flux of volatile Se upon treatment with gluten was as high as 391 μg Se m^{-2} h^{-1}, with an average reading of 44.7 μg Se m^{-2} h^{-1}.

The diurnal activity of volatile Se production indicated that this microbial transformation is highly temperature-dependent. The peak of volatile Se emission was always detected during midday (Figure 5). Seasonal cycles and temperature coefficient data indicated that Se emission rates are optimum during the spring and summer months with warmer temperature (Figure 6).

Soil Analysis

Starting September 11, 1987, all subplots were sampled for sediment Se at monthly intervals. Five sediment samples were collected within each subplot (approximately 1.2 m apart from each other) with a 2.5 cm diameter probe to a depth of 15 cm (Ap, plow layer). The distribution of Se in sediment profiles indicated that approximately 80% of the total Se occurred in the upper 15 cm of the soil profile.

The spatial distribution of Se was studied on an 80-point grid. Fractile diagrams of individual sediment Se concentrations were constructed assuming normal and ln-normal distribution. Initially, the Se inventory data followed the ln-normal distribution, but with time and mixing action (rototilling) a normal distribution was approached. After 3 m of soil management, the standard deviation of the soil Se data was reduced by roughly 50%. The physical effect of tillage, which should be considered as a treatment itself, removed the "hot spots," i.e., dangerously high Se levels, by distributing the Se evenly. The initial Se content ranged from 10 to 229 mg kg^{-1} ($\overline{x} = 50.1$, median = 39.0 mg/kg), indicating high initial spatial variability (Figure 7).

The removal of Se as determined by sediment analyses ranged from 30% (cattle manure) to 69% (casein) (Figure 8). Figure 9 shows the dispersion of the collected sediment data and the striking decline in the Se concentration in the Ap horizon by just tillage and adding water.

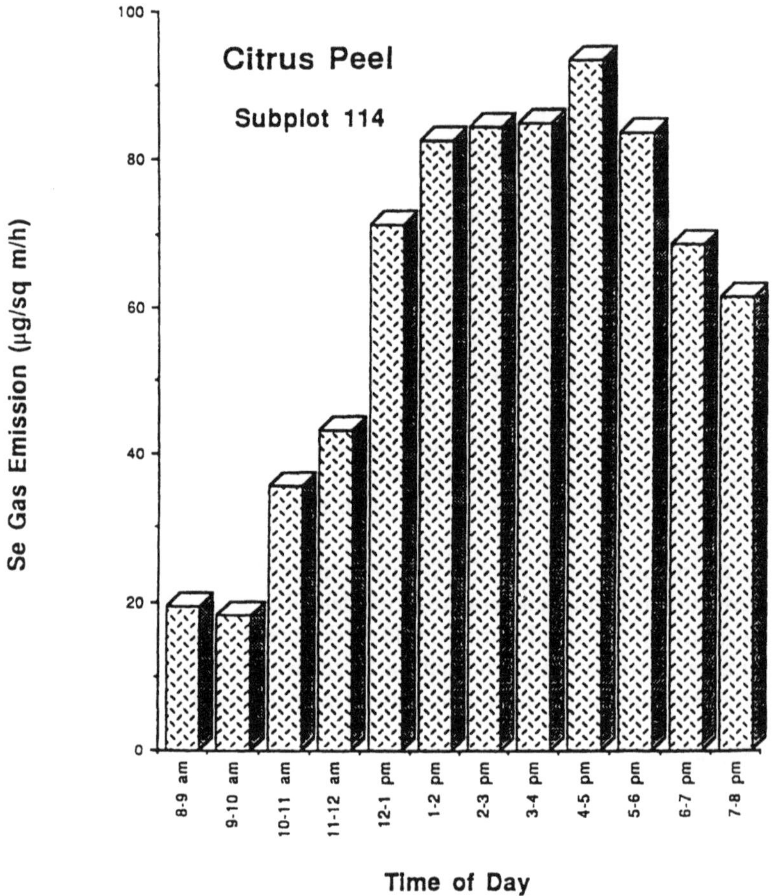

FIGURE 5. Hourly measurements of alkylselenide production from a subplot treated with citrus peel in January 1989.

DISCUSSION

The results of this field study show accelerated rates of volatilization as much as 42-fold over the background controls. Under optimum conditions, approximately 69% of the initial Se inventory in the Ap layer (15 cm deep) of Pond 4 has been dissipated after 23 m of study. Although there is a considerable fraction of Se remaining in the sediment after 2 years, this fraction appears to be insoluble and/or complexed with the organic matter, thus it is unavailable for redistribution in the food chain. It is suggested that rather than setting a specific total concentration for clean-up standards, regulatory agencies should consider a Se speciation (e.g., soluble fraction) concentration that poses little to no threat to the environment. The U.S. Environmental Protection Agency (EPA) has developed a new toxicity characteristic leaching procedure (TCLP) to characterize hazardous waste and requires that all hazardous materials be treated by the best

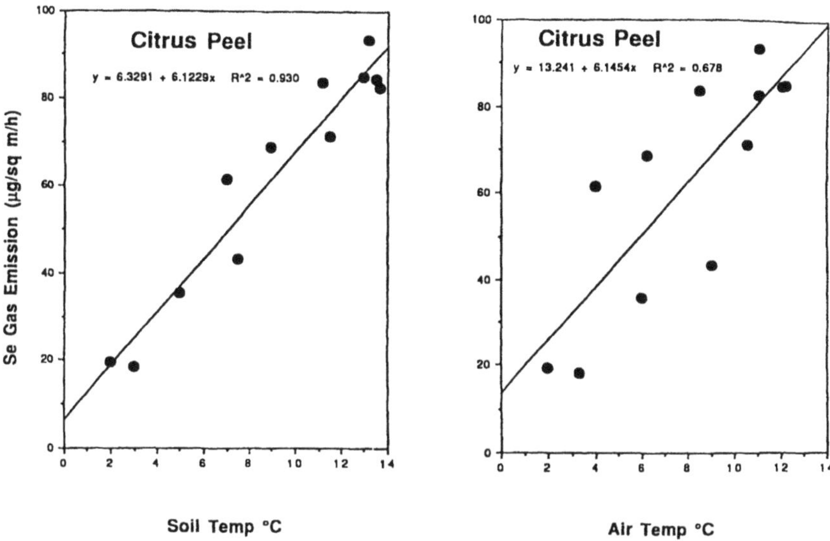

FIGURE 6. Linear regression of alkylselenide production with sediment and air temperatures.

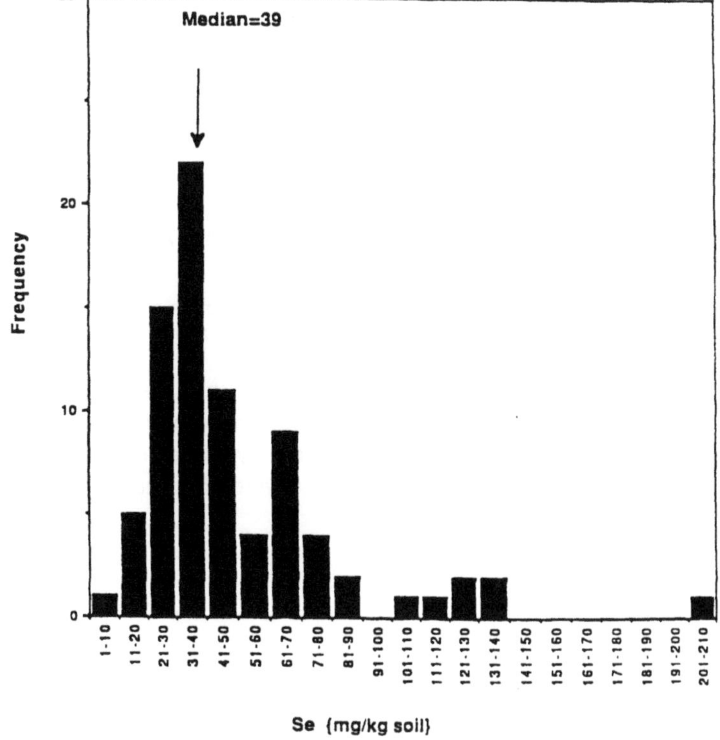

FIGURE 7. Frequency distribution of sediment selenium inventory at Pond 4, Kesterson Reservoir.

POND 4

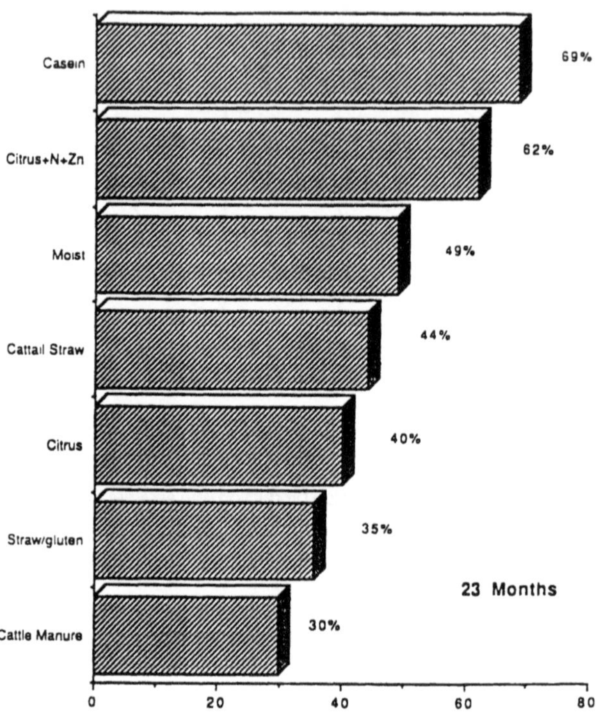

PERCENTAGE OF SELENIUM REMOVAL (%)

FIGURE 8. Influence of various treatments on the percentage of Se dissipated from Pond 4 sediments, Kesterson Reservoir, based on sediment analyses.

demonstrated available technology (BDAT) to reduce the hazardousness of the waste. TCLP is used to judge whether the treatment used is effective. The regulatory concentration for Se is 1.0 mg/L. It is highly recommended that TCLP be considered by State Regulatory Agencies in developing criteria for clean-up standards for seleniferous sediments.

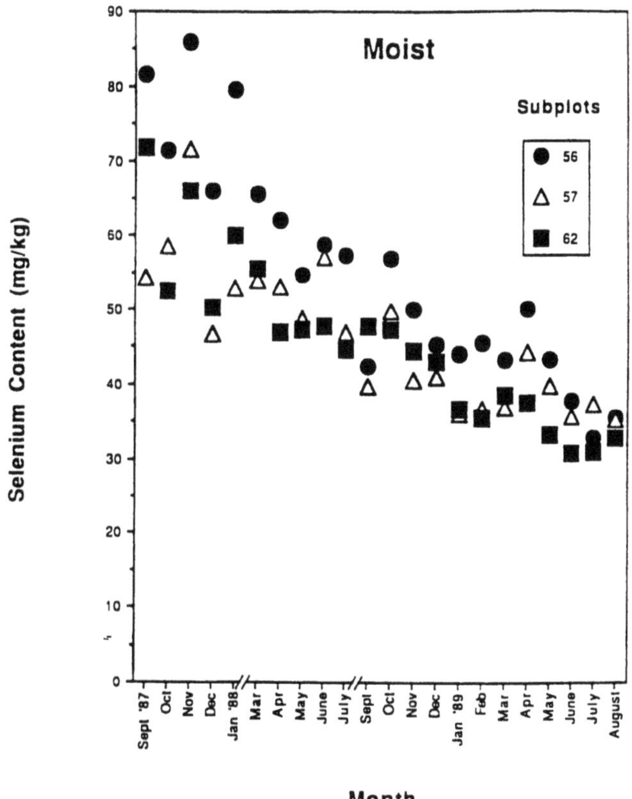

FIGURE 9. Selenium content of Pond 4 sediments, Kesterson Reservoir, vs. time of treatment with water and tillage alone.

REFERENCES

1. Mackenzie, F. T., R. J. Lantzy, and V. Paterson. "Global Trace Metal Cycles and Predictions," *Mathematical Geology* 11:99–142 (1979).
2. Ross, H. B. "An Atmospheric Selenium Budget for the Region 30°N to 90°N," *Tellus Ser. B* 37:78–90 (1985).
3. Ganje, T. J. and E. I. Whitehead. "Evolution of Volatile Selenium from Pierre Shale Supplied with Selenium-75 as Selenite or Selenate," *Proc. S.D. Acad. Sci.* 37:81–84 (1958).
4. Challenger, F. and H. E. North. "The Production of Organometalloidal Compounds by Microorganisms. II. Dimethyl Selenide," *J. Chem. Soc.* 1934:68–71 (1934).

5. Challenger, F. and P. T. Charlton. "Studies on Biological Methylation. X. The Fission of the Mono- and Di-sulfide Links by Moulds," *J. Chem. Soc.* 1947:424–429 (1947).
6. Challenger, F., D. B. Lisle, and P. B. Dransfield. "Studies on Biological Methylation. Part XIV. The Formation of Trimethylarsine and Dimethyl Selenide in Mould Cultures from Methyl Sources Containing ^{14}C," *J. Chem. Soc.* 1954:1760–1771 (1954).
7. Fleming, R. W. and M. Alexander. "Dimethylselenide and Dimethyltelluride Formation by a Strain of *Penicillium*," *Appl. Microbiol.* 24:424–429 (1972).
8. Cox, D. P. and M. Alexander. "Factors Affecting Trimethylarsine and Dimethylselenide Formation by *Candida humicola*," *Microb. Ecol.* 1:136–144 (1974).
9. Barkes, L. and R. W. Fleming. "Production of Dimethylselenide Gas from Inorganic Selenium by Eleven Soil Fungi," *Bull. Environ. Contam. Toxicol.* 12:308–311 (1974).
10. Doran, J. W. and M. Alexander. "Microbial Formation of Dimethylselenide," Abstr. of the American Society for Microbiology, Annual Meeting, New York (1975) p. 188 (N22).
11. Karlson, U. and W. T. Frankenberger, Jr. "Determination of Gaseous Selenium-75 Evolved from Soil," *Soil Sci. Soc. Am. J.* 52:678–681 (1988).
12. Karlson, U. and W. T. Frankenberger, Jr. "Effects of Carbon and Trace Element Addition on Alkylselenide Production by Soil," *Soil Sci. Soc. Am. J.* 52:1640–1644 (1988).
13. Karlson, U. and W. T. Frankenberger, Jr. "Accelerated Rates of Selenium Volatilization from California Soils," *Soil Sci. Soc. Am. J.* 53:749–753 (1989).
14. Frankenberger, W. T., Jr. and U. Karlson. "Environmental Factors Affecting Microbial Production of Dimethylselenide in a Selenium-Contaminated Sediment," *Soil Sci. Soc. Am. J.* 53:1435–1442 (1989).
15. Chan, Y. K., P. T. S. Wong, B. A. Silverberg, P. L. Luxon, and G. A. Bengert. "Methylation of Selenium in the Aquatic Environment," *Science* 192:1130–1131 (1976).
16. Zieve, R. and P. J. Peterson. "Volatilization of Selenium from Plants and Soils," *Sci. Total Environ.* 32:197–202 (1984).
17. Thompson-Eagle, E. T., W. T. Frankenberger, Jr., and U. Karlson. "Volatilization of Selenium by *Alternaria alternata*," *Appl. Environ. Microbiol.* 55:1406–1413 (1989).
18. Thompson-Eagle, E. T. and W. T. Frankenberger, Jr. "Volatilization of Selenium from Agricultural Evaporation Pond Water," *J. Environ. Qual.* 19:125–131 (1989).
19. Doran, J. W. and M. Alexander. "Microbial Transformations of Selenium," *Appl. Environ. Microbiol.* 33:31–37 (1977).
20. Francis, A. J., J. M. Duxbury, and M. Alexander. "Evolution of Dimethylselenide from Soils," *Appl. Microbiol.* 28:248–250 (1974).
21. Hamdy, A. A. and G. Gissel-Nielsen. "Volatilization of Selenium from Soils," *Z. Pflanzenernaehr. Bodenkd.* 6:671–678 (1976).
22. Doran, J. W. and M. Alexander. "Microbial Formation of Volatile Selenium Compounds in Soil," *Soil Sci. Soc. Am. J.* 40:687–690 (1977).
23. Zieve, R. and P. J. Peterson. "Factors Influencing the Volatilization of Selenium from Soil," *Sci. Total Environ.* 19:277–284 (1981).

24. Reamer, D. C. and W. H. Zoller. "Selenium Biomethylation Products from Soil and Sewage Sludge," *Science* 208:500–502 (1980).
25. Abu-Erreish, G. M., E. I. Whitehead, and O. E. Olson. "Evolution of Volatile Selenium from Soils," *Soil Sci.* 106:415–420 (1968).
26. Franke, K. W. and A. L. Moxon. "A Comparison of the Minimum Fatal Doses of Selenium, Tellurium, Arsenic and Vanadium," *J. Pharmacol. Exp. Ther.* 58:434–459 (1936).
27. McConnell, K. P. and O. W. Portman. "Toxicity of Dimethyl Selenide in the Rat and Mouse," *Proc. Soc. Exp. Biol. Med.* 79:230–231 (1952).
28. Frankenberger, W. T., Jr. and U. Karlson. "Dissipation of Soil Selenium by Microbial Volatilization at Kesterson Reservoir," *U.S. Dept. Interior, Bur. of Reclamation* (1989).
29. Karlson, U. and W. T. Frankenberger, Jr. "Volatilization of Selenium from Agricultural Evaporation Pond Sediments," *Sci. Total Environ.* 92:41–54 (1990).
30. Farrell, D. A., E. L. Greacen, and C. G. Curr. "Vapor Transfer in Soil Due to Air Turbulence," *Soil Sci.* 102:305–313 (1966).

14

Mechanism of Manganese Adsorption on Soil Constituents

Riaz A. Khattak and A. L. Page

Department of Soil Science, NWFP Agric. University, Peshawar, Pakistan
and
Department of Soil & Environmental Sciences, University of California, Riverside, California

ABSTRACT

A review of the adsorption of manganese by soils, layer silicates, hydrous oxides of iron, aluminum and manganese, carbonates, fulvic and humic acids, and soil organic matter is presented. In general, the adsorption of manganese by soils increases as the pH, clay, and organic matter content of the soil increase. Adsorption of manganese by hydrous oxides of iron, aluminum, and manganese also increase as the pH of the system increases. Although the mechanism of adsorption by hydrous oxides is not completely understood, it is generally considered to be related to the creation of negative charge through dissociation of protons from surfaces or association of hydroxyl ions with surfaces. At low ionic strength under acidic conditions the free metal ion is considered to be the dominant species adsorbed, while under basic conditions adsorption of $MnOH^+$ is more likely. Studies on the adsorption of Mn^{2+} by manganese oxide show that surface charge reversal accompanies the adsorption of Mn^{2+}. This suggests that an inner sphere complex is formed between the manganese oxide surface and the

Mn^{2+} ion. After removal of exchangeable plus soluble Mn^{2+} from soils by extraction with 1 *M* NH_4OAc, extraction with Cu salts removes additional Mn^{2+}. The Mn^{2+} extracted with Cu following removal of exchangeable plus soluble is considered to be mainly that complexed by soil organic matter. Electron spin resonance spectroscopy evidence suggests that Mn^{2+} is bound to soil organic matter as both inner and outer sphere complexes. At pH values greater than about 6, bonding of Mn^{2+} with soil organic matter is apparently as an inner sphere complex, while at pH values less than about pH 5 bonding of Mn^{2+} with soil organic matter is considered to be as an outer sphere complex.

INTRODUCTION

Trace element solubility and mobility in the soil environment have received considerable attention over the past several years. They are applied in fertilizers, pesticides, and mainly as constituents of waste products derived from different sources. Like other trace elements, Mn is not only important for plant, animal, and human nutrition, but it plays an important role in the chemistry of the terrestrial ecosystem. Its deficiency in calcareous soils and toxicity in acidic and waterlogged soils have been reported all over the world. Because Mn exists in the oxidation state of +2, +3, +4, +6, and +7 [designated Mn(II), Mn(III) Mn(IV), Mn(VI) and Mn(VII)], its chemistry is more complex than other nonoxidizing and nonreducing metals (e.g., Cd, Ni, Zn).

Manganese-adsorption on surfaces of clays and on different oxides and hydroxides has not been studied in detail. The instability of Mn in oxidizing or reducing environments demands more rigorous and careful study. It is important to know the mechanisms of Mn-adsorption to understand its role in influencing the chemical equilibria of soil solutions. Adsorption affects its bioavailability and its mobility in the soil-plant system. Also, Mn oxides adsorb many trace elements[1] and may be useful in sequestering trace elements contained in agricultural and industrial waste products.[2] McKenzie[3] has reviewed the chemistry of Mn oxides and hydroxides while Adriano[2] has presented an excellent review on Mn in soil, plants, and its occurrence in the terrestrial environment. Bartlett[4] has comprehensively reviewed the behavior of Mn in relation to soil redox potential; however, Mn adsorption and its mechanism in soil per se has not been reviewed. The objective of this paper is to review and synthesize the literature concerning Mn adsorption on inorganic and organic soil constituents and to highlight key areas for future research.

ADSORPTION ON SOIL MINERAL CONSTITUENTS

Manganese retention by soil is well recognized, but the exact mechanisms of adsorption and precipitation are not well understood. The crystallographic radii of divalent (II), trivalent (III), and tetravalent (IV) Mn are 0.08, 0.062, and 0.05

nm, respectively.[5] The radius ratio rule dictates that, if the Mn ion is coordinated with O, with a crystallographic radius (effective ionic radius) of 0.14 nm, divalent Mn is most stable in octahedral, trivalent in octahedral or tetrahedral, and tetra-valent in tetrahedral coordination. This would mean if Mn were to enter a clay crystal structure in isomorphic substitution, Mn(II) and Mn(III) could substitute for Al(III) in the octahedral position, and Mn(III) and Mn(IV) could substitute for silicon in the tetrahedral position. But, if the Mn(II) and Mn(III) were to substitute for Al and Si, respectively, the crystal lattice would be rendered unstable, based on the electrostatic valence rule of Pauling.[6]

Hemstock and Low[7] investigated the mechanism of Mn adsorption to a montmorillonite-beidellite type clay from Upton, Wyoming. A stock suspension containing 1.36% clay by weight was equilibrated with $MnCl_2$. Brownish red concretions were formed when the clay was dehydrated at 350°C. At 2θ, X-ray diffraction of the powdered concretions showed peaks at angles of 36.2° and 47.5° corresponding to d spacing of 0.311 and 0.241 nm, respectively, which represent the crystal interplanar spacings of tetragonal MnO_2. Although these results can be regarded as evidence for the presence of MnO_2 in clays heated to 350°C, it is not proof of the presence of this particular oxide in clays prior to heating. Recovery of 0.88 mol/kg clay exchangeable Mn(II) from unheated clays compared to 0.02 mol/kg of clay Mn in the heated clay suggested oxidation of Mn(II) to Mn(IV) during heating.

The dependence of Mn retention on pH eliminates the possibility of isomorphic substitution as the mechanism of Mn adsorption.[7] A significant decrease in suspension pH after equilibration suggested the possibility of the oxidation reaction:

$$a\mathrm{Mn}^{2+} + b\mathrm{O}_2 + c\mathrm{H}_2\mathrm{O} = d\mathrm{MnO}_w\mathrm{O}_x(\mathrm{OH})_y + e\mathrm{H}^+$$

in which the product containing Mn was a precipitate. Assuming the reaction solution was ideal and O_2 and H_2O concentration were constant during the reaction, an approximate equation

$$\log Z = B - (e/a)\ \mathrm{pH}$$

was developed where Z = fraction of initial Mn^{2+} concentration remaining unprecipitated, and B is a constant that depends on temperature, nature of reaction, and initial concentration of Mn, O, and H_2O. The numerical coefficient of pH, e/a, is equal to the ratio of the moles of hydrogen ion produced to the moles of Mn^{2+} ion reacting. From the resulting data Hemstock and Low[7] found that log $Z = 26.375 - 4.26$ pH with $r^2 = 0.98$. Because this equation fit the data well, they emphasized chemical oxidation rather than isomorphic substitution as a mechanism of Mn retention in clay systems. The decrease in Mn retention with decreasing pH seems to be a typical charge balance effect of H^+ and OH^- on clay surfaces and suggests that adsorption could be a possible mechanism of retention along with oxidation.

Reddy and Perkins[8] examined Mn fixation* at four pH values (4.9, 6.2, 7.6, and 8.8), three alternate wetting and drying cycles, and at different levels of moisture. It was found that bentonite and illite fixed significant quantities of Mn under wetting and drying cycles. The fixation was directly related to pH and the amount of Mn added. When subjected to repeated wetting and drying, incubation at moisture saturation resulted in considerably less Mn fixation. Kaolinite fixed relatively small amounts compared to bentonite and illite, regardless of the treatment. Because X-ray diffraction did not reveal a mineralogical change, they excluded the possibility of isomorphic substitution of Mn for Al and Si and attributed Mn fixation to precipitation, physical entrapment in clay "wedge zones," oxidation to higher-valence oxides, and/or strong adsorption at exchange sites. The observed differences in the amounts of Mn fixed by different clays were attributed to variations in cation exchange capacity (CEC) and specific surface area.

The recent studies of Khattak and Jarrell[9,10] indicate that wetting-drying cycles significantly ($p \leq 0.05$) decrease the concentration of Mn in water-saturation extracts of soils. This decrease is accompanied by 50 to 60% increases in exchangeable Mn extracted from the residual soil samples by 10 mmol/L $CuCl_2$ solution. Because other nonoxidizing and nonreducing cations (Ca, Mg, Na, or K) behaved similarly to Mn, the increases in exchangeable Mn were attributed to sorption processes along with oxidation of Mn(II) to Mn(IV) during drying. Use of saline irrigation water can greatly increase the soil solution Mn^{2+} concentration mainly by displacement of the Mn^{2+} sorbed on the solid surfaces as well as by formation of Mn–Cl complexes.[10]

The X-ray spacings and electron spin resonance (ESR) studies of McBride et al.[11] showed that for hectorite under fully hydrated conditions, where $Mn(H_2O)_6{}^{2+}$ is present in approximately 12.5-nm interlayers containing several molecular layers of water, the lifetime of the complex ion between collisions with solvent molecules was estimated to be 30% longer than for $Mn(H_2O)_6{}^{2+}$ in bulk solution. Thermal dehydration of Mn(II) smectite minerals transformed the solution-like ESR spectrum into one characteristic of the solid state as the Mn^{2+} ions moved into ditrigonal cavities of the siloxane surface. Furuhata and Kuwata[12] have reported that the width of the hyperfine (hf) lines of hydrated Mn^{2+} are broader on the exchange sites of montmorillonite than in bulk solution; they attributed this increase in line width to relaxation effects of the more restricted surface-adsorbed ions.

Data from Mn adsorption isotherms on soils have been fit to the Langmuir equation.[13–15] According to Shuman,[13] the data for four acid soils (pH 5.41 to 6.67) conformed to the Langmuir isotherm over the entire range from 0.04 to 1.2 mol/m^3, except in a sandy, low-organic-matter topsoil with low adsorption. Higher adsorption capacities for Mn were noted for the fine-textured, high-

* Fixed Mn refers to Mn that could not be removed from clays by neutral N NH_4OAc.

organic-matter soils; however, the Mn adsorption maximum never exceeded 40% of CEC and was only 6% in the case of sandy soils. At constant pH, the soil properties more closely related to Mn adsorption were clay and organic matter contents. No pattern was established between Mn-adsorption and mineralogical differences between clays.

Curtin et al.[16] studied Mn adsorption and desorption in 20 calcareous soils of Lebanon with varied pH (6.90 to 8.23), $CaCO_3$ equivalent (0.1 to 66.1%), and Fe_2O_3 (0.61 to 8.58%). They noted that when the data were plotted according to the conventional Langmuir equation, the isotherms were curvilinear at low Mn concentration and linear at higher concentrations. When the data were fitted to a "competitive" Langmuir equation, the isotherms were still nonlinear. They inferred that more than one type of adsorption site existed in these soils and the Mn adsorption maximum based on the linear parts of the conventional Langmuir isotherms were highly correlated with CEC, but not with values of organic matter, $CaCO_3$ equivalent, or free iron oxides (Fe_2O_3) of the soils.

The common use of the Langmuir adsorption isotherm is not considered a valid approach explaining the mechanisms of adsorption. Various authors[17–22] have stated that the fitting of data to the Langmuir equation may not be related to mechanisms of adsorption. Harter and Smith[23] emphasized the use of alternate methods, *viz.*, kinetic approaches and application of thermodynamic parameters to study adsorption reaction. Sposito[24] stated that the adherence of experimental sorption data to Langmuir equation gives no information concerning chemical mechanisms of the sorption reaction or of the nature of the solid surfaces in soils.

ADSORPTION ON OXIDE, HYDROXIDE, AND CARBONATE SURFACES

In contrast to the extensive literature on the adsorption of other elements onto Mn-oxide surfaces, information about the adsorption of Mn itself onto soil surfaces is lacking. In this section, the relevant available data on Mn adsorption on surfaces of Fe, Al, and Mn oxides and $CaCO_3$ are summarized.

Iron and Aluminum Hydroxides

Some of the following studies do not deal directly with Mn adsorption on Fe and Al oxides; however, considering the intermediate behavior of Mn between alkali and transition metals, they can be extended to include Mn. It has been shown that Fe and Al hydrous oxide gels and an amorphous aluminum oxide selectively adsorbed traces of Ca and Sr from solutions containing a large excess (8×10^3 mol/m^3) of $NaNO_3$. The fraction of Ca adsorbed depended principally on the suspension pH, the amount of solid present, and to a lesser extent on the $NaNO_3$ concentration.[25] Due to the rapid increase in adsorption at a specific pH ("the adsorption edge"), the adsorption process might be viewed in terms of H^+-M^{2+} exchange, the proton being derived from the weakly acidic surface groups on the hydrated oxide surface.

Adsorption from a mixed solution of eight divalent cations, each in a suspension of freshly precipitated Fe and Al gels, measured as a function of pH in 1×10^3 mol/m^3 $NaNO_3$,[26] showed the selectivity sequence (lower pH = greater selectivity) for the retention of alkaline earth cations by Fe gel: Ba > Ca > Sr > Mg, but for Al gel it was Mg > Ca > Sr > Ba. For transition and alkaline earth metals the selectivity sequence was Pb > Cu > Zn > Ni > Cd > Co > Sr > Mg for Fe gel, and Cu > Pb > Zn > Ni > Co > Cd > Mg > Sr for Al gel.

It was noted that the adsorption occurred even when the extent of cation hydrolysis was much less than 1% and invariably occurred at a pH lower than hydroxide precipitation. These researchers also noted that the selectivity sequence for a fresh Al gel at pH_{50} (the pH at which 50% of the original cation is adsorbed) corresponded to the inverse sequence of ionic radii, i.e., Mg^{2+} (0.065 nm) < Ca^{2+} (0.099 nm) < Sr^{2+} (0.113 nm) < Ba^{2+} (0.135 nm). However, comparison of the observed divalent cation adsorption to published solubility and hydrolysis data (see Sillen and Martell[27,28]) suggested that, although the pH effect on cation adsorption appears to be related to the solubility of the corresponding hydroxide, divalent cation adsorption occurs at a lower pH than that required for hydroxide precipitation. The extent of this difference between pH required for adsorption and that required for hydroxide precipitation depends on the concentration of cations and of gel, and the energy of the particular cation-surface interaction involved.

Grimme[29] studied the adsorption of Mn, Co, Cu, and Zn on synthetically prepared goethite in 100 mol/m^3 KNO_3 containing 0.01 mol/m^3 of each metal and observed that the amount of the metals adsorbed increased with increasing pH. At a given pH, adsorption decreased in the order: Cu > Zn > Co > Mn. This selectivity sequence is in agreement with that reported by Kinniburgh et al.[26] for Fe gel: Cu > Zn > Co. Grimme[29] postulated that iron oxide enrichment occurred through initial metal adsorption on the surface, followed by occlusion and irreversible fixation.

McBride[30] concluded from detailed spectroscopic studies on the retention of Cu^{2+}, Ca^{2+}, Mg^{2+}, and Mn^{2+} by amorphous alumina, that at near neutral pH divalent (Mg^{2+}, Cu^{2+}) and trivalent (Fe^{3+}) cations were capable of coprecipitating with Al, occupying octahedral coordination sites in the alumina. The presence of the coprecipitating cations increases the degree of noncrystallinity of alumina and the capacity of alumina to adsorb other metals. Freshly precipitated, high surface area alumina adsorbed Cu^{2+} at low pH (4.5), but was unable to adsorb Mn^{2+}, Ca^{2+}, and Mg^{2+} suggesting Cu^{2+} bond formation with surface O atoms by displacement of protons from hydroxyl groups.

On the basis of similarity and symmetry of covalency of the bonding environments of adsorbed and coprecipitated Cu^{2+}, it was inferred that adsorption and coprecipitation involved the same mechanism of metal ion substitution into the structure of alumina. In contrast, as evidenced by a strong ESR signal similar to that of Mn^{2+} in aqueous solution or adsorbed on solvated smectite, silica gel retained a large quantity of Mn^{2+} in a soluble-like environment.[11] The earlier work of Furuhata and Kuwata[12] also showed similarity between Mn^{2+}

spectra on adsorbents in moist state and air dry state, and that of aqueous solution, but the hyperfine linewidths for Mn^{2+} on adsorbents were broader, even though the amount adsorbed was too little to cause dipole-dipole interaction. They assumed that the hydrated Mn ions on adsorbents move freely in the surface layer, but the motion of ions is restricted and the rotational time is slower than in dilute solution. The adsorbents were ranked in decreasing order of line-width broadening as follows: silica-alumina gel > montmorillonite > kaolinite > silica alumina catalyst. This ordering could be associated with the differences in the surface reactivity, surface area, charge distribution of the above adsorbents, and with the steric environment at the adsorption site. The gel is porous and montmorillonite contains interlayer sites that can restrict (slow) motion.

Adsorption on Manganese Oxide Surfaces

The ability of colloidal hydrous Mn oxides to adsorb metal has been a continuing subject of investigation. While certain aspects of the subject are well understood, there is still much to be learned about the mechanisms of adsorption on Mn oxide surfaces. A comprehensive study of the general colloid-chemical properties of synthetic MnO_2 closely related to d-MnO_2 or to manganous manganite, clarified some aspects of adsorption.[31] The release of H^+ or binding of OH^- with adsorption of Mn^{2+} and other metals (Ca^{2+}, Mg^{2+}, and Zn^{2+}) was demonstrated. The log of the distribution coefficient (Mn^{2+} adsorbed/Mn^{2+} solution) was linearly related to the pH indicating the pH dependence of cation adsorption on these surfaces. No chemical oxidation was observed between pH 4.0 and 8.0. From the slope of the Langmuir plot, an adsorption capacity of 0.5 and 2 mol of Mn^{2+} per mole of MnO_2 was observed at pH 7.5 and 9.0, respectively.

Under the experimental conditions, the Mn(II) solutions were considerably undersaturated with respect to $Mn(OH)_2(c)$ ($K_{so} = 10^{-13}$) and the possibility of hydrolysis of Mn^{2+} ($K_1 = 10^{-0.6}$) was excluded. The Mn^{2+} showed higher affinity than Zn^{2+}, Mg^{2+}, and Ca^{2+}, in decreasing order, for Mn-oxide surfaces. On the basis of these results, the removal of Mn^{2+} from solution was attributed to surface complexation of Mn^{2+} and H^+-Mn^{2+} exchange.[31] The follow-up work of Murray et al.[32] indicated that the adsorption of Group I and II cations on Mn(II) manganite (PZC 1.8 by electrophoresis) took place in the diffuse part of the double layer at low concentration (0.1 to 1.0 mol/m^3) and low pH (2.0 to 4.0). At higher concentration (10 mol/m^3) the adsorption was associated with some other mechanism, possible incorporation into the disordered layer of manganite lattice. This explanation is based on the MnO_2 structure, which is one of the general class of "manganites" that consists of ordered layers of MnO_2 alternating with disordered layers of metal ions coordinated by H_2O, OH^- and/or other anions.

Murray[33] studied the adsorption of several metal ions at the MnO_2-solution interface using direct atomic absorption spectrometry and indirect alkalimetric titration methods. Both sets of experiments indicated the affinity of the metals for the MnO_2 surface in the order (when compared at equal concentrations)

$$Mg^{2+} < Ca^{2+} < Ba^{2+} < Ni^{2+} < Zn^{2+} < Mn^{2+} < Co^{2+}$$

Indirect evidence for specific adsorption through desorption studies (minimum desorption equivalent to greater specific adsorption) was suggested. Magnesium adsorption was most easily reversible and Co^{2+} desorption was very slow, incomplete, and irreversible, followed by Mn^{2+}. To explain the results, the model of Stumm et al.,[34] which involves (1) the separation of proton from the covalent bond at the surface and (2) the association of a solute cation with this site, was proposed:

$$Mn—OH^{\circ} + Co^{2+} = MnO—Co^{+} + H^{+}$$

or

$$Mn—OH^{\circ} + Mn^{2+} = MnO—Mn^{+} + H^{+}$$

This suggests that Mn^{2+} is specifically adsorbed in the sense that it can enter the Stern layer and can change the sign of the potential of the Helmholtz layer. Murray[35] observed charge reversal of MnO_2 by Mn(II) at 1 mol/m^3 and 0.1 mol/m^3 below pH 8.0 with microelectrophoresis. The ability of Mn to reverse and reduce the potential at the surface below pH 8.0 suggests specific adsorption of Mn on the oxide surface, as it has a very low stability constant for the first hydrolysis species [K_1 ($MnOH^+$) = $10^{-10.6}$] and it will oxidize to hydrous Mn oxide before precipitating as $Mn(OH)_2$ (Morgan and Stumm[31]). Thus, any reversal of charge using Mn(II) will be caused by specific adsorption of the unhydrolyzed metal species.

The chemical properties of Mn oxides and their adsorption potential for trace metals have been studied extensively by McKenzie.[3,36–38] The adsorption of Cu, Co, and Ni, in decreasing order, on Mn nodules was noted, but no mechanism was suggested.[36] Further adsorption studies were made on synthetic Mn oxide[37] using two birnessites (δ-MnO_2) and two cryptomelanes ($K_2Mn_8O_{16}$, i.e., α-MnO_2) prepared by different methods. Adsorption of metals on all the oxides released Mn^{2+}, K^+, and H^+, in decreasing order. The release of H^+ was rapid during the first day of incubation, but diminished slowly over the course of the experiment. This suggests the adsorption sites on Mn oxides are polyfunctional or that the adsorption of metals on Mn oxide is not just an exchange mechanism, as suggested by Morgan and Stumm[31] and Loganathan and Burau.[39] During adsorption of heavy metal ions 1 mol of proton is released for every mole of Pb, Cu, Mn, or Zn adsorbed on the surface of birnessite,[40] but the ratio is higher for cryptomelane.[38] These differences between birnessite and crytomelane are associated with differences in the crystal structure of the synthetic Mn oxides and an ionic strength effect.[38–40] McKenzie[40] confirmed the ionic strength effect on H^+ released/Me^{2+} adsorbed. Two possible reactions were proposed:[38]

$$SOH^{\circ} + MOH^{+} = [SO^{-} - MOH]^{+} + H^{+} \qquad (1)$$

and

$$SOH^\circ + M^{2+} = [SO^- - M]^+ + H^+ \qquad (2)$$

where SOH° represents uncharged surface sites and [ ] represents the adsorbed forms of the metal ion M^{2+}. These reactions show release of 1 mol of proton per mole of metal ion adsorbed. According to the model of James and Healy,[41,42] the adsorption of the first hydrolysis product would be favored because of its lower hydration energy. But at low pH (~4.0) the hydrolysis constants of Pb, Cu, Mn, and Zn vary from 10^{-8} to 10^{-10} and, therefore, the concentration of unhydrolyzed species will be greater than the hydrolyzed species by a factor of 10^4. Second, if the adsorption of MOH^+ is significant at low pH, then the ions are not initially hydrolyzed in solution and both hydrolysis and sorption can take place as follows:

$$SOH^\circ + M^{2+} + H_2O = [SO^- - MOH]^+ + 2H^+ \qquad (3)$$

This reaction would release 2 mol of H^+ in solution for each mole of metal ion adsorbed. It is likely that Reaction 1 may not be important at low pH and Reaction 2 representing the interaction of the unhydrolyzed metal ion with the surface is favored. Since Morgan and Stumm[31] and Murray[33] observed increases in the ratio of H^+ released to the metal ion adsorbed with increasing pH, hence the adsorption of a hydroxy complex could be significant at higher pH.

Traina and Doner[43] studied the adsorption of metals by synthetic birnessite. The adsorption capacity sequence, based on the observed adsorption maxima [Co (1.84) > Cu (1.64) > Mn (1.54) mol/kg] was accompanied by a linear release of H^+. Similar findings were observed by others.[31,35,38,40] Traina and Doner[44] showed that increasing pH increased the adsorption of metal and decreased the release of Mn into solution. This reinforces the involvement of H^+ and OH^- ions as potential-determining ions for hydrous Mn oxide surface chemistry.

Coprecipitation and precipitation can be difficult to distinguish from surface chemisorption. The nature of the reaction will depend upon the ionic strength and background electrolyte effect, the nature of sorbing surfaces, the identity of sorbate ions, and the pH of the system.[45]

Adsorption of Manganese on Calcium and Magnesium Carbonate Surfaces

The information on adsorption of Mn on $CaCO_3$ surfaces is limited to a few studies. Although deficiency of Mn^{2+} in plants has often been reported in calcareous arid zone soils of the world, information on the effect of $CaCO_3$ surfaces on Mn adsorption and on soil solution equilibria is lacking. In many cases, manganese hydroxide or carbonate was much too soluble to account for the low equilibrium concentration of Mn^{2+} found in soil solution.[46] It has generally been found that ions such as Mn^{2+} are strongly adsorbed by $CaCO_3$ and more strongly by $MgCO_3$.[47] However, Curtin et al.[16] reported no correlation between $CaCO_3$ equivalent and adsorption maximum of 20 calcareous soils of Lebanon.

McBride[48] performed a series of experiments to study the Mn adsorption mechanism on calcite surfaces. Significant removal of Mn^{2+} from solution (1.0 to 10^{-3} mol/m^3 $MnCl_2$) by 2 g $CaCO_3$ powder (2×10^3 m^2/kg specific surface area measured by BET) was noticed, whereas the adsorption on 0.2 g fine $CaCO_3$ (4.9 m^2/kg) was much greater, suggesting surface adsorption phenomenon. Based on solubility considerations, neither precipitation as $Mn(OH)_2$ or $MnCO_3$ or oxidation to Mn(IV) could explain the low equilibrium concentration of Mn(II) in the solution. Adsorption was also suggested by the amount of Mn(II) adsorbed in close stoichiometric relation to the Ca^{2+} released into the solution. So the following reactions were suggested:

$$Mn^{2+} + HCO_3^- = MnCO_3(ads) + H^+ \tag{4}$$

$$H^+ + CaCO_3 = Ca^{2+} + HCO_3^- \tag{5}$$

resulting in the net reaction:

$$Mn^{2+} + CaCO_3 = MnCO_3(ads) + Ca^{2+} \tag{6}$$

It has been suggested that in calcareous systems $MnCO_3^\circ$ species are significant in the soil solution.[49] In ESR studies, McBride[48] observed an obvious decrease in the intensity of the six-line spectrum of solution Mn^{2+} when 5 g coarse $CaCO_3$ was added to 25 mL of 0.1 mol/m^3 $MnCl_2$. However, the acidified supernatant solution showed no change in signal intensity that indicates that most of the soluble Mn in this calcareous system was in the form of $Mn(H_2O)_6^{2+}$ rather than $MnCO_3^\circ$.

When 2 g reagent grade (coarse) $CaCO_3$ was equilibrated with 0.5 mol/m^3 $MnCl_2$ solution, no change in the intensity of the solid state spectrum of Mn^{2+} in $CaCO_3$ occurred, but equilibration of 0.2 g of fine $CaCO_3$ with 25 mL of 0.5 mol/m^3 $MnCl_2$ produced a broad resonance superimposed upon the spectrum of structural Mn^{2+}.[48] The addition of a higher concentration of Mn^{2+} resulted in an even more intense broad signal. This phenomenon was attributed to $MnCO_3$ precipitation as a separate phase. Based upon these spectroscopic studies, it was suggested that a continuous process of chemisorption leads to precipitation as adsorption increases.

MANGANESE OXIDE INTERACTION WITH COBALT

In soil, the cobalt concentration generally varies from 1 mg/kg to 50 mg/kg, although higher values have been reported.[50,51] In some Australian soils, Taylor and McKenzie[52] reported an average of 79% of total soil Co associated with Mn oxide minerals ($r^2 = 0.757$ between Co extracted and Mn extracted with H_2O_2). The Co deficiency disease of sheep, "pine", observed on high Mn Irish soils[53] and Co deficiency noticed in clover grown on high Mn soil[54] suggests that Mn oxides control the solubility and bioavailability of Co in Mn-rich soils.

Desorption of Manganese by Cobalt Adsorption

The Co^{2+} ion is strongly adsorbed on ground Mn nodules and on goethite and the extent of adsorption increases with pH.[36,55] The Co^{2+} has very strong affinity for the surface of MnO_2 at low pH (≤4.0) and is more strongly adsorbed than Cu(II) and Ni(II) even below the PZC (pH = 1.8).[32] Generally Co^{2+} is very difficult to desorb, even from positively-charged surfaces due to specific adsorption and possibly because of the surface oxidation of Co^{2+} to Co^{3+}. Murray and Dillard[56] used X-ray photoelectron spectroscopy (XPS) to observe cobalt adsorbed on MnO_2 and suggested Co(II) oxidation to Co(III).

Increases in Co adsorption on synthetic birnessite and MnO_2 as a function of time showed that the initial adsorption was large and very rapid over a period of 1 d and then continued to increase more slowly over a period of 42 d.[37] Similar studies on the lower oxides partridgeite (Mn_2O_3), manganite (MnOOH), and hausmannite ($Mn^{2+}Mn_2{}^{3+}O_4$) suggested that Co was slowly sorbed to a marked degree by partridgeite, but not by manganite or hausmannite. It was speculated[37] that the driving force of this reaction is the high field stabilization energy of the low spin Co^{3+} ion. Of the elements Cu, Co, Ni, and Zn, only Co^{2+} could be oxidized by Mn^{3+}, and only the low-spin Co^{3+} has a higher crystal field stabilization energy than Mn^{3+}. It appears that Co^{2+} is at first adsorbed in a manner similar to other heavy metals, then slowly oxidized to Co^{3+}, and is finally incorporated into the surface layers of the crystal lattice with the resulting Mn^{2+} expelled into solution. This mechanism does not happen in manganite and hausmannite because of crystal field effects related to the greater Mn-O distance (manganite, 0.020 nm and hausmannite octahedral site, 0.205 nm). The greater Mn-O distance in these two minerals may reduce the crystal field strength sufficiently to prevent the oxidation of Co^{3+} to the low-spin state. Supposedly this may make the oxidation of Co^{2+} energetically unfavorable and prevent the replacement of Mn^{3+} by Co^{3+}.[37,38,57]

Although Murray and Dillard[56] showed the presence of Co(III) at the surface of birnessite, the exact mechanism of Co(II) oxidation could not be resolved. Because a small amount of Mn(II) is released (0.05% of total metal adsorbed), it seems that a very small fraction of higher valency Mn is reduced when Co(II) is oxidized to Co(III).[38] Therefore, with Mn(IV) being the dominant form in oxides, it makes it very difficult to detect the presence of Mn(II) by XPS measurement.

Burns[55] proposed a reaction in which Mn(IV) oxidizes Co(II).

$$2Co^{2+}(aq) + 2OH^-(aq) + 2H_2O(l) + MnO_2(s) = 2Co(OH)_3(s) + Mn^{2+}(aq)$$

Assuming pH = 8.0, $(Mn^{2+}) = 10^{-9}$ and $\Delta Gr° = 71.5KJ$.

$$\Delta Gr = \Delta Gr° + 2.303\ RT \log[(Mn^{2+})/(Co^{2+})^2\ (OH^-)^2]$$

This reaction requires an activity of Co^{2+} of 4×10^{-5} that seems high for sea water, but it may not be high for a soil-Mn system. Therefore, the thermodynamic possibility of the above reaction cannot be ruled out for the soil environment.

Traina and Doner[58] detected a release of Mn from birnessite surfaces with the maximum ratio of 0.02 for Mn released/Co(II) sorbed. However, these studies suggest that an agreement exists on the oxidation of Co(II) to Co(III), but it is not clear which Mn species is involved in the oxidation.

ADSORPTION ON SOIL ORGANIC CONSTITUENTS

Divalent Mn may form chelates with alpha hydroxyl and dicarboxylic acids.[59] The enhanced extraction of Mn from soil using various inorganic salts, including $CuSO_4$, $CaCl_2$, and $ZnCl_2$, suggest that some Mn is adsorbed by organic matter functional groups.[60,61] Some researchers believe that the Cu replaceable Mn comes from the Mn-organic complexes after NH_4OAc extractable Mn is removed.[7,62,63] Walker and Barber[64] measured the amount of organically complexed Mn in 12 Indiana soils by displacement with Cu after removing exchangeable Mn with NH_4OAc. The Cu-displaced Mn in these soils varied from none to 0.44 mol/kg of soil. Page[65] emphasized the pH-dependent Mn organic complex formation in the soil solution while Passioura and Leeper[66] considered the formation of nonexchangeable Mn-complexes of little significance in some Australian soils.

Utilizing spectrophotometric and polarographic methods, Geering et al.[67] reported that from 84 to 99% of the Mn in the soil solution exists in an organically bound form. Results of Schnitzer and Skinner[68,69] and Schnitzer[70] suggested the following order of stabilities of complexes between a soil fulvic acid (FA) and divalent ions:

$$Cu > Fe > Ni > Pb > Co > Ca > Zn > Mn > Mg \text{ at pH } 3.5$$

and the order changes to

$$Cu > Pb > Fe > Ni > Mn \approx Co > Ca > Zn > Mg \text{ at pH } 5.0.$$

The Mn-fulvic acid complex log k values reported by Schnitzer and Skinner[69] and Schnitzer[70] are 1.47 at pH 3.5 and 3.78 at pH 5.0. The increase in stability coefficients (log K) at higher pH might be due to increased dissociation of functional groups, especially carboxyls.[71]

The interaction between certain metallic cations and humic acids isolated from different soils of Alberta, showed formation of metal humic acid complexes and release of protons from functional groups of humic acids.[72] The order of magnitude of pH decrease of the humic acids on addition of metal cation was found to be

$$Mn^{2+} < Co^{2+} < Ni^{2+} < Zn^{2+} < Cu^{2+} < Al^{3+} < Fe^{3+}$$

The effectiveness of added cations in coagulating the humic acids followed the same sequence, except Fe^{3+} was less effective than Al^{3+}.

Studies on the mechanism of metal-FA interaction indicates that alcoholic OH groups are relatively unimportant in metal-FA interactions. Two types of reactions occur, a major one, involving simultaneously both acidic COOH and phenolic OH groups, and a minor one, in which only less acidic COOH groups participate.[70] Gamble et al.[73] supported the hypothesis that Mn^{2+} forms outer sphere complexes with fulvic acids. That is, the oxygen atoms of the ligand, probably carboxylates, are excluded from the inner hydration sphere of Mn^{2+} [$Mn(H_2O)_6{}^{2+}$]. The mechanism of Mn^{2+} bonding in humic acid (HA) extracted from a mineral soil using electron spin resonance spectroscopy (ESR) suggested that the Mn ion is retained by organic matter as both inner and outer sphere complexes depending on pH.[74] According to McBride,[75] at a soil pH greater than 5.0 to 6.0 Mn^{2+} is held by soil organic matter functional groups in inner sphere complexes, but bonding to organic matter at more acid pH is by outer sphere complexes.

CONCLUSIONS AND RECOMMENDATIONS

The review indicates a general agreement on some aspects of Mn adsorption on clay minerals, oxides, and hydroxides of Fe, Al, and Mn. Increases in adsorption capacity with pH and occurrence of adsorption on oxide surfaces below the hydrolysis constants are well recognized. Adsorption at higher pH could be affected by Mn(II) oxidation to higher oxides or by precipitation of Mn-carbonate, bicarbonate, and hydroxide. Since these phases are reasonably soluble, it seems important to consider the available data on solubility constants and association constants for such soluble and insoluble complexes.

Several workers using electron spin resonance (ESR) techniques have indicated that the adsorption of Mn^{2+} on clay minerals is via outer sphere complexation mechanisms. However, the ESR spectrum indicated no Mn-adsorption on synthetic alumina at lower pH. The observed strong adsorption of Mn(II) on various Mn oxides and its ability to reverse the charge suggests specific adsorption. The exact mechanism demands more spectroscopic studies.

The specific adsorption of Co on a variety of synthetic Mn oxides has been shown. The oxidation of Co(II) to Co(III) is strongly suggested by XPS, but lack of agreement still exists on whether Mn(III) or Mn(IV) oxidizes Co(II). Some

ESR studies support the chemisorption of Mn as $MnCO_3$ but more work is needed to establish a solid consensus of the sorption mechanism.

Most of the work has been done using synthetic oxides and hydroxides. It is important to acknowledge the complexity and variety of binding sites simultaneously existing in soils when extrapolating these observations to field soils. Also, the methodology used for preparing different oxides and the structural differences among the synthetic polymorphs must be carefully considered.

Adsorption of Mn^{2+} on organic functional groups as outer sphere complexes has been proven at low pH while formation for inner sphere complexes at high pH has been proposed.

Use of kinetic approaches, ESR and XPS, can certainly prove helpful in unraveling the complicated adsorption reactions. A more complete understanding of adsorption will help in predicting plant nutrient uptake, mobility of trace metals, and may help in understanding the potential role Mn oxides can play in trace metal waste management in the terrestrial as well as aquatic ecosystems.

REFERENCES

1. Jenne, E. A. "Controls on Mn, Fe, Co, Ni, Cu and Zn Concentrations in Soils and Water: the Dominant Role of Hydrous Manganese and Iron Oxides," in *Trace Inorganics in Water*, Gould, R. F., Ed. (Advan. Chem. Series. Am. Chem. Soc., Washington, D.C. 1968) p. 337–387.
2. Adriano, D. C. *Trace Elements in the Terrestrial Environment* (New York: Springer-Verlag, 1986).
3. McKenzie, R. M. "Manganese Oxides and Hydroxides," in *Minerals in Soil Environments*, Dixon, J. B. and S. B. Weed Eds. (Madison, WI: Soil Sci. Soc. Am., 1977) pp. 181–192.
4. Bartlett, R. J. "Soil Redox Behavior," in *Soil Physical Chemistry*, Sparks, D. L., Ed. (Boca Raton, FL: CRC Press, 1986) pp. 179–209.
5. Brystrom, A. and A. M. Brystrom. "Crystal Structures of Hollandite, the Related Manganese Oxide Minerals and MnO_2," *Acta Crystallogr.* 3:146–155 (1950).
6. Pauling, L. *The Nature of the Chemical Bond.* (Ithaca, NY: Cornell University Press, 1960).
7. Hemstock, G. Z. and P. F. Low. "Mechanisms Responsible for Retention of Manganese in the Colloidal Fraction of Soil," *Soil Sci.* 76:331–343 (1953).
8. Reddy, M. R. and H. F. Perkins. "Fixation of Manganese by Clay Minerals," *Soil Sci.* 121:21–24 (1976).
9. Khattak, R. A. and W. M. Jarrell. "Effect of Two Moisture Levels and Wetting-Drying Cycles on Manganese Release in NaCl-Amended Soils," *Commun. Soil Sci. Plant Anal.* 20:23–45 (1989).
10. Khattak, R. A., W. M. Jarrell, and A. L. Page. "Mechanism of Native Manganese Release in Salt-Treated Soils," *Soil Sci. Soc. Am. J.* 53:701–705 (1989).
11. McBride, M. B., T. J. Pinnavaia, and M. M. Mortland. "Electron Spin Relaxation and the Mobility of Manganese(II) Exchange Ions in Smectite," *Am. Mineral.* 60:66–72 (1975).

12. Furuhata, A. and K. Kuwata. "Electron Spin Resonance Spectra of Manganese(II) and Copper(II) Adsorbed on Clay Minerals and Silica-Alumina Mixtures," *Nendo Kagaku* 9:19–27 (1969).
13. Shuman, L. M. "Effect of Soil Properties on Manganese Adsorption Isotherms for Four Soils," *Soil Sci.* 124:77–81 (1977).
16. Curtin, D., J. Ryan, and R. A. Chaudhary. "Manganese Adsorption and Desorption in Calcareous Lebanese Soils," *Soil Sci. Soc. Am. J.* 44:947–950 (1980).
17. Hsu, P. H. and D. A. Rennie. "Reactions of Phosphorous in Aluminum Systems. I. Adsorption of Phosphate by X-Ray Amorphous 'Aluminum Hydroxides,'" *Can. J. Soil Sci.* 42:197–209 (1962a).
18. Hsu, P. H. and D. A. Rennie. "Reactions of Phosphorous in Aluminum Systems. II. Precipitation of Phosphates by Exchangeable Aluminum on a Cation Exchange Resin," *Can. J. Soil Sci.* 42:210–221 (1962b).
19. Stumm, W. and J. J. Morgan. *Aquatic Chemistry. An Introduction Emphasizing Chemical Equilibria in Natural Waters* (New York: Wiley-Interscience, 1970).
20. Veith, J. A. and G. Sposito. "On the Use of Langmuir Equation in the Interpretation of "Adsorption" Phenomenon," *Soil Sci. Soc. Am. J.* 41:697–702 (1977).
21. Griffin, R. A. and A. K. Au. Lead Adsorption by Montmorillonite Using a Competitive Langmuir Equation," *Soil Sci. Soc. Am. J.* 41:880–882 (1977).
22. Harter, R. D. and D. E. Baker. "Applications and Misapplications of the Langmuir Equation to Soil Adsorption Phenomenon," *Soil Sci. Soc. Am. J.* 41:1077–1088 (1977).
23. Harter, R. D. and G. Smith. "Langmuir Equation and Alternate Methods of Studying "Adsorption" Reactions in Soils," in *Chemistry in the Soil Environment*, Dowdy, R. H., et al., Eds. ASA Spec. Pub. No. 40 (Madison, WI: Soil Sci. Soc. Am., 1981) pp. 167–182.
24. Sposito, G. "On the Use of the Langmuir Equation in the Interpretation of "Adsorption" Phenomenon. II. The Two-Surface Langmuir Equation," *Soil Sci. Soc. Am. J.* 46:1147–1152 (1982).
25. Kinniburgh, D. G., J. K. Syers, and M. L. Jackson. "Specific Adsorption of Trace Amounts of Calcium and Strontium by Hydrous Oxides of Iron and Aluminum," *Soil Sci. Soc. Am. J.* 39:469–470 (1975).
26. Kinniburgh, D. G., M. L. Jackson, and J. K. Syers. "Adsorption of Alkaline Earth, Transition, and Heavy Metal Cations by Hydrous Oxide Gels of Iron and Aluminum," *Soil Sci. Soc. Am. J.* 40:796–799 (1976).
27. Sillen, L. G. and A. E. Martell. "Stability Constants of Metal-Ion Complexes." Spec. Publ. No. 17 (London: The Chemical Society, 1964).
28. Sillen, L. G. and A. E. Martell. "Stability Constants of Metal-Ion Complexes." Suppl. No. 1, Spec. Publ. No. 25 (London: The Chemical Society, 1971).
29. Grimme, H. "Die Adsorption von Mn, Co, Cu, and Zn durch Goethite aus Verdunnten Losungen," *Z. Pflanzenernaehr Bodenkd.* 121(1):58–65 (1968).
30. McBride, M. B. "Retention of Cu^{2+}, Ca^{2+}, Mg^{2+}, and Mn^{2+} by Amorphous Alumina," *Soil Sci. Soc. Am. J.* 42:27–31 (1978a).
31. Morgan, J. J. and W. Stumm. "Colloidal-Chemical Properties of Manganese Oxide," *J. Colloid Sci.* 19:347–359 (1964).
32. Murray, D. J., T. W. Healy, and D. W. Fuerstenan. "The Adsorption of Aqueous Metal on Collodial Hydrous Manganese Oxide," *Adv. Chem.* 79:74–81 (1968).
33. Murray, J. W. "The Interaction of Metal Ions at the Manganese Dioxide-Solution Interface," *Geochim. Cosmochim. Acta* 39:505–519 (1975a).

34. Stumm, W., C. P. Huang, and S. R. Jenkins. "Specific Chemical Interaction Affecting the Stability of Dispersed Systems," *Croat. Chem. Acta* 42:223–245 (1970).
35. Murray, J. W. "The Interaction of Cobalt with Hydrous Manganese Dioxide," *Geochim. Cosmochim. Acta* 39:635–647 (1975b).
36. McKenzie, R. M. "The Sorption of Cobalt with Manganese Dioxide Minerals," *Aust. J. Soil Res.* 5:235–246 (1967).
37. McKenzie, R. M. "The Reaction of Cobalt with Manganese Dioxide Minerals," *Aust. J. Soil Res.* 8:97–106 (1970).
38. McKenzie, R. M. "The Adsorption of Lead and Other Heavy Metals on Oxides of Manganese and Iron," *Aust. J. Soil Res.* 18:61–73 (1980).
39. Loganathan, P. and R. G. Burau. "Sorption of Heavy Metals Ions by a Hydrous Manganese Oxide," *Geochim. Cosmochim. Acta* 37:1277–1293 (1973).
40. McKenzie, R. M. Proton Release During Adsorption of Heavy Metal Ions by a Hydrous Manganese Dioxide. *Geochim. Cosmochim. Acta.* 43:1855–1857 (1979).
41. James, R. O. and T. W. Healy. "Adsorption of Hydrolyzable Metal Ions at the Oxide-Water Interface. I. Co(II) Adsorption on SiO_2 and TiO_2 as Model Systems," *J. Colloid Interface Sci.* 40:42–52 (1972a).
42. James, R. O. and T. W. Healy. "Adsorption of Hydrolyzable Metal Ions at the Oxide-Water Interface. III. A Thermodynamic Model of Adsorption," *J. Colloid Interface Sci.* 40:65–81 (1972b).
43. Traina, S. J. and H. E. Doner. "Co, Cu, Ni, and Ca Sorption by a Mixed Suspension of Smectite and Hydrous Manganese Dioxide," *Clays Clay Miner.* 33:118–122 (1985a).
44. Traina, S. J. and H. E. Doner. "Copper-Manganese(II) Exchange on a Chemically Reduced Birnessite," *Soil Sci. Soc. Am. J.* 9:307–313 (1985b).
45. Benjamin, M. M. and J. O. Leckie. "Conceptual Model for Metal-Ligand-Surface Interactions During Adsorption," *Environ. Sci. Technol.* 15:1050–1056 (1981).
46. Leeper, G. W. "Factors Affecting Availability of Inorganic Nutrients in Soils with Special Reference to Micronutrient Metals," *Annu. Rev. Plant Physiol.* 3:1–16 (1952).
47. Boischot, P., M. Durroux, and G. Sylvestre. "Etude sur la Fixation du Fer et Manganes dans les Sols Calcaires," *Ann. Inst. Nat. Rech. Agron. Ser A*, 1:307–315 (1950).
48. McBride, M. B. Chemisorption of Mn^{2+} at $CaCO_3$ Surfaces. *Soil Sci. Soc. Am. J.* 43:693–698 (1979).
49. Mattigod, S. V. and G. Sposito. "Estimated Association Constants for Some Complexes of Trace Metals with Inorganic Ligands," *Soil Sci. Soc. Am. J.* 41:1092–1097 (1977).
50. Swaine, D. J. "The Trace Elements Content of Soils," *Soil Sci.* Techn. Comm. No. 48 (York, England: Herald Printing Works, 1955).
51. Vanselow, A. P. "Cobalt," in *Diagnostic Criteria for Plants and Soils*, Chapman, H. D., Ed. (Univ. California, Div. Agric. Sci., 1966) pp. 142–156.
52. Taylor, R. M. and R. M. McKenzie. "The Association of Trace Elements with Manganese Minerals in Australian Soils," *Aust. J. Soil Res.* 4:29–39 (1966).
53. Fleming, G. "Mineral Disorders Associated with Grassland Farming," in Proc. Int. Meeting on Animal Production from Temperate Grassland, Dublin, Ireland (An Foras Taluntais, Dublin, 1977) pp. 88–95.

54. Adam, S. N., J. H. Honeysett, K. G. Tiller, and K. Norrish. "Factors Controlling the Increase of Cobalt in Plants Following the Addition of Cobalt Fertilizer," *Aust. J. Soil Res.* 7:29–42 (1969).
55. Burns, R. G. "The Uptake of Cobalt Into Ferromanganese Nodules, Soils, and Synthetic Manganese(IV) Oxides," *Geochem. Cosmochim. Acta* 40:95–102 (1976).
56. Murray, J. W. and J. G. Dillard. "The Oxidation of Cobalt(II) Adsorbed on Manganese Oxides," *Geochim. Cosmochim. Acta* 43:781–787 (1979).
57. McKenzie, R. M. "The Sorption of Some Heavy Metals by the Lower Oxides of Manganese," *Geoderma* 8:29–35 (1972).
58. Traina, S. J. and H. E. Doner. "Heavy Metal Induced Releases of Manganese(II) from Hydrous Manganese Oxide," *Soil Sci. Soc. Am. J.* 49:317–321 (1985c).
59. Main, R. K. and C. L. A. Schmidt. "Combination of Divalent Manganese with Proteins, Amino Acids and Related Compounds," *J. Gen. Physiol.* 19:127–147 (1935).
60. Heintz, S. G. and P. J. G. Mann. "Soluble Complexes of Manganic Manganese," *J. Agric. Sci.* 37:23–26 (1947).
61. Heintz, S. G. and P. J. G. Mann. "Studies on Soil Manganese," *J. Agric. Sci.* 39:80–95 (1949).
62. Shuman, L. M. "Fractionation for Soil Microelements," *Soil Sci.* 140:11–22 (1985).
63. Shuman, L. M. "Effect of Organic Matter on the Distribution of Manganese, Copper, Iron and Zinc in Soil Fraction," *Soil Sci.* 146:192–198 (1988).
64. Walker, J. M. and S. A. Barber. "The Availability of Chelated Mn to Millet and Its Equilibria with Other Forms of Mn in Soil," *Soil Sci. Soc. Am. Proc.* 24:485–488 (1960).
65. Page, E. R. "Studies in Soil and Plant Manganese. II. The Relationship of Soil pH to Mn-Availability," *Plant Soil* 16:247–257 (1962).
66. Passioura, J. B. and G. W. Leeper. "Available Manganese and the X Hypothesis," *Agrochemica* 8:81–89 (1963).
67. Geering, H. R., J. F. Hodgson, and C. Sdano. "Micronutrient Cation Complexes in Soil Solution. IV. The Chemical State of Manganese in Soil Solution," *Soil Sci. Soc. Am. Proc.* 33:81–85 (1969).
68. Schnitzer, M. and S. I. M. Skinner. "Organo-Metallic Interactions in Soils. 5. Stability Constants of Cu^{++}, Fe^{++}, and Zn^{++}-Fulvic Acid Complexes," *Soil Sci.* 102:361–365 (1966).
69. Schnitzer, M. and S. I. M. Skinner. "Organo-Metallic Interactions in Soils. 7. Stability Constants of Pb^{++}, Ni^{++}, Mn^{++}, Co^{++}, Ca^{++}, and Mg^{++}-Fulvic Acid Complexes," *Soil Sci.* 103:247–252 (1967).
70. Schnitzer, M. "Reaction Between Fulvic Acid, a Soil Humic Compound, and Inorganic Soil Constituents," *Soil Sci. Soc. Am. Proc.* 33:75–81 (1969).
71. Oakes, J. "Magnetic Resonance Studies in Aqueous Systems. Part 3. Electron Spin and Nuclear Magnetic Relaxation Study of Interactions Between Manganese Ions and Micelles," *J. Chem. Soc. Faraday Trans. 2* 9:1321–1329 (1973).
72. Khan, S. U. "Interaction Between the Humic Acid Fraction of Soils and Certain Metallic Cations," *Soil Sci. Soc. Am. Proc.* 33:851–854 (1969).
73. Gamble, G. S., C. H. Langford, and J. P. K. Tong. "The Structure and Equilibria of a Manganese(II) Complex of Fulvic Acid Studied by Ion Exchange and Nuclear Magnetic Resonance," *Can. J. Chem.* 54:1239–1245 (1976).

74. McBride, M. B. "Transition Metal Bonding in Humic Acid — an ESR Study," *Soil Sci.* 126:208–209 (1978b).
75. McBride, M. B. "Electron Spin Resonance Investigation of Mn^{2+} Complexation in Natural and Synthetic Organics," *Soil Sci. Soc. Am. J.* 46:1137–1142 (1982).
76. Rajendran, P. and R. S. Aiyer. "Availability of Added Zinc and Manganese in Relation to Adsorption Characteristics in Different Acid Soil Types of Kerala," *Madras Agric. J.* 69:86–92 (1982).

15

Multi-element Analysis in Plant Materials — Analytical Tools and Biological Questions

Bernd Markert

Systems Research Group, University of Osnabrück, P.O. Box 4469, 4500 Osnabrück, Germany

ABSTRACT

Two thirds of naturally occurring chemical elements in ecosystems are not investigated since they are viewed as nonessential or nontoxic to biota. In view of the important role plants play in most ecosystems, their inorganic chemical characterization, according to modern instrumental multi-element techniques is of high interest. The establishment of "Reference plant", comparable to the "Reference man" by the International Commission on Radiological Protection (ICRP), can be a useful tool for this type of chemical "fingerprinting". In this paper, fingerprints of various plant species will be discussed and their comparability demonstrated. In the future, more attention should be focused on establishing baseline values for "normal" elemental concentrations in ecosystem components, with special emphasis on those elements that are either uncommonly investigated or of undetectable concentrations. Lanthanides and platinum elements are among those for which baseline values in plants will be emphasized. Interelement relationships in plants are covered to provide a clearer understanding of constant element proportions in biological systems.

INTRODUCTION

Trace element analysis is an increasing challenge for research. In the last two decades, approximately every 2 years the status of a trace element changes from inessential to essential, formerly nontoxic elements change to potentially toxic, and elements hitherto not detectable by older techniques have been found in quantitative amounts for organisms by using modern equipment.[1–12] Besides this increase in analytical improvement in trace element research,[13,14] new thinking must be established to clarify objectives and interests for future biological trace element research.

Environmental processes are a complex mixture of biological and abiotic reactions. These processes are largely controlled by the abiotic components of the environment. A major portion are inorganic chemical compounds acting as mineral nutrients, ballast material in biotic entities, or as toxic substances. Environmental chemistry is far from having a complete understanding of the interdependence and possible synergistic interactions of elements.[15] Modern textbooks may give the impression that everything of significance in inorganic chemistry for plant and animal nutrition is already known. This, however, is a misconception, especially for trace elements close to detection limits, which should be accepted by the entire community of scientists involved in biological trace element research.[16,17] For further trends in inorganic environmental chemistry, three objectives must be clarified:

1. What kind of samples should be analyzed?
2. Which chemical elements should be investigated?
3. How can analytical chemistry help to fulfill the previous two demands?

Selection of Samples for Chemical Analysis

In characterizing our environment by its inorganic chemical composition, single ecosystems seem to be the most acceptable units for comparable scientific investigations because

- interactions between organisms themselves (biotic relationships) or between living systems and their physical environment (abiotic relationships) can be studied, and
- comparisons of polluted and nonpolluted areas can be undertaken.

Comparable ecosystems should have similar soil conditions and species composition. Due to the great number of different materials and species found in the field and in view of the limitation of analytical capabilities, a careful selection of samples to be analyzed must be carried out. The sampling of precipitation and water is well described in the literature. The selection of area-specific and ecosystem-specific sets of soils has emerged as one of the most complex and challenging problems facing the introduction of ecosystem research programs, real-time monitoring of the environment, and environmental specimen banks.[19,20]

The selection of living organisms (especially plants and animals) is a difficult step in ecosystem research. From the biological and analytical point of view, the organisms should have the following properties:

- sensitivity to the chemical substances under investigation
- able to accumulate the substances of interest
- distribution in different ecosystems for comparative studies
- representative of the ecosystem investigated
- present in large amounts
- capable of being sampled in an easy and representative manner
- analyzable according to the standard analytical methods currently available

The selection of specific materials for chemical analysis in ecological projects is described, for example, in the Environmental Specimen Bank Project[21–23] or in the Element Concentration Cadasters Project.[16,17] Different species were chosen for biomonitoring purposes.[24–31]

Selection of Elements to be Investigated

The periodic table in Figure 1 gives the elements commonly accepted as essential for living organisms. Our knowledge about their role and fate is limited to about 30 elements.[32,33] Consequently, the majority of the elements are little known and are not considered in evaluations of biological and environmental samples.[34] The chemical elements can be ordered by their physiological function observed in living organisms or by their position in the periodic table, if no physiological role is known. The elements can be classified as structural elements if they participate in the structure of functional molecules of the cell metabolism, e.g., as proteins, lipids, carbohydrates, nucleic acids, etc., or when they are responsible for stabilization effects within the organisms (e.g., Ca and Si). Nitrogen and sulfur are integrated into the carbon chain, which means that after reduction of the high oxidation state (nitrate or sulfate) they will be bound directly to the organic substance. In contrast, phosphorus, boron, and silicon are not reduced, but tend to form esters, especially with sugar molecules.[35] The so-called electrolyte elements serve to build up specific physiological potentials and are responsible for the maintaining of definite osmotic relations within the cell metabolism. The element Ca can be a structural element and an electrolytic element. Several elements, especially metal ions, can act as metal complex compounds in the cell metabolism.[15] These elements are called enzymatic elements. The physiology of these elements is described in detail in various textbooks of plant, animal, and human physiology.

In order to improve our understanding of the elemental composition of ecosystem components and to assist us in predicting the flow directions and effects of chemical elements, a holistic approach was undertaken in the years 1983 to 1986 to determine between 60 to 70 elements in different soils and plants. The data obtained in these investigations were compiled in Element Concentration Cadasters[17,18] in order to facilitate comparisons between different ecosystems and ecosystem compartments. As can be seen in the Element Concentration

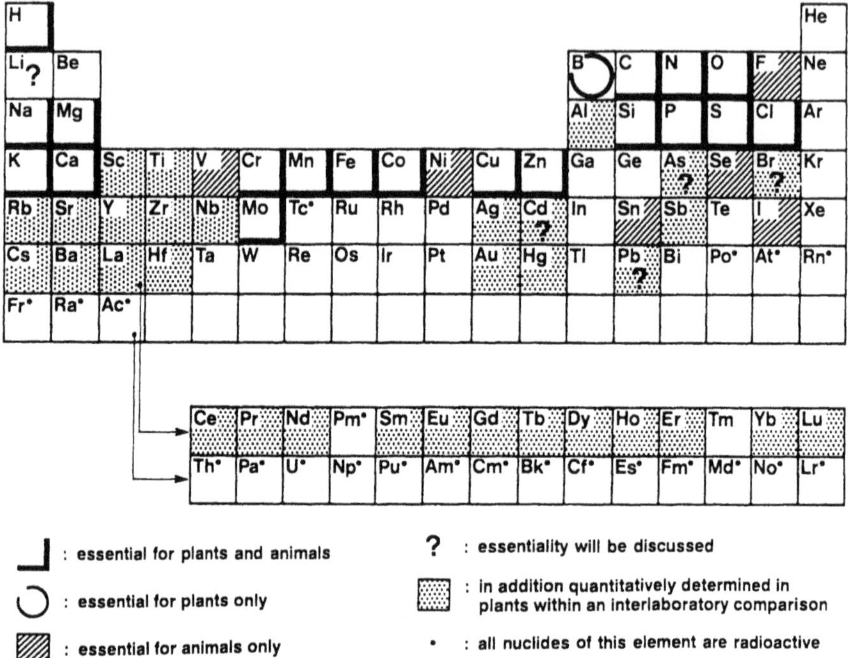

FIGURE 1. The periodic table of the elements with additional information on essentiality and occurrence of single elements in living organisms. (From Markert, B. *Instrumentelle Multielementanalyse an pflanzlichen Systemen* (Weinheim: VCH-Verlagsgesellschaft), in press.)

Cadaster of Figure 2 (leaves of Vaccinium vitis-idaea), a number of so-called inessential elements appear in the same order of magnitude as essential elements. As Horovitz[36] pointed out, the object of future research must be to overcome existing experimental difficulties and limitations in analytical methods. This will enable us to differentiate more accurately between essentiality and inessentiality of chemical elements for life, not forgetting the possibility that such a distinction might not really exist in nature.[36]

One problem of a multielement approach is the comparability of the analytical results in different samples, because a variety of data are produced per sample and these data are mostly distributed over a scale from 10^6 to 10^{-6} mg/kg dry weight. To overcome the problems connected with a logarithmic presentation of the data, a new type of data presentation in the form of chemical fingerprints with normalization to "reference plant" will be discussed.

Beginning in the 1950s the International Commission on Radiological Protection established a so-called "Reference man". This "Reference man" was used to make an estimate of the human body radiation dose, whether from external or internal sources. This estimate required a certain amount of data about the exposed individual. "Reference man" was defined as being between 20 and

ECC of	*Vaccinium vitis-idaea* (leaves) (red whortleberry)	Osnabrück Achmer	E: 7° 50' N: 52° 20' 68 NN	8,8°C 771 mm	rainy
17.6.1983	on Podsol		AAS AES-ICP MAS NAA, EA	D: 48h/105°C W: – H: agate/10 min	De: conc. HNO_3 3h, 170°C 2-4 Torr

10^xmg/kg	−6	−5	−4	−3	−2	−1	0	1	2	3	4	5	6	detection limit	not determined
			Eu	Lu	Gd	Sb	Ti	B	Fe	P	N	O		Bi <0.1	Ac, Ar, At,
				Er	Tb	As	Br	Na	Cl	Mg	H	C		Ga <0.02	Be, F,
				Yb	Sc	La	Ni	Rb	Al	S				Ge <1	Fr, He, I,
				Au	Hg	Cd	Cu	Zn	Mn	Ca				Sn <0.2	In, Ir, Kr,
				Ho	Hf	Ce	Sr	Ba	Si	K				Tm <0.1	Li, Ne, Os,
				Dy	Sm	Zr								U <0.073	Pa, Pd, Pm,
					Th	Cs								W <0.057	Po, Pt, Ra,
					Y	V									Re, Rh, Rn,
					Se	Pb									Ru, Ta, Tc,
					Ag	Cr									Te, Ti, Xe
					Cd										
					Nb										
					Pr										
					Nd										
					Mo										

FIGURE 2. Element Concentration Cadasterl of Vaccinium vitis-idaea (leaves); bar: half order of magnitude; D: drying; W: washing; H: homogenization; De: decomposition; ❍: essential for plants only; ❑ = essential for plants and animals; ■ = essential for animals only. (Modified after Lieth and Markert.[18])

30 years of age, weighing 70 kg, 170 cm in height, and living in a climate with an average temperature from 10° to 20°C. He is a Caucasian and is Western European or North American in habitat and custom. Besides the characteristics and data of different organs, water balance, respiration, etc., the chemical composition of the total body and various tissues were given. By 1959, data on 46 naturally occurring elements found in the adult body and its tissues had been reported. In later years, "Reference man" was extensively modified and additional characteristics were included, which will not be reported here. Up to now, "Reference man" has been a valuable tool to compare data produced by different analysts in various human tissues of different origin.

Similar to "Reference man" is a "Reference plant."[32] Of course, such an undertaking seems to be much more complicated because the plant kingdom is characterized by different types of plants, different plant families, and different plant species, whereas "Reference Man" characterized different races of only one species man. According to earlier work carried out by Duvigneaud and Denaeyer-De Smet[37] and Kinzel,[38] who established "reference values" for macronutrients in plants, and bearing in mind the goal of obtaining a data base system for comparing different analytical data from plant analysis with each other, it seems unimportant for a first approximation whether "Reference Plant" is a moss, a fern, or a higher plant. Table 1 shows the first values of "Reference Plant" for different chemical elements. The data were mainly extracted from our own analyses,[32,39] and when no data were available they were collected from Bowen[4] or Kabata-Pendias and Pendias.[7] No data of accumulator plants or plants with abnormal contents of elements, such as halophytes, were included in Table1. Because no data of natural platinum contents are available at the moment, calculated platinum values given by Markert[32] were included.

For the fingerprints of different plant species, shown in Figures 3 to 5, the values of "Reference plant" were set to zero (normalization) and the data of the element concentration of the plant species under consideration were given as deviations from the values of "Reference Plant". Some fingerprints will be discussed in more detail. For example, Betula pendula (Figure 3) can be characterized as a low mineral plant, which is characterized by high Zn (about 200 mg/kg) and high Mn (300 mg/kg) values. The other elements, especially the alkaline and alkaline earth metals, occur in lower concentrations compared to "Reference Plant". The fingerprint of Vaccinium vitis-idaea (Figure 4) is characterized by the accumulation of the following elements: Mn (674 mg/kg), Sn (2.9 mg/kg), Cs (0.371 mg/kg), and Hf (0.276 mg/kg); the enzymatic elements (Co, Cr, Fe, Mo, Ni, V, and Zn) appear in lower concentrations compared to "Reference Plant". In addition the alkaline (with the exception of Cs) and alkaline earth elements and B are decreased. Low concentrations are also observable for the halogens Cl (100 mg/kg) and Br (0.91 mg/kg). The lanthanide elements appear in almost the same order of magnitude as in "Reference Plant". The fingerprint of Sphagnum (Figure 5) represents accumulations of Co, Cr, and Mo. Mn appears in lower concentration than in "Reference Plant". An enrichment of concentration is observable for the alkali elements Li, Na, and Cs with

Table 1. Data for "Reference Plant"

Element Content of "Reference Plant" (mg/kg dry weight)

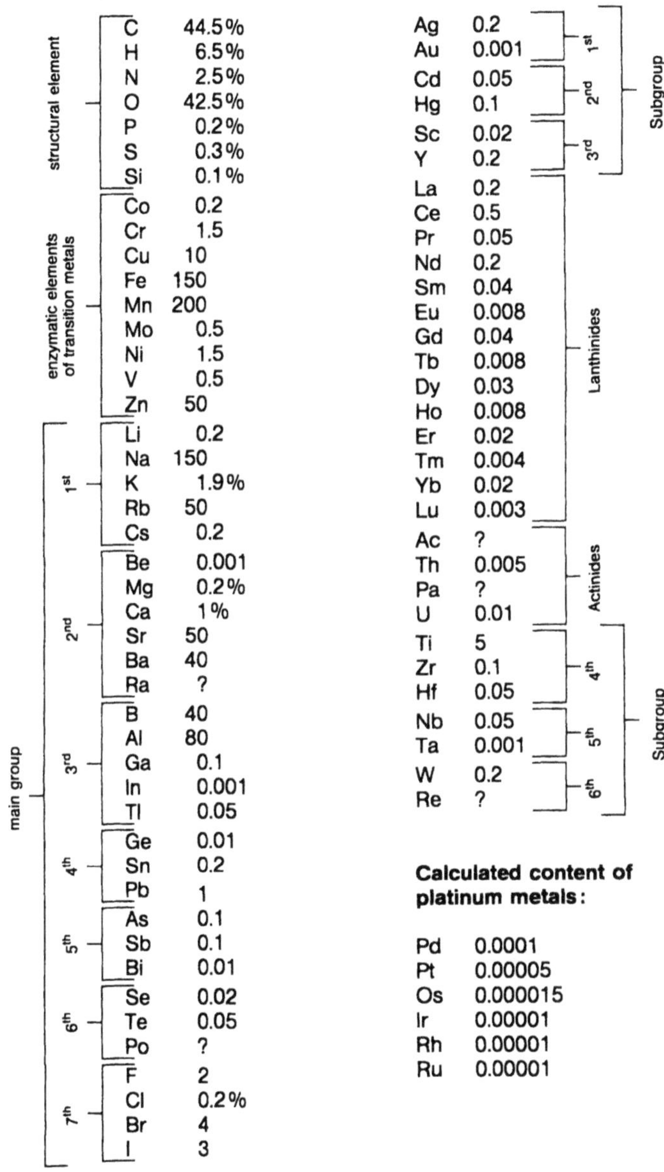

Group		Element	Content
structural element		C	44.5%
		H	6.5%
		N	2.5%
		O	42.5%
		P	0.2%
		S	0.3%
		Si	0.1%
enzymatic elements of transition metals		Co	0.2
		Cr	1.5
		Cu	10
		Fe	150
		Mn	200
		Mo	0.5
		Ni	1.5
		V	0.5
		Zn	50
main group	1st	Li	0.2
		Na	150
		K	1.9%
		Rb	50
		Cs	0.2
	2nd	Be	0.001
		Mg	0.2%
		Ca	1%
		Sr	50
		Ba	40
		Ra	?
	3rd	B	40
		Al	80
		Ga	0.1
		In	0.001
		Tl	0.05
	4th	Ge	0.01
		Sn	0.2
		Pb	1
	5th	As	0.1
		Sb	0.1
		Bi	0.01
	6th	Se	0.02
		Te	0.05
		Po	?
	7th	F	2
		Cl	0.2%
		Br	4
		I	3

Element	Content		Group
Ag	0.2	1st	Subgroup
Au	0.001		
Cd	0.05	2nd	
Hg	0.1		
Sc	0.02	3rd	
Y	0.2		
La	0.2		Lanthinides
Ce	0.5		
Pr	0.05		
Nd	0.2		
Sm	0.04		
Eu	0.008		
Gd	0.04		
Tb	0.008		
Dy	0.03		
Ho	0.008		
Er	0.02		
Tm	0.004		
Yb	0.02		
Lu	0.003		
Ac	?		Actinides
Th	0.005		
Pa	?		
U	0.01		
Ti	5	4th	Subgroup
Zr	0.1		
Hf	0.05		
Nb	0.05	5th	
Ta	0.001		
W	0.2	6th	
Re	?		

Calculated content of platinum metals:

Element	Content
Pd	0.0001
Pt	0.00005
Os	0.000015
Ir	0.00001
Rh	0.00001
Ru	0.00001

Note: No data of typical accumulator or rejector plants were used. Data were mainly extracted from the analytical work by Markert;[32,39] if data for single elements were not available they were collected from Bowen[4] or Kabata-Pendias and Pendias.[7]

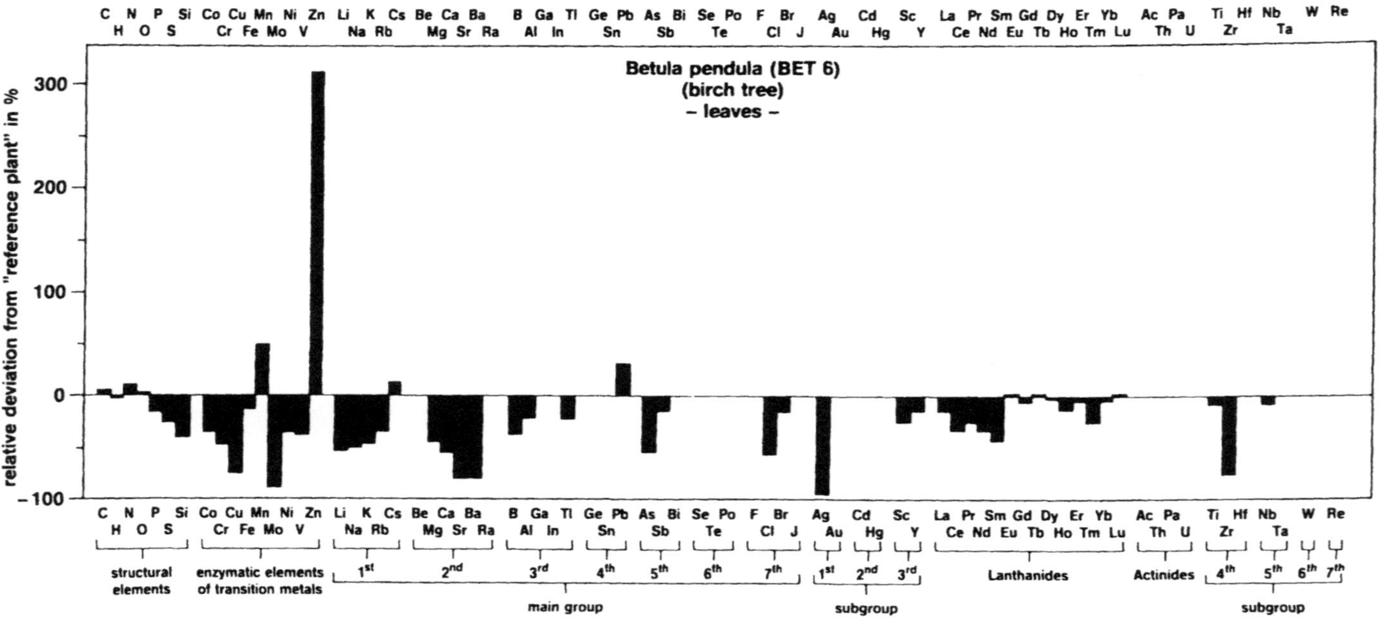

FIGURE 3. Chemical fingerprint of Betula pendula (birch tree, leaves) after normalization against "Reference plant". The samples were collected in the Grasmoor, near Osnabrück, Northwest Germany. (From Markert, B. *Instrumentelle Multielementanalyse an pflanzlichen Systemen* (Weinheim: VCH-Verlagsgesellschaft), in press.)

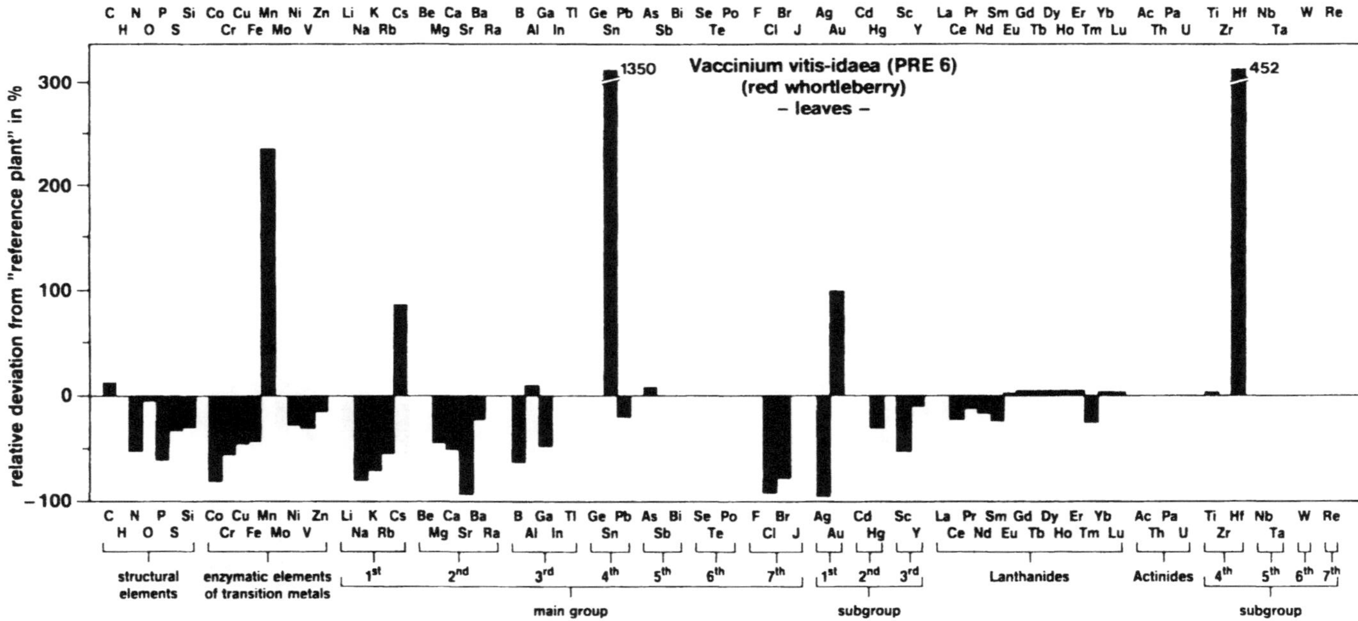

FIGURE 4. Chemical fingerprint of Vaccinium vitis-idaea (red whortleberry, leaves) after normalization against "Reference plant". The samples were collected in the Grasmoor near Osnabrück, Northwest Germany. (From Markert, B. *Instrumentelle Multielementanalyse an pflanzlichen Systemen* (Weinheim: VCH-Verlagsgesellschaft), in press.)

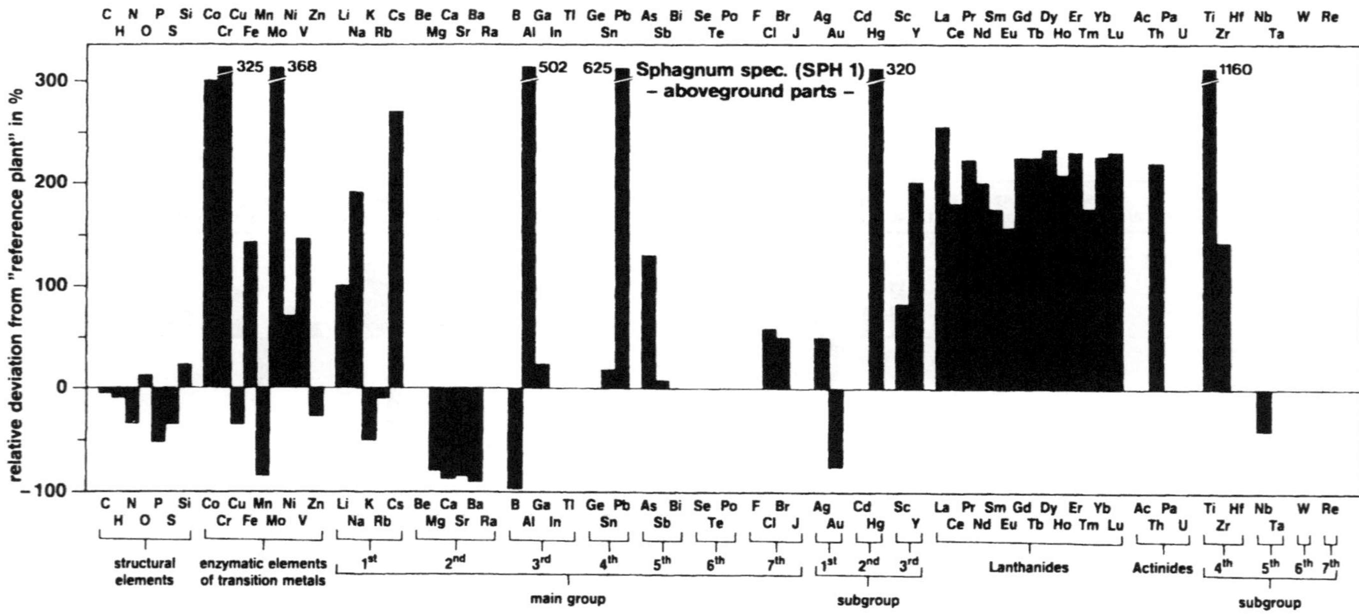

FIGURE 5. Chemical fingerprint of different Sphagnum species (mosses, above-ground parts) after normalization against "Reference plant". The samples were collected in the Grasmoor near Osnabrück, North West Germany. (From Markert, B. *Instrumentelle Multielementanalyse an pflanzlichen Systemen* (Weinheim: VCH-Verlagsgesellschaft), in press.)

decreasing concentrations of K and Rb. All alkaline earth elements and B appear in lower concentrations than in "Reference Plant". The elements Al (482 mg/kg), As (0.23 mg/kg), Hg (0.417 mg/kg), Pb (7.25 mg/kg), and the lanthanides were accumulated by Sphagnum mosses.

As one can see from the examples above, plant species can be characterized by different distribution patterns of inorganic composition. Of interest is the comparison of different fingerprint graphs. For example, plants that belong to the same plant family are compared in Figure 6. Vaccinium vitis-idaea (red whortleberry) and Vaccinium myrtillus (blueberry) are both Ericaceae. With reference to Figure 6, a high degree of agreement between both fingerprint graphs is observable, especially for the elements P, Si, Co, Cr, Cu, Mn, Na, K, Rb, Cs, Mg, Ca, Sr, Ba, B, Hg, Sc, most of the lanthanide elements, and Hf. Different element distributions are only observable for a few elements such as Ga and Pb. This example demonstrates that the close relationship of the two plant species can also be observed in a similar distribution pattern of the chemical elements.

With respect to the examples mentioned above, fingerprint graphs have the following properties:

- A multi-element spectrum of nearly all chemical elements of the periodic table in a plant species can be presented in one figure.
- The normalization of element concentration allows a direct comparison on a percent scale, and prevents a comparison on a logarithmic scale on which slight differences between the element concentration of two samples are not always clearly observable.
- Each plant species can be characterized by a specific element distribution pattern for the accumulation or rejection of elements compared to "Reference Plant".
- Different fingerprint graphs can be compared with each other. Related plant species seem to develop similar distribution patterns for single elements.

Of greatest interest seem to be so-called inter-element relationships found in recent years for different elements within living organisms. A chemical balance of inorganic elements in living organisms seems to be a basic condition for their proper growth and development.[7,40–49] In Figure 7, highly correlated element pairs in different plant reference materials are summarized.[41] In this context, leaves and needles seem of greatest interest, since they represent the productive part of plants and possess the highest metabolic efficiency, which is most important for the active establishment of constant mass fractions by plants. Due to the lack of knowledge of many elements in plant nutrition, a clear-cut statement for this phenomenon cannot be given. Since most interactive processes are controlled by several factors, these mechanisms are still poorly understood.[7]

The call for more data on elements not usually investigated is justified not only by our poor knowledge of biochemical reactions where trace elements are involved. Besides natural sources, human activities release many chemical elements into the environment in forms capable of interacting with biota.[50–64] Year after year, during the past few decades, new problems with inorganic

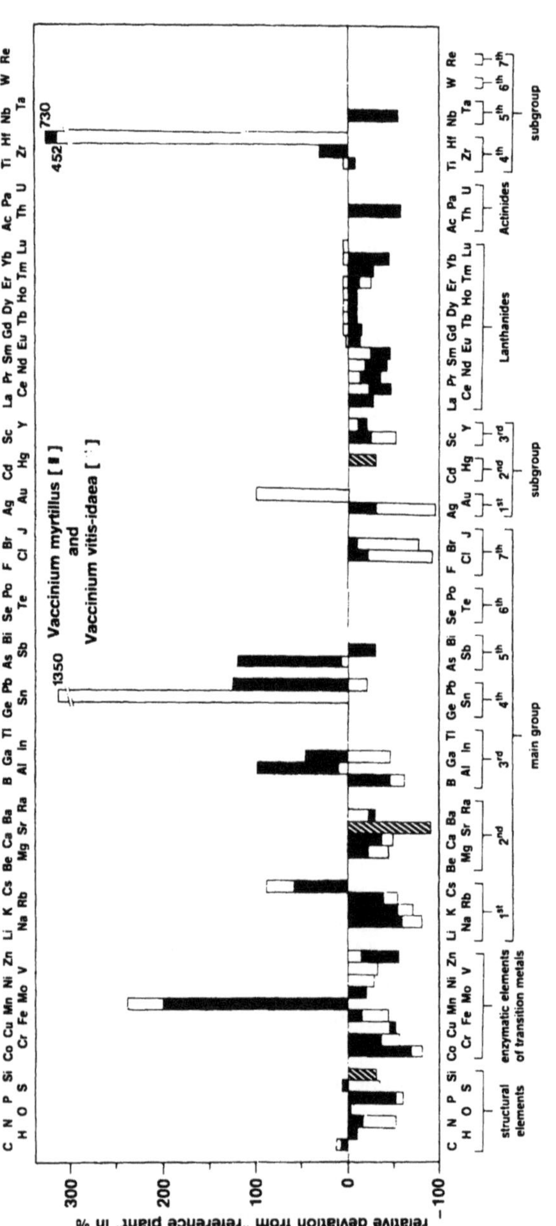

FIGURE 6. Comparison of single fingerprints of Vaccinium vitis-idaea and Vaccinium myrtillus. (From Markert, B. *Instrumentelle Multielementanalyse an pflanzlichen Systemen* (Weinheim: VCH-Verlagsgesellschaft), in press.)

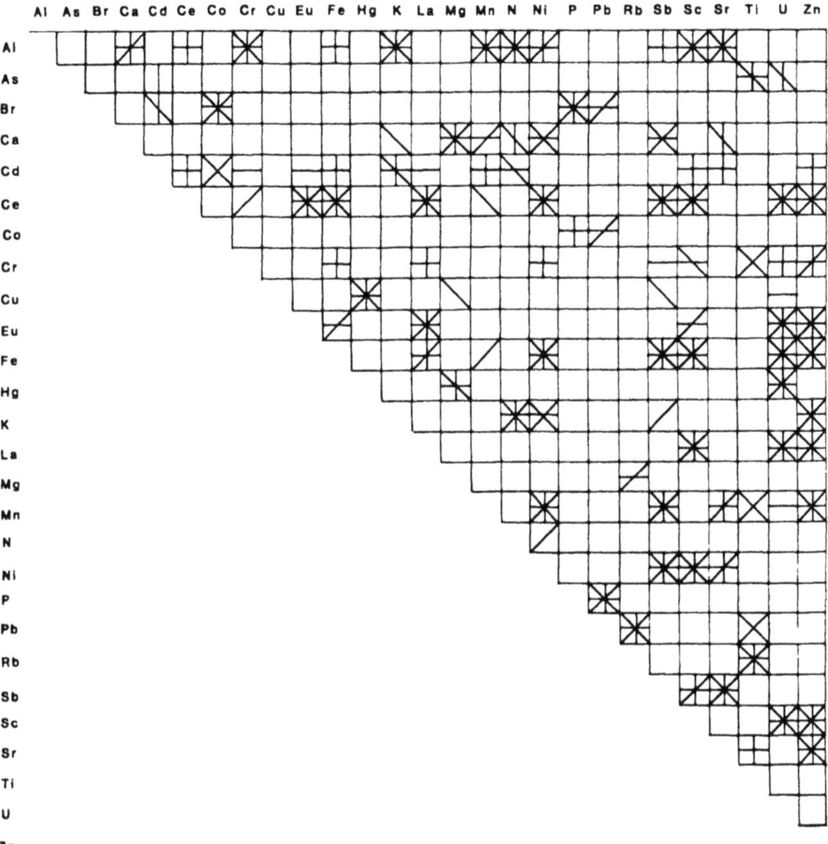

FIGURE 7. Summarizing correlation matrix of highly correlated (>+/-0.9) element pairs. — = highly linearly correlated; | = highly exponentially correlated; / = highly logarithmically correlated; \ = highly potentially correlated.

compounds have surfaced. Heavy metals were among the first concerns, soon followed by mass releases of halogens, amphoteric elements, and selenides. In recent years investigations in different parts of the world have shown that concentrations of platinum metals and lanthanides (rare earth elements) are steadily increasing in the environment.[18,65,66] These inorganic compounds can move along the food chain and gradually influence all biota in the ecosystem. Models describing this movement use detailed information about origin, concentration level, transfer rates, residence times in different parts of the ecosystem, and final sink as well as the type of elemental species in which they exist in different ecosystem compartments. Baseline values for lanthanide elements in a natural forest ecosystem (Figure 8) are especially important for risk assessments and political decisions. Baseline reference values of lanthanide elements in plant material were estimated as follows (all values were given in mg/kg dry weight): La: 0.15–0.2; Pr: 0.030–0.060; Nd: 0.1–0.25; Sm: 0.02–0.04; Eu: 0.005–0.015;

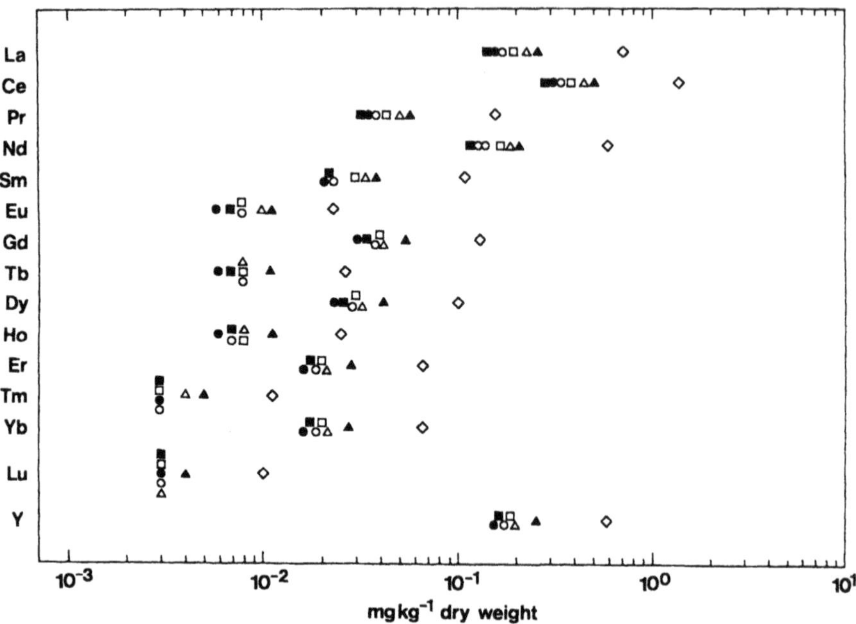

FIGURE 8. Baseline values (background concentrations) for lanthanide elements in a natural forest ecosystem. ○ = Betula alba (leaves); ● = Pinus sylvestris (needles); □ = Vaccinium vitis-idaea (leaves); ■ = Vaccinium myrtillus (leaves); Δ = Deschampsia flexuosa (aboveground parts); ▲ = Polytrichum formosum (above-ground parts); ◇ = Sphagnum species (above-ground parts). (From Markert, B. and Zhang De Li, *Sci. Total Environ.* 103:27–35 (1991). With permission.)

Gd: 0.030–0.060; Tb: 0.005–0.015; Dy: 0.025–0.05; Ho: 0.005–0.015; Er: 0.015–0.030; Tm: 0.0025–0.005: Yb: 0.015–0.030; Lu: 0.0025–0.005; Y: 0.15–0.25.[66]

The measured values shown in Figure 8 were compared with the expected baseline values calculated from the value in the earth's crust using a factor of 100. As can be seen in Figure 9, measured and expected values are in good agreement. It seems that the value, a hundredth part of the concentration of the earth's crust, is a good indicator of the basic reference values in plants.[66] This can be especially helpful for about 20 elements that are not at the moment detectable in plant material due to the very low content of these elements in natural, nonpolluted systems. Platinum metals belong to this group. Platinum metals are a group of 6 metals including platinum, palladium, osmium, iridium, rhodium, and ruthenium. The introduction of catalytic cleaning of motor car exhaust gases has created a new pollution source: platinum emission leading to uncontrolled addition to the environment. Test made in the U.S., where catalytic converters were introduced more than 10 years ago, showed that, depending on the mode of driving, 2 μg platinum and more were emitted per kilometer driven.[67–69] Due to the low natural concentration of platinum and high affinity of these substances to organic molecules, toxic effects can be expected in the future. The analysis of

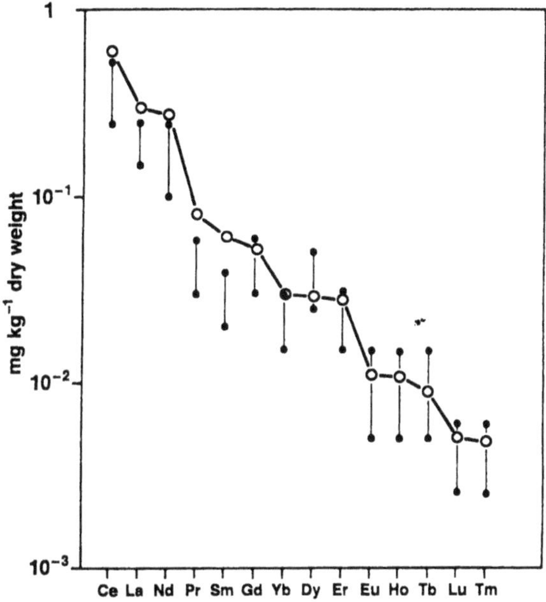

FIGURE 9. Comparison of calculated normal concentration of lanthanides (○) and measured concentrations (●) in plants of the Grasmoor near Osnabrück, Northwest Germany. ●—● gives the range of lanthanide element concentrations quantitatively measured in different plant species investigated. (From Markert, B. in *Modern Ecology: Basic and Applied Aspects.* Esser, G. and D. Overdieck, Eds. (Amsterdam: Elsevier, 1991b). With permission.)

platinum and platinum metals in biological concentrations is just beginning. In order to provide initial information on the natural background concentration of platinum metals and gold in plant material, the normal concentrations calculated against the earth's crust as described above are given in Figure 10. The following background levels were calculated for platinum metals: Pd: 0.0001 mg/kg DW; Pt: 0.00005 mg/kg DW; Os: 0.000015 mg/kg DW; Ir, Rh, and Ru 0.00001 mg/kg DW.

Analytical data of platinum in plants are currently available in two standard reference materials only (citrus leaves, NIST 1572, and orchard leaves, NIST 1571). As can be seen in Figure10, the measured platinum values in citrus leaves are near the expected value. Orchard leaves represent higher measured values. In addition the same procedure was repeated with gold, and similarly good agreement between calculated and measured gold values was observed for citrus leaves. As in the case of platinum in orchard leaves, the gold values determined experimentally are higher than the expected value, which might be related to a contamination during preparation of the material, a higher accumulation factor for these metals by orchard leaves compared with citrus leaves, or pollution of the material by soil or anthropogenic influences during growth. However, as in the case of lanthanide elements, the calculation of background values with the

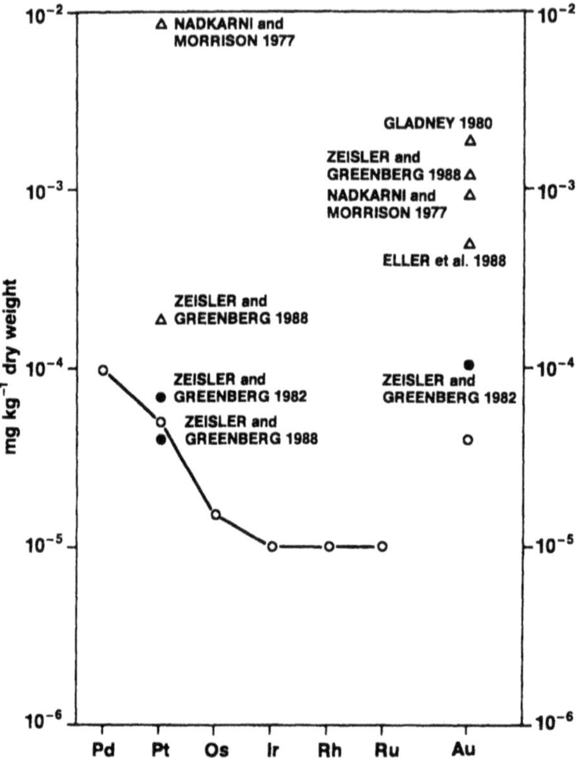

FIGURE 10. Calculated "normal concentrations" of platinum metals and gold in plants (○) compared with measured concentrations of different working groups in NIST-citrus leaves (●) and NIST orchard leaves (Δ). (From Markert, B. in *Modern Ecology: Basic and Applied Aspects.* Esser, G. and D. Overdieck, Eds. (Amsterdam: Elsevier, 1991b). With permission.)

help of normalized earth crust values seems to give a first indication of natural background concentration for plant materials, especially when no analytical data are available.

Use of Instrumental Multielement Methods in Environmental Chemistry

Instrumental techniques available for environmental analysis have reached a high degree of sensitivity for most of the chemical elements of the periodic table.[70–72] The ppm and upper ppb ranges are open for routine analysis, and the lower ppb and ppt ranges are currently a domain for highly specialized laboratories especially equipped with clean room conditions, quartz tools, etc. From an instrumental point of view, the best equipment was used by scientists to perform accurate analytical chemistry. Practical detection limits as given for different instrumental methods for plant analysis (Figure 11) show that about half of the elements can be found directly in plant materials. However, many results

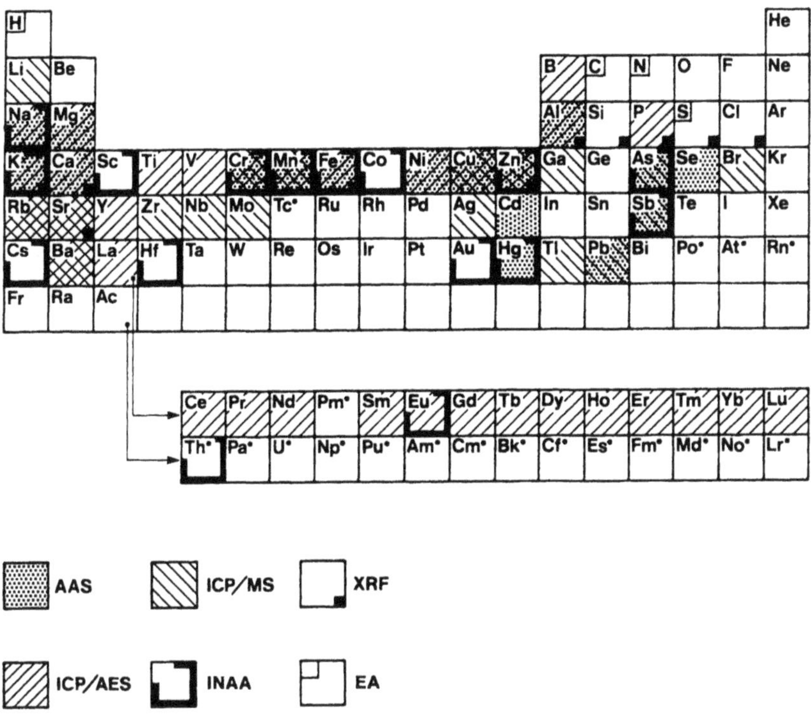

FIGURE 11. Realistic detection limits of different analytical methods in uncontaminated plant materials. (From Markert, B. *Instrumentelle Multielementanalyse an pflanzlichen Systemen* (Weinheim: VCH-Verlagsgesellschaft), in press.)

obtained in the past are no longer satisfactory for present purposes. Interlaboratory comparisons organized by different national and international institutes revealed the analytical discrepancies that have been taking place in the last few decades. Analytical results obtained in these various laboratories often differ by several orders of magnitude. In a comparative study, Horwitz[73] demonstrated that the errors in accuracy increase exponentially in the ppb and ppt ranges. Reasons are due more to errors in sample preparation (drying, weighing, decomposition) than in direct instrumental measurements because the precision (reproducibility of analytical signals) is often acceptable. Only two procedures are possible to control the accuracy of analytical results: (1) use of independent methods for the same sample[74] or (2) use of certified reference materials available from different organizations.[74–85] An example of the certification of the element Cu in rye grass reference material (BCR-CRM 281) by the Bureau Community of Reference in Brussels is given in Figure 12. The use of independent methods is often not feasible for most institutes; the expense may be so high that reference materials

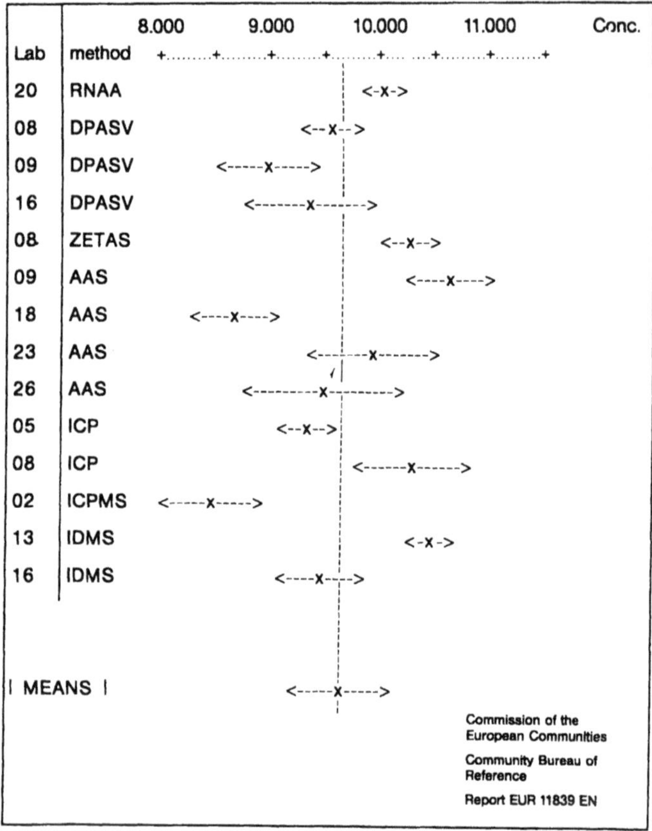

FIGURE 12. Comparison of results obtained by different analytical methods during determination of Cu in rye grass (BCR 281). (From Markert, B. *Instrumentelle Multielementanalyse an pflanzlichen Systemen* (Weinheim: VCH-Verlagsgesellschaft), in press.) RNAA: neutron activation analysis with radiochemical separation, DPASV: differential pulse anodic stripping voltammetry, ZETAS: electrothermal atomic absorption spectrometry with Zeeman-background correction, AAS: atomic absorption spectrometry (flame), ICP: inductively coupled plasma emission spectrometry, ICPMS: inductively coupled plasma (ionization) mass spectrometry, IDMS: isotope dilution mass spectrometry.

seem the best way to reduce costs while producing accurate results. For many elements and matrices, though, either well-characterized reference materials are not available or the concentration ranges of certified elements are not satisfactory. Poor performance is displayed by the whole community of trace element analysts in that they have allowed instruments such as the ICP/mass spectrometer, which is capable of determining nearly the whole spectrum of elements within a few minutes, to be brought onto the market; the accuracy of the results

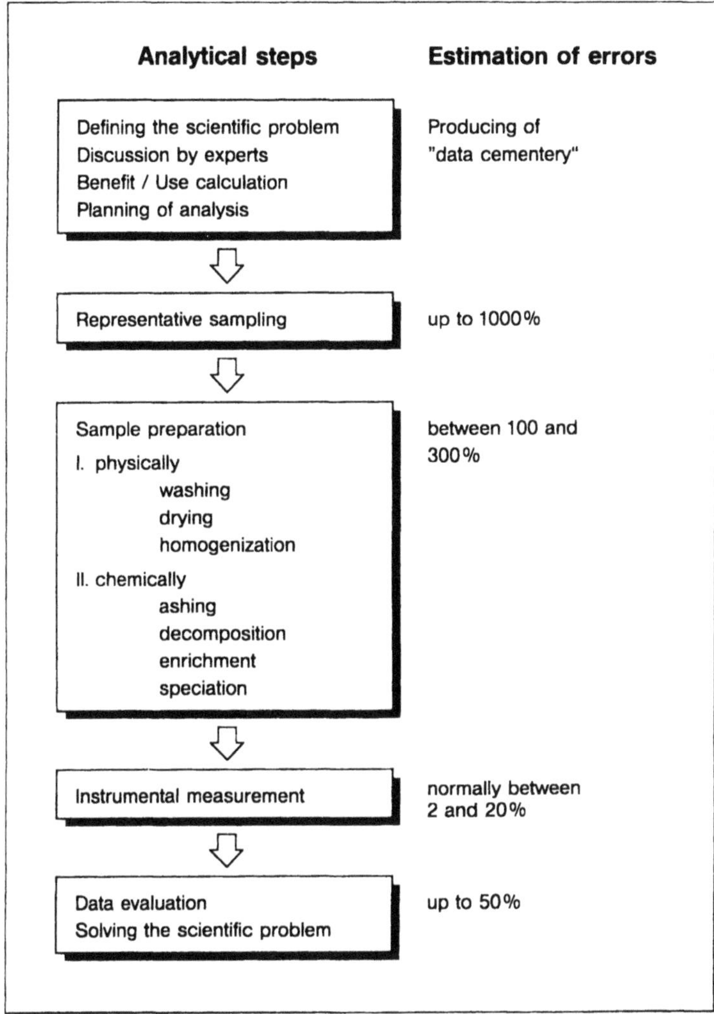

FIGURE 13. Simplified flowchart for different analytical steps during the total procedure of chemical analysis including error estimation. (From Markert, B. *Instrumentelle Multielementanalyse an pflanzlichen Systemen* (Weinheim: VCH-Verlagsgesellschaft), in press.)

cannot be checked because there are no reference materials. It is estimated that incorrect results cause a loss of national wealth in the F.R.G. in the order of $2.5 billion (U.S.) every year.[86]

A second point has been underestimated in recent years. It is not always appreciated that, in general, sampling errors of ecological materials greatly exceed laboratory (analytical) errors in the multi-element analysis of these materials (Figure 13). A major consideration in the reliability of any analytical step is sample quality. Does the collected sample really represent a portion of the

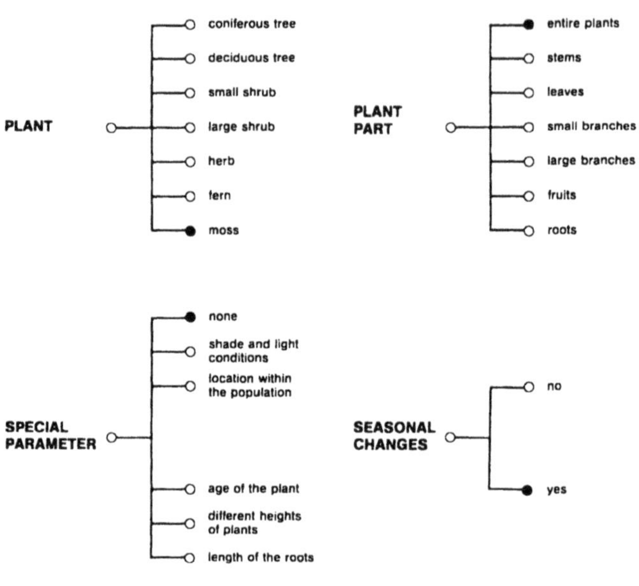

Advice:
Choose a stand of moss after statistical division of your research area. Collect aboveground parts of single and different aged moss plants at weekly intervals (about 50 g fresh weight) over a period of 2 years at least. Transport the samples immediately to the lab to avoid changes in the mineral content produced by microbiological activities.

FIGURE 14. Menu for collection of plant samples with respect to different biological parameters (●). (From Markert, B. and N. Klausmeyer, *Toxicol. Environ. Chem.*, 25:200–212 (1990). With permission.)

whole and is it selected in such a way as to be truly representative of the whole or the question in mind?[87] The analyst often only reports results obtained on a particular test specimen, which may not provide the information desired or needed. This may be due to uncertainties in the sampling process or because the sampling plan itself is so poorly considered that no relation is established between the analytical results and the population from which the sample was drawn. Since incorrect sampling of materials can introduce an error of up to 1000% to chemical analysis, more attention must be paid to this step during the whole procedure of single- or multi-element analysis. Therefore more attention must be paid to the representative sampling procedures, which are in part described in the literature.[88–105]

An example of a menu from a computer-aided sampling procedure for plant materials is given in Figure 14. Different parameters, influencing the chemical composition of plants, can be combined in the procedure. At the end of the procedure, advice is given for collecting the plants under investigation.[104]

The second largest error in chemical multielement analysis can be introduced during the sample decomposition process by dry and wet ashing methods.[106–110] Contamination problems of the acids, vessels, and air; volatilization of elements or element compounds in open systems; and incomplete decomposition are the main reasons for errors during this analytical step. Working in closed systems in quartz vessels at high temperature and high pressure can optimize this procedure.

Application notes of different analytical techniques on environmental samples are given, e.g., for AAS,[111,112] ICP/AES,[113,114] ICP/MS,[115-117] and NAA.[118–124]

Knowledge of the composition of a sample in terms of total trace element concentration is a first step toward complete characterization of an environmental sample, but certainly does not provide all or even the most important information. Synergistic or antagonistic interactions of life-sustaining, biologically important molecules in living organisms occur in most cases, not with elements or their free ions but with specific compounds of these elements (species). To obtain the necessary knowledge, investigations of the toxic or beneficial effects must be followed by studies identifying the various compounds, their transformations, and their interactions with biomolecules.[15] Various methods are available for speciation analysis. Volatile compounds can be determined by gas chromatography with element- or compound-specific detectors, such as mass spectrometers, and atomic absorption or emission spectrometers. Nonvolatile compounds can be separated by high pressure liquid chromatography or ion chromatography and detected by graphite furnace atomic spectrometers or plasma emission spectrometers.[15] Analysts are aware of the difficulties associated with the preservation of samples during and after collection for total trace element analysis, and have at their disposal various methods to ensure stability of their samples. Little is known about the trace element species in environmental samples and even less about methods for the succesful preservation of these compounds during sample storage and sample preparation. It is of vital importance for future work to collect, evaluate, and establish approximate sampling and preservation methodologies for the investigation of natural systems.[15] Speciation analysis seems at present the most exciting challenge for the next century.

ACKNOWLEDGMENTS

Thanks are due to the organizers of the Orlando meeting for their kind invitation and to KFA Jülich for sponsoring the traveling expenses. Prof. Dr. B. Sansoni of KFA and Prof. Dr. H. Lieth (University of Osnabrück) made fruitful remarks on the manuscript. Prof. Dr. D. Adriano and three independent reviewers are thanked for the critical comments made on the manuscript. The "Graphische Betriebe" of KFA are thanked for preparing the figures, and Mrs. J. Carter-Sigglow for polishing the English of the manuscript.

REFERENCES

1. Adriano, D. C. *Trace Elements in the Terrestrial Environment* (New York: Springer-Verlag, 1986).
2. Anke, M., C. Brückner, H. Gürtler, and M. Grün, Mengen- und Spurenelemente Arbeitstagung, Karl-Marx-Universität, Leipzig (1989).
3. Bodek, I., W. J. Lyman, W. F. Reehl, and D. H. Rosenblatt, Eds. *Environmental Inorganic Chemistry, Properties, Processes, and Estimation Methods*, (New York: Pergamon Press, 1988).
4. Bowen, H. J. M. *Environmental Chemistry of the Elements*, (London: Academic Press, 1979).
5. Caroli, S., G. V. Iyengar, and H. Muntau, Eds. "Bioelements: Health Aspects," *Ann. Ist. Super. Sanita* 25:373–378 (1989).
6. Frieden, E., Ed. *Biochemistry of the Essential Ultratrace Elements*, (New York: Plenum Press, 1984).
7. Kabata-Pendias, A. and H. Pendias. *Trace Elements in Soils and Plants*, (Boca Raton, FL: CRC Press, 1984).
8. Kovacs, M. "Chemical Composition of the Lesser Reedmace (Typha angustifolia) in Lake Balaton," *Acta Bot. Acad. Sci. Hung.* 28:297–307 (1982).
9. Kovacs, M., I. Nyary, and L. Toth. "The Microelement Content of Some Submerged and Floating Aquatic Plants," *Acta Bot. Hung.* 30:173–185 (1984).
10. Kovacs, M., G. Turcsanyi, L. Nagy, A. Koltay, L. Kaszab, and P. Szöke. "Element Concentration Cadasters in a Quercetum Petraeae-Cerris Forest," in *Element Concentration Cadasters in Ecosystems*, Lieth, H. and B. Markert, Eds. (Weinheim: VCH Verlagsgesellschaft mbH, 1990) pp. 255–265.
11. Mc Kenzie, H. A. and L. E. Smythe, Eds. *Quantitative Trace Analysis of Biological Materials*, (Amsterdam: Elsevier 1988).
12. Roth-Holzapfel, M. "Multi-Element Analysis of Invertebrate Animals in a Forest Ecosystem (Picea abies L.)," in *Element Concentration Cadasters in Ecosystems*, Lieth, H., and B. Markert, Eds. (Weinheim: VCH Verlagsgesellschaft mbH, 1990) pp. 281–296.
13. Iyengar, V. G. "Biological Trace Element Research: A Multidisciplinary Science," *Sci. Total Environ.* 71:1–5 (1988).
14. Iyengar, V. G. *Elemental Analysis of Biological Systems, Biological, Medical, Environmental, Compositional and Methodological Aspects*, (Boca Raton, FL: CRC Press, 1989).
15. Irgolic, K. J. and A. E. Martell, Eds. *Environmental Inorganic Chemistry*, (Weinheim: VCH Verlagsgesellschaft mbH, 1985).
16. Lieth, H. and B. Markert. "Concentration Cadasters of Chemical Elements in Contrasting Ecosystems," *Naturwissenschaften* 72:322–324 (1985).
17. Lieth, H. and B. Markert. *Aufstellung und Auswertung ökosystemarer Element-Konzentrations-Kataster*, (Berlin: Springer-Verlag, 1988).
18. Lieth, H. and B. Markert, Eds. *Element Concentration Cadasters in Ecosystems, Methods of Assessment and Evaluation*, (Weinheim: VCH Verlagsgesellschaft mbH, 1990).
19. Fränzle, O. Regional Repräsentative Auswahl der Böden für eine Umweltprobenbank, FE-Vorhaben 106 05 028 des Umweltbundesamtes, Berlin (1983).

20. Fränzle, O. "Representative Sampling of Soils in the Federal Republic of Germany and the EC Countries," in *Element Concentration Cadasters in Ecosystems*, Lieth, H. and B. Markert, Eds. (Weinheim: VCH Verlagsgesellschaft mbH, 1990).
21. Stoeppler, M., H. W. Dürbeck, and H. W. Nuernberg. "Environmental Specimen Banking, a Challenge in Trace Analysis," *Talanta* 29:963–972 (1982).
22. Stoeppler, M., F. Backhaus, J. D. Schladot, and N. Commerscheidt. "Environmental Specimen Bank Specific Reference Materials — Main Aims and Preparation," *Fresenius Z. Anal. Chem.* 326:707–711 (1987).
23. Bundesministerium für Forschung und Technologie, Ed. *Umweltprobenbank*, (Berlin: Springer-Verlag, 1988).
24. Arndt, U., W. Nobel, and B. Schweitzer. *Bioindikatoren, Möglichkeiten, Grenzen und neue Erkenntnisse*, (Stuttgart: Verlag Eugen Ulmer, 1987).
25. Glooschenko, W. A. and N. Arafat. "Atmospheric Deposition of Arsenic and Selenium Across Canada Using Sphagnum Moss as a Biomonitor," *Sci. Total Environ.* 73:269–275 (1988).
26. Markert, B. and I. Thornton. "Multi-Element Analysis of an English Peat Bog Soil," *Water Air Soil Pollut.* 49:113–123 (1990).
27. Schubert, R., Ed. *Bioindikation in terrestrischen Ökosystemen*, (Stuttgart: Fischer-Verlag, 1985).
28. Steubing, L. and H. L. Jaeger, Eds. "Monitoring of Air Pollutants by Plants, Methods and Problems," T:VS, No. 7 (Den Haag: Dr. W. Junk Publisher, 1982).
29. Streit, B. "Bioakkumulationen in der Natur," *Biol. unserer Zeit* 3:147–154 (1989).
30. Wiersma, G. B., M. E. Harmon, G. A. Baker, and S. E. Grenne. "Elemental Composition of Hylocomium Splendens in the Hoh Rain Forest Olympic National Park Washington, U.S.A.," *Chemosphere* 16:2631–2645 (1987).
31. Wiersma, G. B., D. A. Bruns, C. Whitworth, C. Boelcke, and L. McAnulty. "Elemental Composition of Mosses from a Remote Nothofagus Forest Site in Southern Chile," *Chemosphere* 20:569–583 (1990).
32. Markert, B. Instrumentelle Multielementanalyse an pflanzlichen Systemen (Weinheim: VCH-Verlagsgesellschaft mbH, in press).
33. Markert, B. "Multi-Element Analysis in Plant Material," in *Modem Ecology: Basic and Applied Aspects*, Esser G. and D. Overdieck, Eds. (Amsterdam: Elseviev, 1991b).
34. Zeisler, R., S. F. Stone, and R. W. Sanders. "Sequential Determination of Biological and Pollutant Elements in Marine Bivalves," *Anal. Chem.* 60:2760–2765 (1988).
35. Mengel, K. *Ernährung und Stoffwechsel der Pflanze*, (Stuttgart: Fischer Verlag, 1984).
36. Horovitz, C. T. "Is the Major Part of the Periodic System Really Inessential for Life?" *J. Trace Elem. Electrolytes Health Dis.* 2:135–144 (1988).
37. Duvigneaud, P. and S. Denaeyer-De Smet. "Phytogeochimie des groupes ecosociologiques forestiers de Haute-Belgique. I. Essai de classification phytochimique des especes herbacees," *Oecol. Plant.* 5:1–32 (1970).
38. Kinzel, H. *Pflanzenökologie und Mineralstoffwechsel*, (Stuttgart: Verlag Eugen Ulmer, 1982).
39. Markert, B. "Aufstellung von Element-konzentrations-katastern in unterschiedlichen Pflanzenarten und Bodentypen in Deutschland, Österreich und Schweden," in *Beiträge zur Umweltprobenbank*, Stoeppler, M. and H. W. Dürbeck, Eds. Jül. Spez., 360 (1986).

40. Fiedler, H. J. and H. J. Rösler, Eds. *Spurenelemente in der Umwelt*, (Stuttgart: Ferdinand Enke Verlag, 1988).
41. Markert, B. "Interelement Correlations in Different Reference Materials," *Fresenius Z. Anal. Chem.* 332:630–635 (1988).
42. Marschner, H. "General Introduction to the Mineral Nutrition of Plants," in *Encyclopedia of Plant Physiology*, New Series, Vol. 15A, Läuchli, A. and R. L. Bieleski, Eds. (1983) pp. 5–60.
43. Marschner, H. *Mineral Nutrition of Higher Plants*, (London: Academic Press, 1986).
44. Cox, R. M. and T. C. Hutchinson. "Metal Cotolerances in the Grass Deschampsia Cespitosa," *Nature (London)* 279:231–233 (1979).
45. Epstein, E. *Mineral Nutrition of Plants, Principles and Perspectives*, (New York: John Wiley & Sons, 1972).
46. Garten, C. T. "Correlations Between Concentrations of Elements in Plants," *Nature (London)* 261:686–688 (1976).
47. Garten, C. T. "Multivariate Perspectives on the Ecology of Plant Material Element Composition," *Am. Nat.* 112:533–544 (1978).
48. Golley, F. B., T. Richardson, and R. G. Clements. "Elemental Concentrations in Tropical Forests and Soils of Northwestern Columbia," *Biotropica* 10:144–151 (1978).
49. Wyttenbach, A., S. Bajo, L. Tobler, and T. Keller. "Major and Trace Element Concentrations in Needles of Picea Abies; Levels, Distribution Functions, Correlations and Environmental Influences," *Plant Soil* 85:313–325 (1985).
50. Chaney, R. L., G. S. Stoewsand, A. K. Furv, C. A. Backe, and D. J. Lisk. "Elemental Content of Tissue of Guinea Pig Fed Swiss Chard Grown on Municipal Sludge Amended Soil," *J. Agric. Food Chem.* 26:994–997 (1978).
51. Ernst, W. H. O. and E. N. G. Joosse Van Damme. *Umweltbelastung durch Mineralstoffe*, (Stuttgart: Gustav Fischer Verlag, 1983).
52. Fortescue, J. A. C. *Environmental Geochemistry*, (Berlin: Springer-Verlag, 1980).
53. Friberg, L., G. F. Norberg, and V. B. Vouk, Eds. *Handbook on the Toxicology of Metals*, (Amsterdam: Elsevier, Vols. 1 and 2, 1981).
54. Grill, E., E. L. Winnacker, and M. H. Zenk. "Occurrence of Heavy Metal Binding Phytochelatins in Plants Growing in a Mining Refuse Area," *Experientia* 44:539–540 (1988).
55. Hamilton, E. I. "The Need for Trace Element Analyses of Biological Materials in the Environmental Sciences," in *Element Analysis of Biological Materials — Current Problems and Techniques with Special Reference to Trace Elements*, International Atomic Energy Agency, Ed. Tech. Rep. Ser. No. 197, Vienna (1980) pp. 39–54.
56. Hemphill, D. D., Ed. "Trace Substances in Environmental Health," Proc. of University of Missouri's Annual Conference on Trace Substances in Environmental Health, University of Missouri, Columbia (1967–1989).
57. Hutchinson, T. C. "Toleranzgrenzen für Pflanzen: Auswahl geeigneter Pflanzen für metallverseuchte Böden," in *Metalle in der Umwelt*, Merian, E., Ed. (Weinheim: VCH Verlagsgesellschaft mbH, 1984) pp. 135–141.
58. Kabata-Pendias, A. and S. Dudka. "Evaluating Baseline Data for Cadmium in Soils and Plants in Poland," in *Element Concentration Cadasters in Ecosystems*, Lieth, H. and B. Markert, Eds. (Weinheim: VCH Verlagsgesellschaft mbH, 1990) pp. 265–280.

59. Likens, G. E., F. H. Bormann, R. S. Pierce, J. S. Eaton, and N. M. Johnson. *Biogeochemistry of a Forested Ecosystem*, (Berlin: Springer-Verlag, 1977).
60. Merian, E., Ed. *Metalle in der Umwelt*, (Weinheim: VCH Verlagsgesellschaft mbH, 1984).
61. Nriagu, J. O. and J. M. Pacyna. "Quantitative Assessment of Worldwide Contamination of Air, Water and Soils by Trace Metals," *Nature (London)* 333:134–139 (1988).
62. Page, A. L. and F. T. Bingham. "Cadmium Residues in the Environment," *Residue Rev.* 48:1–44 (1973).
63. Thornton, I., Ed. "Geochemistry and Health," *Proc. 2nd Int. Symp. on Geochemistry and Health*, (Northwood: Science Review Ltd., 1988).
64. Thornton, I. "A Survey of Lead in the British Urban Environment: An Example of Research in Urban Geochemistry," in *Element Concentration Cadasters in Ecosystems*, Lieth, H. and Markert, B., Eds. (Weinheim: VCH Verlagsgesellschaft mbH, 1990) pp. 221–234.
65. Markert, B., H. Piehler, H. Lieth, and A. Sugimae. "Normalization and Calculation of Lanthanide Element Concentrations in Environmental Samples," *Radiat. Environ. Biophys.* 28:213–221 (1989).
66. Markert, B. and Zhang De Li. "Baseline Levels of Rare Earth Elements in a Natural Ecosystem," *Sci. Total Environ.* 103:27–35 (1991).
67. Alt, F. "Platinum Metals in the Environment — A Challenge to Trace Analysis," in *Trace Element Analytical Chemistry in Medicine and Biology*, Vol. 5, Brätter, P. and P. Schramel, Eds. (Berlin: Walter De Gruyter, 1988) pp. 279–298.
68. Alt, F., U. Jeromo, J. Messerschmidt, and G. Tölg. "The Determination of Platinum in Biotic and Environmental Materials," *Microchim. Acta* 3:299–304 (1988).
69. Hodge, V. F. and M. O. Stallard. "Platinum and Palladium in Roadside Dust," *Environ. Sci. Technol.* 20:1058–1060 (1986).
70. Sansoni, B., Ed. *Instrumentelle Multielementanalyse*, (Weinheim: VCH Verlagsgesellschaft mbH, 1985).
71. Sansoni, B. "Fortgeschrittener chemischer Analysendienst für Elemente, Radionuklide und Phasen, das Jlicher Baukastensystem für Analysenschritte," *Fresenius Z. Anal. Chem.* 323:573–600 (1986).
72. Sansoni, B. "Multi-Element Analysis for Environmental Characterisation," *Pure Appl. Chem.* 59:4:579–610 (1987).
73. Horwitz, W., L. R. Kamps, and K. W. Boyer. "Quality Assurance in the Analysis of Foods for Trace Constituents," *J. Assoc. Off. Anal. Chem.* 63:1344–1354 (1980).
74. Jayasekera, R. and B. Markert, "Multi-Laboratory Chemical Characterization of Ecological Samples," *Fresenius Z. Anal. Chem.* 334:226–230 (1989).
75. ACS Committee on Environmental Improvement "Guidelines for Data Aquisition and Data Quality Evaluation in Environmental Chemistry," *Anal. Chem.* 52:2242–2249 (1980).
76. Caroli, S., M. Mancini, E. Beccaloni, L. Fornarelli, and R. Astrologo, "Reference Values for Elements in Marine Ecosystems," in *Element Concentration Cadasters in Ecosystems*, Lieth, H. and B. Markert, Eds. (Weinheim: VCH Verlagsgesellschaft mbH, 1990) pp. 207–218.
77. Griepink, B., H. Muntau, and E. Colinet. "Certification of the Contents of Cadmium, Copper, Manganese, Mercury, Lead and Zinc in Two Plant Materials of Aquatic Origin and in Olive Leaves," *Fresenius Z. Anal. Chem.* 315:193–196 (1983).

78. Griepink, B. "Quality and Certified Reference Materials," in *Element Concentration Cadasters in Ecosystems*, Lieth, H. and B. Markert, Eds. (Weinheim: VCH Verlagsgesellschaft mbH, 1990) pp. 181–206.
79. Kateman, G. and F. W. Pijpers. *Quality Control in Analytical Chemistry*, (New York: John Wiley & Sons, 1981).
80. International Atomic Energy Agency. "Element Analysis of Biological Materials — Current Problems and Techniques with Special Reference to Trace Elements," Techn. Rep. Ser. No. 197, Vienna (1980).
81. Muntau, H. "The Problem of Accuracy in Environmental Analysis," *Fresenius Z. Anal. Chem.* 324:678–682 (1986).
82. Muramatsu, Y. and R. M. Parr. "Survey of Currently Available Reference Materials for Use in Connection with the Determination of Trace Elements in Biological and Environmental Materials," International Atomic Energy Agency, Vienna, RL, 128 (1985).
83. Muramatsu, Y. and R. M. Parr. "Concentrations of Some Trace Elements in Hair, Liver, and Kidney from Autopsy Subjects — Relationship Between Hair and Internal Organs," *Sci. Total Environ.* 76:29–40 (1988).
84. Tölg, G. "Spurenanalyse der Elemente — Zahlenlotto oder exakte Wissenschaft," *Naturwissenschaften* 63:99–110 (1976).
85. Zimmermann, R. D. "Erste Ergebnisse einer Ringanalyse zur Erstellung eines internen Buchenblatt Referenzmaterials für Ökosystemuntersuchungen," *Fresenius Z. Anal. Chem.* 334:323–325 (1989).
86. Anonymus "Schäden in Milliardenhöhe durch falsche Analysen," *Chem. Rundsch.* 15:132 (1989).
87. Keith, L. H., Ed. *Principles of Environmental Sampling*, ACS Professional Reference Book, American Chemical Society, Washington, D.C. 1988.
88. Anders, O. U. and J. I. Kim. "Representative Sampling and the Proper Use of Reference Materials," *J. Radioanal. Chem.* 39:435–445 (1977).
89. Allen, S. E., Ed. *Chemical Analysis of Ecological Materials*, (Oxford: Blackwell Scientific Publications, 1974).
90. Jones, B. J., Jr., R. L. Large, D. B. Pfleiderer, and H. S. Klosky. "How to Properly Sample for a Plant Analysis," *Crops Soils* 23:15–18 (1971).
91. Brown, D. H. and R. M. Brown. "Reproducibility of Sampling for Element Analysis Using Bryophytes," in *Element Concentration Cadasters in Ecosystems*, Lieth, H. and Markert, B., Eds. (Weinheim: VCH Verlagsgesellschaft mbH, 1990) pp. 55–62.
92. Bryden, G. W. and L. R. Smith. "Sampling for Environmental Analysis, Part 1: Planning and Preparation, American Laboratory, 21, 7:30–39 and Part 2: Sampling Methodology," *Am. Lab.* 21, 9:19–24 (1989).
93. Ernst, W. H. O. "Element Allocation and (Re)Translocation in Plants and its Impact on Representative Sampling," in *Element Concentration Cadasters in Ecosystems*, Lieth, H. and Markert, B., Eds. (Weinheim: VCH Verlagsgesellschaft mbH, 1990) pp. 17–40.
94. Gomez, A., R. Leschber, and P. L'Hermite, Eds. *Sampling Problems for the Chemical Analysis of Sludge, Soils and Plants*, (London: Elsevier, 1986).
95. Iyengar, V. G. "Presampling Factors in the Elemental Composition of Biological Systems," *Anal. Chem.* 54:554a–560a (1982).
96. Jackson, K. W., I. W. Eastwood, and M. S. Wild. "Stratified Sampling, Protocol for Monitoring Trace Metal Concentrations in Soil," *Soil Science* 143:436–443 (1987).

97. Loveland, P. J. "The National Soil Inventory of England and Wales," in *Element Concentration Cadasters in Ecosystems*, Lieth, H. and B. Markert, Eds. (Weinheim: VCH Verlagsgesellschaft mbH, 1990) pp. 73–80.
98. Kratochvil, B. and J. K. Taylor. "Sampling for Chemical Analysis," *Anal. Chem.* 53:924a–938a (1981).
99. Krivan, V. and G. Schaldach. "Untersuchungen zur Probenahme und -vorbehandlung von Baumnadeln zur Elementanalyse," *Fresenius Z. Anal. Chem.* 324:158–167 (1986).
100. Newbould, P. J. *Methods of Estimating the Primary Production of Forests*, (Oxford: Blackwell Scientific, 1967).
101. Wagner, G. "Variability of Element Concentrations in Tree Leaves Depending on Sampling Parameters," in *Element Concentration Cadasters in Ecosystems*, Lieth, H. and B. Markert, Eds. (Weinheim: VCH Verlagsgesellschaft mbH, 1990) pp. 41–54.
102. Markert, B. and V. Weckert. "Fluctuations of Element Concentrations During the Growing Season of Polytrichum Formosum (Hedw.)," *Water Air Soil Pollut.* 43:177–189 (1989a).
103. Markert, B. and V. Weckert. "Use of Polytrichum Formosum (Moss) as Biomonitor for Heavy Metal Pollution — (Cd, Cu, Pb, Zn)," *Sci. Total Environ.* 86:289–294 (1989b).
104. Markert, B. and N. Klausmeyer. "Variations in the Elemental Compositions of Plants and Computer Aided Sampling in Ecosystems," *Toxicol. Environ. Chem.* 25:200–212 (1990).
105. Sansoni, B. and V. Iyengar. "Sampling and Sample Preparation Methods for the Analysis of Trace Elements in Biological Materials," *Forschungszentrum Jülich*, Jül. Spez., 13 (1978).
106. Bock, R. *A Handbook of Decomposition Methods in Analytical Chemistry*, translated and expanded version by Marr, I. L., International Textbook Company Ltd., Glasgow (1979).
107. Kingston, H. M. and L. B. Jassie. *Introduction to Microwave Sample Preparation*, ACS Professional Reference Book, American Chemical Society, Washington, D.C. (1988).
108. Knapp, G. "Der Weg zu leistungsfähigen Methoden der Elementspurenanalyse in Umweltproben," *Fresenius Z. Anal. Chem.* 317:213–219 (1984).
109. Matusiewicz, H. and R. E. Sturgeon. "Present Status of Microwave Sample Dissolution and Decomposition for Elemental Analysis," *Prog. Anal. Spectrosc.* 12:21–39 (1989).
110. Sansoni, B. and V. K. Panday. "Ashing of Trace Element Analysis of Biological Material," in *Analytical Techniques for Heavy Metals in Biological Fluids*, Facchetti, S., Ed. (Amsterdam: Elsevier Science Publisher, 1981) pp. 91–131.
111. Eller, R., F. Alt, G. Tölg, and H. J. Tobschall. "An Efficient Combined Procedure for the Extreme Trace Analysis of Gold, Platinum, Palladium and Rhodium with the Aid of Graphite Furnace Atomic Absorption Spectrometry and Total Reflection X-Ray Fluorescence Analysis," *Fresenius Z. Anal. Chem.* 334:713–739 (1989).
112. Grobecker, K. H. and U. Kurfuerst. "Solid Sampling by Zeeman Graphite-Furnace-AAS, a Suitable Tool for Environmental Analyses," in *Element Concentration Cadasters in Ecosystems*, Lieth, H. and B. Markert Eds. (Weinheim: VCH Verlagsgesellschaft mbH, 1990) pp. 121–138.

113. Sanz-Medel, A., A. Menendez, M. L. Fernandez, A. Lopez, and R. Pereiro. "The Use of Atomic Emission Spectroscopy with Inductively Coupled Plasma in Biological Materials," in *Element Concentration Cadasters in Ecosystems*, Lieth, H. and B. Markert, Eds. (Weinheim: VCH Verlagsgesellschaft mbH, 1990) pp. 139–148.

114. Schramel, P., B. J. Klose, and S. Hasse. "Die Leistungsfhigkeit der ICP-Emissionsspektroskopie zur Bestimmung von Spurenelementen in biologisch medizinischen und Umweltproben," *Fresenius Z. Anal. Chem.* 31:209–216 (1982).

115. Beauchemin, D., J. W. McLaren, S. N. Willie, and S. S. Berman. "Determination of Trace Metals in Marine Biological Reference Materials by Inductively Coupled Plasma Mass Spectrometry," *Anal. Chem.* 60:687–691 (1988).

116. Date, A. and A. Gray. *Applications of Inductively Coupled Plasma Mass Spectrometry*, (Glasgow: Blackie, 1989).

117. Paul, M., K. Prosen, and U. Völlkopf. "The Analysis of Biological Samples by ICP/MS," in *Element Concentration Cadasters in Ecosystems*, Lieth, H. and Markert, B., Eds. (Weinheim: VCH-Verlagsgesellschaft mbH, 1990) pp. 149–170.

118. Bode, P. and M. De Bruin. "Routine Neutron Activation of Environmental Samples," in *Element Concentration Cadasters in Ecosystems*, Lieth, H. and B. Markert, Eds. (Weinheim: VCH Verlagsgesellschaft mbH, 1990) pp. 171–178.

119. De Goeij, J. J. M. "Radiochemical Neutron Activation Analysis of Biological Materials for Intercomparison and Certification of Reference Materials," *Trans. Am. Nucl. Soc.* 60:19–20 (1989).

120. Gladney, E. S. "Elemental Concentrations in NBS Biological and Environmental Materials," *Anal. Chim. Acta* 118:385–396 (1980).

121. International Commission on Radiological Protection, *Report of the Task Group on Reference Man*, (Oxford: Pergamon Press, 1975).

122. Rossbach, M. "Instrumentelle Neutronenaktivierungsanalyse zur standortabhängigen Aufnahme und Verteilung von Spurenelementen durch die Salzmarschpflanze Aster tripolium von Marschwiesen des Scheldeestuars, Niederlande," in *Beiträge zur Umweltprobenbank*, Stoeppler, M. and H. W. Dürbeck, Eds. Jül. Spez. (1986).

123. Zeisler, R. and R. R. Greenberg. "Ultratrace Determination of Platinum in Biological Materials via Neutron Activation and Radiochemical Separation," *J. Radioanal. Chem.* 75:27–37 (1982).

124. Zeisler, R. "Determination of Baseline Platinum Levels in Biological Materials," in *Trace Element Analytical Chemistry in Medicine and Biology*, Vol. 5, Brätter, P. and P. Schramel, Eds. (Berlin: Walter de Gruyter, 1988) pp. 297–303.

16

Metal-Humic Substance Complexes in the Environment. Molecular and Mechanistic Aspects by Multiple Spectroscopic Approach

Nicola Senesi

Istituto di Chimica Agraria, Università di Bari, Via Amendola, 165/A, 70126 Bari, Italy

ABSTRACT

A review on the application of several spectroscopic techniques, including visible-ultraviolet (UV-Vis), infrared (IR), fluorescence, electron spin resonance (ESR), Mössbauer, and nuclear magnetic resonance (NMR), to metal-humic substances studies in environmental and laboratory-modeled systems is provided. The discussion is focused on the molecular and mechanistic aspect of metal-humic substance complexation, i.e., on the physicochemical nature of sites involved and type of binding and formation mechanisms and stability of metal-humic substance complexes. Information generated by spectroscopic techniques, particularly fluorescence quenching spectroscopy, on quantitative aspects of metal-humic substance interactions, such as ligand complexing capacity and equilibrium constants of complexation, are discussed. Advantages and limitations of each technique are also considered. It is inferred that the simultaneous utilization of UV-Vis, IR, fluorescence, and ESR spectroscopies, in combination with chemical derivatization, along with fluorescence and/or ESR probes, and, where possible, with the support of Mössbauer and NMR spectroscopies, represents the most advisable rationale in approaching the complicated chemical problems encountered in investigating metal-humic substance interactions in natural environments.

INTRODUCTION

The behavior of metals in terrestrial and aquatic environments is dependent not only on the total soluble and insoluble metal concentrations, but also on the relative distribution and chemical forms and properties of the various metals in the solution and solid phases. The knowledge of the type and concentration of individual species present in the system is, therefore, essential in assessing the impact of metals on the global ecosystem. The distribution of metal ions in environmental systems is extremely complex and is governed by a variety of reactions that include complexation with organic and inorganic ligands, ion exchange, adsorption and desorption processes, precipitation and dissolution of solids, and acid-base equilibria.

Metal ion speciation is important in determining the general biological and physico-chemical behavior of metal ions. This, in turn, influences bioavailability of metal nutrients to plants and soil microorganisms, toxicity hazard of potentially toxic metals, migration-accumulation phenomena of metals in the soil-water-sediment system, pedogenic processes, and geochemical transfer and mobility pathways.[1,2]

Complexation reactions involving natural organic matter play a key role in establishing the behavior of metal ions, particularly in trace concentrations. Organic compounds in aquatic and terrestrial environments that may form complexes with metal ions may be grouped into three main classes: (1) organic substances of known molecular structure and chemical properties including biochemicals such as simple aliphatic acids, polysaccharides, amino acids, polyphenols; (2) xenobiotic organic chemicals deriving from human, agricultural, industrial, and urban activities; and (3) humic substances.

Humic substances (HS) represent a significant proportion of total organic carbon in the global carbon cycle, constituting the major organic fraction in soils (between 70 and 80%) and the largest fraction of natural organic matter in stream, river, wetland, lake, sea, and ground waters (40 to 60% of dissolved organic carbon).[3,4]

One of the most striking characteristics of HS (see section titled "Humic Substances in the Environment") is their ability to interact with metal ions to form water-soluble, colloidal, and water-insoluble complexes of varying properties and widely differing chemical and biological stabilities.[5,6] The evidence suggests that almost every aspect of the chemistry of heavy metals in soils, waters, and sediments is related in some way to the formation of complexes with HS. The topic is of considerable practical interest because of the continuous and increasing release of various heavy metals to environmental systems by numerous human activities. This has been extensively studied and reviewed in detail,[5,7] and only a brief survey of the practical and theoretical significance of metal-HS complexation in environmental systems will be provided.

The availability of many metal ions, especially trace elements, to higher plants and to soil micro- and macro-faunal organisms is strongly influenced by complexation with the soluble and insoluble fractions of HS in soil. At pH values commonly found in soils, metal ions that would ordinarily be converted to insoluble forms may be maintained in solution by complexation, thus increasing their bioavailability. Soluble complexes can function as metal-carrier in the transport to ground- and surface-water bodies, thereby rendering the water unfit for beneficial uses. The concentration of a toxic metal ion may be reduced to a nontoxic level through complexation to HS, particularly when low-solubility metal complexes are formed with water-insoluble fractions of soil HS. For instance, Bloom et al.[8,9] have found that Al^{3+} ions complexation by soil HS may assume considerable importance in controlling soil solution levels of Al^{3+} in acid soils. Stevenson[6] reviewed complexation reactions with HS that are implicated in the chemical weathering of rocks and minerals, as well as in concentration of metals in a variety of biogenic deposits such as peat, coal, and sediments, thus altering the geochemical cycles. Saar and Weber[7] considered that dissolved HS can cause the release into solution of metal ions adsorbed in sediments, whereas sedimentary HS can sequester metal ions from the aquatic systems. Dissolved HS may decrease metal removal efficiency in water treatment processes. For instance, Truitt and Weber[10] found that the fraction of Cu^{2+}, Cd^{2+}, and Zn^{2+} ions removed by alum from aqueous solutions is increased by soluble fractions of HS. Complexation of metal ions in natural matrices also represents a challenging problem for analytical methodology and procedures of metal determinations in environmental samples.

In this review, attention is focused on the molecular and mechanistic aspects of the complexation of metal ions by humic substances as revealed by the application of spectroscopic methods, including UV-Vis, IR, fluorescence, ESR, Mössbauer, and NMR. The topical discussion is preceeded by brief considerations on some general aspects of the environmental chemistry and reactivity of metal ions and humic substances, with emphasis to their complexing capacity for metal ions.

METAL-HUMIC SUBSTANCE COMPLEXATION IN THE ENVIRONMENT

Metal Reactivity with Environmental Organic Complexants

Inorganic cations may be subdivided into three categories according to their typical reactivity with natural organic ligands.[2,11] "Hard" cations of Group I participate preferentially in electrostatic interactions where the change in free energy results mainly from gain in entropy due to changes in orientation of hydration water molecules. "Soft" cations of Group III tend to form more covalent bonds, for which the free energy of complexation is typically of enthalpic origin. "Borderline cations" of Group II have character intermediate between hard and soft metals.

Hard cations of Group I include alkaline and alkaline-earth metals that form rather weak, outer-sphere complexes with only hard oxygen ligands. Cations such as Ca^{2+}, Mg^{2+}, and Na^{+} can combine significantly with natural organic ligands. The interactions have no significant influence on speciation of Group I metals in the environment, since the natural concentration of these ions is much higher than that of organic ligands. Group I cations, however, may have two important indirect effects on the complexation of other metals: (1) a competitive effect for oxygen ligands, with respect to cations of Groups II and III, and (2) a counterion effect for negatively charged polyelectrolyte complexing agents, thus influencing their reactivity by modifying their charge and/or conformation, as well as their degree of aggregation and dispersion-coagulation.

Borderline metals of Group II, which include Fe^{2+}, Co^{2+}, Ni^{2+}, Cu^{2+}, Zn^{2+}, and Pb^{2+}, possess appreciable affinity for both hard and soft ligands. The concentration of these cations in natural soil and aquatic systems is generally a little lower or similar to that of the organic complexants. These metals can, therefore, compete for hard ligands with Group I metals, which are less strongly bound but at higher concentration in the environment, and with Group III metals, which are at lower concentration but more strongly bound. In addition, some Group II metals, such as Mn^{2+} and Fe^{2+}, can be oxidized to higher, strongly hydrolyzed, oxidation states (Mn(III) and (IV) and Fe^{3+}) that can form organic-reactive colloids or polymers.

The metal ions of Group III, of which the most environmentally representative are Cd^{2+} and Hg^{2+}, possess a strong affinity for intermediate (nitrogen) and soft (sulfur) ligands and are generally found in very low concentration associated to natural organic matter. This implies that Group III metals can be complexed to N and S sites despite the low concentration of these sites in natural organic matter.

In practice, the metals of most immediate concern, in terms of their general environmental impact, toxicity for plant and aquatic organisms, and translocation from soils and waters through the human and animal food chain, are Hg, Pb, Cd, Cu, Ni, Mn, and Zn.

The most important organic complexing functional groups present in natural organic matter, classified according to their affinity for hard, borderline, and soft metals, are listed in Table 1.[2,12,13] For soft metals, the following order of donor atom affinity is observed: $O < N < S$, whereas a reverse order is observed for hard cations. For bidentate ligand sites, affinity for a given soft metal increases with the overall softness of the donor atoms, in the order: $(O,O) < (O,N) < (N,N) < (N,S)$, whereas this order is reversed for the hardest metals.[2] In general, the competitive reactions for a given ligand essentially involve Group I and Group II metals for O sites, and metals of Groups II and III for N and S sites, with competition between metals of Group I and III being weak.[2]

The typical affinity sequence of soil organic matter for divalent metal ions (at pH 5) generally parallels the metal electronegativity values by Pauling, whereas

Table 1. Most Important Organic Complexing Functional Groups Encountered in Natural Systems, Classified According To Their Preference for Hard, Borderline, and Soft Metals

Ligands Preferred by Hard Cations (Hard Bases)	Ligands Preferred by Borderline Cations	Ligands Preferred by Soft Cations (Soft Bases)
$-C(=O)O^{\ominus}$, $-C(=O)O-$ (Carboxylate) (ester)	$-NH2$, $=NH$, $\equiv N$ (primary, sec., tert. amino groups)	$R^{\ominus}$ (Alchil anion)
–OH (alcoholic and phenolic	$-NH-C(=O)-$ (amide)	$-SH$, $-S^{\ominus}$ (Sulfydril, sulfide)
$>C=O$, $-O-$ (Carbonyl) (ether)		–S–S–, –S– (disulfide) (thioether)
$-O-P(=O)(O-)-O^{\ominus}$, $-O-P(=O)(O^{\ominus})-O-$ (Phosphate) (phosphate esther)		
$-O-SO_3^{\ominus}$ (sulfate)		

the stabilities of aquatic organic matter-metal complexes follow the Irving-Williams series.[14–16] Stevenson and Ardakani[16] considered, however, that the relative affinities are often dependent on the method used to measure metal bonding and the pH. The type, source, and concentration of organic matter in the various systems can affect metal binding affinity. For instance, the stability constants of Cu^{2+} complexes with organic matter from different environments followed the sequence: marine sediments > lake and river water > seawater > peat > soil. Mantoura et al.[17] found that organic matter in freshwaters bind more than 90% of total Cu, whereas those in seawater bind only 10%, probably because of large concentrations of Ca and Mg that displace Cu from organic matter in the latter case.

Selectivity coefficients for metal binding vary with the amount of metal bound. Davies et al.[18] found that the strength of binding of Zn^{2+} and Cu^{2+} by soil humic acids increases with a decrease in the metal amount available. At low ligand concentrations, the relatively soft cations Cd^{2+} and Pb^{2+} are highly preferred by "soft", sulfur-containing ligands in HS, and could compete successfully with "hard" cations, e.g., even the much abundant Ca^{2+} ions. The reverse is true, however, for the Cd^{2+}/Ca^{2+} competition at high concentration where carboxylate ligands become dominant.

Humic Substances in the Environment

Isolation and Definitions

Natural humic substances comprise a wide class of naturally occurring, biogenic, structurally complex and heterogeneous, refractory, acidic, yellow- to black-colored organic polyelectrolytes of relatively high molecular weight, that occur in all soils, sediments, fresh waters, and seawaters.[3,5,6] HS are the products of the "humification" process that consists in the chemically and biologically mediated synthesis of compounds originating from the degradation of terrestrial and/or aquatic plant and animal residues and from synthetic activities of microorganisms.

In the environment humic substances are part of a complex chemical system including interactions with metal ions, clay colloids, and nonhumified organic materials. The ultimate objective in the study of HS is to relate structural and chemical information obtained to the environmental roles and biogenesis of HS. Thus, an extraction step that allows separation of HS from other components, followed by a fractionation step that decreases heterogeneity of bulk HS, must be achieved before detailed structural and chemical studies can be performed. The choice of extractant is limited by the complex heterogeneity of the material and the rigorous demands placed upon the extractant. An ideal extractant should ensure universal applicability for the complete isolation of an unaltered and uncontaminated material.[6] Further, in order that the structural and chemical information obtained on HS truly reflect the processes in natural systems, it is of great importance that the compounds fractionated are representative of those in the environment. Unfortunately, the isolation process itself involves unavoidable alterations in the chemistry of HS, such as disruption of interactions with metal ions, and coextraction of tightly bound inorganic and nonhumified organic materials. For studies of metal-HS complexes, it would be advantageous to extract the complexes intact, but removal of the metals is generally a declared objective of the isolation procedure.

The earliest reported reagent used for extracting HS from soil is NaOH, and to this day, with only minor modifications of the procedure, it remains the most widely used. Alkaline extractants for HS have been criticized for many reasons including coextraction processes and artifact productions. Numerous other inorganic reagents and organic solvents, also used in sequential extraction procedures, and physical methods of extraction, such as ultrasonic dispersion and super-critical fluid techniques, have been tried, but none have proved as effective as NaOH. Following alkaline extraction of HS from soil and sediments, an acidification step is classically included to precipitate the humic acid (HA) fraction. Alternative fractionation methods are available, including salting-out, use of metal ions or organic solvents, gel permeation chromatography, ultrafiltration, ultracentrifugation, electrophoresis and electrofocusing, and ion-exchange methods. The crude HA fraction is contaminated with metal ions and may be coprecipitated with both fine clay particles and nonhumic organic materials.

Recovery of HA is, therefore, followed by a purification stage, traditionally consisting in alkaline redissolution and acid precipitation and then washing, or dialyzing against dilute acid followed by water.[6] This step invariably leads to some loss of humic material with failure in complete removal of mineral and organic impurities. The fulvic acid (FA) fraction, that does not precipitate by acidification and remains in solution, is heavily contaminated and of more difficult recovery and purification than the HA fraction. Numerous materials and various methods, including polyvinylpyrrolidone and Amberlite XAD-resins, are used with different success to separate and purify the FA fraction of soil and sediment HS.

Presently, there is no universally accepted method of isolating aquatic HS, and no consensus exists as to the operational definitions of these materials. Column chromatographic methods using either the nonionic XAD resin or the anionic resin, Duolite A-7, are currently the most extensively used methods for isolating HS from water. Thurman and Malcolm[19] have published a procedure using XAD-8 that is commonly used for isolating HS from freshwaters. According to this method, the sample is initially filtered through a 0.45 μm silver filter. After acidification to pH 2 with HCl, the sample is passed through a column of XAD-8 resin, during which HS adsorb to the resin. The HS are then back eluted from the column with NaOH, hydrogen saturated, and lyophilized. Similar procedures are followed with XAD-2 resin for seawater (Mantoura and Riley, 1975).[20] While these methods are highly efficient, they have important limitations, since some chemical alteration and contamination of the isolated HS is unavoidable. However, isolation procedures can be designed that minimize these problems. It is important to fully understand the basic principles of the isolation process to minimize the possibility of producing artifacts, and to choose standard procedures that ensure the isolation of products that are comparable and can produce analytical results that can be compared on a rational basis.

Tremendous efforts have been made in the last few years by the International Humic Substances Society (IHSS) in order to standardize the isolation and fractionation methods of HS from soil, sediment, and water by introduction of a procedure that is highly encouraged to be adopted universally. On these basis, Thurman et al.[21] have recently proposed a comprehensive conceptual view of the isolation procedure and terminology for HA and FA of any source (Figure 1). The right side of Figure 1 refers to solid materials (soil, sediments, etc.) and the left side to aquatic samples (freshwater, marine, pore waters, etc.). During pretreatment and extraction (Steps I and II), samples are treated in different ways. Pretreatment involves preparation of the sample for the extraction of HA and FA. For soil materials, typical pretreatments are demineralization by HCl and/or HF, flotation to separate plant fragments, etc. Water samples are commonly filtered and acidified with HCl to pH 2.0. Samples from other sources, e.g., sediments, may receive different pretreatments, e.g., freeze-drying. The typical extraction for soil involves the removal of the alkaline soluble fraction, commonly by NaOH and pyrophosphate, or dipolar aprotic solvents, such as dimethylsulphoxide. Water samples can be treated by column sorption techniques, including XAD-

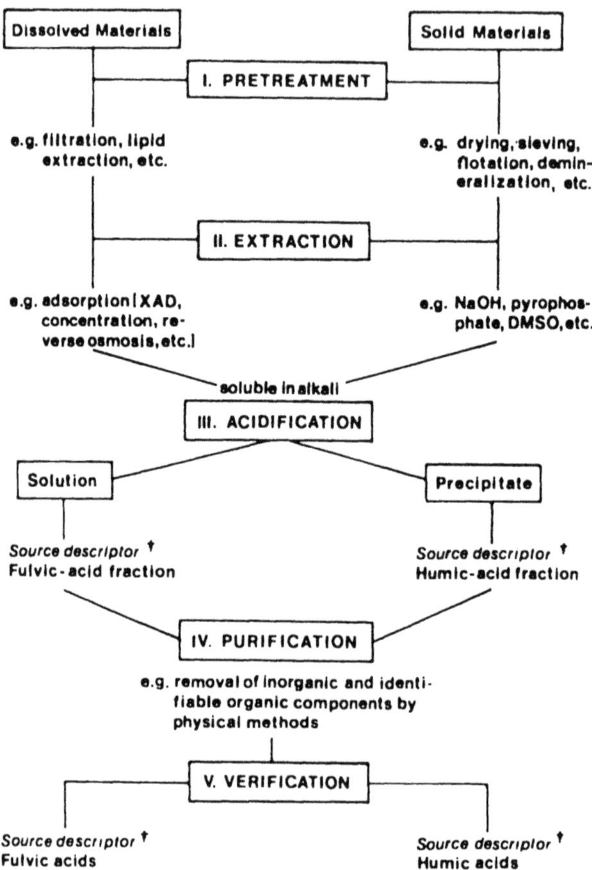

FIGURE 1. Comprehensive scheme proposed for the isolation and nomenclature of humic and fulvic acids from various sources. The source from which the sample is isolated should be included in the specific nomenclature. For instance, fulvic acid isolated from seawater would be named "marine aquatic fulvic acid". (From Thurman, E. M. in *Humic Substances and Their Role in the Environment*, Frimmel, F. H. and R. F. Christman, Eds. (Chichester: Wiley-Interscience, 1988) pp. 31–43. With permission.)

8 or XAD-2 resins, or other concentration processes, such as ion-exchange, freeze-drying, roto-evaporation, etc. In Step III (acidification) the procedures join, thus permitting the use of a final, common terminology, i.e., the use of the terms humic acids and fulvic acids. This step consists in the acidification of the alkali-soluble extract to pH ≤ 2 with HCl and precipitation. Centrifugation is needed to separate the sample into two fractions, a fulvic acid fraction and a humic acid fraction. These two fractions are not solely humic and fulvic acids but may contain specific compound classes such as polysaccharides and aminoacids, as well as inorganic ions, which are removed in the purification step. Step IV describes the "purification" of HA and FA fractions to yield the final products,

HA and FA. Since the structure of HS is not known precisely, the term "purification" refers essentially to the removal by physical methods of substances that are "physically" but not "chemically" bound in humic extracts, e.g., salts, metals, lipids, polysaccharides, fatty acids, etc. Thurman et al.[21] conclude their scheme by proposing a final "verification" step (Step V) for the isolated HA and FA. This is done by comparing the identity of the products isolated on the basis of the simple elemental and acid functional group composition, or on a more complex basis, e.g., NMR, molecular weight determinations, etc. The efforts of IHSS to publish values and ranges for average soil and aquatic HA and FA of its standard and reference collection will be greatly useful for such comparative verification. Hayes[22] and Aiken[23] have recently provided comprehensive and detailed reviews on isolation procedures of HA and FA from soil and aquatic samples, whereas fractionation techniques have been reviewed, respectively, by Swift[24] and Leenheer.[25]

Another difficulty encountered in the study of HS relates to terminology. Since HS do not conform to a unique chemical entity, they cannot be described in unambiguous chemical terms. This implies that they can be described only operationally, as a complex mixture of organic acids. The terms used to describe the principal fractions of HS, HA and FA, originate in the field of soil science and are accepted for HS from sediment and other solid materials. For aquatic HS a classification based on the chromatographic methods used to extract them from water is more reasonable, but problems with this type of terminology exist as well. Since no consensus exists on the method of extraction, no consensus exists as to what part of the dissolved organic carbon pool is classified under the definitions of fulvic and humic acids.

With reference to the isolation procedure shown in Figure 1 and to the recommendations of IHSS, the two major fractions of HS, once isolated from the environment, are operationally defined in terms of their solubilities as follows: (1) humic acid, the fraction of HS that is not soluble in water under acid conditions (below pH 2), but soluble at greater pH; and (2) fulvic acid, the fraction of HS that is soluble under all pH conditions.[21,26] A third fraction, not discussed in this text, is "humin", i.e., the fraction of HS that is not soluble in water solution at any pH value.[21,26]

Molecular Structure and Chemical Properties

Since HS consist of a heterogeneous mixture of compounds, no single structural formula is valid, and it is virtually impossible to describe uniquely the molecular configuration of HA and FA.[3,5,6] According to current concepts it is possible, however, to depict the general structure of a "typical" molecule of HA or FA on the basis of available compositional, structural, functional, and behavioral data. Table 2 summarizes average data on the major elemental composition and principal oxygen functional group content of soil, water, and sedimentary HA and FA.[2,3,5,27–32] In Figure 2 some model structures proposed for natural HA and FA are presented.[6,33–37] Although these "type" structures

Table 2. Ranges of Major Elemental Composition (%) and Functional Group Content (meq/g) of Soil, Aquatic, and Sedimentary Humic Acids (HA) and Fulvic Acids (FA)

	Soil[5,27,28]		Freshwater[a] [3,29]		Seawater[30]	Sedimentary[b]	
Parameter	HA	FA	HA	FA	FA	HA	FA
C	53.8–58.7	40.7–53.1	50.2–62.1	41.6–59.7	50.0	48.4–53.7	41.1–48.9
O	32.7–38.7	39.7–49.8	23.5–44.8	31.6–51.6	36.4	32.3–40.8	37.6–47.0
H	3.2–6.2	3.2–7.0	3.1–5.1	2.7–5.9	6.8	5.0–6.3	3.9–6.4
N	0.8–5.5	0.9–3.3	0.5–3.2	0.5–2.2	6.4	4.6–6.7	7.1–8.2
S	0.1–1.5	0.1–3.6	0.6–1.0	0.4–4.3	0.5	—	—
Total acidity	5.6–7.7	6.4–14.2	7.1–8.9[c]	9.6–16.6[c]	—	3.0–5.5	2.0–5.5
COOH	1.5–5.7	6.1–11.2	4.0–5.9	4.0–8.9	5.5	2.0–4.0	1.0–4.0
Phenolic OH	2.1–5.7	0.3–5.7	2.0–3.0	0.8–3.0	—	0.5–2.5	0.0–1.5
Alcoholic OH	0.2–4.9	2.6–9.5	—	—	—	0.0–3.0	—
Quinonic CO	1.4–2.6	0.3–2.0 }	4.3–5.1c	4.3–7.4c	—	3.0–5.0	5.0–6.0
Ketonic CO	0.3–1.7	1.6–2.7 }			—		
OCH_3	0.3–0.8	0.3–1.2	—	—	—	—	—

[a] Origin: river, lake, wetland (marsh, swamp, bog), groundwater.

a

b

Kleinhempel's model

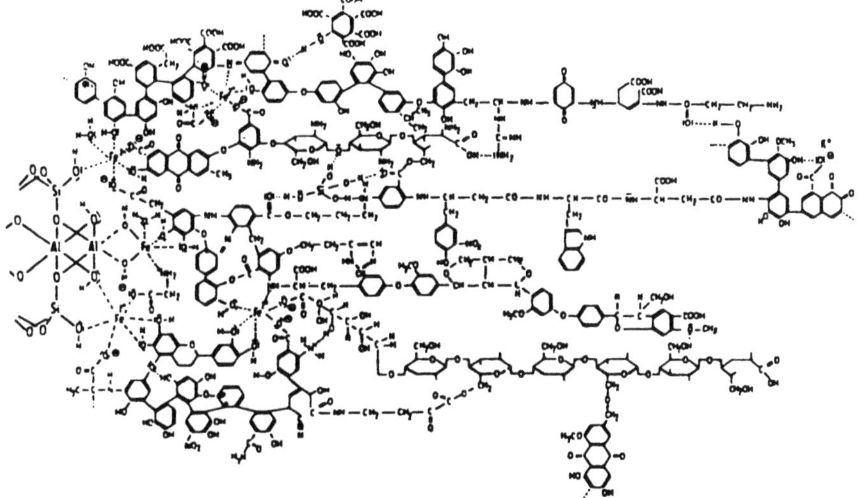

FIGURE 2. Model structures proposed for natural humic acids (HA) and fulvic acids (FA) of various origin: (a) soil HA[6]; (b) soil HA[33]; (c) soil FA[34]; (d) ocean pelagic FA[35]; (e) freshwater FA[36]; (f) sedimentary HA[37].

possess certain common features and may explain acceptably several chemical and functional properties of HA and FA, none of them can be considered entirely satisfactory, and they should probably be considered complementary since each emphasizes certain particular properties but neglects others.

Kleinhempel[33] proposed a model (Figure 2b) ($M_w \sim 1.3 \times 10^4$) that combines realistically the different properties resulting from experimental data for soil HA

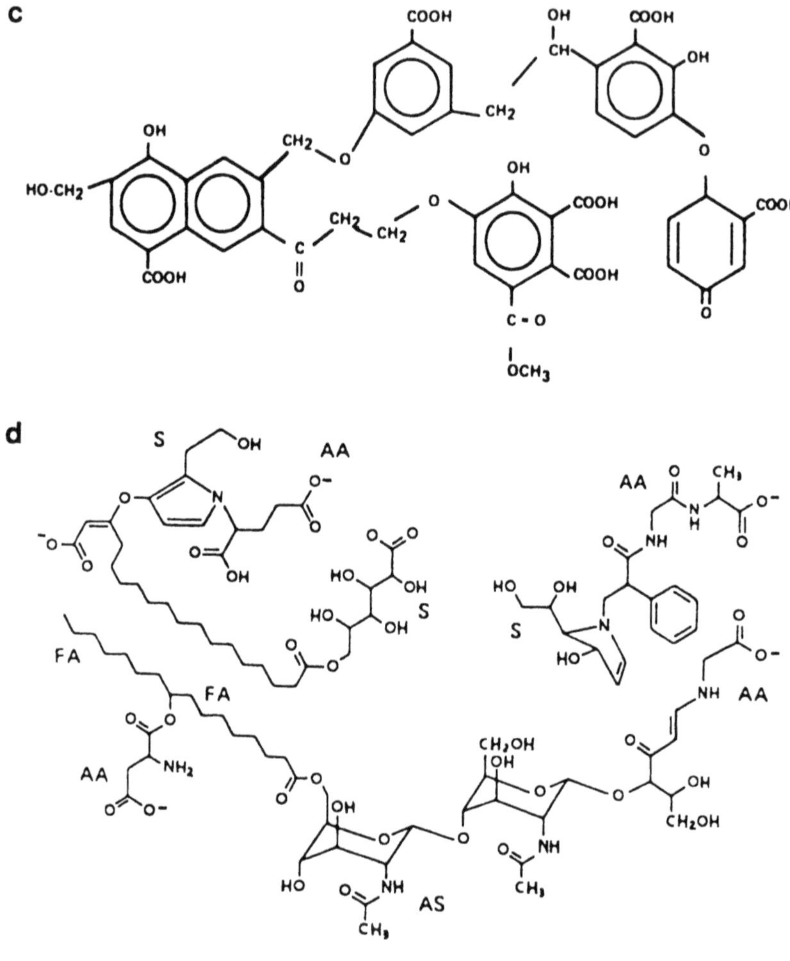

FIGURE 2. (Continued).

(Table 2). The model features a largely three-dimensional structure, a highly polyelectrolytic character, several metal complexing sites, and intra- and intermolecular association to aluminosilicate minerals surfaces. Also apparent are aliphatic, polysaccharide and amino-acidic bridges and chains, heterocyclic nitrogen, and an acceptable degree of aromaticity. Langford et al.[34] suggest a model (Figure 2c) and composition (Table 2) for soil FA that are simpler than those of soil HA (Figure 2a, b), featuring a lower aromatic and higher aliphatic character, more carboxylic sites, and lower molecular weight.

Gagosian and Stuermer[35] proposed a molecular model for the FA fraction isolated from an ocean pelagic zone (Figure 2d) consistent with the corrispondent

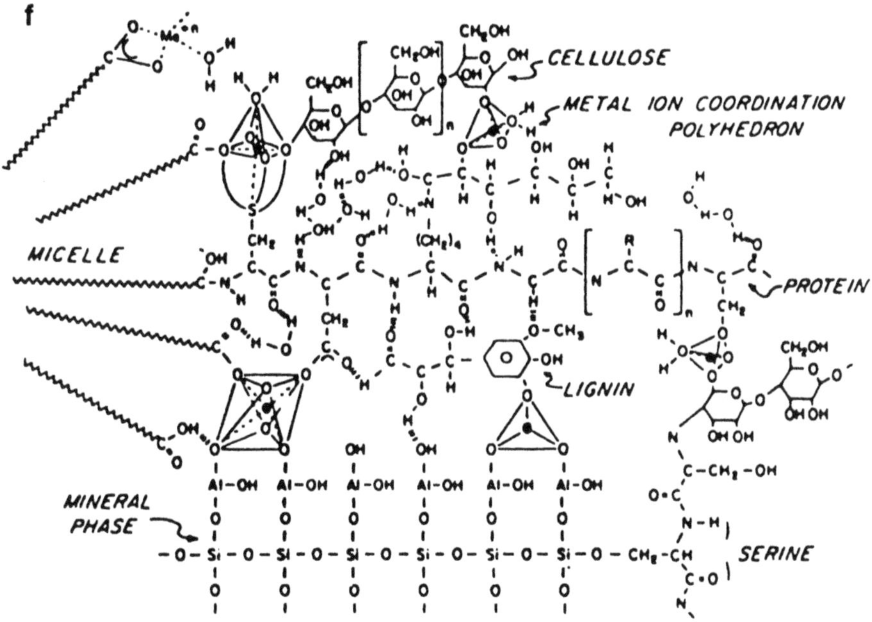

FIGURE 2. (Continued).

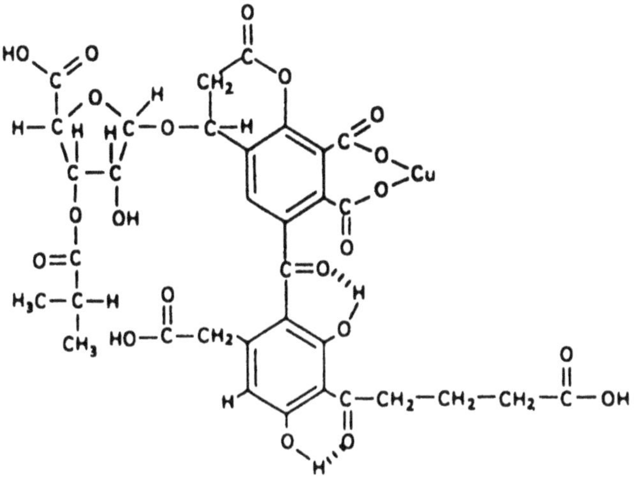

analytical data (Table 2) and clearly distinguishing this type of HS from soil FA and HA. Nitrogen content and aliphaticity are higher, total acidity much lower, and lignin and polysaccharide are absent. Leenheer et al.[36] suggest a structure and composition of HS from freshwaters and coastal and estuarine areas that is intermediate between those of soil and open ocean HS (Figure 2e) (Table 2) and includes components typically originated from both terrestrial and aquatic organisms.

Degens and Mopper[37] found that sedimentary FA and HA also show differences in compositional and structural properties with respect to soil FA and HA, but some similarities to ocean aquatic HS (Figure 2f) (Table 2). In particular, the content of carboxylic and phenolic OH groups is low, whereas that of carbonyl groups, aliphatic structures, carbohydrates, and uronic acids is high. Metals are complexed in sedimentary HS to a significant extent and contribute to the overall structure by forming bridges across complexing molecules reinforcing the refractory character.

Nature of Complexing Sites and Complexing Capacity

Humic substances are described as natural complexing agents like proteins and metal oxides, which are distinguished from "simple" ligands such as inorganic anions and amino acids.[2] This distinctive character of HS is based on the fact that they contain a large number of complexing sites per molecule. In an aqueous system the total molar concentration of these sites can be comparable to the concentrations of many simple ligands, but the characteristics of "multiligand" HS molecules differ appreciably from those of the simple ligands.[39]

According to Buffle,[38] the principal common molecular characteristics that influence the complexing ability of HS are *polyfunctionality, polyelectrolyte character, hydrophilicity,* and the *capacity to form intermolecular associations* and *change molecular conformation.*

The major ligand sites for a metal ion in HA and FA are oxygen-containing functional groups, including carboxylic, phenolic, alcoholic and enolic hydroxyls, and carbonyls of various types. Amino groups and sulfur- and phosphorus-containing groups may also participate in complex formation.[40]

Carboxyl groups are believed to play a prominent role in the binding of metal ions by soil HA and FA. Mixed complexes are probably formed, the most prominent one involving both phenolic and COOH groups, i.e., a salicylate-type site[41,42] (Equation 1).

$$\text{(C(=O)O}^-\text{, OH site)} + M^{2+} \rightleftharpoons \text{(C(=O)–O–M–O– chelate)} + H^+ \qquad (1)$$

Other possible combinations include two COOH, for example a phthalic-type site (Equation 2), two phenolic OH, quinone groups, NH_2 groups, sulfhydryl groups, and conjugated ketonic structures[41–44] (Equation 3).

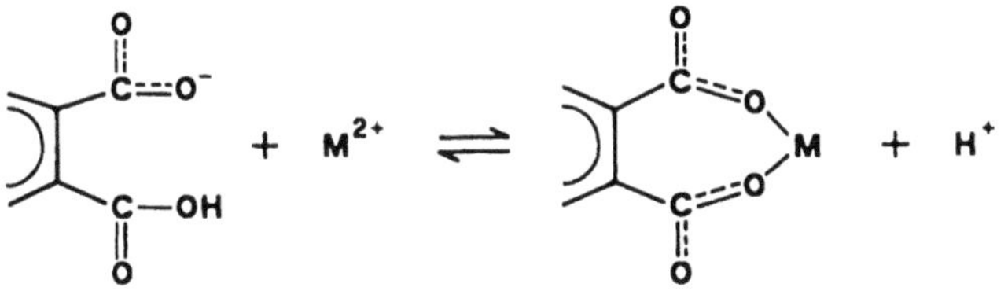

$$\text{(2)}$$

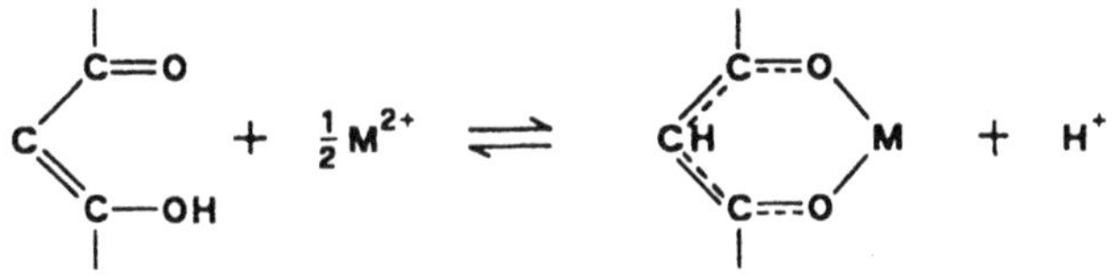

$$\text{(3)}$$

The most stable complex is believed to involve the more strongly acidic COOH groups, whereas the less stable complexes are believed to be associated with weakly acidic COOH and phenolic groups.[45]

Data presented in Table 2 show that the total acidities contributed by COOH and phenolic OH in soil FA are considerably higher than for soil HA, whereas the contents of both groups are much lower in aquatic and sedimentary FA and HA. Nonaromatic carboxyl and hydroxyl sites are speculated to be important in metal ion binding in the aquatic environment. The monomeric analogues of some of these sites, pyruvate and glycolic acid (Equations 4 and 5), have binding constants similar to phtalic and salicylic acid.

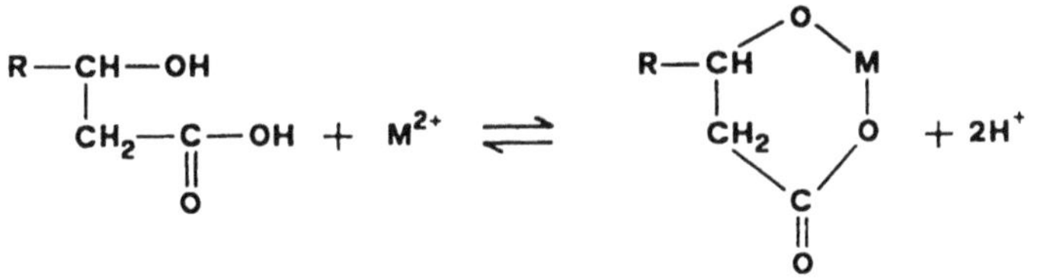

$$\text{(4)}$$

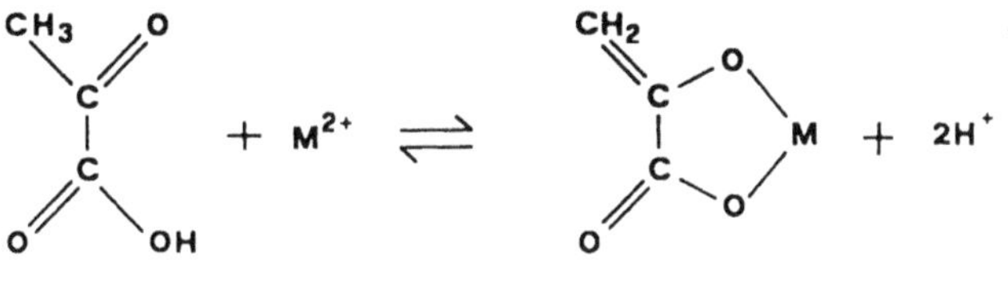

$$\text{(5)}$$

Metal ions may also coordinate with ligands belonging to two (or more) HS molecules, forming 2:1 complexes (Equations 6 and 7) and/or chelates (Equations 8 and 9), and eventually producing an aggregate structure (Equation 10) that may result in the precipitation as the chain grows at high metal to HS ratios.[43]

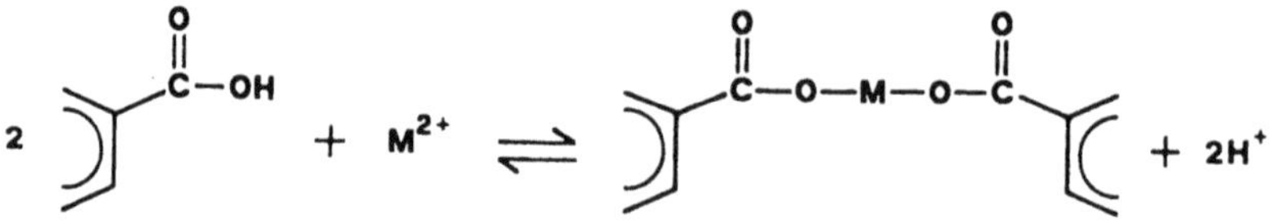

$$2\ \text{HS–C(=O)–OH} + M^{2+} \rightleftharpoons \text{HS–C(=O)–O–M–O–C(=O)–HS} + 2H^{+} \quad (6)$$

$$2\ R\text{–}C(=O)O^{-} + M^{2+} \rightleftharpoons R\text{–}C(=O)\text{–O–M–O–}C(=O)\text{–}R \quad (7)$$

$$2\ \text{HS(–C(=O)–O}^{-})\text{(–C(=O)–OH)} + M^{2+} \rightleftharpoons [\text{HS(COO)}_2\text{M(OOC)}_2\text{HS}]^{2-} + 2H^{+} \quad (8)$$

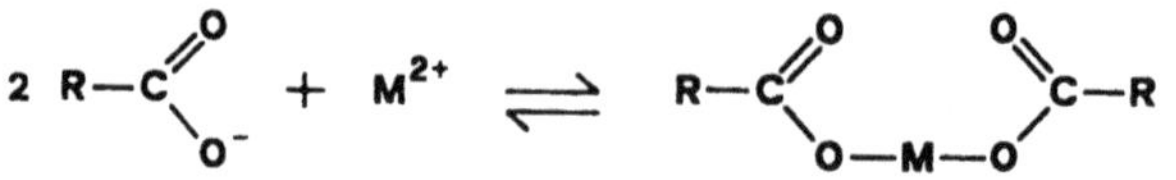

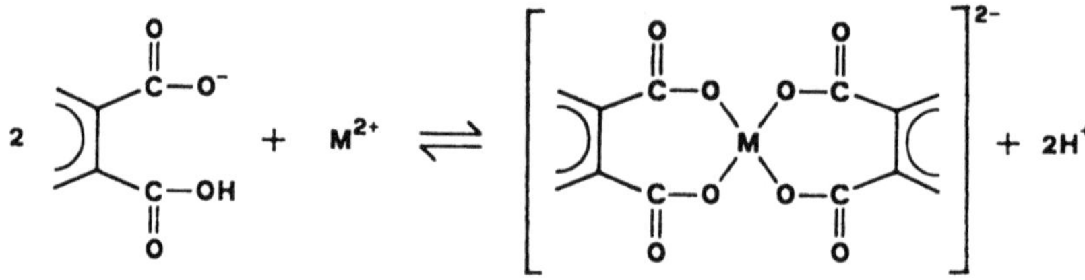

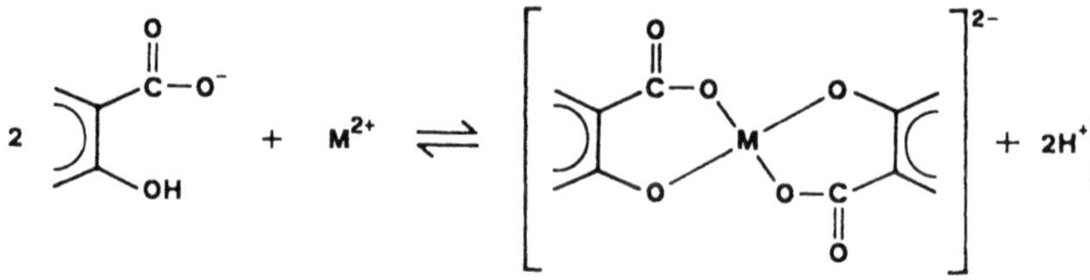

$$2\ \text{HS(–C(=O)–O}^{-})\text{(–OH)} + M^{2+} \rightleftharpoons [\text{HS(COO)(O)M(O)(OOC)HS}]^{2-} + 2H^{+} \quad (9)$$

$$\text{HS(COO)}_2\text{M(OOC)}_2\text{HS} \ \vert\ n \text{ units of HS} \ \vert\ \text{HS(COO)}_2\text{M(H}_2\text{O)}_n \quad (10)$$

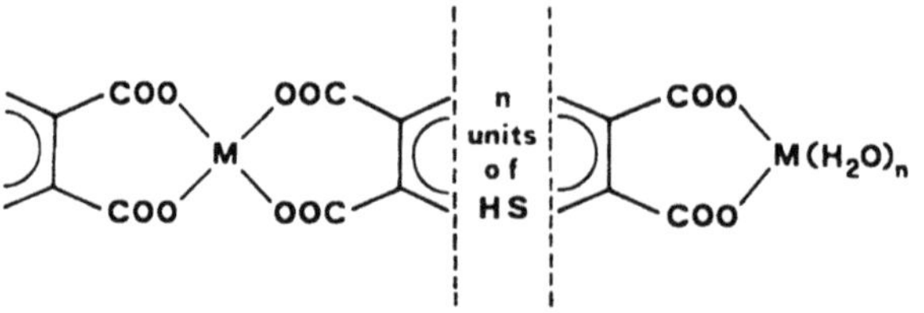

The 2:1 complexes (Equations 6 to 10) may be formed simultaneously with 1:1 complexes described previously (Equations 1 to 5).

Two main types of complexes may be formed between metal ions and HS, that are (1) inner-sphere complexes, resulting in the formation of bonds with some covalent character between the ligand atom(s) and the metal ion, both completely or partially dehydrated; and (2) outer-sphere complexes that result in the electrostatic attraction between the ligand(s) and the metal ion that remains hydrated. For simplicity, all reaction schemes described in Equations 1 to 10 show formation of inner-sphere complexes, but they also represent outer-sphere HS complexes with a solvated cation (Equation 11).

$$\text{HS}(\text{COO}^-)(\text{OH}) + M^{2+}(H_2O)_n \rightleftharpoons \text{HS}(\text{O}^-)\text{C=O}\cdots(H_2O)_n M^{2+}(\text{O}^-) \quad (11)$$

The electronic and steric environment of the ligand site, as well as various physical and chemical characteristics of the surrounding medium, can exert a considerable influence on the overall complexation process.[2] In a given HS macromolecule, identical coordinating groups can be bound to different types of aliphatic chains and aromatic rings of various structure, which can exert differing electronic effects. The steric microenvironment of the binding site, and particularly its size, will depend on the geometry, steric conformation, and flexibility of the whole complexant molecule. This is influenced by the formation-disruption of hydrogen bonds and metal bridges, which can vary with pH, ionic strength, and concentration of the metal to be complexed. The hydration of hydrophilic sites and electrostatic effects — the electric field determined by the extent of ionization of major acidic complexing groups — also can influence the formation process and stability of complexes.[2] In hydrated macromolecules, "tertiary structure" due to the formation of inter- and intramolecular bonds is important.[46] This structure can render the hydrated particle formed sufficiently rigid to be considered a "gel-like" phase distinct from a solution phase.[2] Metal complexation properties of HS may also be affected by these phenomena in that effective retention of cations inside these "gels" can decrease with the increasing hydration.[47,48] The relative importance of these different effects varies with the degree of site occupation by metallic cations and represents the fundamental difference between HS complexants and "simple" ligands.

Measurement Methods. The Spectroscopic Approach

The quantitative measurement and interpretative modeling of trace metal complexation by HS have been widely investigated and constitute the subject of several recent reviews.[7,39,40,49–52] Methods used to determine the binding capacities of HS for metal ions and stability constants of metal-HS complexes formed include (1) separation techniques, such as coagulation, proton release titration, ion-exchange, metal ion competition with cation-exchange resins, equilibrium dialysis, liquid chromatography, and ultrafiltration; and (2) nonseparation techniques, such as hydrogen-ion and ion-selective electrode (ISE) potentiometry, anodic-stripping voltammetry (ASV), fluorescence spectrometry, and electron spin resonance spectrometry.[7,40] Bioassay techniques have also been used, mainly in toxicologycally oriented studies.[39]

Quantitative results obtained by different researchers using different procedures are often not comparable. Complexing capacities of HS are dependent on the method of measurement.[7] Also the source of HS and the procedure used for its isolation, plus many experimental factors, including concentration of HS, ionic strength of solution, pH, temperature, and the method of data manipulation for computation of stability constant can influence the results.[7]

The knowledge of the molecular structure and the binding mechanisms of metal-HS complexes is limited. Progress in the ecotoxicology of trace metals in natural aqueous systems requires a more precise and extended conceptual knowledge of molecular and mechanistic aspects of trace metal-HS reactions. A greater emphasis on basic information about chemical aspects is needed compared to quantitative monitoring studies.[53]

Spectroscopic techniques, e.g., UV-Vis, IR, fluorescence, ESR, Mössbauer, and NMR, offer a great potential to provide a number of important insights at the molecular and mechanistic level of the metal complexation reactions with HS, as well as unique information about the chemical and physical nature of the binding sites.

The principal objective of the present review is to discuss the results available on metal-HS interactions by the application of these spectroscopic methods.

ULTRAVIOLET-VISIBLE SPECTROSCOPY

Principle and Methodology

The absorption bands in the UV-Vis region arise from electronic transitions from bound states (outer valence orbitals) to excited electron states.[54] The wavelength limits of these regions of the electromagnetic spectrum are 200 to 400 nm for the UV and 400 to 800 nm for the visible. In organic molecules such as HS, these exceptionally low energy transitions are associated with the presence of chromophores, i.e., conjugated double-bonds, aromatic, and related molecules with delocalized electronic orbitals. The strongest UV-Vis absorption bands are associated with $\pi \rightarrow \pi^*$ transitions, and weaker ones with $n \rightarrow \pi^*$ transitions, where π and π^* refer to bonding and antibonding p-type orbitals and n refers to lone-pair orbitals.[54] Electronic transitions can occur within the molecular orbitals of chromophores or involve the transfer of an electron from one chromophore to another chromophore or to a nonchromophore (electron- or charge-transfer excitation). For metal-organic complexes, absorption can occur in the UV-Vis region from $d \rightarrow d$ orbital transitions associated with transition metals.

The two most important parameters determined in UV-Vis spectroscopy are the wavelength(s) of maximum absorption and the absorptivity, ε. A change in both parameters of the chromophores may result by ionization of carboxyl and hydroxyl groups.[55] Interaction of some metal ions such as Cu^{2+} with polycarboxylates in dilute aqueous solution leads to typical spectral perturbations in the UV region. They occur: (1) in the 195 to 200 nm range, corresponding to the internal ligand transition band typical of bound (di)carboxylic groups; and (2) between 245 and 265 nm, that is the strong electron transfer band due to electron transition between the central metal ion and the electronic systems of the ligands (charge-transfer process). With increasing the ratio of Cu^{2+} ions to polyelectrolyte ligand, a red-shift and a strengthening is observed for the charge transfer band.[56]

The Job's method has been applied to determine the stoichiometry of metal ion-HS complex formation in aqueous solution. Complex composition can be obtained by measuring the variation of optical densities in the visible range (400 to 800 nm) of solutions containing different ratios of metal ion to complexing agent, while simultaneously maintaining a constant total concentration of both reactants.[57]

Metal-Humic Substance Complexes

In early studies, Broadbent and Ott[58] suggest the formation of more than one type of chelate for the complexation of soil FA with Cu^{2+} ions. Pavel[59] showed that a peat FA forms at least one type of chelate complex with Cu^{2+}, whereas it combines with Co^{2+} in at least three to four, and with Cr^{3+} in at least four different types of chelate complexes. Schnitzer and Skinner[60] plotted the difference between optical density measured for various metal-soil FA complexes and the value corresponding to no complexation vs. system composition. The maximum (or minimum) occurs at the metal to ligand ratio corresponding to the composition of the metal complex. This suggests the formation of 1:1 complexes between FA and Cu^{2+}, Fe^{3+}, and Al^{3+} at pH 3, whereas at pH 5 Cu^{2+} and Fe^{3+} form 2:1 molar complexes with the FA, while Al^{3+}-FA complex composition remains at 1:1. Ram and Raman[61] obtained Job plots for Zn^{2+}, Mn^{2+}, and Ca^{2+} complexes with soil FA samples indicating that, in any case, 1:1 metal-FA complexes form.

Wershaw et al.[62] observed a well-defined absorption band at approximately 245 nm, consistent with a Cu^{2+} charge transfer complex with carboxylate groups of a river FA. The structure and conformation of the FA molecule are a limiting factor for the ability of carboxylate groups to form bidentate complexes with copper. The UV spectra also showed particulate scattering, probably caused by the binding of partially hydrated copper ions to two or more FA molecules, that produces aggregation. Wershaw et al.[63] suggest a fraction of the FA molecules possessing accessible, strong S- and/or N-binding sites as responsible of aggregation.

Conclusions

The UV-Vis spectra of HS show little structure, appearing largely as broad, coarsely structured absorption bands. This is because of heterogeneous substitution that results in chromophores with overlapping bands and spectral shifts due to slight differences in the macromolecular structures.

This lack of distinctive structure in electronic absorption spectroscopy is a severe drawback when being used in a qualitative manner. UV-Vis spectroscopy is largely restricted in its application to providing evidence for carboxylic complexing sites in HS involved in charge transfer complexes with metal ions. On the other hand, most useful applications of the technique are found in quantitative measurements of the stoichiometry of metal-HS complexes based on the Job's method.

INFRARED SPECTROSCOPY

Basic Principles and Methodology

The most interesting portion of the infrared (IR) spectrum for the structural and analytical study of molecules is the medium IR region, between 4000 and 400 cm^{-1} (2.5 and 25 nm). Energy absorbed by an organic molecule in this region is converted into energy of molecular vibration, and the IR spectrum of the organic molecule consists of vibrational bands. There are two types of bond vibration modes in simple molecules: stretching, that involves changes in the bond length between the atoms along the bond axis, and bending or deformation, which involves a change of bond angles.

The characteristic wavenumber (or frequency) ranges for the vibrations of particular functional groups, such as CH, NH, OH, C=O, C=C, aromatic rings, etc., depend on the vibrational mode, the strength of the bonds involved, and the masses of the atoms.[63] Intensities in IR spectra depend on dipole-changes occurring during the vibrations and so the polar bonds and groups frequently give the strongest absorptions. In organo-metallic chemistry particular metal-atom-bonded ligands can be identified from their IR absorptions.[64] The general range of applicability of IR spectroscopy is much greater than that of UV-Vis spectroscopy because saturated as well as unsaturated, unconjugated as well as conjugated, molecules all have vibration frequencies. Lists of band frequencies occurring in very complex organic molecules, with correlation tables and empirical rules of comparison, are very useful in the interpretation of IR spectra for the assignment of experimentally observed absorption bands to specific functional groups.[63,64]

The biggest improvement in performance of IR spectroscopy came about when dispersive, diffration grating spectrometers were replaced by interferometers, together with the necessary computing facilities for Fourier-transform (FT) conversion of the interferogram into a normal (intensity vs. wavenumber) spectrum.

The very high sensitivity brought about by FT methods has also led to the development and use of a number of additional nontransmission techniques including diffuse reflectance.[65] Besides its high sensitivity, the great practical advantage of diffuse reflectance infrared Fourier-transform (DRIFT) method is that powdered samples can be studied with little or no preparation. Compared with transmission spectra, the strongest absorption features in the spectra of organic molecules are relatively reduced in intensity by selective reflection processes. The weaker absorptions can be, therefore, very well recorded with little distortion by DRIFT, and the result is generally adequate for structure and identification purposes.

Infrared Spectra of Humic Substances

The most striking feature of the IR spectra of HS is their overall *simplicity*,[5,6] but this is more apparent than real for a complex macromolecule such as that of HS. The observed broadness of the bands generally results from the extended overlapping of very similar absorptions arising from individual functional groups of the same type, with different chemical environments.

Another remarkable characteristic of the IR spectra of HS of different origin is their overall apparent *similarity*, more than their diversity. This is in part due to the fact that most groups of atoms vibrate with almost the same frequency irrespective of the molecule to which they are attached. This does not mean, however, that HS displaying similar spectra must have similar structures, but only that the functional groups and structures are similar.[66]

A complete and detailed description and interpretation of IR spectra of HS in the solid and aqueous solution phase is beyond the limits of the present discussion. Several detailed and comprehensive reviews of the application of IR spectroscopy to HS have been recently provided.[6,28,66–68]

Metal-Humic Substance Complexes

Since any change in the strength of the bonds and masses in a given system alters the vibrational frequency, the formation of metal-HS complexes may be studied by the shifts observed for vibrational frequencies. These can be used to identify the absorbing functional groups involved in the complexation and, possibly, provide information on the type of interaction occurring between the metal and HS in the complex formed.

Carboxylate Ligands

Results of IR spectroscopy have provided evidence of the prominent role played by COOH groups in metal ion complexation by HA and FA. The C=O stretching absorption band at about 1720 cm^{-1} and the C–O stretching and O–H deformation absorptions at about 1200 cm^{-1} disappear upon ionization of the COOH groups and new bands appear near 1600 and 1380 cm^{-1}, arising,

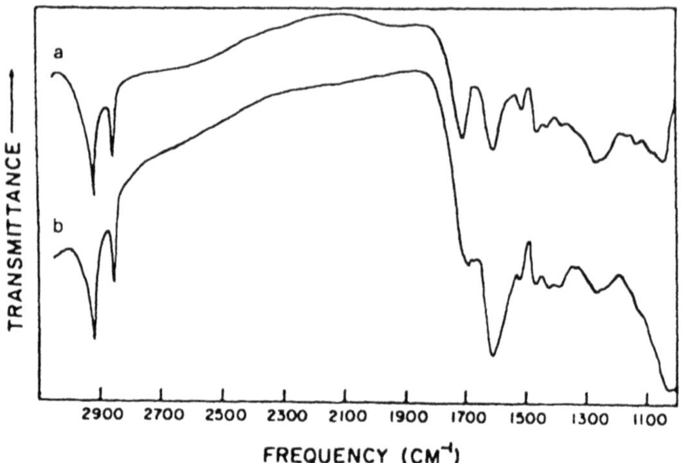

FIGURE 3. Fourier-transform infrared (FT-IR) spectra in H_2O (CaF_2 cells) of a soil humic acid (HA) (a) and its complex with Cu^{2+} (b). (From MacCarthy, P., H. B. Mark, Jr., and P. R. Griffiths. *J. Agric. Food Chem.* 23:600–602 [1975]. With permission.)

respectively, from the asymmetric and symmetric stretching vibrations of the COO^- groups.[6]

Schnitzer and Skinner[60] showed that both Fe^{3+}- and Al^{3+}-soil FA complexes prepared with increasing metal:FA ratios feature a decrease in intensity of the IR bands at 1725 and 1200 cm^{-1}, with respect to those at 1625 and 1400 cm^{-1}, that indicates conversion of COOH to COO^- groups. Similar results were obtained in several successive IR studies on the complexation of FA and HA from different soils and aquatic and sedimentary sources to several divalent and trivalent metal ions including Cu^{2+}, Mn^{2+}, Zn^{2+}, Pb^{2+}, Co^{2+}, Ni^{2+}, Ca^{2+}, Mg^{2+}, Sr^{2+}, Fe^{3+}, Al^{3+}.[67,69–86] MacCarthy and Mark[85] and MacCarthy et al.[86] confirmed the participation of carboxyl groups of HS in metal complexation by IR spectra of deuterated HA and Cu^{2+} and Fe^{3+}-HA complexes in D_2O and FT-IR spectra of HA and its Cu^{2+}-HA complex in aqueous solution (Figure 3). Deuterium oxide does not absorb in the 1600 cm^{-1} region and Fourier-transform method allows for the subtraction of the water spectrum and detection of relatively weak absorptions from solutes. The IR spectra of metal-FA and metal-HA fractions isolated from a number of natural, cultivated, and sludge-amended soils of widely differing pedogeographical origin also showed the presence of stable metal-carboxylate complexes formed by HS in natural conditions.[87,89]

Juste and Delas[71] showed that the COO^- to the COOH absorption ratio in the IR spectra of a number of metal-HA complexes prepared at the same HA:metal ratio is dependent on the nature of the metal ion complexed and vary in the decreasing order: $Fe^{3+} > Cu^{2+} > Al^{3+} > Ca^{2+} > Mg^{2+}$. This indicates a decreasing conversion of COOH groups to COO^- groups involved in metal complexation. A successive IR study by Banerjee and Mukherjee[77] suggested the following order, Cu > Zn > Fe > Co > Mn, for metal interaction with carboxylate in various HA complexes.

The frequency of the asymmetric and symmetric stretching vibrations of COO^- may help to solve the controversy existing as to whether the COO^- linkages are ionic or covalent. With increasing covalency in metal carboxylate complexes, the COO^- asymmetrical stretching band shifts to a higher frequency and the symmetrical stretching band to lower frequency.[64] Frequency shifts measured by conventional IR spectroscopy on various metal-HA complexes, however, are variable and slight and do not produce any conclusive evidence of metal binding character. This may be due to either the formation of complexes of mixed, electrovalent, and covalent nature or the limitations imposed by relatively poor resolution of peaks. Vinkler et al.[90] studied the influence of several metal cations on the COO^- asymmetrical stretching in HA complexes. Absorption bands that range from 1585 cm^{-1} for Ca^{2+} to 1625 cm^{-1} for Al^{3+} indicate a different degree of bonding covalency for the various metals examined. Piccolo and Stevenson[79] found that covalent bonds for Cu^{2+} in HA and FA are preferentially formed at low levels of metal ion, whereas bonding becomes increasingly ionic as the system is saturated with the metal. Interpretations in the 1620 cm^{-1} region are, however, complicated by interference from other groups, as will be discussed further in this section. More recently, Baes and Bloom[91] obtained evidence by DRIFT spectroscopy of the covalent character of Cu-HA bonds, indicated by the much larger separation between the asymmetrical and symmetrical COO^- stretching band of Cu-HA complex with respect to Na-humate.

The nature of the carboxylate binding site in HS can be evaluated by measuring the separation between the two frequencies of the antisymmetric and symmetric stretching vibrations for the metal complexed COO^- group (near 1600 and 1400 cm^{-1}, respectively), with respect to the uncomplexed carboxylate ion. This separation is larger in unidentate complexes (Structure I), smaller in bidentate (chelate) complexes (Structure II) and comparable in bridging complexes (Structure III)[64] (Equation 12).

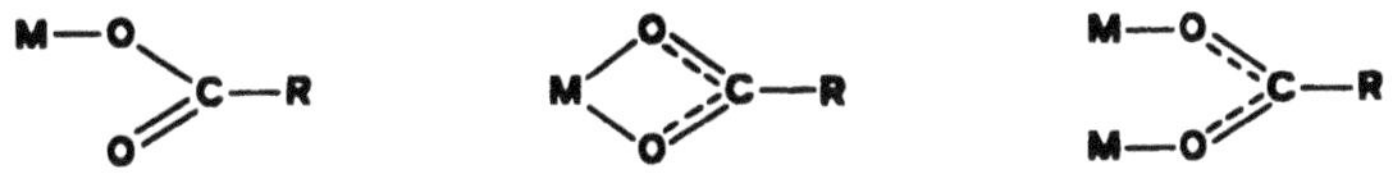

(12)

I II III

Boyd et al.[92] found that a large separation occurred (Figure 4) for Cu^{2+} and Fe^{3+} complexes of a soil HA, thus suggesting the formation of unidentate metal complexes. These observations were later confirmed by Prasad et al.[83] for Fe^{3+}-, Co^{2+}-, and Zn^{2+}-FA complexes. These results are also consistent with metal chelation involving either two adjacent COOH groups (phtalate type) or a COOH and adjacent phenolic OH group (salicylate type), each forming single bonds with the metal ion.[83,92] Piccolo and Stevenson[79] observed the disappear-

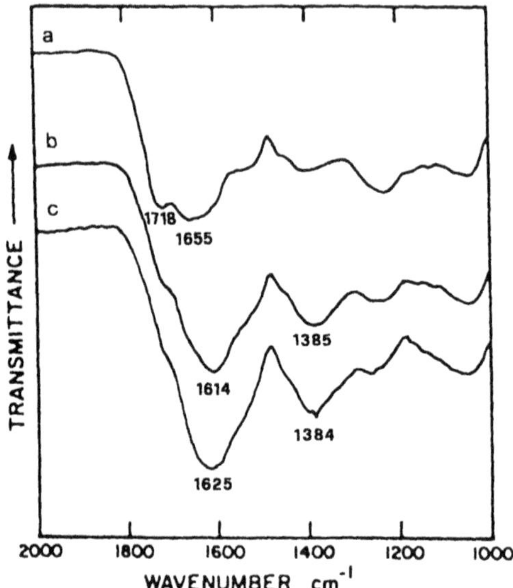

FIGURE 4. Infrared (IR) spectra of a soil humic acid (HA) (a) and its complexes with Cu^{2+} (b) and Fe^{3+} (c). (From Boyd, S. A., L. E. Sommers, and D. W. Nelson. *Soil Sci. Soc. Am. J.* 45:1241–1242 [1981]. With permission.)

ance of a small band at 1855 cm^{-1} attributed to cyclic anhydrides, in highly loaded (metal:HA ratio > 1) metal-HA complexes that would confirm the formation of phtalate-like metal chelates.

Phenolic and/or Alcoholic Hydroxyl Ligands

A decrease in intensity and/or a shift of the IR absorption near 3440 cm^{-1} is expected to occur where oxygen of phenolic OH groups are involved in metal binding, alone or together with COOH groups. Schnitzer and Skinner[93] and Schnitzer and Hoffman[69] failed in measuring these changes, although selective blocking of COOH and OH groups showed that both groups react simultaneously with Cu^{2+}, Fe^{3+}, Al^{3+}. Absence of changes is ascribed to increased absorption contributed to the 3400 cm^{-1} band by OH groups of metal ions complexed as partially hydroxylated forms by soil FA at pH 3.

Successively, several authors[76–80] observed a net shift in the OH band, from 3500 to 3400 cm^{-1} in HA or FA, to 3300 to 3200 cm^{-1} in HA- or FA-complexes with Zn^{2+}, Fe^{3+}, Cu^{2+}, Pb^{2+}, and Mn^{2+}. Phenolic and/or alcoholic OH groups may thus form covalent bonds with the metal ion. Griffith and Schnitzer[87] obtained similar results on naturally occurring metal-HA and -FA fractions isolated from soil.

Banerjee and Mukherjee[77] found that the extent of the OH stretching shift toward lower frequencies depends on the type of metal and follows the order: Mn (3383) < Co (3354) < Cu (3334) < Fe (3325), whereas an unespected increase (3442) is measured for Zn.

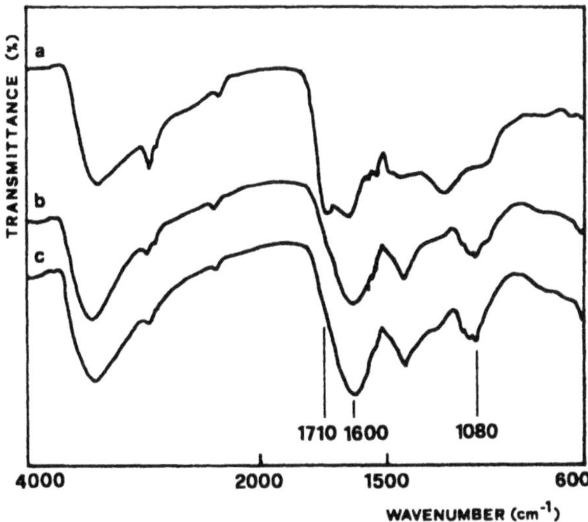

FIGURE 5. Infrared (IR) spectra of a soil humic acid (HA) (a) and its complexes with Cu^{2+} (Cu^{2+}:HA = 0.04) (b), and Cu^{2+} and Fe^{3+} [(Cu^{2+} + Fe^{3+}):HA = 0.08] (c). (Adapted from Senesi, N., G. Sposito, and J. P. Martin. *Sci. Total Environ.* 55:351–362 [1986].)

Prasad and Sinha[80] observed modifications for the typical band at 1070 cm^{-1} suggesting that polysaccharide structures in FA can also be involved in metal complexation.

Complexes of Hydroxylated and/or Hydrated Metals

Dupuis et al.[74] observed a strong and broad band at 3200 cm^{-1} in the IR spectra of Fe^{3+}-FA complexes prepared at pH > 4, that may be attributed to partially hydroxylated Fe^{3+} ions bound electrovalently to the FA. Piccolo and Stevenson[79] measured a progressive enhancement of the strong band centered at 3420 cm^{-1} with increasing the amount of Cu^{2+} or Pb^{2+} ions added to a peat FA. This result, together with the resistance of this IR band to heating at 100°C for 24 h, suggests that at least part of this absorption is due to hydration water of complexed metal.

Several authors[78,79,81,82,84] attributed the very sharp IR bands observed in the region 1130 to 1080 cm^{-1} (Figure 5) and at about 890 cm^{-1} for HA or FA complexes with Cu^{2+}, Zn^{2+}, Al^{3+}, or Fe^{3+}, to metal-oxygen vibration of hydroxylated and/or hydrated metal ion bound to HA or FA. On the contrary, Tan[78] confirmed these findings measuring IR spectra for physical mixtures of HA and $Zn(OH)_2$ or $Al(OH)_3$ resembling those of original HA and different than those of metal-HA complexes.

Juste and Delas[71,72] observed additional IR bands at low wavenumbers (853, 817, 781, 697 cm^{-1}) in Cu^{2+}-HA complexes prepared from HA and $Cu(OH)_2$ at pH 7, that are assigned to Cu^{2+}-OH vibrations. Higashi[94] measured a regular increase in IR absorptions of Al-HA complexes near 730 and 980 cm^{-1}, that is attributed to formation of additional Al–O bonding with increasing the pH value during complexation.

Conjugated Ketonic Complexing Sites

Stevenson and Goh[95] assigned absorption in the 1660 to 1600 cm^{-1} region in IR spectra of HS in part to conjugated ketonic, or β-diketone structures. Although frequency shifts in this region are hardly detectable due to the broad absorption, Piccolo and Stevenson[79] observed a shift to a lower frequency for the band at 1610 cm^{-1} in Cu^{2+}-FA complexes, that is attributed to the C=O group vibration in conjugated ketones weakened by resonance between C–O–Cu and C=O–Cu in the complexes. Further IR findings led to the conclusion that ketonic structures are among the first in FA to be involved in metal complexation.[79]

In an IR study of FA-metal-clay complexes, Schnitzer and Kodama[96] observed that a band at 1525 cm^{-1} appears when the bridging metals between FA and the clay are Cu^{2+} or Al^{3+}. This band is absent in the spectra of untreated FA and of the clay, and also in ternary complex formed with FA having C=O groups selectively blocked. These results provide evidence that β-diketone groups in the FA may form metal chelates similar to acethylacetonates.[64] The chelate ring is planar and symmetrical and the two C–O bonds are equivalent as are the two C–C bonds in the ring, thus providing strong possibilities for resonance and stability (Equation 3). Since this type of reaction is not observed in aqueous solution and in the absence of clay, Schnitzer and Kodama[96] conclude that the presence of the clay, e.g., montmorillonite, is essential in affecting the conformation of the FA polymer, thus favoring the complexation of metal ions by C=O groups in the FA.

Nitrogen and Sulfur Ligands in Sewage Sludge Humic Substances

IR evidence has been obtained of metal binding sites involving amide N (and possibly amide C=O) or sulfonic groups (SO_3H) in nitrogen- and sulphur-rich HS isolated from sewage sludges.[83,97–99] Boyd et al.[97] observed a shift of the amide-I band (1644 cm^{-1}) to higher frequencies (~1655 cm^{-1}) on removal of indigenous metal cations from a sludge FA, which can be reversed by addition of Cu^{2+} ions to the extracted material. This suggests the occurrence of binding amide sites in sewage sludge FA for metals like Cu^{2+} and Ni^{2+}, whereas Zn^{2+} is expected to bind preferentially to amide O-donors.

Prasad et al.[83] observed that the band at 1920 cm^{-1}, attributed to –NH group vibrations, disappears in complexes of a soil FA with Fe^{3+}, Zn^{2+}, and Co^{2+} ions, thus suggesting the formation of coordinate covalent bonds between –NH groups and metal ions.

Sposito et al.[98] and Baham et al.[99] obtained IR evidence of the involvement of sulphonic groups in metal-sludge FA interactions by the presence of typical bands of metal sulfonates at 1120 to 1130, 1070, and 620 cm^{-1}.

Conclusions

The complexity and heterogeneity of HS is still cause of some ambiguity and uncertainty in the interpretation of IR spectra of their metal complexes. IR spectrometry can, however, provide useful information about the nature and reactivity of HS structural components that interact with metal ions and on the molecular arrangement of the binding sites in HS involved in metal complexation.

Application of IR spectroscopy to the study of HS-metal complexes in deuterated water solution and of FT-IR technique in common aqueous solution shows very promising in providing unique information on the interacting species that may be observed in the equilibrium state, in low perturbed system, and in close-to-environmental conditions.

The use of DRIFT spectroscopy applied to metal-HS complex powders will surely facilitate future studies of metal ion binding by HS.

The utilization of IR spectroscopy in conjunction with selective group blocking techniques and, especially, together with chemical derivatization methods has the potential for considerably enhancing the quality of IR spectra and facilitating their interpretation, thus providing IR data more informative on metal-HS complexation.

FLUORESCENCE SPECTROSCOPY

Basic Principles and Methodology

The absorption of visible and ultraviolet radiation raises a molecule from the ground electronic and vibrational state to excited states. The most important ways in which excited electronic states decay are radiative processes, i.e., fluorescence and phosphorescence, which occur when excited electrons return to ground state, and nonradiative processes, i.e., internal conversion and intersystem crossing.[100] Fluorescence is a radiative photoprocess that occurs between two energy levels of the same multiplicity and consists in the emission of less energetic (longer wavelength) photons than the photons absorbed to produce the excited state. Phosphorescence is a radiative photoprocess that occurs between two energy levels of different multiplicity and is most exclusively preceeded by intersystem crossing. Fluorescence is generally a rapid process that takes place within 10^{-9} to 10^{-6} s, while phosphorescence lifetimes are in the range 10^{-4} to 10^{-2} s.[100] The fluorescence process always competes, however, with nonradiative decay processes and photodecomposition. The dominating process is dependent on the relative rates of decay.[100]

The fluorescence behavior of a molecule, especially in the solution state, is affected to various extent by several molecular properties and environmental factors. In particular, an electronically excited molecule can lose its energy by interacting with another solute (a quencher) and thus its fluorescence reduced. Fluorescence quenching process may be ascribed to two distinct mechanisms: (1) static quenching, which occurs when a quencher interacts with the ground state of the fluorescent molecule, and (2) dynamic quenching, which involves the excited-state molecule.[100] Commonly encountered quenchers are (1) dissolved molecular oxygen, whose effect can be easily eliminated by flushing it out from the solution by bubbling nitrogen; and (2) metal ions, especially paramagnetic ions, even if they do not form complexes with the ground state of the fluorescent solute. There is another circumstance in which fluorescence quenching is apparently observed: this is under experimental conditions that introduce the "inner filter" effect. This effect arises generally from light absorption by the solvent or high concentration of solute.[100]

Two types of conventional fluorescence spectra can be obtained: emission and excitation. Emission spectrum is recorded by measuring the relative intensity of radiation emitted as a function of wavelength for a fixed excitation wavelength. Excitation spectra are obtained by measuring the emission intensity at a fixed wavelength while varying the excitation wavelength.

Fluorescence efficiency, or quantum yields, Φ_f, measures the efficiency with which the absorbed energy is re-emitted. It is defined as the ratio of total energy emitted as fluorescence per total energy absorbed. The fluorescence efficiency depends on the rate of fluorescence emission relative to the rates of radiationless processes. Any factor related to molecular structure and environment that affects the rate of any of these processes will influence fluorescence efficiency. This is, therefore, a primary factor in determining the sensitivity of the fluorescence method.

Fluorescence lifetime, τ, is related to the probability of finding a given molecule that has been excited in the excited state at time t after the excitation source is turned off, and refers to the mean lifetime of the excited state. Fluorescence lifetimes measured for most organic molecules are typically of the order of nanosecond and reflect the overall rate at which the excited state is deactivated, including both radiative and nonradiative processes.

The intensity of fluorescence, I_f, depends on the concentration of absorbing species in solution and the efficiency of the fluorescence process:

$$I_f = \Phi_f I_o [1 - \exp(-\varepsilon bC)] \tag{13}$$

where Φ_f is the fluorescence efficiency, I_o is the intensity of incident radiation, ε is the molar absorptivity at the excitation wavelength, b is the path length of the cell, and C is the molar concentration. For very dilute solutions, where εbC is sufficiently small, Equation 13 reduces to a linear relationship between measured fluorescence intensity and concentration:

$$I_f = \Phi_f I_o \varepsilon bC \tag{14}$$

When εbC values are low the fluorescence intensity is essentially homogeneous throughout the sample.

Fluorescence of Humic Substances

Humic substances of any source and nature fluoresce because of conjugated double bonds or aromatic rings bearing various functional groups. In general, only a small fraction of HS molecules that absorb radiation actually undergo fluorescence.

Fluorescence emission spectra of HS are generally characterized by a broad band with a wavelength and intensity of the maximum that are variable with the nature and origin of the HS sample. Excitation spectra feature, in general, structured lineshapes with a number of peaks and shoulders of varying relative intensity and wavelength. The fluorescence properties of FA have been recently reviewed by Senesi.[101]

Quenching of Humic Substance Fluorescence by Metal Ions

In general, metal ions, especially paramagnetic ions, are able to quench the fluorescence of organic ligands by enhancing the rate of some nonradiative processes that compete with fluorescence, such as intersystem crossing. Paramagnetic transition metal ions such as Cu^{2+}, Fe^{3+}, Fe^{2+}, Co^{2+}, Ni^{2+}, Cr^{3+}, and VO^{2+}, which possess d levels of energy lower than the excited singlet state, may effectively quench the fluorescence of HS ligands via intramolecular energy transfer.[102,103] Although the quenching effect for diamagnetic metal ions is expected to be much less pronounced than for paramagnetic ions, Pb^{2+} and Al^{3+} quench fluorescence of HS, whereas Cd, which forms much weaker complexes, does not quench.[104,105] Wavelength shifts of fluorescence emission maximum and/or excitation peaks are often observed upon interaction of HS with some paramagnetic metal ions.[101] This suggests that some fluorescent metal-HS complexes can involve a transition either between two metal energy levels or between a metal energy level and π energy level in the ligand, i.e., a charge-transfer transition.

Titration curves of HS fluorescence quenching vs. concentration of added quencher have been used to obtain complexing capacity of HS ligands and the stability constant of HS-metal complex. Quantitative treatment of fluorescence data, however, will not be discussed in this text.

Early investigations showed that Cu^{2+}, Fe^{2+}, and Ni^{2+} are strong quenchers of soil and seawater FA and HA fluorescence, Co^{2+} and Mn^{2+} ions have a small effect, whereas cations like K^+, Na^+, Ca^{2+}, and Ba^{2+} show no quenching effect.[106–108] This suggests that Cu^{2+} and Fe^{2+} form more stable complexes than Mn^{2+} and Co^{2+} with soil HA and FA.[106] Adhikari and Hazra[108] measured a shift toward the shorter wavelength for excitation peaks of soil FA in the presence of Cu^{2+} or Ni^{2+}, whereas no shift in the emission wavelength maxima was observed by Almgren et al.[107] for seawater FA with added Cu^{2+}.

Ghosh and Schnitzer[109] showed that the position and intensity of the peak at higher wavelength (465 nm) of a soil FA is unchanged as more of either Cu^{2+} or Fe^{3+} is complexed, whereas the intensity of the lower wavelength peak (360 nm) decreases and shifts to longer wavelength (390 nm) at both pH 4 and 6 (Figure 6). Since fluorescence characteristics of synthetic metal complexes of a low-ash content (<3%) FA are similar to those observed with high ash contents FA samples (9.4 to 35.5%), Ghosh and Schnitzer[109] conclude that similar fluorescent aromatic structures are involved in metal-complexation of FA in both natural and laboratory conditions. Boto and Isdale[110] observed that addition of excess Fe^{3+} to FA solutions isolated from inclusions in the crystal structure of inshore-growing coral skeletons also produces a shift of the fluorescence emission maximum of the FA 360 nm peak toward longer wavelength (+25 nm), with a 30% quenching of fluorescence intensity.

Several investigators found that fluorescence quenching of FA and HA from different sources by various metal ions, including Cu^{2+}, Pb^{2+}, Co^{2+}, Ni^{2+}, and Mn^{2+}, increases as pH increases[104,111–114] (Figure 7). Saar and Weber[104] measured almost no Cu^{2+} quenching at pH below 2.1, or below 4 for Co^{2+}, Ni^{2+}, and Mn^{2+}, for a soil FA and a river aquatic FA at the same metal ion/FA ratio. At pH 6 and 7, Cu^{2+} quenches nearly 80% of FA fluorescence, Co^{2+} is less than half effective, and Mn^{2+} quenches about one third of FA fluorescence.[111,112] Further, Cu^{2+}, Pb^{2+}, and Mn^{2+} binding to FA, as measured by ISE potentiometry, was found to be proportional to FA fluorescence quenching by these ions,[104,115] whereas no proportionality was found for Ni^{2+} and Co^{2+}.[104] Underdown et al.[115] conclude in favor of a static quenching mechanism, depending primarily on the fraction of FA ligand sites complexing the metal ions, and exclude a dynamic quenching mechanism occurring by collision of FA fluorophores with the metal ions. The latter would depend exclusively on the concentration of quenching ion. Underdown et al.[115] also observed an increase of FA fluorescence, with increasing pH up to the equivalence point, that indicates protons are quenching cations similar to Cu^{2+} and Mn^{2+} and confirms quenching is not collisional since the concentration of free protons is very small.

Cabaniss and Shuman[116] found that ISE potentiometry and fluorescence spectroscopy used in combination give comparable results for Cu^{2+} binding by lake FA at low, natural levels of Cu^{2+} loading. The discrepancy observed at high Cu^{2+}-loading and high pH can be due to either stoichiometry changes, or weak fluorescence of ligands bound, or removal of Cu^{2+} from solution in absorbed or colloidal forms.

Pb^{2+} and Mn^{2+} behave as less effective quencher than Cu^{2+} at the same level of FA binding, i.e., the Pb^{2+} and Mn^{2+} required a concentration much higher than

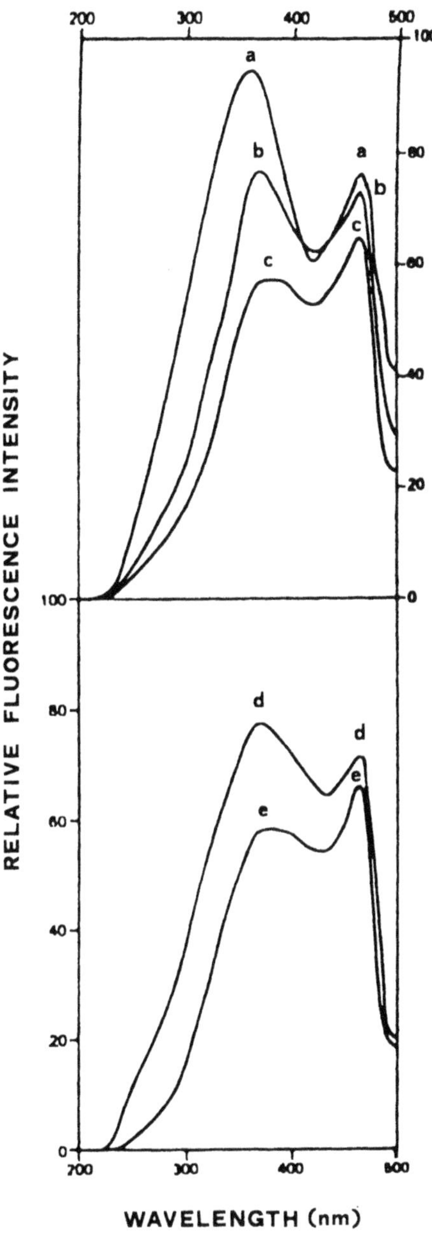

FIGURE 6. Fluorescence excitation spectra at pH 6 of a soil fulvic acid (FA) (a) and its complexes with Cu^{2+} (% Cu: 3.51 [b] and 7.22 [c]) and Fe^{3+} (% Fe: 3.47 [d] and 6.27 [e]). (Adapted from Ghosh, K. and M. Schnitzer. *Soil Sci. Soc. Am. J.* 45:25–29 [1981].)

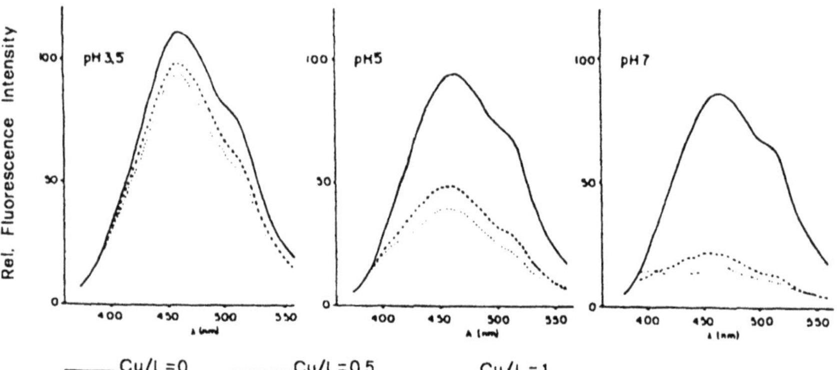

FIGURE 7. Fluorescence emission spectra (λ_{exc} = 355 nm) of a soil fulvic acid (FA) and its complexes with Cu^{2+} obtained at three pHs and two different Cu^{2+}:FA ligand (L) ratios. (From Bartoli, F., A. Hatira, J. C. Andre, and J. M. Portal. *Soil Biol. Biochem.* 19:355–362 [1987]. With permission.)

Cu^{2+} to give the same degree of quenching.[104,115] This confirms that the "close" association of metal ion and ligand is a necessary requirement for effective quenching. Strongly bound paramagnetic Cu^{2+} and diamagnetic Pb^{2+} are more effective than weakly bound paramagnetic Ni^{2+}, Co^{2+}, and Mn^{2+} for quenching FA fluorescence, whereas the weakly bound and diamagnetic Cd^{2+} has no effect on fluorescence of FA.

Ryan and Weber[111] attributed the greater quenching ability of Cu^{2+} to its capacity to form strong inner-sphere complexes that involve in binding several weak sites causing conformational changes making available additional, internal binding sites of FA. Olson[113] suggests that this can occur for a peat HA, especially at pH 10 where dissolved HA would be relatively flexible and open-structured, whereas at pH 7 intense intra- and inter-molecular H-bonding would reduce the access of Cu^{2+} to some sites. Ryan et al.[112] suggest that Mn^{2+}, and likely Co^{2+}, would form outer-sphere complexes with FA not involving weakly acidic phenolic sites, where the ion is farther from the fluorophore, and thus exhibits low quenching ability. Additional data by Underdown et al.[115] indicate that the most strongly acidic carboxyl groups react first with Cu^{2+}, as would occur in bidentate chelation sites with salicylic or phthalic acid type structures. Shestakov et al.[117] observed that fluorescence quenching by Mn^{2+} decreases as the molecular weight of soil FA increases, as a function of the FA nature.

Bartoli et al.[114] proposed two binding mechanisms to explain adsorption of Cu^{2+} by a soil FA at pH 3.5, 5, and 7 and various Cu:FA ratio (Figure 7): (1) a protolytic reaction (strictly complexation), occurring up to a Cu:FA ratio of 0.6; and (2) a charge neutralization, from a Cu:FA ratio of 0.6 to 1.1. Above a 1:1 ratio, the almost constant fluorescence suggests copper is no longer adsorbed on FA.

Gregor et al.[118] showed by a comparative ISE, ASV, and fluorescence quenching study that no single ligand adequately describes Cu^{2+}-FA complexing over the entire pH range. A mixed mode of coordination is proposed, with the dominant binding sites varying with pH and metal to ligand ratio. Comparison of FA-metal binding curves with those for model ligands indicates that potentially fluorescing groups such as salicylates or phthalates seem unlikely to be directly involved in Cu^{2+} complexing at low pH. Aromatic amino acid groups, such as tyrosine and phenylamine moieties, and citrate and malonate moieties in FA may provide important binding sites for Cu^{2+} at pH 3 to 7. At high pH, either polydentate fluorescing moieties are directly involved in complexing, or weakly complexing fluorescent centers become involved because of their close proximity to a strong donor site, i.e., within 1 or 2 carbon atoms of the complexing moieties.[118]

Ryan et al.[112] measured no or a relatively small scattering increase during the fluorescence titration of FA with Mn^{2+} and Co^{2+}, whereas a large increase was observed, especially at higher pH, with Cu^{2+} addition. This suggests that cation-induced coagulation is less for Mn^{2+} and Co^{2+} because of their weaker association to FA. Light-scattering measurements by Underdown et al.[115] indicate, however, that little, if any, aggregation occurs below 6 mmol Cu^{2+} per gram FA. Thus, quenching is apparently not related to aggregation, that is, only the uncomplexed FA fraction fluoresces extensively and only the fraction already quenched by bound metal aggregates.

Frimmel and Hopp[119] showed that fluorescence quenching of HS isolated from lake and river waters at pH 6.8 parallels the relative stabilities of HS complexes, in the order $Cu^{2+} > Fe^{3+} > Fe^{2+} > Co^{2+} = Ni^{2+} > Mn^{2+}$. Diamagnetic metal ions such as Ca^{2+} may compete with Cu^{2+} giving a decreased quenching effect, whereas paramagnetic, uncomplexed metal ions do not quench the fluorescence of HS. In most cases, average complexation capacities determined by fluorescence quenching are of the same order of magnitude of those determined by polarographic titration.[119]

Waite and Morel[120] studied by ligand exchange kinetics and fluorescence quenching the molecular speciation of iron in the presence of a river FA and a grassy pond FA. The existence is shown of different operationally defined iron-FA groupings, that are dependent on pH and light conditions. At pH 3.9, only the portion of iron that is not reduced and is present as small (ultrafiltrable) strongly bound Fe^{3+}-FA complexes is effectively quenching fluorescence. At this pH, the degree of quenching rapidly declines on light irradiation because of the increasing reduction of iron to FA-unbound, not quenching forms. At pH 6.5, almost all the iron is present in large (nonultrafiltrable), relatively strongly bound Fe^{3+}-FA groupings that behave as slightly effective quenchers and are not affected significantly by light. Quenching ability of Fe^{3+} in small complexes with FA is due to the close proximity of metal-ion d orbitals to the excited energy levels of the FA molecule, that allows efficient electron transfers.[120]

The fluorescence quenching method has been also used for the study of metal binding by unfractionated HS contained in untreated natural waters. Ryan and Weber[121] observed that Cu^{2+} may complex to HS dissolved in freshwaters and marine waters probably by displacing more weakly bound metal ions such as

Zn^{2+}, Cd^{2+}, Ca^{2+}, and Mg^{2+}. More strongly bound ions such as Al^{3+}, Fe^{3+}, and Pb^{2+} appear to compete successfully with Cu^{2+} for sites, at least until Cu^{2+} is present in large excess.[121] Tuschall and Brezonik[49] confirmed that Cu^{2+} is more effective than Cd^{2+} and Zn^{2+} in binding to HS dissolved in swamp waters. The relatively low value measured in all cases for maximum quenching suggests, however, that not all HS-fluorophores are involved in Cu^{2+} complexation. A similar result, obtained for a leaf leachate FA and a brownwater FA by Frimmel and Bauer,[122] suggests a low interaction between metal complexing sites and fluorescent units in these materials. Boussemart et al.[123] measured by fluorescence quenching and differential pulse anodic stripping voltammetry (DPASV) a 1:1 stoichiometry for the Cu^{2+} complexation by HS in interstitial water from estuarine marine sediments, together with the formation of large aggregates. A discrepancy was observed, however, between the complexing capacity measured by fluorescence quenching and DPASV. This is ascribed either to the presence of nonfluorescent complexing organic ligands or to exchange reactions of added Cu^{2+} with weakly bound fluorescence quenching metals.

The Lanthanide Ion Fluorescence Probe

Eu^{3+} has hypersensitive emission transition that is extremely sensitive to ligation and whose intensity is enhanced in metal ligand complexes. The degree of signal enhancement depends on the specific system studied. The large disparity in fluorescence lifetimes, that is of several milliseconds for Eu^{3+} vs. the nanoseconds observed in HS, allows the relatively weak Eu^{3+}-fluorescence to be temporarily resolved from the strong background fluorescence of the HS by means of a fairly simple, time-gated response.[124]

The emission spectrum of the Eu^{3+} ion features two peaks, a nonhypersensitive peak at 592 nm and a hypersensitive peak at 616 nm. The relative intensities of these two peaks vary in the presence of FA at different Eu^{3+} concentrations in relation to the overall extent of Eu^{3+} complexation. Dobbs et al.[125] obtained a sigmoidal shaped curve by plotting the ratio of the integrated spectral intensity of the nonhypersensitive transition to the hypersensitive transition, R, as a function of log C_M, the total concentration of the metal ion. Dobbs et al.[125] applied a continuous, multiligand pH-dependent ligand model[51] to describe the binding characteristics of Eu^{3+} with FA in terms of mean binding strengths, distributions, and concentrations.

Conclusions

Fluorescence spectroscopy has several advantages over most other methods for speciation studies of metal-HS complexation in aqueous media. The method is relatively rapid since no separation is required between bound and free metal ion. Errors associated with the separation step in most speciation methods are avoided. Neither supporting electrolyte nor buffer nor adsorbing material are required to be added to the samples. The method is even more sensitive than

DPASV and ISE potentiometry and is sensitive enough for application to unmodified, natural organic ligands without preconcentration. Unlike most other methods, fluorescence spectroscopy allows the direct measurement of the complexing capacity of the ligand through the determination of the concentration of free ligands, thus differentiating free and bound ligands. It represents, therefore, an excellent complement to other, indirect complexing capacity measurement techniques based on the determination of free metal ion concentration with respect to the bound metal fraction.

The possibility of simultaneous monitoring of light scattering, with the same instrumentation used for fluorescence measurement, provides a means to obtain information on the aggregation and metal ion colloids and precipitates that can occur under certain conditions. It also allows to distinguish these phenomena, that are very likely to occur near the end of the titrations, from true complexation in the solution phase.

The major disadvantage of fluorescence spectroscopy is that it is very effective only with strongly binding, paramagnetic metal ions such as Cu^{2+}. This limitation can be, however, overcome by the use of fluorescent probes, such as the lanthanide ion probe, that is conveniently applied in biochemical studies and shows particularly promising for studies of the metal binding sites in HS in environmental conditions and at natural concentrations of both metal and HS.

ELECTRON SPIN RESONANCE SPECTROSCOPY

Basic Principles and Methodology

The electron is a charged particle with angular momentum and, in consequence, it also possesses a magnetic moment μ_e that can be detected by its interaction with a magnetic field. In zero field, the magnetic moments of unpaired electrons in a sample have randomized directions, but in the presence of a magnetic field, H, the electron moments are aligned parallel or antiparallel to the field, giving rise to two discrete energy states. The interaction between the magnetic moment of the electrons and the applied magnetic field producing the splitting of the energy levels of unpaired electrons is referred to as the Zeeman effect, which is the basic physical phenomenon underlying electron spin resonance (ESR) spectroscopy, also called electron paramagnetic resonance (EPR) spectroscopy.

The energy of an unpaired electron in a magnetic field of strength H, applied parallel to the z-axis, is given by

$$E = -\mu_z H = g \beta H M_z \tag{15}$$

where $\mu_z = -g \beta M_z$ is the electron magnetic moment with respect to the conventional reference z-axis, g is the Landè or spectroscopic splitting factor (or magnetogyric ratio, i.e., the ratio of the magnetic moment to the angular momentum possessed by the electron), β is the Bohr magneton, and M_z is the component of the electron spin angular momentum in the direction of the z-axis of the applied magnetic field, H. M_z may assume discrete values of +1/2 and −1/2, in dependence of the alignment of the spin-magnetic moment either with the magnetic field direction (high energy) or against it (low energy). Thus, the spin states are not energetically equivalent and there are two energy levels with an energy difference that increases linearly with the intensity of H, given by

$$\Delta E = g \, \beta \, H \tag{16}$$

In a sample containing unpaired electrons in the thermodynamic equilibrium in a magnetic field of value H_0, there will be a population difference between the two energy levels given by the Boltzmann law:

$$\frac{N+}{N-} = \exp \frac{-g\beta H_o}{KT} \tag{17}$$

where T is the temperature, K the Boltzmann constant, and N_+ and N_- refer to the population in the upper ($M_z = +1/2$) and lower ($M_z = -1/2$) levels, respectively, leading to an excess population in the lower level. If an incident electromagnetic radiation is supplied to the sample, e.g., by applying an alternating magnetic field of frequency ν perpendicular to the static magnetic field H_0, absorption occurs provided that the energy of each incident quantum equals the difference in energy between the electron states, that is

$$h \, \nu = \Delta E = g \, \beta \, H_o \tag{18}$$

where h is the Planck's constant. This is known as the "resonance condition".

When Equation 18 is satisfied, electrons in the lower level absorb radiation energy and are excited to the upper level, immediately followed by emission of radiation quanta of the same frequency ν by electrons in the upper level that fall to the lower level. A net absorption of radiation occurs, however, only when the population of the lower level is maintained greater than that of the upper level. Provided the magnitude of the alternating field is not too large, this population difference (i.e., steady-state conditions) is maintained by a mechanism known as "spin-lattice relaxation", by which electrons that have been excited to the upper level can dissipate energy to their surroundings (the lattice) and return to the lower level. The measurement of this absorption of energy is the basis of ESR spectroscopy.

The ESR signal is highly dependent on the nature of the local environment about the absorbing electron. The effective magnetic field experienced by an electron (H_{eff}) is thus the result of two terms: the applied magnetic field generated by the spectrometer magnet (H_{appl}) and the local perturbations of the field produced by the electron's environment (H_{loc}), that is

$$H_{eff} = H_{appl} \pm H_{loc} \tag{19}$$

In definitive, H_{eff} determines the separation of the Zeeman energy levels, so that the resonant field strength H_o, that is the position of the ESR signal, and the overall ESR spectral pattern depend on the environment conditions in the vicinity of the electron. The most important types of interactions in the spin system that affects the position and pattern of the ESR spectrum are the "electron Zeeman", "nuclear hyperfine", and "ligand superhyperfine" interactions.

The "electron Zeeman" effect arises from interaction of unpaired electrons with external magnetic field and determines the position at which resonance occurs, i.e., the deviation of the g-factor from the free electron value (g = 2.00232). The main source of local magnetic fields affecting the g-value is orbital magnetic moment originating from spin coupling to excited electronic states (spin-orbit coupling). In the solid state, for most paramagnetic metal species the spin-orbit coupling is anisotropic, i.e., dependent on the orientation of the molecule relative to the external magnetic field, so that g results orientation dependent. Species with "axial symmetry" such as Cu^{2+} and V^{4+}, i.e., with one principal axis of symmetry, conventional the z-axis, and equivalent x- and y-axes, exhibit two g-values, usually labeled $g_{\parallel}$ (= g_{zz}, i.e., the g-value along the z or symmetry axis) and $g_{\perp}$ (= g_{xx} = g_{yy}, i.e., the g-value perpendicular to the z-axis in the x-y plane). "Rigid-limit" spectra with resolved $g_{\parallel}$ — and $g_{\perp}$ — values are systematically obtained for single crystals, whereas for powdered samples with randomly oriented paramagnetic particles, resonances are broadened and some loss of information may occur. Anisotropy of the g-tensor is often averaged for paramagnetic species in solution by rapid rotation of the metal ion and a single isotropic g-value is exhibited, g_{iso} or $g_o = 1/3(g_{xx} + g_{yy} + g_{zz})$. "Rigid-limit" spectra with resolved $g_{\parallel}$ — and $g_{\perp}$ — values can be, however, observed even in solution for paramagnetic species of high molecular weight that cannot undergo rotational motion rapidly enough to average the g-factor anisotropy.

A "nuclear hyperfine" interaction arises from the magnetic moments of the unpaired electron and its nucleus. This occurs only with magnetic nuclei, i.e., nuclei with nonzero spin ($I \neq 0$) such as Cu ($I = 3/2$), Mn ($I = 5/2$), and V ($I = 7/2$). Nuclear spin causes a splitting into $2I + 1$ components; four lines for Cu, six for Mn, and eight for V. The splitting of the hyperfine components is, in general, approximated by $A/g\beta$, where A is the magnitude of the nuclear hyperfine interaction, the so-called "hyperfine coupling constant". The parameter A, like g, exhibits an orientation dependence.

The "ligand superhyperfine" interaction can occur if the ligand atoms have nuclear spin, such as ^{14}N ($I = 1$). A number of components may result for each ligand nucleus, thus leading to very complex spectra, particularly if the ligands are not identical.

In order to determine accurately and rigorously the g-values and the hyperfine and superhyperfine coupling constants, A, the experimental ESR spectrum should be compared to a computer-simulated spectrum calculated using trial parameters. A convenient mathematical representation of the ESR spectrum involves the operator, "spin Hamiltonian". The mathematical description and application of spin Hamiltonians to obtain the ESR parameters of paramagnetic species and a discussion of the physicochemical methods used to relate the obtained ESR parameters to structural and chemical properties of the sample are outside the scope of this review and can be found elsewhere.[126–131]

In practice, the values of g-factors and hyperfine and superhyperfine coupling constants may be obtained relatively simply, though not rigorously, by direct computation from data accurately derived from the experimental ESR spectrum, and from spectrometer settings used in the measurement, according to the following standard equations:[129]

$$g = (h\,\nu)/(\beta\,H_o) = 0.714484\,\nu/H_o \tag{20}$$

and

$$A\ (\mathrm{cm}^{-1}) = (2.80247\,a\,g)/(c\,g_e) = 0.469766\ 10^{-4}\,g\,a \tag{21}$$

where ν (MHz) is the microwave frequency used, H_o is the value of the field at which the resonance is centered, $g_e = 2.00232$ is the g-value for the free electron, a is the hyperfine splitting computed on the experimental spectrum, expressed in gauss (1 gauss = 10^{-4} mT), and h, β, and c are known constants.

Metal-Humic Substance Complexes

Since ESR is a technique of absorption spectroscopy for detecting paramagnetism due to the magnetic moments of unpaired (odd) electrons, the method is applicable to transition metal ions in both the free state and organic complexed forms in which they maintain unpaired electron(s), in either solid or solution state.

The ESR technique has been extensively applied to elucidate the chemical and geometrical properties of naturally occurring and laboratory-prepared synthetic complexes formed by HA and FA of various origin and nature with paramagnetic transition metal ions of great chemical and biological importance to agriculture and environment, including Fe, Cu, Mn, V, and Mo.[132,133] ESR analysis of metal-HS associations can provide useful and, in some cases, unique information about binding mechanisms of metals to HS, oxidation states of metals bound, symmetry and type of coordination sites in HS, identity of ligand atoms and groups involved in metal complexing, and degree of mobility of HS-bound metals, i.e., stability of metal-HS complexes formed.

The method of paramagnetic metal "probe" addition has also been widely applied to the study of the "residual" binding capacity of natural HA and FA, in order to ascertain some molecular and quantitative aspects of the complexation. They include the nature of HS binding sites involved in the various experimental conditions and the stability of complexes formed towards competitive physical and chemical treatments, including proton and metal ion-exchange.[132,133]

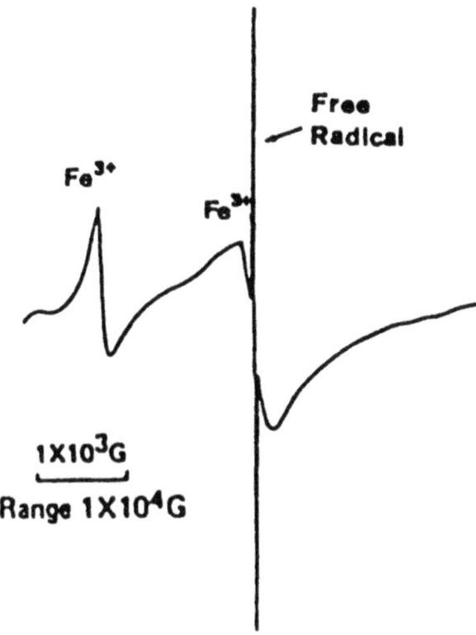

FIGURE 8. Electron spin resonance (ESR) spectrum (scan range, 10,000 G) recorded at RT of a natural soil fulvic acid. (Adapted from Senesi, N., S. M. Griffith, M. Schnitzer, and M. G. Townsend. *Geochim. Cosmochim. Acta* 41:969–976 [1977].)

The following is intended to provide the reader with the principal structural and chemical information that the direct analysis and comparative interpretation of ESR spectra and parameters derived therefrom can provide about paramagnetic metal-HS complexes.

Iron Complexes

ESR spectroscopy suggests Fe^{3+} is complexed in at least two nonequivalent oxygen sites in HA and FA of various nature and origin. A common feature of the ESR spectra of almost any type of HS is the asymmetric resonance line occurring at g-values ranging from 4.1 to 4.3, that is consistent with high-spin Fe^{3+} ions bound to carboxylic groups and, possibly, polyphenols of HA or FA in tetrahedral or octahedral sites in low symmetry (rhombic) ligand field[81,82,84,89,134–157] (Figure 8). Senesi et al.[137] showed that this form of iron exhibits considerable resistance to proton and metal exchange and to reduction, suggesting that Fe^{3+} is strongly bound and protected by HS in inner-sphere complexes.

Another generally observed feature in the ESR spectra of HS is the broad resonance signal centered near g = 2 that probably consists of an envelope of several resonances arising from extended spin-spin coupling interactions between various paramagnetic metal ions absorbing in this region. This resonance is derived prevalently from neighboring high-spin Fe^{3+} ions in octahedral sites with no or only small axial distortion from cubic symmetry[136,137,139,140,144,148,149]

(Figure 8). Senesi et al.[137] showed that Fe^{3+} in such sites is easily complexed and reduced by chemical agents, thus suggesting Fe^{3+} ions loosely held onto HS surfaces.

Weaker resonances have been observed to occur in some samples of natural HS and Fe^{3+}-added HS at g = 8.5 to 9.0 and at g = 5.8 to 6.0. The former resonance has been generally attributed to Fe^{3+} ions in sites with near-orthorhombic symmetry and the latter to high-spin Fe^{3+} ions in largely distorted sites in strong axially symmetric crystal fields.[84,136,140,146,147,150,151,153–156] Filip et al.[144,149] suggested that Fe^{3+} ions responsible for the resonance at g = 6.0 may be bound to four nitrogen atoms in a planar porphyrin configuration possibly deriving from a residual chlorophyll structure incorporated in the HS macromolecule, particularly suited to the geometric and chemical features of Fe^{3+} ion that replaces Mg^{2+} ion.

HS of any origin possess a high residual binding capacity toward Fe^{3+} ion that can form complexes stable against various physical and chemical treatments. Senesi et al.[137] found that the intensity of the resonance at g = 2 relative to that at g = 4.2, however, increases with increasing Fe^{3+} addition to a soil FA, indicating that most of the added Fe is bound to surface octahedral sites. Schnitzer and Ghosh[142] confirmed that, at pH 4 and 6, Fe^{3+} can form inner-sphere complexes mostly with carboxylic and phenolic OH groups of soil FA and that the Fe in all complexes occurs in the trivalent form. Senesi[84] obtained ESR spectra of aquatic FA with Fe^{3+} ions featuring a strong enhancement of the Fe^{3+} resonance at g = 4.2 associated with a weaker signal at g = 8.8. Senesi and collaborators[150,152–155] obtained similar results for FA and HA of various origin when treated simultaneously with Fe^{3+}, Cu^{2+}, and Mn^{2+} ions.

No direct ESR evidence has been obtained for the existence of Fe^{2+} species complexed by HA and FA. Goodman and Cheshire[148] observed an increasing intensity and decreasing width of the g = 2 feature with decrease in pH below 2 in Fe^{3+}-FA solutions at a ratio 1:50. This is consistent with Fe^{2+} ions not contributing directly to the resonance but involved in ferromagnetic exchange interactions with Fe^{3+} ions, that is, some of the Fe^{2+} may form magnetically exchanging ion clusters involving oxide or hydroxide bridges between Fe ions.

Copper Complexes

ESR analysis confirms the presence of inner-sphere complexes for Cu^{2+} in HS, also suggested by other spectroscopic and physicochemical studies. ESR spectra of Cu^{2+}-HA or -FA complexes feature a rigid-limit (anisotropic) pattern of the axial type consisting of a major unresolved absorption at higher field ($g_\perp$) associated with a lesser absorption at lower field ($g_\parallel$). Both $g_\perp$ and $g_\parallel$ can be partially resolved into a quadruplet (nuclear spin for Cu, I = 3/2) (Figure 9b). These results are consistent with a $d_{x^2-y^2}$ ground state for Cu^{2+} ions held in inner-sphere complexes in HS, with ligands arranged in a square planar (distorted octahedral) coordination around the central ion (tetragonal symmetry).[81,82,84,89,136,139,140,142,144,146,147,149,152–155,158–167]

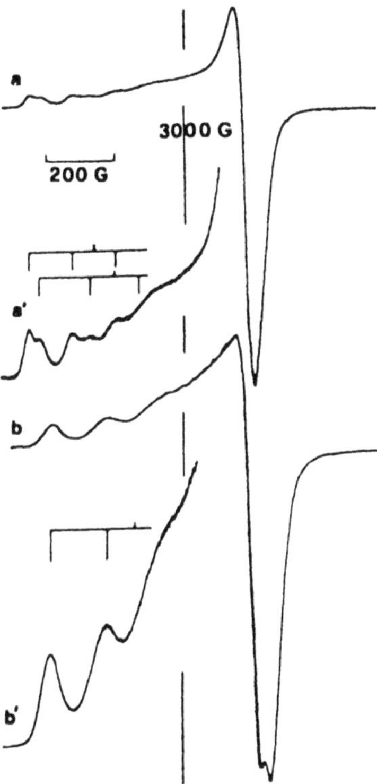

FIGURE 9. Electron spin resonance (ESR) spectra (scan range, 2,000 G) recorded at 77 K of laboratory-prepared Cu^{2+} complexes of a river aquatic fulvic acid (a; a′, higher gain) and humic acid (b; b′, higher gain). (From Senesi, N. in *Pollution in the Urban Environment, Polmet 88*, Hills, P., R. Keen, K. C. Lam, C. T. Leung, M. A. Oswell, M. Stokes, and E. Turner, Eds. (Hong Kong: Vincent Blue Copy, 1988) pp. 607–612. With permission.)

Information regarding the nature of the binding and the kind of ligand atoms coordinating to the Cu^{2+} centers can be derived from the interpretation of ESR parameters, especially $g_{\parallel}$ and $A_{\parallel}$ values, calculated from the experimental spectra (Equations 20 and 21). The ESR parameters exhibited by some representative Cu^{2+}-HS complexes are listed in Table 3. Data obtained suggest a generally high covalent bond contribution consistent with either fully oxygenated binding sites and/or sites involving one or more nitrogen ligand atoms. Carboxyls, phenolic hydroxyls, carbonyls, and eventually water molecules are generally indicated as the main oxygenated functional groups participating in the Cu^{2+} binding by HS.[81,84,89,139,142,146,147,150,152–155,161–167]

ESR studies of Bresnahan et al.[161] confirmed the existence of two classes of oxygenated binding sites of different affinities for Cu^{2+} ion in soil and aquatic FA, as suggested by titration experiments. At low metal-to-FA ratio, ESR data suggest that sites of high binding strength coordinate to Cu^{2+}, with four (or three) oxygen atoms of carboxylate, phenolate, or carbonyl groups. At large Cu-to-FA

Table 3. ESR Spectral Parameters for Some Representative Naturally Occurring and Laboratory-Prepared Cu^{2+} Complexes with HA and FA of Different Origin and Nature

Sample	Measurement Condition	$g_{\parallel}$	$\lvert A_{\parallel}\rvert$ ($10^{-4}cm^{-1}$)	$g_{\perp}$	Ligand Atoms	Ref.
FA, aquatic (river)	Lab. prep. solut. 77 K	2.407[a]	141	2.081	4 O[b]	84
		2.355	157		4 O[b]	
FA, soil	Natural, solut. 77 K	2.344[a]	158	2.074	4 O	165
		2.291	180		3 O, 1 N	
FA, sludge	Lab. prep. solut. 77 K	2.300[a]	169/184[c]		4 O	167
		2.275[a]	184/196[c]	2.055[d]	2 O, 2 N	
		2.249	197/212[c]	2.038[e]	1 O, 3 N	
HA, soil	Natural, powder 77 K	2.269	180	2.070	2 O, 2 N	89
HA, soil fungal	Natural, powder 77 K	2.242[a]	190	2.058	1 O, 3 N	147
		2.283	164		3 O, 1 N	
		2.226	196		1 O, 3 N	
HA, peat	Lab. prep. powder RT	2.171	187	2.047	4 N	158

[a] Main resonance.
[b] Water oxygen involved.
[c] Two values for $\lvert A_{\parallel}\rvert$ are referred, respectively, to two hyperfine quadruplets of closely spaced peaks deriving from hyperfine coupling of the unpaired electron to the two copper isotope nuclei ^{63}Cu and ^{65}Cu.
[d,e] The $g_{\perp}$ component is resolved into two groups of 5 (d) and 7 (e) lines deriving from superhyperfine coupling of the copper unpaired electron to two or three equivalent N ligand nuclei, respectively.

ratios, however, numerous weaker sites predominate and probably only two FA donor atoms are bonded to each Cu^{2+} ion.

ESR patterns displaying two or more partially overlapping components in the $g_{\parallel}$ Cu^{2+}-feature indicate that more than one type of binding sites in HS may be involved in Cu^{2+} complexation at high Cu^{2+} loading.[132,133] For example, in aquatic FA, two different fully oxygenated sites, the one major, the other minor, are simultaneously involved in the binding of added Cu^{2+} ions[84] (Figure 9a, Table 3).

Nitrogen-containing sites can be important in Cu^{2+}-complexes of some soil, peat, sludge and compost HA and FA[81,146,155,160,167] and particularly in HA-type polymers synthesized by soil fungi,[147,164] due to proteinaceous components that are rich in these materials. Evidence for participation of nitrogen in the binding of Cu^{2+} by nitrogen-rich HS has been found in the $g_{\perp}$ region of the Cu^{2+} ESR spectrum.[144,147,149,165,167] Senesi et al.[167] tentatively resolved the ESR pattern at $g_{\perp}$ of a Cu^{2+}-sludge FA complex (Figure 10) into two groups of five lines each, attributed to a superhyperfine coupling of the unpaired copper electron to two different sets of two equivalent nitrogen ligand nuclei. ESR evidence of Cu^{2+}-tetraporphyrine complexes highly resistant to acid washing in HA isolated from peat, mineral soils, and salt marsh lake mud is provided by the very structured pattern observed at $g_{\perp}$.[136,140,149,159] This implies that biologically stable porphyrine structures, originating from plant chlorophyll and incorporated into HS, may be extensively involved in copper fixation by HS in peat and soil.

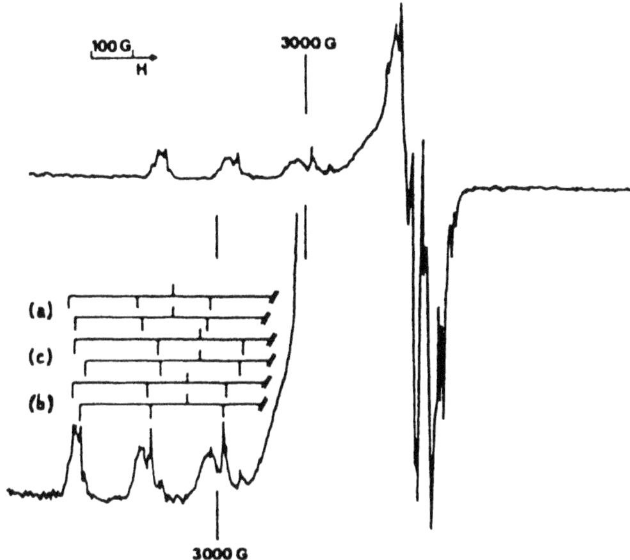

FIGURE 10. Electron spin resonance (ESR) spectrum (scan range, 1,000 G) recorded at 77 K of a Cu^{2+}-sludge fulvic acid complex in aqueous solution, laboratory-prepared at Cu:FA = 8. (From Senesi, N., D. F. Bocian, and G. Sposito. *Soil Sci. Soc. Am. J.* 49:119–126 [1985]. With permission.)

According to McBride,[157] low $g_{\parallel}$ values measured for Cu^{2+}-HA complexes at low levels of Cu^{2+} loading support formation of bonding of high covalent character involving amino groups. In contrast, high g-values measured for Cu^{2+} bonded to HS at high Cu^{2+} loading would indicate a preferential coordination to O-containing ligands in sites with low covalency, where the metal ion exhibits a high degree of mobility. McBride[157] also illustrated the correlation between the rigid-limit $g_{\parallel}$ ESR parameter and the nature of the coordination site around Cu^{2+} ion for model compounds with varying N- and O-ligands and natural HS. Care must be exercized, however, in this kind of interpretation since some O-containing ligands, such as polyphenolic compounds, may generate small values of $g_{\parallel}$ and large hyperfine constants, $A_{\parallel}$, by covalent bonding with Cu^{2+}.

Large ESR evidence has been obtained of the capacity of Cu^{2+} ions to replace Mn^{2+}, VO^{2+}, and Fe^{3+} from FA and HA, when these materials are treated with excess Cu^{2+} ions, and to form complexes of various stability toward protons or competing cations.[81,82,84,137,150,153,154,166–170]

Manganese Complexes

Naturally occurring and laboratory-prepared complexes of Mn^{2+} (nuclear spin, I = 5/2) with aquatic, terrestrial, and sedimentary HA and FA exhibit ESR spectra featuring a six-line isotropic pattern (Figure 11) characterized by g_{iso} and A_{iso} values typical of hexahydrated, high-spin Mn^{2+} ion. This is consistent with Mn^{2+} bound by electrostatic forces in outer-sphere complexes in highly sym-

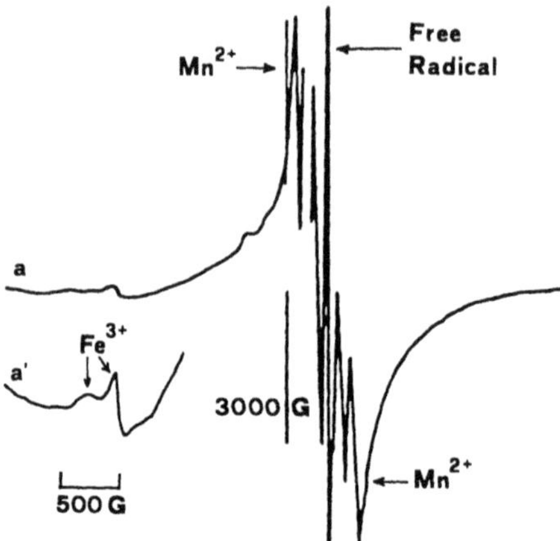

FIGURE 11. Electron spin resonance (ESR) spectrum (scan range, 8,000 G) recorded at 77 K of a laboratory-prepared Mn^{2+} complex of a paleosol humic acid, acid-leached after metal saturation. (Adapted from Senesi, N. and G. Calderoni. *Org. Geochem.* 13:1145–1152 [1988]. With permission.)

metrical, undistorted octahedral sites, probably involving carboxylate and/or phenolate groups.[134,136,139,140,149,150,152–154,168,171,172] Manganese complexed in this form would result easily available to plant roots and microorganisms.

On the contrary, Lakatos et al.[160] measured ESR parameters consistent with inner-sphere complexes of Mn^{2+} ions in octahedral coordination with six oxygen-ligands of carboxyl, phenolic hydroxyl, and/or carbonyl groups, for some laboratory-prepared Mn^{2+}-peat HA complexes. Carpenter[173] showed, however, that in marine and lacustrine pore waters only 8 to 13% of the Mn^{2+} is present as inner-sphere organic or carbonate complexes. The relatively low portion of Mn^{2+} organically complexed in these systems is ascribed to the low concentration of dissolved organic matter and the competition for organic ligands by other abundant cations such as Ca^{2+} and Mg^{2+}.

McBride[174] found that most Mn^{2+} is adsorbed by a soil FA at low pH in outer-sphere, but at pH > 8, or T > 50°C, Mn^{2+} may enter inner-sphere multiligand complexation sites. These results indicate that the type and stability of Mn^{2+}-HS complexes and, in turn, its ease of exchangeability and bioavailability in natural systems are strongly dependent on pH and temperature.

Senesi and collaborators[150,152–154,168] have shown that HS isolated from various sources exhibit a high residual complexing capacity for added Mn^{2+}. Mn^{2+} is bound in water-stable forms, but, unlike Fe^{3+} and Cu^{2+}, it may be completely displaced by protons or strongly complexed metal ions.

Gamble et al.[172] determined by ESR spectroscopy stability constants of water soluble Mn^{2+}-FA complexes on the basis of the linear functionality existing between the height of Mn^{2+}-hyperfine peaks and the concentration of free Mn^{2+}. The increasing relaxation line broadening, observed in the Mn^{2+}-ESR spectrum

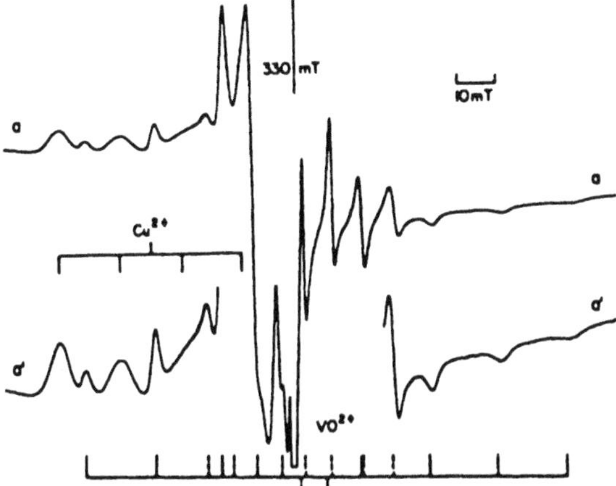

FIGURE 12. Electron spin resonance (ESR) spectrum (scan range, 2000 G) recorded at 77 K of a natural soil humic acid showing VO^{2+} complexes (a′ refers to a higher gain spectrum). (From Senesi, N., G. Sposito, K. M. Holtzclaw, and G. R. Bradford. *J. Environ. Qual.* 18:186–194 [1989]. With permission.)

with increasing addition of complexing ligand, can be directly related to increasing complexation of Mn^{2+}. K_c values measured by ESR are in excellent agreement with K_c values determined by an ion-exchange method. The ESR method is faster, more sensitive, and more convenient than the ion-exchange procedure.

Vanadium Complexes

The formation of complexes of vanadyl (VO^{2+}) ions occurs in both natural and vanadium-enriched HA and FA of various origin. A quite-well resolved anisotropic ESR spectrum of the axial type, consisting of two partially overlapping hyperfine multiplets of eight lines each, corresponding to the parallel and perpendicular components of a VO^{2+} ion (nuclear spin of V, $I = 7/2$) (Figure 12), has been observed.[89,134,136,139,140,146,147,152,153,160,175–178] The ESR parameters of this spectrum are consistent with the vanadyl group held in inner-sphere complex in an equatorial plane by either four oxygen donor atoms, possibly including water molecules, or, more rarely, by both oxygen and nitrogen ligands.

ESR spectroscopy provided evidence of the ability of HS to reduce the metavanadate ion, VO_3^-, the most stable form of vanadium under typical soil conditions, to VO^{2+} in acidic media.[136,160,177,178] The VO^{2+} ion can be complexed, partially or predominantly depending on the pH of the medium, in forms that are protected from oxidation even at very high pH values. Templeton and Chasteen[176] have suggested that catechol groups may be responsible for the observed reduction of VO_3^- to VO^{2+} by FA.

Templeton and Chasteen[176] have studied by ESR the dynamics of motion, molecular conformation, aggregation properties, stability constants, and stoichiometry in aqueous solution of VO^{2+} complexes with two gel-filtrated FA fractions. The approximate molecular weights of the two FA fractions, calculated by rotational correlation times obtained from ESR data, agree well with estimates obtained by gel permeation chromatography. ESR results showed that, in the presence of VO^{2+}, the higher molecular weight fraction forms a complex approximated as $(VO)_2(FA\text{-}I)_6$, whereas the lower molecular weight fraction forms a simpler complex, VO^{2+}-FA-II. Both FA fractions involve similar binding sites of low symmetry consisting of four oxygen ligands bound in the first coordination sphere (inner-sphere) VO^{2+} ions. Comparative evaluation of ESR parameters suggest that the ligand fields existing about the metal can be modeled by the complexes bis(phtalato)(salicylato)-oxovanadium (IV) and mono(salicylato)-oxovanadium (IV), respectively.

Templeton and Chasteen[176] also proposed a classification into three groups of binding sites for VO^{2+} ions in HS of different nature and source. ESR parameters of VO^{2+} ion complexes in a number of soil and peat HA and FA fractions[136,175,178] suggest relatively strong ligand fields and high covalency, consistent with phenolate or possibly nitrogen donor groups. This renders VO^{2+} ions difficult to remove from these sites, even with acid leaching. Templeton and Chasteen[176] suggest that catechols represent a model ligand that can form extremely stable complexes with VO^{2+}. In contrast, naturally occurring and laboratory-prepared VO^{2+} complexes of HA and FA of different origin[89,139,146,147,152,153,176] are characterized by weak ligand fields and low covalency, probably involving carboxylate groups and, perhaps, water molecules. Thus, VO^{2+} ions in this type of complexes are relatively labile and exchangeable.[81,152–154] Finally, some laboratory-prepared VO^{2+}-peat HA complexes[160] exhibit values of ESR parameters indicating the presence of binding sites having properties intermediate between those previously discussed, probably arising from a combination of phenolate and carboxylate ligands.

Molybdenum Complexes

ESR studies suggest that HA can reduce molybdate to Mo(V) and complex Mo(V) species in strongly acidic media.[160,179] Goodman and Cheshire[179] obtained an ESR spectrum of a peat HA complex with Mo(V) enriched in ^{95}Mo (nuclear spin, $I = 5/2$) that features two distinct components, each split into two six-line hyperfine patterns at $g_{\parallel}$ and $g_{\perp}$, consistent with two different axially symmetric Mo(V)-HA complexes.

The sample obtained after 0.1 *M* HCl treatment of the ^{95}Mo(V)-enriched-HA gave a low-intensity six-line ESR spectrum that probably arises from a Mo(III) species.[179] Although the Mo(III) accounts for only a very small proportion of the Mo originally added to the HA, its formation and detection by ESR is highly significant because there are little definite evidence of its formation in natural systems. These findings also indicate that Mo(III) species can be formed and remain stable in the solid state even in aerobic conditions when protected in HS complexes.

Conclusions

Major advantages of ESR spectroscopy are its high sensitivity and the ability to measure spectra directly with minimal or no pretreatment. ESR evidence suggests that small quantities of metal ions can bond selectively in inner-sphere complexes at the most preferred sites for the metal, whereas in the presence of high amounts of metal added, the high degree of site occupation generally results in a loss of relative selectivity. The ESR data also show that high pH values, generating a greater availability of negatively charged O-ligands, favors inner-sphere complexation for metals that are retained as hydrated ions at lower pH. Inner-sphere coordination is also preferred when competing water ligands are removed by dehydration, thus forcing the metal to enter into direct bonding with HS ligands.

Most ESR evidence indicates that Cu^{2+} and VO^{2+}, together with Fe^{3+}, tend to form inner-sphere complexes with HS at all experimental conditions used, whereas Mn^{2+}-HS complexes are condition dependent. ESR data confirm that the more electronegative the metal ion, the stronger the metal bound to HS, the higher the degree of bond covalency. The ESR method also allows, in principle, the determination of the free ion concentration and, therefore, of the degree of complexation of a paramagnetic metal ion.

The intrinsic limitation to the ESR technique is that it is applicable only to paramagnetic metal ions that give a detectable ESR signal. No data are available in the literature on the detection by ESR of paramagnetic metal ions other than Fe, Cu, Mn, V, and Mo complexed in HS. Lakatos et al.[160] obtained an unresolved ESR spectrum for a peat HA doped with either paramagnetic Cr^{3+} ions or diamagnetic dichromate, which is apparently reduced to Cr^{3+} by the HA. Lakatos et al.[160] failed to observe ESR spectrum for Co^{2+} and Ni^{2+} ions doped in the same peat HA. It should be noted, however, that the absence of a detectable ESR signal in a sample does not prove the absence of paramagnetic species, since relaxation and other effects can prevent its observation by ESR.

The major limitation of the ESR experiment is the inability to resolve signal component lines that may overlap to such an extent that information is lost. The width of the resonance line is determined by two mechanisms, homogeneous and inhomogeneous broadening. Homogeneous line-width broadening arises from "microwave-power saturation" effects that produce broad spectra for some transition metal ions, such as Fe^{3+} and Ni^{2+}, particularly at room temperature. The choice of power is, thus, critical as paramagnetic species with long relaxation times become "saturated" at high power, with a reduction in intensity or loss of signal, as previously noted (section titled "Basic Principles and Methodology"). Inhomogeneous broadening arises from nonuniformities of the magnetic field throughout the sample, caused by neighboring paramagnetic species, or magnetic moments of neighboring nuclei, or dipolar interactions between unlike species. Any of these effects may result in merging of the individual resonant lines or spin packets into a single overall line or envelope, with a loss of information.

Temperature is also a critical parameter in the ESR experiment, since sensitivity for paramagnetic species increases with lowering sample temperature, according to Curie's law. ESR measurements are often made at either liquid nitrogen (77 K) or liquid helium (4.2 K) temperature, that may reduce some type of line broadening.

The technique of ESR has been underutilized for the study of metal-HS complexation chemistry in soil and aquatic systems. Further extension of ESR investigations to these systems will provide new information on metal speciation in solution and on metals at liquid-solid interfaces, as well as on mechanisms of HS-induced metal reduction in natural systems. ESR can be used to monitor metal redox reaction kinetics and to correlate ESR data and thermodynamic data of paramagnetic metal-HS complexation. Finally, modeling chemical bonding in binding sites using suitable paramagnetic metal ion probes and/or model ligand compounds will add to the knowledge of the nature of complexation reactions of nonparamagnetic metal ions of environmental and/or agricultural importance with natural humic substances.

MÖSSBAUER SPECTROSCOPY

Basic Principles and Methodology

Mössbauer spectroscopy measures the resonant absorption of nuclear gamma rays involved with transitions between the ground and excited state of atomic nuclei with nonzero angular momenta. The precise energy of such transitions is influenced by the chemical environment of the nuclei and any external magnetic field and electric field gradients. Mössbauer technique thus provides a method of observing energy changes of the order of those produced by interactions of the nuclear states with the surrounding electronic charge distribution. The Mössbauer effect is highly isotope-specific since it refers to transitions of very narrow energy widths between ground and excited states of the nucleus. Approximately 30 isotopes are Mössbauer-active, including iron, nickel, zinc, and mercury, but the most easily studied metal is iron.[180] All investigations so far reported on metal-HS complexes have been with ^{57}Fe.

In the Mössbauer experiment with iron, the gamma source consists of ^{57}Co, which decays by electron capture to ^{57}Fe, with the emission of gamma rays that are then absorbed by ^{57}Fe nuclei in the sample. The source (emitter) is moved alternatively toward and away from the sample (absorber) by a velocity transducer. A velocity range of ± 10 mm s^{-1} is suitable. The first-order relativistic Doppler shift is used to produce small changes in the gamma ray energy emitted by the source and adsorbed by the sample. The change in energy, E, due to a relative velocity v of the source relative to absorber (or vice versa) is given by:

$$\triangle E = \nu E\, \gamma / c \qquad (22)$$

where E_γ is the energy of the radiation and c is the velocity of light. An iron foil is generally used as a calibration standard absorber. The energy transmitted by the absorber is detected by a synchronized counter of high effectiveness and long life. Spectral data are accumulated automatically in a multichannel analyzer and computer-fitted to a series of Lorentzian or Gaussian lines. Widths and areas of doublet components are constrained to be equal and areas of sextets are held to a 3:2:1:1:2:3: ratio. Mössbauer data are referred as percent absorption (or transmission) as a function of source velocity (±) in mm s^{-1}. Spectra are generally obtained on solid samples over a temperature range 4 to 300 K.

The principal energy-dependent parameters that can be obtained from a Mössbauer spectra are the isomer shift, the quadrupole coupling constant, the magnetic field, and the peak width.

The magnitude of the isomer shift, δ, is proportional to the difference in electron density at the Fe nucleus in the sample and in reference, usually metallic iron. A nonzero isomer shift is observed as a shift in center of gravity of the spectrum. Factors influencing the isomer shift include localization of electron density in 4s orbitals and the total population of 3d orbitals, which can modify the s-electron density at the nucleus by a shielding effect.

The quadrupole splitting, or magnetic hyperfine interaction, Δ, is determined by the interaction between the quadrupole moment of the excited state and the electric field gradient at the iron nucleus that mostly derives by a combination of charges originating from the electronic environment (valence electrons) of the iron and from surrounding atoms. The nonzero electric field gradient occurring in ^{57}Fe nucleus, having spins of ground and excited states 1/2 and 3/2, respectively, results in a partial removal of degeneracy in the nuclear excited state that gives rise to a doublet in the Mössbauer spectrum.

The magnetic hyperfine field, H_{eff}, is proportional to any magnetic field experienced by the nucleus. The interaction between the nuclear magnetic dipole moment and any magnetic field at the nucleus causes the complete removal of degeneracy of nuclear energy levels, producing six peaks in the spectrum with intensity ratios 3:2:1:1:2:3, if the magnetic domains are randomly oriented. In a paramagnetic material, magnetic hyperfine structure may also be observed if the rate of relaxation between the electronic spin states is slow compared to the timescale of the Mössbauer transition (about 10^{-8} sec).

Finally, the value of the peak width can provide useful information on the possible presence of unresolved components in the Mössbauer spectrum.

In the high-spin state, the ferric ion has a $3d^5$ configuration (s = 5/2), each d orbital containing a single electron, that results in a spherical and symmetrical electronic charge distribution in either octahedral or tetrahedral coordination for Fe^{3+}. This results in a zero contribution from valence electrons to the quadrupole splitting and an electric field gradient can arise, for example, where of nonidentical ligands distort the cubic symmetry. For the high-spin ferrous ion, the additional electron present in the d orbitals (S = 2) makes the electronic contributions to electric field gradients generally larger than lattice contributions. Therefore, high-spin Fe^{2+} ion in an octahedral or tetrahedral arrangement in an ionic complex experiences an asymmetrical electron distribution due to the distortion of the first coordination sphere of ligands, that may determine a high value of quadrupole splitting. In conclusion, in the high-spin state, ferrous ion usually has a much larger value of Δ than does the ferric ion, except in some very distorted environments. The presence of an additional electron in the ferrous ion (d^6 configuration) also results in an increased screening of the s-electrons, that is, a decreased d-electron density at the nucleus, producing a marked increase in the isomer shift, δ, with respect to the ferric ion (d^5 configuration). In paramagnetic materials, the magnitude of H_{eff} is proportional to the unpaired electron density at the nucleus, whereas in magnetically ordered (ferromagnetic) materials it may be decreased by exchange interactions.

In the low spin state ($S = 1/2$ and $S = 0$, respectively for Fe^{3+} and Fe^{2+}), trends in Mössbauer parameters are much more difficult to be rationalized because of the high degree of bond covalency. Little evidence exists, however, of the occurrence in nature of low-spin iron forms, except where coordinated to S-containing groups and in some nitrogenated heterocycles.

Iron-Humic Substance Complexes

Ferric Iron Complexes

The Mössbauer spectra recorded at either room temperature (RT) or 77 K of iron complexes with HA or FA prepared by reaction of HS with the ^{57}Fe isotope at various Fe:HS ratios and at pH values 4 to 6, generally consist of a paramagnetic doublet exhibiting isomer shift and quadrupole splitting values typical of high-spin Fe^{3+} ion complexes.[135,137,142,143,145,148,181–185] Hansen and Mosbaek[181] and Lakatos et al.[182] analyzed comparatively the spectra of Fe^{3+}-FA complexes and those of a variety of structurally similar Fe^{3+} chelates. Data obtained suggest that the relatively large value of the quadrupole splitting exhibited by the Mössbauer doublet measured for the Fe^{3+}-FA complex is inconsistent with mononuclear iron species, but more likely with a trinuclear complex featuring an oxygen atom at the center and three high-spin Fe^{3+} ions at the vertices of an equilateral triangle. Each Fe^{3+} is also bound to a carboxylate and a phenolate, or two carboxylate groups. Babanin et al.[135] gave an alternative explanation for a similar spectrum of HA, that is, Fe^{3+} are bound to HS in the high-spin state in several, similar, mononuclear, six-ligand coordination positions, regardless of the nature of the sample.

Senesi et al.[137] used computer-fitting procedures to analyze the Mössbauer spectrum obtained at RT on a natural soil HA containing 0.57% Fe. The spectrum was fitted to three doublets exhibiting Mössbauer parameters suggesting, together with ESR data, the presence of three sites for Fe^{3+} ions in the HA, two with octahedral and one with tetrahedral coordination (Figure 13a). Goodman and Cheshire[183] computer-separated the Mössbauer spectrum of a $^{57}Fe^{3+}$-peat HA

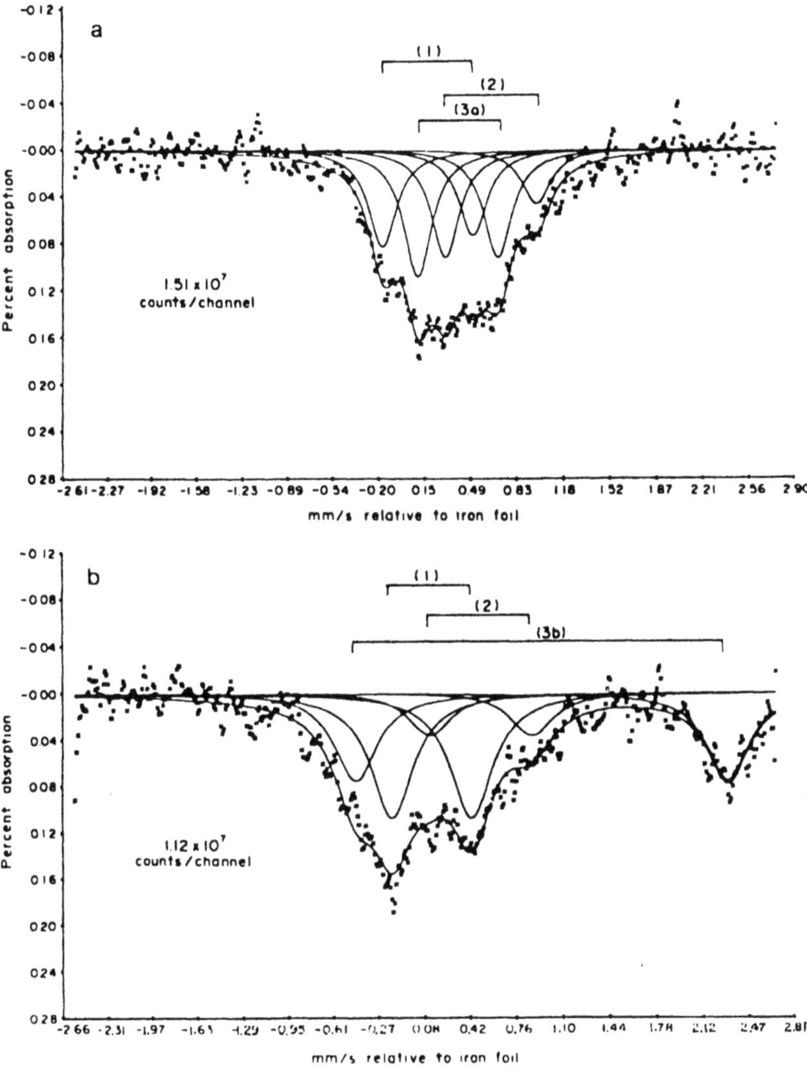

FIGURE 13. Mössbauer spectra of a natural soil humic acid (HA) (a) and the same HA after reduction with hydrazine (b). Doublets (1), (2), (3) refer to three different sites for iron in HA. Doublets (3a) and (3b) refer, respectively, to Fe^{3+} and Fe^{2+} ions in similar octahedral sites. (From Senesi, N., S. M. Griffith, M. Schnitzer, and M. G. Townsend. *Geochim. Cosmochim. Acta* 41:969–976 [1977]. With permisssion.)

slurry at pH values greater than 3 into two doublets consistent with Fe^{3+} ions in more than one type of environment, and an additional six-peak component typical of a magnetically-ordered species that becomes dominant at pH 4 to 6. The peak positions and relative intensities of the latter component are very close to those obtained from microcrystalline β-FeOOH. In a later study on a $^{57}Fe^{3+}$-soil FA at pH values from 5 to 1, Goodman and Cheshire[184] found that only small amounts of a magnetically ordered phase is found at pH 5, where virtually all the iron is in the ferric form.

Cressey et al.[143] showed that the Mössbauer spectra recorded at 82 K on naturally occurring or laboratory-prepared Fe^{3+}-FA complexes at pH 4.0 can be computer fitted alternatively to one and two Fe^{3+} doublets exhibiting Mössbauer parameters similar to those obtained for Fe^{3+}-HA complexes, and one additional Fe^{2+} doublet, which is discussed later in this section. Kallianou and Yassoglou[145] computer modeled Mössbauer spectra of some natural HA with (1) a strong Fe^{3+} doublet, suggesting one type of environment for Fe^{3+} in HA complexes; (2) a subordinate magnetically-split sextet, attributed to finely sized, coextracted inorganic iron forms; and (3) a weak Fe^{2+} doublet.

McBride et al.[185] observed that the doublet measured at temperatures above 20 K on soil extracted Fe^{3+}-HA complexes gives hyperfine splitting consistent with magnetically ordered iron when the temperature is lowered to 4.2 K. This suggests most Fe^{3+} is in the form of oxide and hydroxide cores surrounded by large organic molecules.

Goodman and Cheshire[148] showed, however, that in soil FA solutions with added $^{57}Fe^{3+}$ at Fe:FA ratios 1:5 to 1:500 and at pH values 5.0 to 0.5, the hyperfine splitting, H_{eff}, is too large to correspond to any known oxide or oxihydroxide that can be present in the system. The presence of the hyperfine structure over the pH range examined excludes the possibility that it is due to solvated Fe^{3+} ions in solution and indicates that these complexes are mononuclear, with Fe^{3+} ions probably 6-coordinated and bound to oxygen atoms.

Kodama et al.[186] showed that the Mössbauer spectrum of a soil FA (Fe^{3+}, 0.1 to 0.5%) recorded at 4.2 K can be fitted to a broad asymmetric doublet, plus a sextet that may be attributed in part to Fe^{3+} bound to the FA. A similar Mössbauer spectrum is obtained at 78 K for the $^{57}Fe^{3+}$-doped-FA, whereas at 4.2 K the spectrum shows two distinct magnetic sextets. This suggests that the added Fe occurs partly as Fe-FA complexes and partly as inorganic species.

Ferrous Iron Complexes

Mössbauer studies suggest that Fe^{3+} bound to HA and FA may not be stable. At pH 4 and 5 some natural and $^{57}Fe^{3+}$-added soil HA and FA show, as previously noted in this section, an additional weak Mössbauer peak consistent with high-spin Fe^{2+} ions.[143,145,148] Goodman and Cheshire[184] found that this peak may account for as much as 20% of the iron in $^{57}Fe^{3+}$-FA solutions at pH 4. This supports the concept that iron in soil may be transported as FA complexes of ferrous ions formed by reduction of ferric iron.

Senesi et al.[137] showed that chemical reduction of a naturally-occurring soil Fe^{3+}-HA with hydrazine results in a Mössbauer spectrum in which one of the two doublets assigned to Fe^{3+} ions in octahedral sites disappears (3a in Figure 12a) and a new quadrupole doublet appears featuring much higher values for δ and Δ parameters, typical of Fe^{2+} ions in the same type of sites (3b in Figure 12b).

In an extended Mössbauer analysis of several Fe^{2+}-HA complexes, Lakatos et al.[182] showed that in samples dried at 130°C, the Fe^{2+} ions are held in elongated octahedral arrangement in high spin, ionic, inner-sphere chelate-type complexes with carboxylate and phenolic or alcoholic groups and nitrogen-containing groups. Values of quadrupole splitting for Fe^{2+}-HA complexes in the dehydrated state lower than those measured in aqueous suspension indicates that the Fe^{2+} ions are at least partially hydrated. Analysis of the 4-doublet fit of the Mössbauer spectrum obtained at pH 1 for a $^{57}Fe^{3+}$-HA slurry by Goodman and Cheshire[183] indicates the presence of two components (amounting to about one half of the iron) with quadrupole splitting values consistent with Fe^{2+} ions in two types of environment: a completely hydrated Fe^{2+} (outer-sphere complexes with HA) and Fe^{2+} partly associated with oxygen ligands in HA. Kallianou and Yassoglou[145] arrived to similar conclusions for naturally occurring Fe^{2+}-HA complexes and synthetic Fe^{2+}-HA and -FA complexes prepared at pH 2.8 and 4.0. These results demonstrate that Fe^{2+} is only weakly bound in HA.

Goodman and Cheshire[148,183] also showed that when the pH of $^{57}Fe^{3+}$-FA or -HA slurry is decreased below 2, the Mössbauer spectrum changes dramatically and exhibits an asymmetrical doublet with a large quadrupole splitting consistent with a high-spin ferrous ion, which tends to dissolve. By raising the pH above 3, Fe^{2+} can be reoxidized to Fe^{3+} that precipitates as inorganic β-Fe-OOH, thus demonstrating the reversibility of the process. Previously, Lakatos et al.[182] obtained Mössbauer evidence of the partial oxidation of Fe^{2+} to Fe^{3+} on exposure to air of some Fe^{2+}-HA complexes.

Goodman and Cheshire[148] found that Fe^{2+} ions in $^{57}Fe^{3+}$-FA systems decrease as the Fe:FA ratio decreases from 1:20 to 1:200, suggesting that reduction and complexation reactions are competitive, with the latter being favored but limited to a very small fraction of active functional groups (complexing sites) in the FA.

Conclusions

The Mössbauer effect has been used successfully for studying iron-complexes with HS, commonly in conjunction with other spectroscopic techniques, particularly ESR spectroscopy. The intrinsic limitation of the technique is its high specificity for the ^{57}Fe isotope. With ^{57}Fe, about 1 µmol of the isotope is required to permit good accuracy with several hours of acquisition of data. Although the low natural abundance of ^{57}Fe (2.19%), it is possible to obtain data with unenriched, natural Fe-HS complexes. The analysis of isomer shift and quadrupole splitting data taken together permits evaluation of the role of covalency and back-binding in iron complexes. Caution must be exercised, however, in interpretation. Model compounds must be investigated for comparison, as there often are several binding sites that may give rise to similar spectra.

Much Mössbauer and ESR evidence suggest that both Fe^{3+} and Fe^{2+} ions occur mostly in high-spin forms in combination with HS. Ferric iron exhibits little tendency to be reduced by HS at the most common natural pH values (from 3 up to 6) and environmental concentration. Mössbauer spectroscopy is able to distinguish unambiguously between high spin Fe^{3+} and Fe^{2+} ions on the basis of major differences observed in both the isomer shift and quadrupole splitting. Further, values of experimental Mössbauer parameters may provide information on the type of coordination, symmetry, and chemical nature of the groups bound to the high-spin iron forms.

NUCLEAR MAGNETIC RESONANCE SPECTROSCOPY OF CATION SOLVATION WATER

Principle

The binding of water in the inner-sphere solvation shell of paramagnetic ions has the effect of shortening relaxation times of proton-NMR spectra. When nuclei (protons) of coordinated water molecules exchange rapidly with free molecules of the solvent, the water NMR signal expresses an average between the two environments with an observed value of the relaxation time, $T_{1,2(obs)}$— for either longitudinal (T_1) or transverse (T_2) relaxation in the proton NMR signal that is governed by the McConnell's equation for fast exchange:[187]

$$1/T_{1,2(obs)} = P_A/T_{1,2A} + P_M/T_{1,2M} \tag{23}$$

where P_A and P_M are the probabilities that a proton may be found in the bulk solvent environment (A) or in the solvation shell of the paramagnetic ion (M). In a dilute solution, P_A is equal to unity to a good approximation, whereas P_M may be expressed in terms of C_M, the concentration of paramagnetic ions, and q, the number of water molecules coordinated to the paramagnetic ion (six in the case of metal ions such as Mn^{2+}, Cu^{2+}, and Fe^{3+}). Thus the following relation of proportionality is held:[187]

$$(1/T_{1,2(obs)} - 1/T_{1,2A}) \propto qC_M/T_{1,2M} \tag{24}$$

The coordination of the metal ion to a nonsolvent ligand, such as FA, lowers q and may simultaneously alter the value of $T_{1,2M}$. The factors that control relaxation parameters, T_{1M} or T_{2M}, are determined by fluctuating magnetic fields that can influence the magnetic state of the nucleus under study. These factors are described by τ_R, the rotational correlation time of the molecule containing the nucleus, τ_S, the paramagnetic electron relaxation correlation time, and τ_M, the chemical exchange correlation time. While the latter parameter has no control in the case of fast exchange, as in Equation 23, the effect exerted by τ_S and/or τ_R on observed T_1 or T_2 depends upon the fact that both τ_S and τ_R are short times compared to the reciprocal of the NMR frequency. Observed relaxation processes are, therefore, governed by the most rapid correlation time.

As a result of the effects described above, coordination of a paramagnetic ion to a nonsolvent ligand may lead to either a decrease or an increase of observed relaxation rates ($1/T_{1,2}$). A decrease may be observed as a consequence of two possible mechanisms: (1) reduction of q due to replacement of coordinated water by the ligand; and/or (2) increase of electronic relaxation times ($T_{1,2M}$). The latter effect is usually associated with removal of symmetry occurring in high spin ions, such as Mn^{2+} and Fe^{3+}, in highly symmetrical environments. Alternatively, an increase in relaxation rates may be observed because of a reduction of $T_{1,2M}$, that occurs in situations where τ_R is most important, i.e where rotation of a metal ion is slowed down by complexation to a large molecule such as FA.

Since the experimental value of $1/T_{2(obs)}$ is proportional to the width at half height, $\Delta v_{1/2}$, of the Lorentzian NMR absorption line for water, addition of an organic ligand to a water solution of a paramagnetic metal ion may produce either line narrowing, when one of the two mechanisms described in (1) and (2) occurs, or enhanced line broadening, when the alternate situation described previously occurs. As long as magnetic dipolar coupling dominates the interaction between the paramagnetic electron and the nucleus under study, T_1 equals T_2, whereas T_1 is different from T_2 only in the fast exchange where τ_S is the important factor and "scalar" coupling is important between the paramagnetic electron and the nucleus.

Manganese, Copper, and Iron Complexes with Humic Substances

The effect of coordination to an organic ligand on NMR line broadening of water has been used successfully to explore the type of binding of Mn^{2+}, Cu^{2+}, and Fe^{3+} ions to HS.[187,188]

Gamble et al.[188] interpreted the slight but definite decrease in line widths of the water NMR-signal observed upon addition of FA to a Mn^{2+} water solution, as an indication that either a small amount of water is displaced from the coordination sphere of Mn^{2+}, and/or a slight but significant distortion from octahedrical symmetry occurs on association of the hexahydrated Mn^{2+} to FA, thus changing electron relaxation times. Comparison made with NMR data of Mn^{2+} complexes with known simple ligands supports outer-sphere binding of hexahydrated Mn^{2+} ion onto FA sites.

In a later more detailed study, Deczky and Langford[187] found that the relaxation mechanism in the presence of FA is only slightly different from that in the absence of FA. Further, FA has a nearly zero effect on T_1, that is controlled by τ_R, thus suggesting τ_S is more sensitive than τ_R to Mn^{2+} complexing by the large FA molecule. These results are consistent with a fully hydrated Mn^{2+} ion rotating, nearly as freely as in aqueous solution, on electrostatic binding sites of the FA ligand (outer-sphere Mn^{2+}-FA complex). The picture suggests a "caged" hexahydrated Mn^{2+} ion surrounded by a not fully symmetric environment, thus altering τ_S, and consequentely T_2, moderately.

NMR data by Gamble et al.[188] suggest a different situation for Fe^{3+} ion complexing by FA. Line width of the water NMR signal is greatly reduced upon addition of FA to Fe^{3+} solution, i.e., the remaining paramagnetic broadening is small. This effect, later confirmed by the large decrease measured for both relaxation rates, $1/T_1$ and $1/T_2$, is too large to be accounted for by changes in the q parameter, i.e., displacement of water from hexahydrated Fe^{3+} ions. The effect is more likely attributed to a change in τ_S, that is, to a substantial alteration of electron relaxation time by the FA ligands, consequent on reduction of symmetry of the high spin d^5-center, implying inner-sphere coordination of Fe^{3+}.

Deczky et al.[187] showed that the mechanism of proton relaxation does not change significantly on addition of FA to a Cu^{2+} solution. Longitudinal relaxation rate ($1/T_1$) dependence on the Cu^{2+} ion concentration increases markedly in the presence of FA, when compared to the Cu^{2+}-aquo ion and to the small Cu^{2+}-bipyridine complex. This is consistent with τ_R remaining important for an inner-sphere complex of Cu^{2+} where rotation requires tumbling of the entire large FA entity.

Conclusions

The NMR technique appears to be underutilized so far in the study of metal-HS complexation. Pommery et al.[189] applied ^{113}Cd-NMR spectroscopy to study complexing of Cd^{2+} ions to HA, providing evidence of two kinds of Cd^{2+}-HA complexes involving either 1 or 10 Cd ions per HA molecule. The potential applicability of NMR water probes and of metal nuclei-NMR deserves a higher attention for investigations of metal-HS complexation chemistry in natural systems.

CONCLUDING REMARKS AND RECOMMENDATIONS

Spectroscopic methods represent a valuable means for the study of metal ion interaction and binding to natural humic substances. In general, spectroscopic techniques can furnish valuable information on the metal-HS complexes at the environmental level as it concerns either molecular and mechanistic aspects, that is chemical and physical nature of sites involved and type of binding, and formation mechanisms and stability of metal-HS complexes. Spectroscopic techniques, particularly fluorescence quenching spectroscopy, appear also promising for the study of quantitative aspects of the metal-humic substance interaction, such as ligand complexing capacity and equilibrium constants of complexation or both.

When used alone, none of the spectroscopic methods, however, is able to describe satisfactorily the nature of complex species formed in metal-HS complexation. When solution of a complicated chemical problem such as this is attempted, the adoption of a variety of experimental approachs is strongly recommended. The simultaneous utilization of UV-Vis, IR, fluorescence, and ESR spectroscopies, in combination with chemical derivatization, along with fluorescence and/or ESR probes, and, where possible, with the support of Mössbauer and NMR spectroscopies, represents the most advisable rationale in approaching the complicated chemical problems encountered in investigating metal-humic substance interactions in natural environments.

REFERENCES

1. Stevenson, F. J. *Cycles of Soil. Carbon, Nitrogen, Phosphorus, Sulfur, Micronutrients* (New York: Wiley-Interscience, 1986).
2. Buffle, J. *Complexation Reactions in Aquatic Systems: An Analytical Approach* (Chichester: Ellis Horwood, 1988).
3. Thurman, E. M. *Organic Geochemistry of Natural Waters* (Dordrecht: Nijhoff, 1986).
4. Schnitzer, M. "Binding of Humic Substances by Soil Mineral Colloids," in *Interactions of Soil Minerals with Natural Organic Microbes*, Huang, P. M. and M. Schnitzer, Eds. (Madison, WI: Soil Science Society of America, 1986) pp. 77–101.
5. Schnitzer, M. "Humic Substances: Chemistry and Reactions," in *Soil Organic Matter*, Schnitzer, M. and S. U. Khan, Eds. (Amsterdam: Elsevier, 1978) pp. 1–64.
6. Stevenson, F. J. *Humus Chemistry. Genesis, Composition, Reactions* (New York: Wiley-Interscience, 1982).
7. Saar, R. A. and J. H. Weber. "Fulvic Acid: Modifier of Metal-Ion Chemistry," *Environ. Sci. Technol.* 16:510A–517A (1982).
8. Bloom, P. R., M. B. McBride, and R. M. Weaver. "Aluminum Organic Matter in Acid Soils: Buffering and Solution Aluminum Activity," *Soil Sci. Soc. Am. J.* 43:488–493 (1979a).
9. Bloom, P. R., M. B. McBride, and R. M. Weaver. "Aluminum Organic Matter in Acid Soils: Salt Extractable Aluminum," *Soil Sci. Soc. Am. J.* 43:813–815 (1979b).
10. Truitt, R. E. and J. H. Weber. "Influence of Fulvic Acid on the Removal of Trace Concentrations of Cadmium(II), Copper(II), and Zinc(II) from Water by Alum Coagulation," *Water Res.* 13:1171–1177 (1979).
11. Pearson, R. G. "Hard and Soft Acids and Bases," *J. Am. Chem. Soc.* 85:3533–3539 (1963).
12. Chaberek, S. and A. E. Martell. *Organic Sequestering Agents* (New York: John Wiley & Sons, 1959).
13. Schnitzer, M. and S. U. Khan. *Humic Substances in the Environment* (New York: Dekker, 1972).
14. Schnitzer, M. and S. I. M. Skinner. "Organo-Metallic Interactions in Soils. 5. Stability Constants of Cu^{2+}, Fe^{2+}, and Zn^{2+}-Fulvic Acid," *Soil Sci.* 102:361–365 (1966).
15. Schnitzer, M. and S. I. M. Skinner. "Organo-Metallic Interactions in Soils. 7. Stability Constants of Pb^{2+}, Ni^{2+}, Mn^{2+}, Co^{2+}, Ca^{2+}, and Mg^{2+} Fulvic Acid Complexes," *Soil Sci.* 103:247–252 (1967).
16. Stevenson, F. J. and M. S. Ardakani. "Organic Matter Reactions Involving Micronutrients in Soils," in *Micronutrients in Agriculture*, Mortvedt, J. J., et al., Eds. (Madison, WI: Soil Sci. Soc. Am., 1972) pp. 79–114.
17. Mantoura, R. F. C., A. Dickson, and J. P. Riley. "The Complexation of Metals with Humic Materials in Natural Waters," *Estuarine Coastal Mar. Sci.* 6:387–408 (1978).
18. Davies, R. I., M. V. Cheshire, and I. J. Graham-Bryce. "Retention of Low Level of Copper by Humic Acids," *J. Soil Sci.* 20:65–71 (1969).

19. Thurman, E. M. and R. L. Malcolm. "Preparative Isolation of Aquatic Humic Substances," *Environ. Sci. Technol.* 15:463–466 (1981).
20. Mantoura, R. F. C. and J. P. Riley. "The Analytical Concentration of Humic Substances from Natural Waters," *Anal. Chim. Acta* 76:97–106 (1975).
21. Thurman, E. M. "Isolation of Soil and Aquatic Humic Substances," Group Report, in *Humic Substances and Their Role in the Environment*, Frimmel, F. H. and R. F. Christman, Eds. (Chichester: Wiley-Interscience, 1988) pp. 31–43.
22. Hayes, M. H. B. "Extraction of Humic Substances from Soil," in *Humic Substances in Soil, Sediment and Water*, Aiken, G. R., D. M. McKnight, R. L Wershaw, and P. MacCarthy, Eds. (New York: Wiley-Interscience, 1985) pp. 329–362.
23. Aiken, G. R. "Isolation and Concentration Techniques for Aquatic Humic Substances," in *Humic Substances in Soil, Sediment and Water*, Aiken, G. R., D. M. McKnight, R. L Wershaw, and P. MacCarthy, Eds. (New York: Wiley-Interscience, 1985) pp. 363–385.
24. Swift, R. S. "Fractionation of Soil Humic Substances," in *Humic Substances in Soil, Sediment and Water*, Aiken, G. R., D. M. McKnight, R. L Wershaw, and P. MacCarthy, Eds. (New York: Wiley-Interscience, 1985) pp. 387–408.
25. Leenheer, J. A., "Fractionation Techniques for Aquatic Humic Substances," in *Humic Substances in Soil, Sediment and Water*, Aiken, G. R., D. M. McKnight, R. L Wershaw, and P. MacCarthy, Eds. (New York: Wiley-Interscience, 1985) pp. 409–429.
26. Aiken, G. R., D. M. McKnight, R. L. Malcolm, and P. MacCarthy. "An Introduction to Humic Substances in Soil, Sediment, and Water," in *Humic Substances in Soil, Sediment and Water*, Aiken, G. R., D. M. McKnight, R. L Wershaw, and P. MacCarthy, Eds. (New York: Wiley-Interscience, 1985) pp. 1–9.
27. Schnitzer, M. "Recent Findings on the Characterization of Humic Substances Extracted from Soils from Widely Differing Climatic Zones," in *Soil Organic Matter Studies, Vol. 2* (Vienna: I. A. E. A., 1977) pp. 117–132.
28. Stevenson, F. J. and J. H. A. Butler. "Chemistry of Humic Acids and Related Pigments," in *Organic Geochemistry*, Eglinton, G. and M. T. J. Murphy, Eds. (New York: Springer-Verlag, 1969) pp. 534–557.
29. Stimberg, C. and U. Munster. "Geochemistry and Ecological Role of Humic Substances in Lakewater," in *Humic Substances in Soil, Sediment and Water. Geochemistry, Isolation, and Characterization*, Aiken, G. R., D. M. McKnight, R. L. Wershaw, and P. MacCarthy, Eds. (New York: Wiley-Interscience, 1985) pp. 105–145.
30. Stuermer, D. H. and J. R. Payne. "Investigation of Seawater and Terrestrial Humic Substances with Carbon-13 and Proton Nuclear Magnetic Resonance," *Geochim. Cosmochim. Acta* 40:1109–1114 (1976).
31. Ishiwatari, R. "Geochemistry of Humic Substances in Lake Sediments," in *Humic Substances in Soil, Sediment and Water. Geochemistry, Isolation, and Characterization*, Aiken, G. R., D. M. McKnight, R. L. Wershaw, and P. MacCarthy, Eds. (New York: Wiley-Interscience, 1985) pp. 147–180.
32. Rashid, M. A. and L. H. King, "Major Oxygen-Containing Functional Groups Present in Humic and Fulvic Acid Fractions Isolated from Contrasting Marine Environments," *Geochim. Cosmochim. Acta* 34:193–201 (1970).
33. Kleinhempel, D. "Ein Beitrag zur Theorie des Huminstoffezustandes," *Albrecht Thaer Arch.* 14:3–10 (1970).

34. Langford, C. H., D. S. Gamble, A. W. Underdown, and S. Lee. "Interaction of Metal Ions with a Well Characterized Fulvic Acid," in *Aquatic and Terrestrial Humic Materials*, Christman, R. F. and E. T. Gjessing, Eds. (Ann Arbor, MI: Ann Arbor Science, 1983) pp. 219–237.
35. Gagosian, R. B. and Stuermer, D. H. "The Cycling of Biogenic Compounds and Their Biogenetically Transformed Products in Seawater," *Mar. Chem.* 5:605–632 (1977).
36. Leenheer, J. A., D. M. McKnight, E. M. Thurman, and P. MacCarthy. "Structural Components and Proposed Structural Models of Fulvic Acid from the Suwannee River," in *Humic Substances in the Suwannee River, Georgia: Interactions, Properties, and Proposed Structures*, Averett, R. C., J. A. Leenheer, D. M. McKnight, and K. A. Thorn, Eds. (Denver: U.S. Geological Survey, Open-File Report 87–557, 1989) pp. 335–359.
37. Degens, E. T. and K. Mopper. "Factors Controlling the Distribution and Early Diagenesis of Organic Materials in Marine Sediments," in *Chemical Oceanography, Vol. 6*, Riley, J. P. and R. Chester, Eds. (London: Academic Press, 1976) pp. 60–114.
38. Buffle, J. "Natural Organic Matter and Metal-Organic Interactions in Aquatic Systems," in *Metal Ions in Biological Systems, Vol. 18, Circulation of Metals in the Environment*, Sigel, H., Ed. (New York: M. Dekker, 1984) pp. 165–221.
39. Mantoura, R. F. C. "Organo-Metallic Interactions in Natural Waters," in *Marine Organic Chemistry*, Duursma, E. K. and R. Dawson, Eds. (Amsterdam: Elsevier/North Holland, 1981) pp. 179–223.
40. Stevenson, F. J. and A. Fitch. "Chemistry of Complexation of Metal Ions with Soil Solution Organics," in *Interactions of Soil Minerals with Natural Organic Microbes*, Huang, P. M. and M. Schnitzer, Eds. (Madison, WI: Soil Science Society of America, 1986) pp. 29–58.
41. Schnitzer, M. "Reaction Between Fulvic Acid, a Soil Humic Compound, and Inorganic Soil Constituents," *Soil Sci. Soc. Am. Proc.* 33:75–81 (1969).
42. Gamble, D. S., M. Schnitzer, and I. Hoffman. "Cu^{2+}-Fulvic Acid Chelation Equilibrium in 0. 1 M KCl at 25. 0 °C," *Can. J. Chem.* 48:3197–3204 (1970).
43. Stevenson, F. J. "Binding of Metal Ions by Humic Acids," in *Environmental Biogeochemistry, Vol. 2, Metals Transfer and Ecological Mass Balances*, Nriagu, J. O., Ed. (Ann Arbor, MI: Ann Arbor Science, 1976) pp. 519–540.
44. Chen, Y. and F. J. Stevenson. "Soil Organic Matter Interactions with Trace Elements," in the *Role of Organic Matter in Modern Agriculture*, Chen, Y. and Y. Avnimelech, Eds. (Dordrecht: Nijhoff, 1986) pp. 73–116.
45. Randhawa, N. S. and F. E. Broadbent. "Soil Organic Matter-Metal Complexes: 5. Reactions of Zinc with Model Compounds and Humic Acid," *Soil Sci.* 99:295–300 (1965).
46. Van Dijk, H. "Colloidal Chemical Properties of Humic Matter," in *Soil Biochemistry, Vol. 2*, McLaren, A. D. and J. Skujins, Eds. (New York: Dekker, 1971) pp. 16–35.
47. Marinsky, J. A., A. Wolf, and K. Bunzl. "The Binding of Trace Amount of Lead(II), Copper(II), Cadmium(II), Zinc(II), and Calcium(II) to Soil Organic Matter," *Talanta* 27:461–468 (1980).
48. Marinsky, J. A., S. Gupta, and P. W. Schindler. "The Interaction of the Cu(II) Ion with Humic Acid," *J. Colloid Interface Sci.* 89:401–411 (1982).

49. Tuschall, J. R. and P. L. Brezonik. "Complexation of Heavy Metals by Aquatic Humus: A Comparative Study of Five Analytical Methods," in *Aquatic and Terrestrial Humic Materials*, Christman, R. F. and E. T. Gjessing, Eds. (Ann Arbor, MI: Ann Arbor Science, 1983) pp. 275–294.
50. Weber, J. H. "Metal Ion Speciation Studies in the Presence of Humic Materials," in *Aquatic and Terrestrial Humic Materials*, Christman, R. F. and E. T. Gjessing, Eds. (Ann Arbor, MI: Ann Arbor Science, 1983) pp. 315–331.
51. Perdue, E. M. and C. R. Lytle "A Critical Examination of Metal-Ligand Complexation Models: Application to Defined Multiligand Mixtures," in *Aquatic and Terrestrial Humic Materials*, Christman, R. F. and E. T. Gjessing, Eds. (Ann Arbor, MI: Ann Arbor Science, 1983) pp. 295–313.
52. Sposito, G. "Sorption of Trace Metals by Humic Materials in Soils and Natural Waters," *CRC Crit. Rev. Environ. Control* 16:193–229 (1986).
53. Stumm, W., R. Schwarzenbach, and L. Sigg. "From Environmental Analytical Chemistry to Ecotoxicology — A Plea for More Concepts and Less Monitoring and Testing," *Angew. Chem. Int. Ed. Engl.* 22:380–389 (1983).
54. McCoustra, M. R. S. "Electronic Absorption Spectroscopy: Theory and Practice," in *Perspective in Modern Chemical Spectroscopy*, Andrews, D. L., Ed. (Berlin: Springer-Verlag, 1990) pp. 88–101.
55. Scott, A. I. *Interpretation of Ultraviolet Spectra of Natural Products* (New York: Pergamon Press, 1964).
56. Campanella, L., V. Crescenzi, M. Dentini, C. Fabiani, F. Mazzei, A. I. Nero Scheffino. "Polyelectrolytic Metal Ions Sequestrants," in *Metal Speciation, Separation, Recovery, Vol. II*, Patterson, J. W. and R. Passino, Eds. (Chelsea, MI: Lewis Publishers, Inc., 1990) pp. 359–375.
57. Vosburg, W. C. and G. R. Cooper. "Complex Ions. 1. The Identification of Complex Ions in Solution by Spectrophotometric Measurements," *J. Am. Chem. Soc.* 63:437–442 (1941).
58. Broadbent, F. E. and J. B. Ott. "Soil Organic Matter-Metal Complexes. I. Factors Affecting Retention of Various Cations," *Soil Sci.* 83:419–427 (1957).
59. Pavel, L. "On the Knowledge of Humic Substances. 4. Watersoluble Metal-Chelates Formed by the Dark-Coloured Fulvic Acid Fraction," *Sb. Cesk. Akad. Zemed. Ved. Rostl. Vyroba* 32:639–650 (1959).
60. Schnitzer, M. and S. I. M. Skinner. "Organo-Metallic Interactions in Soils. I. Reactions Between a Number of Metal Ions and the Organic Matter of a Podzol Bh Horizon," *Soil Sci.* 96:86–93 (1963).
61. Ram, N. and K. V. Raman. "Characterization of Metal-Humic and -Fulvic Acid Complexes," *Pedologie* 33:137–145 (1983).
62. Wershaw, R. L., D. M. McKnight, and D. J. Pinckney. "The Speciation of Copper in the Natural Water System. I. Evidence of the Presence of a Copper(II)-Fulvic Acid Charge Transfer Complex," in *Proceedings 2nd Int. Symp. Peat in Agriculture and Horticulture*, Schallinger, K. M., Ed. (Rehovot-Bet Dagan: the Hebrew University of Jerusalem, Faculty of Agriculture, 1983) pp. 205–222.
63. Bellamy, L. J. The *Infrared Spectra of Complex Molecules* (New York: John Wiley & Sons, 1975).
64. Nakamoto, K. *Infrared and Raman Spectra of Inorganic and Coordination Compounds* (New York: Wiley-Interscience, 1986).
65. Willis, H. A., J. H. Van Der Maas, and R. G. J. Miller. *Laboratory Methods in Vibrational Spectroscopy* (Chichester: John Wiley & Sons, 1988).

66. MacCarthy, P. and J. A. Rice. "Spectroscopic Methods (Other Than NMR) for Determining Functionalities in Humic Substances," in *Humic Substances in Soil, Sediment and Water. Geochemistry, Isolation, and Characterization*, Aiken, G. R., D. M. McKnight, R. L. Wershaw, and P. MacCarthy, Eds. (New York: Wiley-Interscience, 1985) pp. 527–559.
67. Orlov, D. S. *Humus Acids of Soils* (New Delhi: Oxonian Press, 1985).
68. Bloom, P. R. and J. A. Lenheer. "Vibrational, Electronic, and High-Energy Spectroscopic Methods for Characterizing Humic Substances," In *Humic Substances II. In Search of Structure*, Hayes, M. H. B., P. MacCarthy, R. L. Malcolm, and R. Swift, Eds. (Chichester: John Wiley & Sons, 1989) pp. 409–446.
69. Schnitzer, M. and I. Hoffman. "Thermogravimetric Analysis of the Salts and Metal Complexes of a Soil Fulvic Acid," *Geochim. Cosmochim. Acta* 31:7–15 (1967).
70. Juste, C. "Modifications de la Mobilité Électrophorétique et du Spectre Infrarouge d'un Acide Humique Artificiellement Enrichi en Fer ou en Aluminium," *C.R. Acad. Sci.* 262:2692–2695 (1966).
71. Juste, C. and J. Delas. "Influence de l'Addition d'Aluminium, de Fer, de Calcium, de Magnésium ou de Cuivre sur la Mobilité Électrophorétique, le Spectre d'Absorption Infrarouge et la Solubilité d'un Composé Humique," *Ann. Agron.* 18:403–427 (1967).
72. Juste, C. and J. Delas. "Etude de Quelques Propriétés des Complexes Formés par les Acides Humiques et les Cations," *Bull. Assoc. Fr. Etude Sol.* 4:39–49 (1970).
73. Juo, A. S. R. and S. A. Barber. "Reaction of Strontium with Humic Acid," *Soil Sci* 108:89–94 (1969).
74. Dupuis, T., P. Jambu, and J. Dupuis. "Sur les Formes de Liaisons Entre le Fer et les Acides Fulviques de Sols Hydromorphes," *C.R. Acad. Sci.* 270:2264–2267 (1970).
75. Rashid, M. A. "Role of Humic Acids of Marine Origin and Their Different Molecular Weight Fractions in Complexing Di- and Tri-Valent Metals," *Soil Sci.* 111:298–306 (1971).
76. Tan, K. H., L. D. King, and H. D. Morris. "Complex Reactions of Zinc with Organic Matter Extracted from Sewage Sludge," *Soil Sci. Soc. Am. Proc.* 35:748–752 (1971).
77. Banerjee, S. K. and S. K. Mukherjee. "Studies of the Infrared Spectra of Some Divalent Transitional Metal Humates," *J. Indian Soc. Soil Sci.* 20:91–94 (1972).
78. Tan, K. H. "Formation of Metal-Humic Acid Complexes by Titration and Their Characterization by Differential Thermal Analysis and Infrared Spectroscopy," *Soil Biol. Biochem.* 10:123–129 (1978).
79. Piccolo, A. and F. J. Stevenson. "Infrared Spectra of Cu^{2+}, Pb^{2+}, and Ca^{2+} Complexes of Soil Humic Substances," *Geoderma* 27:195–208 (1982).
80. Prasad, B. and M. K. Sinha. "Physical and Chemical Characterization of Molecularly Homogeneous Fulvic Acid Fractions and Their Metal Complexes," *J. Indian Soc. Soil Sci.* 31:187–191 (1983).
81. Senesi, N., G. Sposito, and J. P. Martin. "Copper(II) and Iron(III) Complexation by Soil Humic Acids: An IR and ESR Study," *Sci. Total Environ.* 55:351–362 (1986).
82. Senesi, N., G. Sposito, and J. P. Martin. "Copper(II) and Iron(III) Complexation by Humic Acid-Like Polymers (Melanins) from Soil Fungi," *Sci. Total Environ.* 62:241–252 (1987).

83. Prasad, B., G. D. Dkhar, and A. P. Singh. "Cobalt(II), Iron(III) and Zinc(II) Complexation by Fulvic Acids Isolated from North-Eastern Himalayan Forest and Cultivated Soils," *J. Indian Soc. Soil Sci.* 35:194–197 (1987).
84. Senesi, N. "Role of Aquatic Fulvic and Humic Acids in Copper and Iron Speciation in River Waters," in *Pollution in the Urban Environment, Polmet 88*, Hills, P., R. Keen, K. C. Lam, C. T. Leung, M. A. Oswell, M. Stokes, and E. Turner, Eds. (Hong Kong: Vincent Blue Copy, 1988) pp. 607–612.
85. MacCarthy, P. and H. B. Mark, Jr. "Infrared Studies on Humic Acid in Deuterium Oxide. I. Evaluation and Potentialities of the Technique," *Soil Sci. Soc. Am. Proc.* 39:663–668 (1975).
86. MacCarthy, P., H. B. Mark, Jr., and P. R. Griffiths. "Direct Measurement of the Infrared Spectra of Humic Substances in Water by Fourier Transform Infrared Spectroscopy," *J. Agric. Food Chem.* 23:600–602 (1975).
87. Griffith, S. M. and M. Schnitzer. "The Isolation and Characterization of Stable Metal-Organic Complexes from Tropical Volcanic Soils," *Soil Sci.* 120:126–131 (1975).
88. Chen, Y., N. Senesi, and M. Schnitzer. "The Chemical Degradation of Humic and Fulvic Acid Extracted from Mediterranean Soils," *J. Soil Sci.* 29:350–359 (1978).
89. Senesi, N., G. Sposito, K. M. Holtzclaw, and G. R. Bradford. "Chemical Properties of Metal-Humic Acid Fractions of a Sewage Sludge-Amended Aridisol," *J. Environ. Qual.* 18:186–194 (1989).
90. Vinkler, P., B. Lakatos, and J. Meisel. "Infrared Spectroscopic Investigations of Humic Substances and Their Metal Complexes," *Geoderma* 15:231–242 (1976).
91. Baes, A. U. and P. R. Bloom. "Diffuse Reflectance and Transmission Fourier Transform Infrared (DRIFT) Spectroscopy of Humic and Fulvic Acids," *Soil Sci. Soc. Am. J.* 53:695–700 (1989).
92. Boyd, S. A., L. E. Sommers, and D. W. Nelson. "Copper(II) and Iron(III) Complexation by the Carboxylate Group of Humic Acid," *Soil Sci. Soc. Am. J.* 45:1241–1242 (1981).
93. Schnitzer, M., and S. I. M. Skinner. "Organo-Metallic Interactions in Soils. 4. Carboxyl and Hydroxyl Groups in Organic Matter and Metal Retention," *Soil Sci.* 99:278–284 (1965).
94. Higashi, T. "Characterization of Al/Fe-Humus Complexes in Dystrandepts Through Comparison with Synthetic Forms," *Geoderma* 31:277–288 (1983).
95. Stevenson, F. J. and K. M. Goh. "Infrared Spectra of Humic Acids and Related Substances," *Geochim. Cosmochim. Acta* 35:471–483 (1971).
96. Schnitzer, M. and H. Kodama. "Reactions Between Fulvic Acid and Cu^{2+}-Montmorillonite," *Clays Clay Miner.* 20:359–367 (1972).
97. Boyd, S. A., L. E. Sommers, and D. W. Nelson. "Infrared Spectra of Sewage Sludge Fractions: Evidence for an Amide Metal Binding Site," *Soil Sci. Soc. Am. J.* 43:893–899 (1979).
98. Sposito, G., K. M. Holtzclaw, and J. Baham. "Analytical Properties of the Soluble, Metal-Complexing Fractions in Sludge-Soil Mixtures. II. Comparative Structural Chemistry of Fulvic Acid," *Soil Sci. Soc. Am. J.* 40:691–698 (1976).
99. Baham, J., N. B. Ball, and G. Sposito. "Gel Filtration Studies of Trace Metal-Fulvic Acid Solutions Extracted from Sewage Sludge," *J. Environ. Qual.* 7:181–188 (1978).
100. Creaser, C. S., and J. R. Sodeau. "Luminescence Spectroscopy," in *Perspective in Modern Vibrational Spectroscopy*, Andrews, D. L., Ed. (Berlin: Springer-Verlag, 1990) pp. 103–136.

101. Senesi, N. Molecular and Quantitative Aspects of the Chemistry of Fulvic Acid and its Interactions with Metal Ions and Organic Chemicals. Part II. The Fluorescence Spectroscopy Approach," *Anal. Chim. Acta* 232:77–106 (1990).
102. Seitz, W. R. "Fluorescence Methods for Studying Speciation of Pollutants in Water," *Trends Anal. Chem.* 1(16):79–83 (1981).
103. Weber, J. H. "Binding and Transport of Metals by Humic Materials," in *Humic Substances and Their Role in the Environment*, Frimmel, F. H. and R. F. Christman, Eds. (Chichester: Wiley-Interscience, 1988) pp. 165–178.
104. Saar, R. A. and J. H. Weber. "Comparison of Spectrofluorometry and Ion-Selective Electrode Potentiometry for Determination of Complexes Between Fulvic Acid and Heavy-Metal Ions," *Anal. Chem.* 52:2095–2100 (1980).
105. Blaser, P. and G. Sposito. "Spectrofluorometric Investigation of Trace Metal Complexation by an Aqueous Chestnut Leaf Litter Extract," *Soil Sci. Soc. Am. J.* 51:612–619 (1987).
106. Banerjee, S. K. and S. K. Mukherjee. "Physico-Chemical Studies of the Complexes of Divalent Transitional Metal Ions with Humic and Fulvic Acids of Assam Soil," *J. Indian Soc. Soil Sci.* 20:13–18 (1972).
107. Almgren, T., B. Josefson, and G. Nyquist. "A Fluorescence Method for Studies of Spent Sulfite Liquor and Humic Substances in Sea Water," *Anal. Chim. Acta* 78:411–422 (1975).
108. Adhikari, M. and G. G. Hazra. "Humus-Metal Complex: Spectra Studies," *J. Indian Chem. Soc.* 53:513–516 (1976).
109. Ghosh, K. and M. Schnitzer. "Fluorescence Excitation Spectra and Viscosity Behavior of a Fulvic Acid and its Copper and Iron Complexes," *Soil Sci. Soc. Am. J.* 45:25–29 (1981).
110. Boto, K. and P. Isdale. "Fluorescent Bands in Massive Corals Result from Terrestrial Fulvic Acid Inputs to Nearshore Zone," *Nature (London)* 315:396–397 (1985).
111. Ryan, D. K. and J. H. Weber. "Fluorescence Quenching Titration for Determination of Complexing Capacities and Stability Constants of Fulvic Acid," *Anal. Chem.* 54:986–990 (1982a).
112. Ryan, D. K., C. P. Thompson, and J. H. Weber. "Comparison of Mn^{2+}, Co^{2+} and Cu^{2+} Binding to Fulvic Acid as Measured by Fluorescence Quenching," *Can. J. Chem.* 6:1505–1509 (1983).
113. Olson, B. M. "The Use of Fluorescence Spectroscopy to Study Herbicide-Humic Acid Interactions: Preliminary Observations," *Can. J. Soil Sci.* 70:515–518 (1990).
114. Bartoli, F., A. Hatira, J. C. Andre, and J. M. Portal. "Proprietes Fluorescentes et Colloidales d'une Solution Organique de Podzol au Cours du Processus de Complexation par le Cuivre," *Soil Biol. Biochem.* 19:355–362 (1987).
115. Underdown, A. W., C. H. Langford, and D. S. Gamble. "The Fluorescence and Visible Absorbance of Cu(II) and Mn(II) Complexes of Fulvic Acid: The Effect of Metal Ion Loading," *Can. J. Soil Sci.* 61:469–474 (1981).
116. Cabaniss, S. E. and M. S. Shuman. "Combined Ion Selective Electrode and Fluorescence Quenching Detection for Copper-Dissolved Organic Matter Titrations," *Anal. Chem.* 58:398–401 (1986).
117. Shestakov, E. I., A. I. Karpukhin, V. V. Fadeev, and V. V. Chubarov. "Fluorescence Intensity of Organic Compounds Containing Manganese in Podzolic Soils," *Izv. Timiryazevsk.* 2:82–85 (1987).

118. Gregor, J. E., H. K. J. Powell, and R. M. Town. "Evidence for Aliphatic Mixed Mode Coordination in Copper(II)-Fulvic Acid Complexes," *J. Soil Sci.* 40:661–673 (1989).
119. Frimmel, F. H. and W. Hopp. "Fluorimetric Investigation of the Interactions of Heavy Metals with Humic Substances," volunteered papers 2nd Int. Conf. International Humic Substances Society, Birmingham, U. K. (1984) pp. 200–202.
120. Waite, T. D. and F. M. M. Morel. "Ligand Exchange and Fluorescence Quenching Studies of the Fulvic Acid-Iron Interaction," *Anal. Chim. Acta* 162:263–274 (1984).
121. Ryan, D. K. and J. H. Weber. "Copper(II) Complexing Capacities of Natural Waters by Fluorescence Quenching," *Environ. Sci. Technol.* 16:866–872 (1982b).
122. Frimmel, F. H. and H. Bauer. "Influence of Photochemical Reactions on the Optical Properties of Aquatic Humic Substances Gained from Fall Leaves," *Sci. Total Environ.* 62:139–148 (1987).
123. Boussemart, M., C. Benamou, M. Richou, and J. Y. Benaïm. "Comparison of Differential Pulse Anodic Stripping Voltammetry and Spectrofluorometry for Determination of Complexes Between Copper and Organic Matter in Interstitial Waters Extracted from Marine Sediments," *Mar. Chem.* 28:27–39 (1989).
124. Horrocks, W. DeW. and M. Aibin, *Progress in Inorganic Chemistry*, Vol. 31, Lippard, S. I., Ed. (New York: Wiley-Interscience, 1984).
125. Dobbs, J. C., W. Susetyo, F. E. Knight, M. A. Castles, L. A. Carrera, and L. V. Azarraga. "A Novel Approach to Metal-Humic Complexation Studies by Lanthanide Ion Probe Spectroscopy," *Int. J. Environ. Anal. Chem.* 37:1–17 (1989).
126. Carrington, A. and A. D. McLachlan, *Introduction to Magnetic Resonance with Applications to Chemistry and Chemical Physics* (New York: Harper & Row, 1967).
127. Ingram, D. J. E. *Biological and Biochemical Applications of Electron Spin Resonance* (New York: Hilger, 1969).
128. Abragam, A. and B. Bleaney, *Electron Paramagnetic Resonance of Transition Ions* (Oxford: Clarendon Press, 1970).
129. Wertz, J. E. and J. R. Bolton. *Electron Spin Resonance: Elementary, Theory and Practical Applications* (New York: McGraw-Hill, 1972).
130. Swartz, H. M., J. R. Bolton, and D. C. Borg. *Biological Applications of Electron Spin Resonance* (New York: Interscience, 1972).
131. Vänngård, T. "Copper Proteins," in *Biological Applications of Electron Spin Resonance*, Swartz, H. M., J. R. Bolton, and D. C. Borg, Eds. (New York: John Wiley & Sons, 1972) pp. 411–447.
132. Senesi, N. "Molecular and Quantitative Aspects of the Chemistry of Fulvic Acid and its Interactions with Metal Ions and Organic Chemicals. Part I. The Electron Spin Resonance Approach," *Anal. Chim. Acta* 232:51–75 (1990).
133. Senesi, N. "Application of Electron Spin Resonance (ESR) Spectroscopy in Soil Chemistry," in *Advances in Soil Science*, Vol. 14, Stewart, B. A., Ed. (New York: Springer-Verlag, 1990) pp. 77–130.
134. Hall, P. L., B. R. Angel, and J. Braven. "Electron Spin Resonance and Related Studies of Lignite and Ball Clay from South Devon, England," *Chem. Geol.* 13:97–113 (1974).
135. Babanin, V. F., A. D. Voronin, G. M. Zenova, L. O. Karpachevskiy, A. S. Manucharov, A. A. Opalenko, and T. N. Pochatkova. "Study of Fe-Organic Soil Compounds by Nuclear Gamma-Resonance Spectroscopy," *Sov. Soil Sci.* 7:128–134 (1976).

136. Cheshire, M. V., M. L. Berrow, B. A. Goodman, and C. M. Mundie. "Metal Distribution and Nature of Some Cu, Mn and V Complexes in Humic and Fulvic Acid Fractions of Soil Organic Matter," *Geochim. Cosmochim. Acta* 41:1131–1138 (1977).
137. Senesi, N., S. M. Griffith, M. Schnitzer, and M. G. Townsend. "Binding of Fe^{3+} by Humic Materials," *Geochim. Cosmochim. Acta* 41:969–976 (1977).
138. Eltantawy, I. M. and M. Baverez. "Structural Study of Humic Acids by X-Ray, Electron Spin Resonance, and Infrared Spectroscopy," *Soil Sci. Soc. Am. J.* 42:903–905 (1978).
139. McBride, M. B. "Transition Metal Binding in Humic Acids: An ESR Study," *Soil Sci.* 126:200–209 (1978).
140. Abdul-Halim, A. L., J. C. Evans, C. C. Rowlands, and J. H. Thomas. "An EPR Spectroscopic Examination of Heavy Metals in Humic and Fulvic Acid Soil Fractions," *Geochim. Cosmochim. Acta* 45:481–487 (1981).
141. Senesi, N. "Spectroscopic Evidence of Organically-Bound Iron in Natural and Synthetic Complexes with Humic Substances," *Geochim. Cosmochim. Acta* 45:269–272 (1981).
142. Schnitzer, M. and K. Ghosh. "Characteristics of Water-Soluble Fulvic Acid-Copper and Fulvic Acid-Iron Complexes," *Soil Sci.* 134:354–363 (1982).
143. Cressey, P. J., G. R. Monk, H. K. J. Powell, and D. J. Tennent. "Fulvic Acid Studies: Evidence for a Polycarboxylate Co-Ordination Mode at Soil pH," *J. Soil Sci.* 34:783–799 (1983).
144. Filip, Z., M. V. Cheshire, B. A. Goodman, and D. B. McPhail. "The Occurrence of Copper, Iron, Zinc and Other Elements and the Nature of Some Copper and Iron Complexes in Humic Substances from Municipal Refuse Disposed in a Landfill," *Sci. Total Environ.* 44:1–16 (1985).
145. Kallianou, Ch. S. and N. J. Yassoglou. "Bonding and Oxidation State of Iron in Humic Complexes Extracted from Some Greek Soils," *Geoderma* 35:209–221 (1985).
146. Senesi, N., G. Sposito, and J. P. Martin. "Complexation of Some Transition Metal Ions in Soil Humic Acids: An ESR Study," in *Proceedings of 5° Int. Conference Heavy Metals in the Environment* (Edimburgh: CEP Consultants, 1985) pp. 478–480.
147. Senesi, N., G. Sposito, and J. P. Martin. "Intrinsic Copper, Iron and Vanadyl Complexes in Humic Acid-Type Polymers (melanins) from Soil Fungi: An IR and ESR Study," in *Current Perspectives in Environmental Biogeochemistry*, Giovannozzi Sermanni, G. and P. Nannipieri, Eds. (Roma: C.N.R. — I.P.R.A., 1987) pp. 295–308.
148. Goodman, B. A., and M. V. Cheshire. "Characterization of Iron-Fulvic Acid Complexes Using Mössbauer and EPR Spectroscopy," *Sci. Total Environ.* 62:229–240 (1987).
149. Filip, Z., J. J. Alberts, M. V. Cheshire, B. A. Goodman, and J. R. Bacon. "Comparison of Salt Marsh Humic Acid with Humic-Like Substances from the Indigenous Plant Species *Spartina Alterniflora* (Loisel)," *Sci. Total Environ.* 71:157–172 (1988).
150. Senesi, N. and G. Calderoni. "Structural and Chemical Characterization of Copper, Iron and Manganese Complexes Formed by Paleosol Humic Acids," *Org. Geochem.* 13:1145–1152 (1988).

151. Senesi, N., T. M. Miano, M. R. Provenzano, and G. Brunetti. "Spectroscopic and Compositional Comparative Characterization of Some I.H.S.S. Reference and Standard Fulvic and Humic Acids of Various Origin," *Sci. Total Environ.* 81/82:143–156 (1989).

152. Senesi, N. and F. Sakarelladiou. "Structural, Chemical and Functional Characterization of Humic Acids from Coastal River, Brakish-Water Lake and Marine Sediments," Proc. 9° Int. Symposium on Environmental Biogeochemistry, Moscow 1989, in press (1991).

153. Senesi, N., T. M. Miano, and G. Sposito. "Molecular and Metal Chemistry of Leonardite Humic Acid in Comparison to Typical Soil Humic Acids," in *Proceedings of Int. Conference Peat 90* (Jyskä, Finland: The Association of Finnish Peat Industries, 1990) pp. 412–421.

154. Senesi, N., G. Sposito, G. R. Bradford, and K. M. Holtzclaw. "Residual Metal Reactivity of Humic Acids Extracted from Soil Amended with Sewage Sludge," *Water Air Soil Pollut.* 55: 409–425 (1991).

155. Senesi, N., C. Saiz-Jimenez, and T. M. Miano. "Spectroscopic Characterization of Metal-Humic Acid-Like Complexes of Earthworm-Composted Organic Wastes," *Sci. Total Environ.* in press (1992).

156. Cheshire, M. V., M. L. Berrow, B. A. Goodman, and C. M. Mundie. "The Nature and Origin of the Organic Matter-Metal Complexes of Soil," Coll. Int. C.N.R.S. No. 303, Migrations Organo-Minérales dans les Sols Temperés (1979) pp. 241–246.

157. McBride, M. B. "Reactions Controlling Heavy Metal Solubility in Soils," in *Advances in Soil Science*, Vol. 14, Stewart, B. A., Ed. (New York: Springer-Verlag, 1989) pp. 1–56.

158. Goodman, B. A. and M. V. Cheshire. "Electron Paramagnetic Resonance Evidence That Copper Is Complexed in Humic Acid by the Nitrogen of Porphyrin Groups," *Nature (London New Biol.)* 244:158–159 (1973).

159. Goodman, B. A. and M. V. Cheshire. "The Occurrence of Copper-Porphyrin Complexes in Soil Humic Acids," *J. Soil Sci.* 27:337–347 (1976).

160. Lakatos, B., T. Tibai, and J. Meisel. "ESR Spectra of Humic Acids and Their Metal Complexes," *Geoderma* 19:319–338 (1977).

161. Bresnahan, W. T., C. L. Grant, and J. H. Weber. "Stability Constants for the Complexation of Copper (II) Ions with Water and Fulvic Acids Measured by an Ion Selective Electrode," *Anal. Chem.* 50:1675–1679 (1978).

162. Boyd, S. A., L. E. Sommers, D. W. Nelson, and D. X. West. "The Mechanism of Copper (II) Binding by Humic Acid. An Electron Spin Resonance Study of a Copper (II)-Humic Acid Complex and Some Adducts with Nitrogen Donors," *Soil Sci. Soc. Am. J.* 45:745–749 (1981).

163. Boyd, S. A., L. E. Sommers, D. W. Nelson, and D. X. West. "Copper (II) Binding by Humic Acid Extracted from Sewage Sludge: An Electron Spin Resonance Study," *Soil Sci. Soc. Am. J.* 47:43–46 (1983).

164. Saiz-Jimenez, C. and F. Shafizadeh. "Iron and Copper Binding by Fungal Phenolic Polymers: An Electron Spin Resonance Study," *Curr. Microbiol.* 10:281–286 (1984).

165. Senesi, N. and G. Sposito. "Residual Copper(II) Complexes in Purified Soil and Sewage Sludge Fulvic Acids: An Electron Spin Resonance Study," *Soil Sci. Soc. Am. J.* 48:1247–1253 (1984).

166. Senesi, N., D. F. Bocian, and G. Sposito. "Electron Spin Resonance Investigation of Copper(II) Complexation by Soil Fulvic Acid," *Soil Sci. Soc. Am. J.* 49:114–119 (1985).
167. Senesi, N., D. F. Bocian, and G. Sposito. "Electron Spin Resonance Investigation of Copper(II) Complexation by Fulvic Acid Extracted from Sewage Sludge," *Soil Sci. Soc. Am. J.* 49:119–126 (1985).
168. Senesi, N. and G. Sposito. "Manganese(II) Complexation by Humic Acids from Soils and Soil Fungi," in *Proceedings of 6° Int. Conference Heavy Metals in the Environment* (Edinburgh: CEP Consultants, 1987) pp. 330–333.
169. Senesi, N. and G. Sposito. "Characterization and Stability of Transition Metal Complexes of Chestnut (*Castanea sativa* L.) Leaf Litter," *J. Soil Sci.* 40:461–472 (1989).
170. Senesi, N., G. Sposito, and G. R. Bradford. "Iron, Copper and Manganese Complexation by Forest Leaf Litter," *For. Sci.* 35:1040–1057 (1989).
171. Alberts, J. J., J. E. Schindler, D. E. Nutter, Jr., and E. Davis. "Elemental, Infra-Red Spectrophotometric and Electron Spin Resonance Investigations of Non-Chemically Isolated Humic Material," *Geochim. Cosmochim. Acta* 40:369–372 (1976).
172. Gamble, D. S., M. Schnitzer, and D. S. Skinner. "Mn(II)-Fulvic Acid Complexing Equilibrium Measurements by Electron Spin Resonance Spectrometry," *Can. J. Soil. Sci.* 57:47–53 (1977).
173. Carpenter, R. "Quantitative Electron Spin Resonance (ESR) Determinations of Forms and Total Amounts of Mn in Aqueous Environmental Samples," *Geochim. Cosmochim. Acta* 47:875–885 (1983).
174. McBride, M. B. "Electron Spin Resonance Investigation of Mn^{2+} Complexation in Natural and Synthetic Organics," *Soil Sci. Soc. Am. J.* 46:1137–1143 (1982).
175. Goodman, B. A. and M. V. Cheshire. "The Bonding of Vanadium in Complexes with Humic Acid: an Electron Paramagnetic Resonance Study," *Geochim. Cosmochim. Acta* 39:1711–1713 (1975).
176. Templeton, G. D. and N. D. Chasteen. "Vanadium-Fulvic Acid Chemistry: Conformational and Binding Studies by Electron Spin Probe Techniques," *Geochim. Cosmochim. Acta* 44:741–752 (1980).
177. McBride, M. B. "A Comparative Electron Spin Resonance Study of VO^{2+} Complexation in Synthetic Molecules and Soil Organics," *Soil Sci. Soc. Am. J.* 44:495–499 (1980).
178. Wilson, S. A. and J. H. Weber. "An EPR Study of the Reduction of Vanadium(V) to Vanadium(IV) by Fulvic Acid," *Chem. Geol.* 26:345–354 (1979).
179. Goodman, B. A. and M. V. Cheshire. "Reduction of Molybdate by Soil Organic Matter: EPR Evidence for Formation of Both Mo(V) and Mo(III)," *Nature (London)* 299:618–620 (1982).
180. Gutlich, P., R. Link, and A. Trautwein. *Mössbauer Spectroscopy and Transition Metal Chemistry* (New York: Springer-Verlag, 1978).
181. Hansen, E. H. and K. H. Mosbaek. "Mössbauer Studies of an Iron (III) Fulvic Acid Complex," *Acta Chem. Scand.* 8:3083–3084 (1970).
182. Lakatos, B., L. Korecz, and J. Meisel. "Comparative Studies on the Mössbauer Parameters of Iron Humates and Polyuronates," *Geoderma* 19:149–157 (1977).
183. Goodman, B. A. and M. V. Cheshire. "A Mössbauer Spectroscopic Study of the Effect of pH on the Reaction Between Iron and Humic Acid in Aqueous Media," *J. Soil Sci.* 30:85–91 (1979).

184. Goodman, B. A. and M. V. Cheshire. "A Mössbauer Effect Study of the Reduction of Iron by Fulvic Acid," volunteered papers 2nd Int. Conf. International Humic Substances Society, Birmingham, U.K. (1984) pp. 187–188.
185. McBride, M. B., B. A. Goodman, J. D. Russell, A. R. Fraser, V. C. Farmer, and D. P. E. Dickson. "Characterization of Iron in Alkaline EDTA and NH_4OH Extracts of Podzols," *J. Soil Sci.* 34:825–840 (1983).
186. Kodama, H., M. Schnitzer, and E. Murad. "An Investigation of Iron (III)-Fulvic Acid Complexes by Mössbauer Spectroscopy and Chemical Methods," *Soil Sci. Soc. Am. J.* 52:994–998 (1988).
187. Deczky, K. and C. H. Langford. "Application of Water Nuclear Magnetic Resonance Relaxation Times to Study of Metal Complexes of the Soluble Soil Organic Fraction Fulvic Acid," *Can. J. Chem.* 56:1947–1951 (1978).
188. Gamble, D. S., C. H. Langford, and J. P. K. Tong. "The Structure and Equilibria of a Manganese (II) Complex of Fulvic Acid Studied by Ion Exchange and Nuclear Magnetic Resonance," *Can. J. Chem.* 54:1239–1245 (1976).
189. Pommery, J., J. P. Ebenga, M. Imbenotte, G. Palavitt, and F. Erb. "Determination of the Complexing Ability of a Standard Humic Acid with Cadmium Ions," *Water Res.* 22:185–189 (1988).

Index